Springer Series in
SOLID-STATE SCIENCES 127

Springer
Berlin
Heidelberg
New York
Barcelona
Hong Kong
London
Milan
Paris
Singapore
Tokyo

Springer Series in

SOLID-STATE SCIENCES

---

126 **Physical Properties of Quasicrystals**
Editor: Z.M. Stadnik

127 **Positron Annihilation in Semiconductors**
Defect Studies
By R. Krause-Rehberg and H.S. Leipner

Volumes 1–125 are listed at the end of the book.

R. Krause-Rehberg H. S. Leipner

Positron Annihilation in Semiconductors

Defect Studies

With 212 Figures and 20 Tables

Springer

Dr. Reinhard Krause-Rehberg
Dr. Hartmut S. Leipner
Martin-Luther-Universität Halle-Wittenberg, Fachbereich Physik, Friedemann-Bach-Platz 6
D-06108 Halle, Germany
e-mail: krause@physik.uni-halle.de
leipner@physik.uni-halle.de

Series Editors:

Professor Dr., Dres. h. c. Manuel Cardona
Professor Dr., Dres. h. c. Peter Fulde*
Professor Dr., Dres. h. c. Klaus von Klitzing
Professor Dr., Dres. h. c. Hans-Joachim Queisser
Max-Planck-Institut für Festkörperforschung, Heisenbergstrasse 1, D-70569 Stuttgart, Germany
* also at Max-Planck-Institut für Physik komplexer Systeme, Nöthnitzer Strasse 38
D-01187 Dresden, Germany

Professor Dr. Roberto Merlin
Department of Physics, 5000 East University, University of Michigan
Ann Arbor, MI 48109-1120, USA

Professor Dr. Horst Störmer
Dept. Phys. and Dept. Appl. Physics, Columbia University, New York, NY 10023 and
Bell Labs., Lucent Technologies, Murray Hill, NJ 07974, USA

ISSN 0171-1873

ISBN 3-540-64371-0 Springer-Verlag Berlin Heidelberg New York

Library of Congress Cataloging-in-Publication Data applied for

Die Deutsche Bibliothek – CIP-Einheitsaufnahme
Krause-Rehberg, Reinhard: Positron annihilation in semiconductors: defect studies; with 20 tables/ R. Krause-Rehberg; H. S. Leipner.– Berlin ; Heidelberg ; New York ; Barcelona ; Hong Kong ; London ; Milan ; Paris ; Singapore ; Tokyo : Springer 1999
(Springer series in solid-state sciences ; 127)
ISBN 3-540-64371-0

Typesetting: camera-ready copies by the authors
Cover concept: eStudio Calamar Steinen
Cover production: *design & production* GmbH, Heidelberg

SPIN: 10576794 57/3144/ba - 5 4 3 2 1 0 – Printed on acid-free paper

Foreword

Positron annihilation studies of semiconductors have for a long time been considered a very complex, indirect way to investigate lattice defects in covalently bonded solids. The results were open to diverging interpretations, and quantitative analysis of defect concentrations was considered with great caution. However, in the last decade this situation has changed drastically. Combining positron annihilation studies with other spectroscopic analysis tools such as optical absorption or DLTS has allowed us in many cases to unambiguously solve the question of defect identification, and positron annihilation was able to make decisive contributions to the solution of longstanding questions such as the nature of the metastable state of the EL2 (As_{Ga}) defect in GaAs. Moreover, even quantitative analysis can today be considered as well established so that defect concentrations deduced from these measurements have to be taken seriously.

In this respect the present volume serves as an important milestone, summarizing our present knowledge of positron annihilation studies in semiconductors and providing a basis for the use of these results to solve different defect-related problems of the future. The reader will find it instructive to follow this treatise chapter-by-chapter or to use it as an encyclopedic source, summarizing the state of the art in this field. Thus the volume will be of equal interest to students who want to familiarize themselves with an advanced semiconductor characterization technique and to scientists who want to assess whether this technique might contribute to the solution of a specific problem.

The idea of preparing this volume was stimulated by a comprehensive review seminar given by Reinhard Krause-Rehberg at the Max Planck Institute of Microstructure Physics in Halle/Saale in the Spring of 1994. This presentation distinguished clearly between well-founded conclusions derived from positron annihilation studies and more speculative assumptions that had given this technique a mixed reputation in the past. This critical spirit was to be the guiding principle in writing this volume, and the reader can attest to how far this goal actually has been achieved.

Berkeley, June 1998 *Eicke R. Weber*

Preface

Lattice defects strongly influence or even determine most of the electrical and optical properties of semiconductors and are thus of essential technological importance. However, there are only a few experimental techniques of structural sensitivity on an atomic scale. Positron annihilation has become one of these techniques during the last decade. Positrons have assisted the structural investigation of eminent defects, such as EL2 in gallium arsenide and DX centers in semiconductor compounds.

The first review of positron annihilation was published by West (1973). The interaction of positrons with metals and ionic crystals was treated by Hautojärvi (1979). Two Enrico Fermi schools dealt with the basics and recent applications of positron annihilation (Brandt and Dupasquier 1981; Dupasquier and Mills 1995). A further monograph treated the application of positron annihilation in chemistry (Mogensen 1995). A few review papers on positron annihilation in solids are available: The special topics of these were the slow-positron-beam technique (Schultz and Lynn 1988) and the theory of positron annihilation in solids (Puska and Nieminen 1994). III–V compound semiconductors (Dlubek and Krause 1987) and silicon (Asoka-Kumar et al. 1994) were the subjects of two other review papers. The results on gallium arsenide obtained in the Helsinki positron group were summarized by Saarinen et al. (1998).

The first workshop on positron studies of semiconductor defects was organized by the authors of this book as the 132th WE-Heraeus seminar in Halle in 1994. Up to now, there has been no comprehensive treatise dealing with positron annihilation in semiconductors. On the occasion of a series of lectures on crystal defects in semiconductors at Halle University, Prof. Eicke Weber, Berkeley, encouraged us to compile a monograph. We are grateful to Springer-Verlag who gave us the opportunity to realize this project in the form of this book.

The book is dedicated to experimental investigations of crystal imperfections in elemental and compound semiconductors by means of positron annihilation. It represents the state of the art reported at the 11th International Conference on Positron Annihilation (Kansas City) and the 19th International Conference on Defects in Semiconductors (Aveiro) in 1997. In the meanwhile, the field of positron annihilation in semiconductors has become very extended. No complete bibliography can be given in this book. Moreover, many investigations are presented exemplarily with selected outstanding results.

This book is written both for specialists working on positron annihilation and, particularily, for scientists and students interested in semiconductors and their defect structure in general. The different methods and the theoretical basics of positron annihilation are therefore extensively explained. Basic knowledge in solid-state physics should be sufficient to understand the book. A brief comparison is made between positron annihilation and other relevant techniques with regard to the kind of defects detectable and the sensitivity range of the methods.

In the 1970s, the Halle positron group was built up by Prof. Günter Dlubek, who was one of the first to intensively study defects in compound semiconductors with positrons. We are indebted to him for the introduction to the secrets of positron annihilation. The stimulating atmosphere and the intensive work of many students and scientists in our positron group gave the human and scientific background and enabled us to write this book. We are especially grateful to Dr. Angelika Polity for her committed work in our lab. The sections on irradiation-induced defects benefited much from her intensive research in this field.

Many suggestions for the improvement of the text were given by Dr. David Britton, Cape Town. Special thanks are due to Dr. Torsten Staab who wrote essential parts of Sect. 3.3 and Appendix A2 and to Jörg Gebauer for writing Sect. 5.2.5. We want to thank Katrin Lehmann for the careful preparation of the figures and Dr. Claus Ascheron at Springer-Verlag for the excellent editorial support.

Halle, July 1998

Reinhard Krause-Rehberg
Hartmut S. Leipner

Contents

Abbreviations and Symbols

A	empirical parameter of the positron implantation profile
A_0	integral of the 0.511-MeV annihilation line
A_w	integral of the 0.511-MeV annihilation line in the wing region
ACAR	angular correlation of annihilation radiation
ADC	analog–digital converter
AES	Auger electron spectroscopy
AMOC	age–momentum correlation
c	speed of light
C	concentration
CVD	chemical vapor deposition
Cz	Czochralski
D	positron decay spectrum
D_f	decay spectrum folded by the time resolution function
D_+	positron diffusion constant
DLTS	deep level transient spectroscopy
DOBS	Doppler-broadening spectroscopy
e	elementary charge
E	energy
E_{cond}	energy of lower edge of the conduction band
E_d	ionization energy level of defect d
	electron–positron correlation energy
E_F	Fermi level
EPR	electron paramagnetic resonance
E_R	positron binding energy to Rydberg states
E_{val}	energy of upper edge of the valence band
f	occupation probability of a defect level
f_s	fraction of positron diffusing back to the surface
F	positronium fraction
F_t	time resolution function
f_s	fraction of positrons diffusing back to the surface
G	Gaussian resolution function
GGA	generalized gradient approximation
FWHM	full width at half maximum
FZ	float zone
H	angular-correlation peak-height parameter
I	intensity of positron lifetime components
IR	infrared
k	number of different defect types
k_B	Boltzmann constant
K_i	equilibrium constant of electrons and holes
K_{Hg}	equilibrium constant of V_{Hg} in $Hg_{1-x}Cd_xTe$
$\langle l \rangle$	mean free path of positrons
L_+	positron diffusion length
LDA	local density approximation
LEC	liquid encapsulated Czochralski
LEED	low-energy electron diffraction

LEPD	low-energy positron diffraction
LMTO-ASA	linear muffin-tin orbital in the atomic sphere approximation
LT	low-temperature-grown
MCA	multi-channel analyzer
MCDA	magnetic circular dichroism of absorption
MELT	maximum entropy of lifetime
MOS	metal oxide semiconductor
m_e	electron mass
m^*	effective positron mass
m_0	electron or positron rest mass
n	density of free electrons
n_+	positron density
n_-	total electron density
n_b	number of positrons in the bulk
N	positron lifetime spectrum
N_0	number of positrons in the sample at the time $t = 0$
N_{at}	number of atoms per unit volume
N_c	coincidence counting rate in ACAR measurements
N_D	Doppler-broadening spectrum
p	free-hole density
$\mathbf{p}$	electron momentum
P	positron penetration depth profile
	positron implantation profile for unmoderated positrons
PAC	perturbed angular correlation
PAES	positron annihilation-induced Auger electron spectroscopy
p_{77}	hole density measured at 77 K
P_{Hg}	mercury vapor pressure
PM	photomultiplier
POLIS	positron lifetime spectroscopy
Ps	positronium
r	empirical parameter of the positron implantation profile
$\mathbf{r}$	position vector
r_0	classical electron radius
r_c	capture radius
r_d	defect radius
R	“relative” line shape parameter
RBS	Rutherford backscattering
S	annihilation line shape parameter
SCA	single-channel analyzer
SEM	scanning electron microscopy
SIMOX	separation of implanted oxygen
STM	scanning tunneling microscopy
t	time
t_0	time zero
T	temperature
TAC	time-to-amplitude converter
T_m	melting temperature
TEM	transmission electron microscopy
TPM	transmission positron microscopy
TRIM	transport of ions in matter
v	thermal positron velocity
V_+	positron potential
V_{ext}	external potential
VEPFIT	variable energy positron fit
W	“wing” parameter of the annihilation line
x	composition or coordinate
z	depth measured from the sample surface
	mean penetration depth of positrons
	information depth of positron annihilation
Z	atomic number

δ	positron detrapping rate
ΔV_+	positron potential step
ε_0	dielectric constant
ε_r	relative permitivity
ϕ	fingerprint spectrum of Doppler difference curves
Φ	implantation dose
Φ_+	positron work function
γ	enhancement factor
η	annihilation fraction positrons
κ	positron trapping rate
λ	annihilation rate
λ_{eff}	effective annihilation rate
μ	positron trapping coefficient
μ^0	positron trapping coefficient of a neutral defect
μ^+	positron trapping coefficient of a positively charged defect
ρ	density
σ	electron momentum distribution
σ_s	standard deviation
ϑ	transition rate from a shallow to a deep positron trap
Θ	angular deviation of annihilation quanta
τ	positron lifetime
	average positron lifetime

Indices

a	annealing
b	bulk
d	defect
dis	dislocation
i, j	counting index
s	surface
st	shallow trap
t	trap
v	vacancy
2v	divacancy

1 Introduction

The positron as the antiparticle of the electron was predicted by Dirac (1928). First experimental indications of an unknown particle were found in cloud-chamber photographs of cosmic rays (Anderson 1932). This particle was identified later as the positron, which was thus the first antiparticle in physics. The annihilation of the positron with electrons in matter was first studied in the 1940s. It was discovered early that the energy and momentum conservation during the annihilation process could be utilized to study properties of solids. The bound state of a positron and an electron, as the lightest known atom to be formed, is analogous to a hydrogen atom, where the proton is replaced by the positron. This e^+–e^- state is called positronium (Ps). It was predicted by Mohorovicic (1934) and discovered by Deutsch (1951).

The early experiments with positrons were dedicated to the study of the electronic structure, e.g. the Fermi surface in metals and alloys (Behringer and Montgomery 1942; de Benedetti et al. 1950). The various experimental techniques of positron annihilation based upon the equipment of nuclear spectroscopy were developed strongly in the two decades after 1945. In addition to the angular correlation of annihilation γ-quanta, Doppler broadening of the annihilation line and positron lifetime spectroscopy were established as independent methods. It was realized by the end of the 1960s that the annihilation parameters are sensitive to lattice imperfections. It was discovered that the positron may be trapped in crystal defects, i.e. the wave function of the positron is localized at the defect site until annihilation. This behavior of positrons was clearly demonstrated by several authors, e.g. MacKenzie et al. (1967) for thermal vacancies in metals, Brandt et al. (1968) in ionic crystals, and Dekhtyar et al. (1969) after plastic deformation of semiconductors. The investigation of crystal defects has become the dominant issue of positron annihilation studies. This development is illustrated in Fig. 1.1, which shows the number of published papers on defect studies with positrons. The number of papers has increased continuously. Up to the mid-1980s, defect studies in solids were mainly carried out in metals and alloys. The experience obtained in this field was applied to semiconductors. The study of defects in compound and elemental semiconductors amounts now to about the half of the total number of papers on defect studies with positrons.

The trapping of the positron in defects is based on the formation of an attractive potential at open-volume defects, such as vacancies, vacancy agglomerates, and dislocations. The main reason for this potential is the lack of a repulsive positively charged nucleus in such a defect. The sensitivity range for the vacancy detection

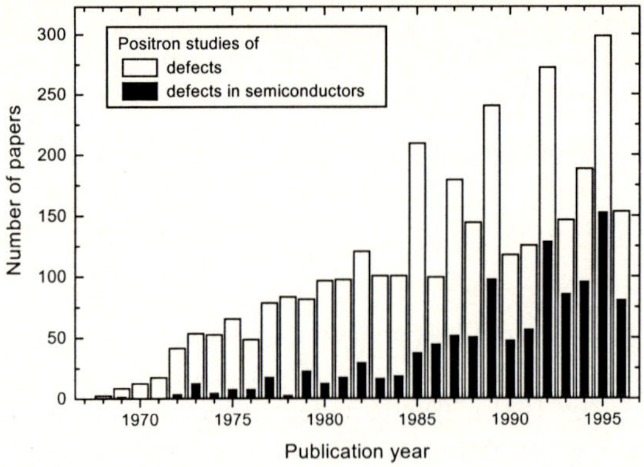

Fig. 1.1. Number of papers on positron studies of defects based on the INSPEC database. The full height of the columns represents the total number of defect studies in the respective year. The fraction of studies on semiconductor defects is drawn in black. The strong fluctuations starting in 1985 are due the international conferences on positron annihilation held every three years.

in metals starts at about one vacancy per 10^7 atoms. This enormous sensitivity is caused by the fact that the positron diffuses about 100 nm through the lattice and probes a high number of atoms until annihilation. When the positron is trapped in an open-volume defect, the annihilation parameters are changed in a characteristic way. The positron lifetime increases in an open-volume defect due to the lower electron density. Momentum conservation during 2γ-annihilation leads to a small angular spread of the collinear γ-quanta or a Doppler shift of the annihilation energy. Both properties, the density and the momentum distribution of electrons participating in the annihilation, result in observables to be detected in a positron experiment. The analysis of the annihilation radiation thus gives the possibility of defect detection. Clustering of vacancies as a typical defect reaction may be observed as the increase in the defect-related lifetime due to the further decrease in the electron density.

The positron trapping behavior in semiconductors is more complex compared with metals. This is mainly due to the fact that the positron as a charged particle is sensitive to the charge state of defects. The defect identification is more difficult in compound semiconductors since the variety of point defects increases. The simple transfer of the knowledge obtained in metals gave rise to some confusing results and misinterpretations at the beginning of semiconductor studies with positrons. The trapping coefficient as the sensitivity-scaling factor taken from metals gave vacancy concentrations that were too high. Since then, reliable experiments have been carried out to determine the trapping coefficient in semiconductors.

The improvement of the theoretical understanding of positron trapping, due especially to M. Puska and R. Nieminen (for a review see Puska and Nieminen

1994), the refinement of positron lifetime spectroscopy as the main technique for defect studies, and the increasing amount of data have made it possible to give reliable qualitative and quantitative information on defects in semiconductors. A large number of investigations have been carried out for differently grown bulk semiconductor crystals. Irradiation with high-energy electrons at low temperature represents an important tool for defined generation of point defects. Many positron studies on the defect structure of semiconductors have been carried out after such irradiation. Practically all positron lifetimes for the important bulk semiconductor materials and many lifetimes for various vacancy-type defects have been experimentally determined. In comparison with theoretical calculations, the defects can be characterized by their specific positron lifetimes. A promising new development is the background-reduced measurement of positron annihilation with core electrons by Doppler-broadening coincidence spectroscopy. It seems to be possible to obtain chemical information on the surroundings of the annihilation site by comparison of measured and theoretically calculated Doppler spectra in the high-momentum region. The sublattice of the defect and eventually the type of impurities stabilizing the defect complexes may be determined. A newer development is the use of ratio or difference curves of the Doppler-broadened annihilation lines as fingerprints for specific annihilation sites.

The charge state of vacancies in semiconductors strongly influences the sensitivity range of positron annihilation. Positively charged vacancies repel positrons and are thus invisible to the different positron-related defect detection methods. The sensitivity range for neutral vacancies in semiconductors is similar to that for metals. The attractive potential of the vacancy is superimposed on the Coulomb potential of negatively charged vacancies. This results in a higher sensitivity of positrons at low temperatures and in a pronounced temperature dependence of positron trapping.

In principle, dislocations are also positron traps in semiconductors. However, the sensitivity range is comparable to that of metals and, hence, dislocations can be observed only after plastic deformation. Precipitates were found to be positron traps in some alloys. Their density is normally too low to be detected in semiconductors.

In addition to the positron traps known from metals, negatively charged non-open-volume defects, e.g. acceptor-type impurities or negatively charged antisites in a compound, represent positron trapping centers at low temperatures. The Coulomb potential of these defects is responsible for the formation of shallow Rydberg states. Positrons may be bound with a small binding energy to such centers, which are called shallow positron traps.

In summary, neutral and negative vacancy-type defects, as well as negative ions, are the dominant positron traps in semiconductors. Temperature-dependent lifetime measurements may distinguish between both defect types. However, it should be noted that the main advantage of positron annihilation is the sensitive detection of open-volume defects. This defines the power of the positron methods in contrast to most electrical and optical spectroscopies. On the other hand, it may also be a disadvantage, as important defect types, such as interstitials and neutral or positive impurities or antisites, remain in the dark.

Modern semiconductor devices are based on layered structures, such as epitaxially grown layers, oxides, or implanted layers. The disadvantage of conventional positron techniques directly utilizing positrons from the β^+-decay of a radioactive source is the broad energy distribution and the large penetration depth of positrons in the sample. Consequently, only bulk material can usually be investigated. This drawback has been overcome by the development of the slow-positron-beam technique. This is based on the moderation of positrons. Monoenergetic positrons are obtained at surfaces exhibiting a negative positron work function, e.g. tungsten. The emitted positrons are guided as a beam to the sample. The variation of the incident positron energy in an accelerator stage enables the measurement of near-surface defect profiles. The positron beam can be pulsed and thus the positron lifetime can be measured too. The focusing of the positron beam may in future allow the three-dimensional probing of defect structures, i.e. the development of a positron microscope and microprobe.

The book is organized as follows. First, the different positron techniques are described in Chap. 2. As far as is necessary for the understanding of the experimental results, the fundamentals of positron annihilation are summarized in Chap. 3. The following chapters review the experimental results on the different elemental and compound semiconductors (Chaps. 4 to 7). Prospective technological applications of positron annihilation are described in Chap. 8. The comparison of positron annihilation with other techniques sensitive to point defects in semiconductors is presented in Chap. 9. Important data on positron annihilation in semiconductors are collected in the Appendix.

2 Experimental Techniques

Positrons diffusing through matter may be captured in special trapping sites. As was shown in the early studies on semiconductors, these trapping centers are crystal imperfections, e.g. vacancies and dislocations. In this sense, the behavior of positrons in semiconductors is similar to that in metals, which was reviewed by Hautojärvi (1979). The wave function of a positron captured in such a defect is localized until it annihilates with an electron of the immediate surrounding into γ-rays. Since the local electron density and the electron momentum distribution are changed with respect to the defect-free crystal, the annihilation radiation can be utilized to obtain information on the localization site. Hence, the character and the concentration of the trapping centers, i.e. lattice defects, can be investigated by positrons.

The different positron techniques dealt with in this chapter are based on the analysis of the annihilation radiation. It appears through the mass–energy transformation of electron–positron pairs, mostly into two γ-quanta of 511 keV. In special cases, e.g. at surfaces, a positron may form a bound state with an electron, which is called positronium (Ps). The annihilation of Ps can lead to two or three annihilation γ-quanta.

The principles of the different positron techniques are illustrated in Fig. 2.1. They can be classified into two principal groups which are distinguished by the sensitivity of positrons to the electron density (positron lifetime measurement) and to the electron momentum distribution in the sample (Doppler-broadening spectroscopy and angular correlation of annihilation radiation). Since the electron density and the electron momentum distribution in defects capable of trapping positrons differ from the perfect lattice, these techniques represent a tool to study the type and number of these crystal defects. The following abbreviations are used for these techniques throughout the book: POLIS (positron lifetime spectroscopy[1]), DOBS (Doppler-broadening spectroscopy), and ACAR (angular correlation of annihilation radiation.

The positron lifetime is registered as the time difference between the emission of the 1.27-MeV γ-quantum generated almost simultaneously with the positron in a ^{22}Na isotope, which is the most commonly used positron source, and one of the 0.51-MeV annihilation γ-quanta (Sect. 2.2). The two other techniques are based on the principle of momentum conservation in the annihilation process (Sect. 2.3).

[1] In a narrower sense, it is not a "spectroscopic" technique, but the term has become established in the literature and we will use it as well.

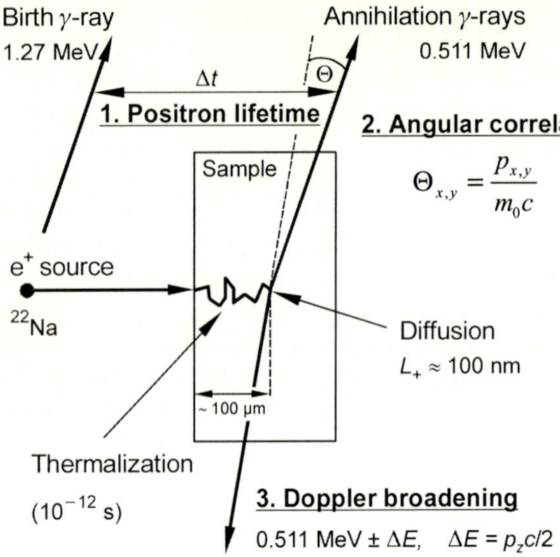

Fig. 2.1. Scheme of different positron experiments. Positrons from an isotope source (e.g. ^{22}Na) penetrate the sample, thermalize within a few ps, and diffuse before annihilation. The thermalization distance is of the order of magnitude of 100 μm and is much longer than the positron diffusion length L_+, which amounts to about 100 nm. The positron lifetime can be measured as the time difference Δt between the emission of the 1.27-MeV and the 0.511-MeV γ-quanta. Doppler-broadening spectroscopy records the broadening ΔE of the 0.511-MeV annihilation line in an energy-dispersive spectrometer. ΔE can be calculated from the electron momentum component in the propagation direction, p_z, $\Delta E = p_z c/2$. The angular deviation Θ of the annihilation quanta from collinearity is measured by the method of angular correlation of annihilation radiation. The angular deviations in the x and y directions $\Theta_{x,y}$ are related to the components of the electron momentum $p_{x,y}$, $\Theta_{x,y} = p_{x,y}/(m_0 c)$, where m_0 is the electron rest mass and c is the speed of light.

The momentum distribution of the electron–positron pair is conserved in the annihilation radiation. The momentum component in the γ-ray propagation direction leads to a small Doppler shift of the annihilation energy. Several million annihilation events are summed up as a Doppler-broadened spectrum of the 0.51-MeV annihilation line in an energy-dispersive spectrometer. This method is called Doppler-broadening spectroscopy (Sect. 2.3.2). The momentum components in the other two space directions result in the deviation Θ from collinearity of the two 0.51-MeV annihilation γ-quanta. This deviation can be measured in a γ–γ-coincidence arrangement in one or two dimensions. These setups are called one- and two-dimensional angular correlation of annihilation radiation (Sect. 2.3.3).

The different positron techniques introduced above, described in more detail in the following sections, are limited to defect studies in bulk material of a thickness of at least 100 μm, since they use positrons directly from isotope sources. Due to the broad energy distribution of the positrons resulting from the β^+-decay, the penetration depth is also broadened and thus these methods are not suitable to investigate near-surface defects or for defect profiling. The slow-positron-beam technique (Sect. 2.4), developed in the early 1980s, uses monoenergetic positrons from moderators (Schultz and Lynn 1988). The energy of such a positron beam can be set in a wide range to allow near-surface defect profiling. Doppler-broadening and angular correlation measurements, as well as positron lifetime spectroscopy, can be carried out with a slow-positron beam too. In contrast, the methods applying unmoderated positrons directly are referred to as conventional positron techniques throughout the book.

In addition to the positron lifetime spectroscopy and momentum distribution techniques, there are other positron techniques, such as low-energy positron diffraction (Sect. 2.5.4), positron-annihilation-induced Auger electron spectroscopy (Sect. 2.5.3), and age–momentum correlation (Sect. 2.5.2). They are, however, less important for defect studies in semiconductors. The development of the positron microprobe/microscope is reviewed in Sect. 2.5.1.

2.1 Positron Sources

Positrons can be obtained from the β^+-decay of radioactive isotopes, e.g. from ^{22}Na according to the decay reaction $^{22}\mathrm{Na} \rightarrow {}^{22}\mathrm{Ne} + \beta^+ + \nu_e + \gamma$. The ^{22}Na isotope gives a relatively high positron yield of 90.4 % (Fig. 2.2) and has several other advantages. First of all, the appearance of a 1.27-MeV γ-quantum almost simultaneously with the positron enables positron lifetime measurement by a start–stop coincidence γ-spectrometer (see Sect. 2.2.1). Moreover, the manufacture of laboratory sources is simple, due to the easy handling of different sodium salts in aqueous solution, such as sodium chloride or sodium acetate. The half-life of 2.6 years and a reasonable price make this isotope the most used source material in positron research. The accidental contamination of laboratory personnel is less harmful, since the biological half-life is only a few days. In addition to ^{22}Na,

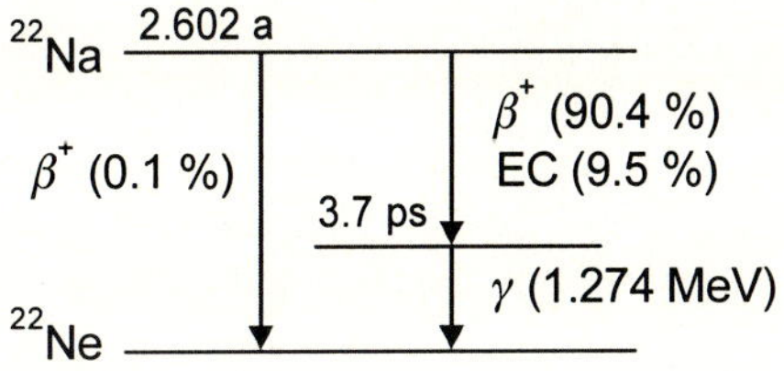

Fig. 2.2. Decay scheme of the radioactive isotope ^{22}Na. 90.4 % decays by emission of a positron and an electron neutrino to the excited state of ^{22}Ne. The ground state is reached after 3.7 ps by emission of a γ-quantum of 1.274 MeV. Competitive processes with lower probabilities are electron capture (EC) and direct transition to the Ne ground state.

other isotopes (^{64}Cu, ^{58}Co, etc.) can be used, but are less common. The generation of electron–positron pairs from high-energy photons is an alternative way to obtain positrons for use in beam systems. No broad application has been reported so far for applications in semiconductor research due to the need for appropriate accelerators.

The positrons generated in the β^+-decay reaction mentioned above exhibit a broad energy distribution up to an energy of 540 keV (see Fig. 2.17) and can penetrate deep into a sample. In electron-irradiation experiments, a particle energy of 0.5 MeV may be sufficient to generate irradiation defects such as Frenkel pairs in many materials. This should be similar for positrons. However, the dose amounts only to $\sim 1\times10^{11}$ cm^{-2} for a rather long positron experiment of several days. Supposing a defect introduction rate of 1 cm^{-1}, the density of introduced defects is $\sim 1\times10^{11}$ cm^{-3} and thus several orders of magnitude lower than the sensitivity limit of positron annihilation.

Sources of weak activity are needed for conventional positron lifetime and Doppler-broadening measurements. Even activities as low as 8×10^5 Bq (20 μCi) are sufficient. The sources are usually prepared by evaporating a solution of a ^{22}Na salt on a thin metal or polymer foil. The most common foil materials are Al, Ni, and mylar or kapton. In order to ensure the almost complete annihilation of positrons in the specimen volume, the foil source is placed between two identical samples ("sandwich arrangement", Fig. 2.3). A minimum sample thickness is required to ensure that the essential fraction of positrons annihilates in the sample pair. The

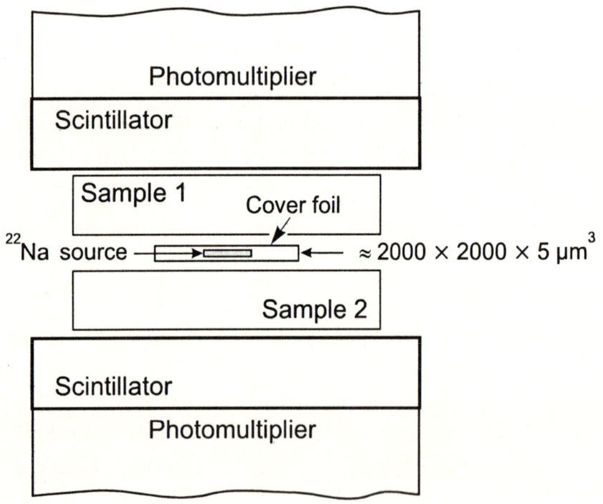

Fig. 2.3. Scheme of the sample–source sandwich arrangement for the registration of the birth and annihilation γ-rays as start and stop signals in the conventional positron lifetime experiment. The γ-quanta are detected as light flashes in the scintillators. These photons are converted into electrical pulses in fast photomultipliers. ^{22}NaCl is typically used as the source material.

scintillators should be placed as close as possible to the sample–source sandwich. If cooling of the sample is required, the design of the cryostat should allow the separation between the scintillators to be as small as possible in order to avoid a considerable reduction in the counting rate. A small fraction of the positrons annihilates in the source. This amounts to 2 to 15 % depending on the foil thickness and the atomic number of the sample, which determines the backscattering and, thus, the multiple passing of positrons through the source. For the analysis of positron lifetime spectra, this fraction must be carefully determined and subtracted. The so-called source correction is an essential task of the lifetime spectra evaluation. Details can be found e.g. in the paper of Staab et al. (1996). To reduce the source contribution, the isotope solution can be directly deposited onto the sample surface. The disadvantage of this procedure is radioactive contamination of the samples, which can thus hardly be used for further investigations.

The positron beam and angular correlation techniques require much stronger sources. Source capsules of about 4 GBq (100 mCi) are commercially available, but are sometimes also prepared in the laboratory (see e.g. Huomo et al. 1989). In contrast to the weaker foil sources of lifetime and Doppler-broadening measurements, radiation protection becomes more important. Carefully arranged lead shielding is necessary.

Another possibility for strong positron sources is to generate positrons by pair production (Schultz and Lynn 1988). In this case, an electron beam is stopped in an absorber of high atomic number creating bremsstrahlung γ-rays. Provided that the energy of the primary electron beam is high enough, the generation probability of electron–positron pairs is sufficiently high. The bunching of the primary electrons leads to a pulsed positron beam.

Furthermore, positrons may be obtained from nuclear reactions. It was proposed by Stein et al. (1974) to direct high-energy protons at a boron target in order to obtain an intense flux of positrons. Another concept for positron production is to be realized in the Munich research reactor (Triftshaeuser 1995). During the $^{113}\mathrm{Cd}(\mathrm{n},\gamma)^{114}\mathrm{Cd}$ reaction, three γ-quanta with a total energy of 9.041 MeV are released and can be used for pair production. The advantage of this technique is the formation of a continuous positron beam of high intensity.

2.2 Positron Lifetime Spectroscopy

The positron lifetime τ is a function of the electron density at the annihilation site. The annihilation rate λ, which is the reciprocal of the positron lifetime τ, is given by the overlap of the positron density $n_+(\mathbf{r}) = |\psi^+(\mathbf{r})|^2$ and the electron density $n_-(\mathbf{r})$ (Nieminen and Manninen 1979),

$$\lambda = \frac{1}{\tau} = \pi r_0^2 c \int \left|\psi^+(\mathbf{r})\right|^2 n_-(\mathbf{r})\,\gamma\,\mathrm{d}\mathbf{r}. \tag{2.1}$$

r_0 is the classical electron radius, c the speed of light, and $\mathbf{r}$ the position vector. The correlation function $\gamma = \gamma[n_-(\mathbf{r})] = 1 + \Delta n_-/n_-$ describes the increase Δn_- in the electron density due to the Coulomb attraction between a positron and an electron. This effect is called enhancement.

When positrons are trapped in open-volume defects, such as in vacancies and their agglomerates, the positron lifetime increases with respect to the defect-free sample. This is due to the locally reduced electron density of the defect. Thus, a longer lifetime component, which is a measure of the size of the open volume, appears. The strength of this component, i.e. its intensity, is directly related to the defect concentration. In principle, both items of information, i.e. the kind and concentration of the defect under investigation, can be obtained independently by a single measurement. This is the major advantage of positron lifetime spectroscopy compared with angular correlation of annihilation radiation or Doppler-broadening spectroscopy with respect to defect issues.

2.2.1 Basics of the Measurement

The conventional positron lifetime measurement is possible since a γ-quantum with an energy of 1.27 MeV is emitted almost simultaneously with the positron in the ^{22}Na source. The positron energy, which extends up to 540 keV, decreases in the sample within a few picoseconds by non-elastic interactions. The mean positron penetration depth of this so-called thermalization process is of the order of 100 μm. The thermalization time usually amounts to a few picoseconds. It is thus small compared with the positron lifetime and can be neglected. The details of the thermalization are described in Sect. 3.1.3. On reaching thermal energies, the positron diffuses in the periodic lattice potential before it is possibly trapped in a lattice defect. The diffusion length is in of order of 100 nm. This distance determines the number of atoms to be probed for positron traps during the positron lifetime. Hence, the diffusion length strongly determines the sensitivity of the positron methods to detect defects. The processes of diffusion and positron trapping are described in detail in Sects. 3.1.4 and 3.1.5.

The positron lifetime of a single event can be measured by detecting the time difference between the birth γ-quantum of the β^+-decay in the source and one of the annihilation γ-quanta of an energy of 511 keV. The activity of the source must be sufficiently low in order to ensure that on average only one positron is in the sample. This avoids the intermixing of start and stop quanta originating from different annihilation events. A special "sandwich" arrangement of foil source, samples, and detectors guarantees that all positrons emitted from the source are penetrating the sample material (Fig. 2.3). The γ-rays are converted by scintillator–photomultiplier detectors into analog electrical pulses. The pulses are processed by discriminators. Their output pulses start and stop a time-to-amplitude converter as an "electronic stopwatch". The amplitude of the output pulse is proportional to the time difference between the birth and the annihilation γ-quanta and, thus, represents a measure of the positron lifetime. The single annihilation event is stored after analog–digital conversion in the memory of a multi-channel analyzer. The channel numbers represent the time scale. In order to obtain the complete lifetime spectrum, more than 10^6 annihilation events must be recorded.

The scheme of the positron lifetime measurement is shown in Fig. 2.4. BaF_2 or plastic scintillators and photomultipliers with a short pulse rise-time are used to obtain a high time resolution. The discriminators suppress noise and generate standard timing pulses by the constant-fraction discrimination principle. This principle is favored over leading-edge discrimination in order to ensure stable time markers independent of the pulse height. Another task is to guarantee that the 1.27-MeV and 0.51-MeV quanta are accepted only in the appropriate channels. The discriminators are of differential type (single-channel analyzer) and accept input pulses within an adjustable energy window.

The timing pulses are used to start and stop the charging of a capacitor in the time-to-amplitude converter (TAC). The time linearity is ensured there by constant-current charging that is stopped at the arrival of the stop pulse originating from the annihilation γ-quantum. The stop pulse is coax-cable delayed in order to shift the time spectrum into a linear region of the TAC. The spectrum is stored in a multi-channel analyzer.

This experimental arrangement is called "fast–fast coincidence" setup. The term is related to the fact that the time measurement as well as the energy selection is performed in a fast channel. A slow channel was used for energy selection when fast differential discriminators were not available at the beginning of positron lifetime experiments. This arrangement is called a fast–slow setup.

Inexpensive multi-channel plug-in boards for personal computers with 2000 channels are sufficient for storing the spectra. The time resolution of the spec-

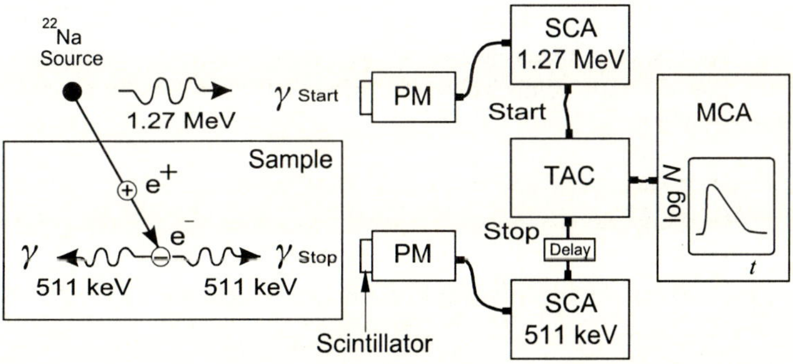

Fig. 2.4. Scheme of the positron lifetime experiment in fast–fast coincidence. The lifetime is measured as the time difference between the appearance of the start and stop γ-quanta (PM—photomultiplier, SCA—single-channel analyzer). The amplitude of the time-to-amplitude converter (TAC) analog output pulse is proportional to this time difference. The whole lifetime spectrum $N(t)$ is stored in a multi-channel analyzer (MCA).

trometer is determined mainly by the scintillator–multiplier part and ranges between 180 and 280 ps. The practical consequence of this relatively poor resolution is the limitation of the determination of positron lifetime components larger than about 50 ps. The determination of positron lifetimes can, however, be carried out with an accuracy of about 1 ps.

2.2.2 Data Treatment

The time-dependent positron decay spectrum $D(t)$ in the sample is given by

$$D(t)=\sum_{i=1}^{k+1} I_i \exp\left(-\frac{t}{\tau_i}\right). \tag{2.2}$$

k different defect types contributing to the positron trapping are related to $k+1$ components in the spectra with the individual lifetimes τ_i and intensities I_i. If no positron traps are present in the sample, (2.2) is reduced to $D(t)=\exp(-t/\tau_b)$, where τ_b is the positron lifetime in the defect-free bulk of the sample. The positron lifetime spectrum $N(t)$ is the absolute value of the time derivative of the positron decay spectrum $D(t)$,

$$N(t)=\sum_{i=1}^{k+1} \frac{I_i}{\tau_i} \exp\left(-\frac{t}{\tau_i}\right). \tag{2.3}$$

The spectrum is shifted on the time scale by the delay cable (Fig. 2.4) to time zero t_0, and consequently the time t in (2.2) and (2.3) has to be replaced by $t-t_0$.

Typical positron lifetime spectra obtained in as-grown silicon and in Si after plastic deformation are shown in Fig. 2.5 (for more details see Sect. 4.1.9). The experimentally obtained spectra differ from the analytical description (2.3) mainly by the convolution with the time resolution function $F_t(t)$. In general, it is approximated by a sum of several Gaussians or one Gaussian with exponential tails. In the case of the use of plastic scintillators, it is often sufficient to take a single Gaussian $G(t)$ centered at t_0,

$$G(t)=\frac{1}{\sigma\sqrt{\pi}} \exp\left[-\left(\frac{t-t_0}{\sigma}\right)^2\right]. \tag{2.4}$$

The time resolution is characterized by the full width at half maximum, FWHM,

$$\mathrm{FWHM}=2\sigma_s\sqrt{\ln 2}\,. \tag{2.5}$$

with the standard deviation σ_s. The convolution of the spectra (2.2) with $F_t(t)$ gives the convolved decay spectrum

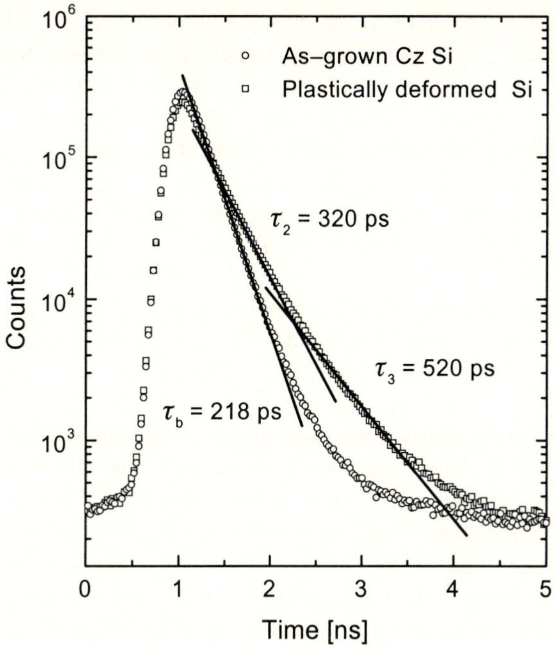

Fig. 2.5. Experimental positron lifetime spectra obtained in as-grown and in plastically deformed Czochralski-grown (Cz) silicon (Hübner et al. 1997b). The curve of the deformed sample (3 % strain, deformation temperature 775 °C), is located significantly higher, indicating the occurrence of long-lived lifetime components. After the decomposition of the upper spectrum, the obtained lifetime components τ_2 and τ_3 are added as straight lines in the semi-logarithmic plot for illustration. The τ_1 component is not indicated as a straight line ($\tau_1 = 120$ ps). Only one lifetime component corresponding to the positron bulk lifetime τ_b is found in the as-grown sample. The deviations from the straight line at higher times are due to annihilations in the source and the background contribution. The Gaussian-like shape of the left part of the curve is mainly caused by the resolution function.

$$D_f(t) = \int_{-\infty}^{\infty} D(t-t')F_t(t')\mathrm{d}t' . \tag{2.6}$$

If $F_t(t) = G(t)$, i.e. a single Gaussian is taken as the resolution function, the positron decay spectrum is given as

$$D_f(t) = \sum_{i=1}^{k+1} \frac{I_i}{2} \exp\left[-\frac{t - t_0 - \sigma_s^2/(4\tau_i)}{\tau_i} \right] \left[1 - \operatorname{erf}\left(\frac{1}{2\sigma_s \tau_i} - \frac{t - t_0}{\sigma_s} \right) \right] . \tag{2.7}$$

Standard computer programs based on Gauss–Newton non-linear fitting routines are available for the decomposition of the spectra (e.g. LIFSPECFIT by

Puska 1978, POSITRONFIT or PATFIT by Kirkegaard et al. 1989). A model spectrum with a given number of decay components and a given resolution function is used for the least squares fit to the measured spectrum by the variation of the parameters of the lifetime components. The linear parameters, i.e. the intensities and the background, are fitted independently of the non-linear ones, which are the lifetimes τ_i and the time-zero channel t_0.

Only a one-component lifetime spectrum is obtained in a defect-free semiconductor. The deformed sample of Fig. 2.5 contains two defect types with open volume, viz. vacancies and voids, and thus, three distinct components appear as a superposition of straight lines in the semi-logarithmic plot, which are to be folded by the time resolution function to give the lifetime spectrum. The spectra additionally contain background contributions as well as annihilation events in the positron source.

The source correction has to be performed after background subtraction. This means subtracting the characteristic lifetime spectrum of the source. The determination of this source spectrum is rather complicated and usually carried out in such a way that a one-component spectrum of a defect-free sample is analyzed. The source components are varied using a single-component fit to get the best fit, i.e. the smallest variance. The problem of the source correction was investigated in detail on measured and simulated spectra (Staab et al. 1996). It was found that sources prepared from Al foils provide a three-component source spectrum. The fraction of the annihilations in the source is not only determined by the source itself but it is also a function of the atomic number of the sample and increases due to multiple positron backscattering through the source.

The determination of the number of components in samples with unknown defect populations is started by a one-component fit of the spectrum. Additional components are added as long as the variance of the fit decreases. However, the possibility of obtaining a reliable multi-component fit is restricted due to the limited time resolution of the spectrometer and the statistics of the measurements. A number of one to four components can be resolved, depending on the intensity and the mutual separation of the individual lifetime components, the time resolution, and the statistics. These effects were studied with Monte-Carlo-simulated as well as experimentally obtained spectra (Somieski et al. 1996). It was found that the possibility of decomposing the positron lifetime spectra depends essentially on the number of components in the spectrum, the distance of the individual lifetimes, and the statistics of the measurement. At least five million events should be collected for a reliable lifetime separation, even if only two components are present. If only the average positron lifetime is to be measured, a lower number of counts is sufficient.

The decomposition of spectra described above is related to spectra having discrete lifetime components, as is expected in most cases for semiconductors. A nearly continuous distribution of lifetime components, as found in polymers (Jean 1995), may occur in semiconductors if vacancy clusters of different sizes are present. In this case, and if the number of components is too high to be resolved indi-

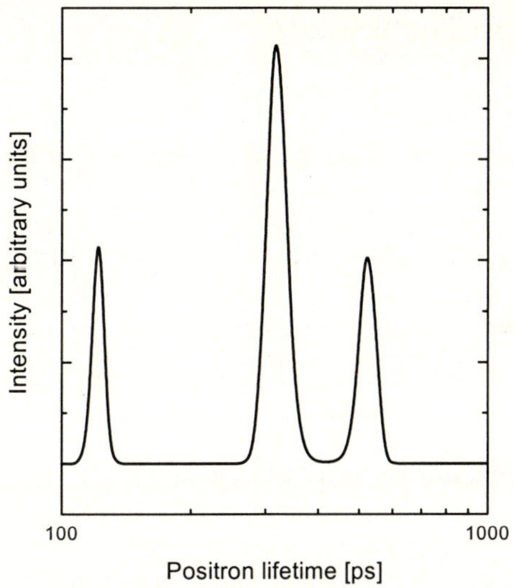

Fig. 2.6. Result of a maximum entropy of lifetime (MELT) analysis of the positron lifetimes of plastically deformed Czochralski-grown silicon (Hübner et al. 1997b). The sample is the same as in Fig. 2.5 (3.5 % deformation at 775 °C). The peaks obtained correspond well to the discrete lifetimes from the decomposition of the spectrum in Fig. 2.5, $\tau_1 = 120$ ps, $\tau_2 = 320$ ps, and $\tau_3 = 520$ ps.

vidually, a continuous distribution of lifetimes can be fitted to the experimental data and an intensity versus lifetime plot is obtained. Such a data analysis is based on Laplace transformation of the measured spectrum (CONTIN program, Provencher 1982).

Another approach has been made by Shukla et al. (1993) using linear filtering and the method of maximum entropy (MELT program). Principally, both methods can also be applied to discrete lifetime spectra. The output is a graph displaying the intensity versus lifetime. Knowledge of the number of components is not required. This is an advantage for multi-component spectra, but it has not found a broad application to semiconductors up to now. The result of a MELT[1] analysis of a plastically deformed silicon sample is shown in Fig. 2.6. The same lifetimes were obtained from the decomposition by a non-linear fit according to (2.3) taking into account the resolution function.

[1] Maximum entropy of lifetime.

2.3 Momentum Distribution Techniques

Conservation of momentum during the annihilation process is the reason for the fact that the annihilation radiation contains information on the electron momentum distribution at the annihilation site. This can be used for the study of the electron structure in solids and for the investigation of defects. There are two basic techniques studying the momentum distribution: Doppler-broadening spectroscopy and angular correlation of annihilation radiation. The basics of both methods are described in Sect. 2.3.1. A survey of the respective experimental setups and the data treatments are given in Sects. 2.3.2 and 2.3.3.

2.3.1 Principle of the Momentum Distribution Techniques

As a result of momentum conservation during the annihilation process, the momentum of the electron–positron pair, $\mathbf{p}$, is transferred to the photon pair. The momentum component p_z in the propagation direction z of the γ-rays results in a Doppler shift ΔE of the annihilation energy of 511 keV, which amounts approximately to

$$\Delta E = p_z c/2. \tag{2.8}$$

Since numerous annihilation events are measured to give the complete Doppler spectrum, the energy line of the annihilation is broadened due to the individual Doppler shifts in both directions, $\pm z$. This effect is utilized in Doppler-broadening spectroscopy, which is described in Sect. 2.3.2.

The momentum components $p_{x,y}$ perpendicular to the propagation direction lead to angular deviations $\Theta_{x,y}$ of the collinearity of the annihilation γ-rays (Fig. 2.1) according to

$$\Theta_{x,y} = \frac{p_{x,y}}{m_0 c} \tag{2.9}$$

m_0 is the rest mass of the electron. These equations hold for small angles. $\Theta_{x,y}$ can be registered simultaneously in both x and y directions by a coincidence measurement using position-sensitive detection of the γ-quanta. This technique is known as the angular correlation of annihilation radiation and is the subject of Sect. 2.3.3.

The momentum conservation during the annihilation process gives a tool to study the momentum distribution of electrons in the solid. The momentum of the positron after thermalization is significantly smaller than that of most electrons. The reason is the Pauli principle and the resulting distribution of the electron momenta up to the Fermi momentum. Although the positron is also a Fermi particle, it can thermalize, because there is no more than one positron in the sample at a given time. This allows the probing of the electron distribution in the electron structure in the momentum space. The effect of the localization of positrons in lattice defects enables such a study, not only in the bulk, but also for open-volume defects. Due to the limited energy resolution of Doppler-broadening spectroscopy, electron structure investigations are carried out mainly by angular correlation of annihilation radiation.

In the case of localization of positrons in open-volume defects, the fraction of valence electrons taking part in the annihilation process increases compared with that of core electrons. Because the momentum of valence electrons is significantly lower, the momentum distribution of annihilating electrons shifts to smaller values. This means a smaller angular deviation for ACAR and a smaller Doppler broadening for DOBS. The curve of defect-rich material is thus higher and narrower than that of defect-free reference material, when both curves are normalized to equal area.

Typical Doppler-broadening spectra measured for as-grown and plastically deformed GaAs are compared in Fig. 2.7 with the resolution function measured as the 514-keV γ-line of ^{85}Sr. The comparison of the measured GaAs curves to this reference γ-line illustrates the effect of Doppler broadening of the annihilation line. Typical values of the FWHM of the ^{85}Sr curve are between 1 and 2 keV.

As an example of the application of momentum techniques for defect investigations, the Doppler spectra of as-grown and plastically deformed GaAs are compared in Fig. 2.7. If the Doppler curves are normalized to equal area, the curve of the deformed sample is higher due to the occurrence of vacancy-type defects. This effect can be used to get quantitative information about these defects (see Sects. 2.3.2.2 and 3.2.2).

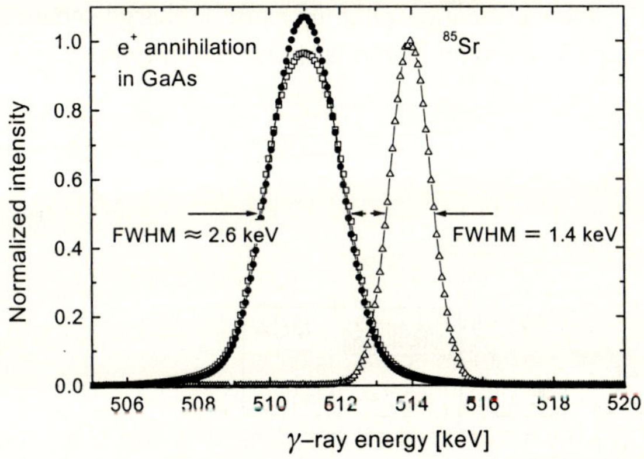

Fig. 2.7. Doppler-broadening spectra of as-grown (□) and plastically deformed (•) gallium arsenide (Hübner et al. 1997b). The Doppler lines of GaAs are broadened compared with the reference line of ^{85}Sr at 514 keV. The full width at half maximum (FWHM) of this reference line characterizes the Gaussian resolution function of the spectrometer. The intensity is normalized to the peak height of the strontium curve (△). The lines are drawn to guide the eye.

Similar investigations providing the kind and concentration of defects in the sample are in principle also possible with angular correlation of annihilation radiation. However, 2D-ACAR is very time consuming for defect investigations and cannot be carried out with positron lifetime spectroscopy at the same time. ACAR has thus rarely been used for defect studies in semiconductors.

An advantage of the momentum techniques is their sensitivity to the chemical environment of the annihilation site, which is higher compared with positron lifetime spectroscopy. This is because the momentum distribution information is more influenced by the chemical surroundings than the electron density, which is the basis of the positron lifetime technique.

2.3.2 Measurement of Annihilation-Line Doppler Broadening

The Doppler-broadening technique requires an energy-dispersive system. Compared with the angular correlation technique (Sect. 2.3.3), a compact and relatively simple setup is possible and, thus, such spectrometers are used in almost all positron laboratories.

2.3.2.1 Experimental Setup

The energy broadening of the annihilation line described above is measured by a high-resolution energy-dispersive detector system (Fig. 2.8). Liquid-nitrogen-cooled pure Ge crystals of high efficiency (about 20 %) are used. Under the applied high voltage of several kV, the annihilation photons cause a charge separation that is converted by a preamplifier into an electrical pulse. Its amplitude is a

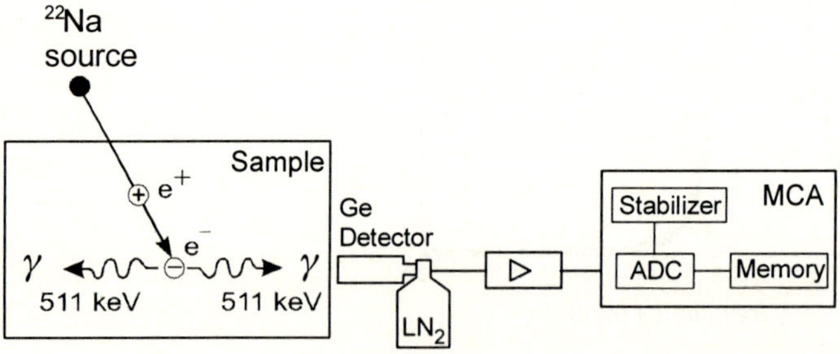

Fig. 2.8. Experimental arrangement of Doppler-broadening spectroscopy. The energy distribution of the annihilation line is measured with a liquid nitrogen (LN_2) cooled germanium detector. The signal of the pure-Ge detector is processed by a preamplifier integrated in the detector. A spectroscopy amplifier provides the necessary pulses for the subsequent digitally stabilized analog–digital converter (ADC). The event is stored in the memory of the multi-channel analyzer (MCA).

measure of the photon energy and can be registered after main amplification in a multi-channel analyzer (MCA). A digital peak-stabilizing system as part of the MCA allows the long-term collection of several million counts. The time of the measurement is comparable to the collection time of a positron lifetime spectrum. Both techniques can easily be performed at the same time because the Ge detector should be separated sufficiently from the sample in order to avoid pile-up effects in the detector system.

The background can be drastically reduced by the application of the Doppler-broadening coincidence technique, which registers both γ-quanta. This technique is especially important for the measurement of the high-momentum part as far as 9 keV away from the center of the annihilation line (Sect. 2.3.2.2). The experimental setup for this technique is shown in Fig. 2.9. The second annihilation γ-quantum, being almost collinear to the primarily monitored quantum, can be detected in a second Ge detector. Alternatively, a cheaper scintillator/photomultiplier detector may be used. The main advantage of the system with two Ge detectors is the improvement of the energy resolution, ideally by a factor of $\sqrt{2}$ (MacDonald et al. 1978).

The coincident γ-quantum serves to reduce the background arising mainly from the 1.27-MeV γ-quanta of the β^+-decay in the source. The background can be suppressed by at least two orders of magnitude in this way. The first experimental realizations of such a coincidence setup were reported by Lynn et al. (1977),

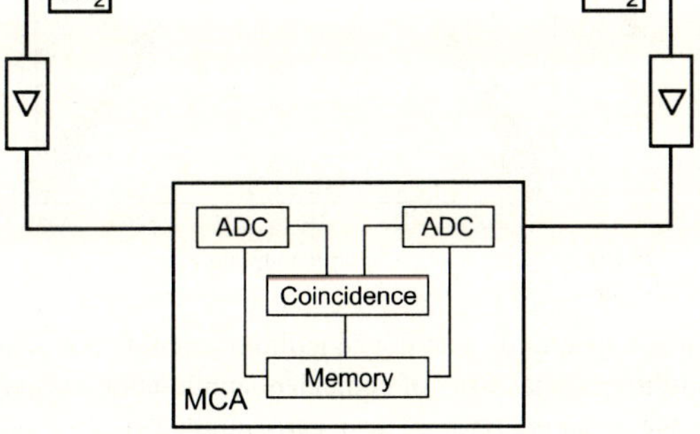

Fig. 2.9. Coincidence setup for the measurement of background-reduced Doppler-broadening spectra. Two liquid nitrogen (LN_2) cooled germanium detectors are used for the registration of the collinear coincidence γ-quanta (ADC—analog–digital converter, MCA—multi-channel analyzer).

MacDonald et al. (1978), and Troev et al. (1979). The first application of this technique to semiconductors was given by Iwase et al. (1985b), who studied the lateral distribution of grown-in defects in an InP wafer. A major step forward in the evaluation of coincidence spectra was the combination with theoretical calculations of the high-momentum part of the annihilation line (Alatalo et al. 1995, 1996; Asoka-Kumar 1997; Asoka-Kumar et al. 1996).

The result is a two-dimensional array of counting rates, where the dimensions represent the energy scales of the respective detectors. An example is shown in Fig. 2.10. A coincidence spectrum of both Ge detectors is obtained as the intensity profile along the diagonal from the upper left to lower right of the measured array. This diagonal profile can be explained by momentum conservation during the annihilation process. The increase in the annihilation γ-ray energy in one detector according to the Doppler shift (2.8) leads to a simultaneous reduction of the γ-ray energy in the second detector, i.e. the sum of the annihilation γ-ray energies remains constant at 1022 keV. The second diagonal is a measure of the resolution of the system. The profiles parallel to the axes at the energy of 511 keV are also coincident spectra. The second detector provides in this case only the gate pulse for the detection of the γ-quanta. Although these spectra also have a reduced background, they are still asymmetrical (closed squares in Fig. 2.10b). Such spectra are usually also obtained if, instead of a second Ge detector, a scintillator is used. A

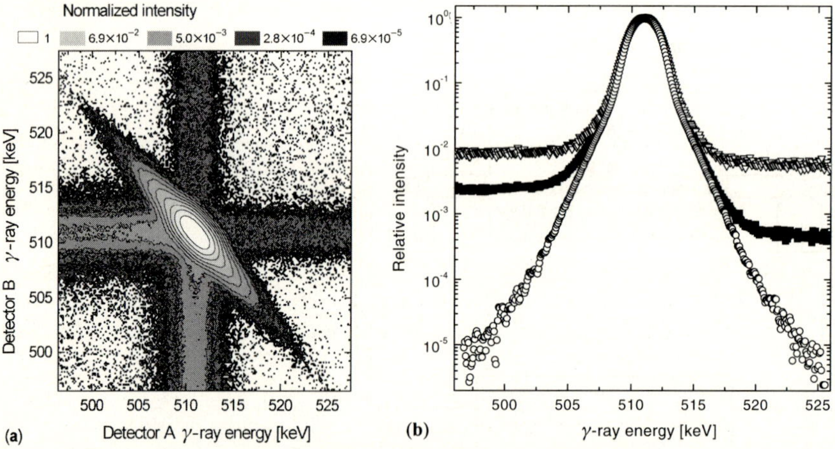

Fig. 2.10. Doppler-broadening coincidence spectra of zinc-doped gallium arsenide (Gebauer and Krause-Rehberg 1998). **(a)** Two-dimensional array of coincident annihilation events. The vertical and horizontal axes are the γ-ray energies of two germanium detectors arranged according to Fig. 2.9. The gray value represents the counting rate normalized to peak maximum. **(b)** Demonstration of the background reduction. The Doppler-broadening spectrum (▽) was measured conventionally (Fig. 2.8) with one Ge detector. The coincidence spectra were obtained from (a) as intensity profiles parallel to the horizontal axis at an energy of 511 keV (■) and diagonal (○).

background reduction of better than one order of magnitude is still possible with a scintillation detector with an optimized setup. The diagonal coincidence spectrum is almost symmetrical and has the improved energy resolution mentioned above. This becomes obvious in Fig. 2.10b, where the improved peak-to-background ratio of 10^{-5} and the better resolution are visible.

2.3.2.2 Data Treatment

The effect of positron trapping in defects on the Doppler-broadening spectrum $N_D = f(E)$ is demonstrated in Fig. 2.11. The quantitative evaluation can be carried out with specific line shape parameters. The S parameter[1] is defined as the area of the central low-momentum part of the spectrum, A_s, divided by the area below the whole curve A_0 after background subtraction,

$$S = \frac{A_s}{A_0}, \quad A_s = \int_{E_0 - E_s}^{E_0 + E_s} N_D \, dE \tag{2.10}$$

The W parameter[2] is taken in a high-momentum region far from the center, as indicated in Fig. 2.11. It is calculated as the area of the curve in a fixed energy interval, A_w, divided by A_0,

$$W = \frac{A_w}{A_0}, \quad A_W = \int_{E_1}^{E_2} N_D \, dE \,. \tag{2.11}$$

The interval limits are chosen symmetrically around the energy of $E_0 = 511$ keV for the calculation of the S parameter, $E_0 \pm E_s$. The energy limits E_1 and E_2 for the W parameter must be defined in such a way as to have no correlation effects with the S parameter. These chosen limits are kept constant for all spectra to be compared. For the determination of the S parameter, the curves of a defect-rich sample and a sample free of positron traps are plotted in Fig. 2.11 normalized to equal area. The S parameter limits should be taken as the intersection points of both curves to obtain the largest sensitivity for a defect-induced change in the line shape. Often, these limits were simply taken to give an S parameter of 0.5 for defect-free material. This means that the determination of W must start at a limit significantly separated from the intersection points of the curves. The limits were set to (511 ± 0.8) keV for the determination of S and to $(511 + 2.76)$ and $(511 + 4)$ keV for W in the example of Fig. 2.11. The ratio of the S parameters for the plastically deformed and the GaAs reference sample was found to be 1.0695. The corresponding ratio of the W parameters was 0.7726.

---

[1] "Shape" parameter; recently also called the valence annihilation parameter.

[2] "Wing" or core annihilation parameter.

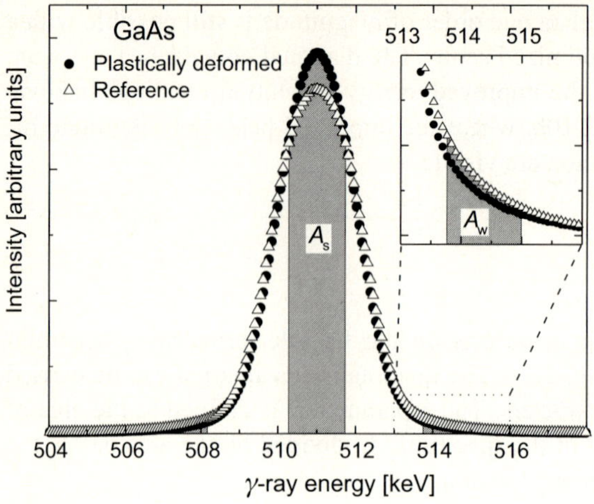

Fig. 2.11. Doppler-broadening spectra of as-grown zinc-doped gallium arsenide showing no positron trapping (reference) compared with plastically deformed GaAs (Hübner et al. 1997b). The line shape parameters S and W are determined by the indicated areas A_s and A_w divided by the area below the whole curve. The curves are normalized to equal area.

The background correction is often performed as the subtraction of a straight line. More sophisticated treatments use a realistic background distribution modeled by a non-linear function. This function takes into account that the background at a given γ-ray energy is proportional to the sum of annihilation events with higher energies. Despite such a background reduction, the Doppler curve remains slightly unsymmetrical. The calculation of the W parameter is, therefore, frequently carried out only in the high-energy wing of the Doppler curve. High-quality Doppler-broadening spectra are obtained by the coincidence technique, preferably with a setup using two Ge detectors. In this case, the W parameters can be taken on both sides of the curve.

The line shape parameters S and W are commonly normalized to their respective bulk values, S_b and W_b. The advantage of such a normalization is that, independent of the chosen limits for the determination of S and W from the Doppler spectrum, the obtained values can be compared. This does not hold for the comparison of values measured with different spectrometers. The reason is the dependence of the normalized line shape parameters on the resolution of the spectrometer (Fig. 2.12). The deviations of the S parameter resulting from differing spectrometer resolutions are much larger than the statistical errors. Resolutions ranging from 1.2 to 2.2 eV were reported. Fig. 2.12 clearly demonstrates that S parameter values reported in the literature and obtained with different spectrometers cannot be directly compared.

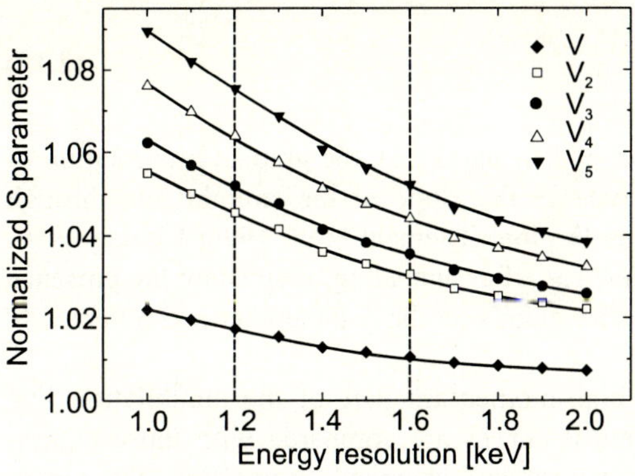

Fig. 2.12. Dependence of the S parameter on the energy resolution of the Doppler-broadening spectrometer. This dependence was obtained from theoretical spectra calculated for different vacancy clusters in silicon (Hakala et al. 1998). It was convolved by Gaussians of different full widths at half maximum simulating the energy resolution (Eichler 1997). The S parameters were each normalized to the respective bulk value convolved by the same Gaussian. The dashed lines indicate the resolution limits of frequently used spectrometers.

A further method for the quantification of the annihilation line shape is the calculation of so-called RID curves[1] (Coleman 1979). For the calculation, the difference of the normalized Doppler curves and its running integral is determined. Information on the type and density of defects may be obtained from the resulting curve. Difficulties arise when more than one defect type contributes significantly to the spectrum. No broad applications of this method are known. Another possibility of defect identification and quantification is the evaluation of difference or ratio curves. A reference spectrum, e.g. a Gaussian function or the spectrum from the defect-free bulk, may be used to reveal hidden details in the Doppler-broadening spectra in order to obtain a fingerprint of the defect structure (see Sect. 3.2.2).

The W parameter is more sensitive to the chemical surrounding of the annihilation site than the S parameter, because the core electrons having a high momentum are contributing mainly in the region of large energy deviations from the annihilation energy of 511 keV.

Both the S and W parameters are sensitive to the concentration and kind of the defect. In order to obtain the defect concentration, one has to presume the defect type, i.e. the defect-related parameters S_d and W_d. A new parameter, R, can be defined, which depends only on the defect type, but not on the concentration (Mantl and Triftshäuser 1978),

[1] Running integrals of difference curves.

$$R=\left|\frac{S-S_{\mathrm{b}}}{W-W_{\mathrm{b}}}\right|=\left|\frac{S_{\mathrm{d}}-S_{\mathrm{b}}}{W_{\mathrm{d}}-W_{\mathrm{b}}}\right|. \tag{2.12}$$

Instead of the numerical calculation of R from (2.12), the graphic presentation of $S = S(W)$ gives R for one defect type as the slope of the straight line through (W_b, S_b) and (W_d, S_d). Such S versus W plots (Clement et al. 1996; Liszkay et al. 1994) for differently treated samples, e.g. after annealing, may show the presence of more than one defect type (for more details on the R parameter and S versus W plots see Sect. 3.2.2.2).

It is possible to calculate the high-momentum part of the annihilation line (compare Sect. 3.3.2). Such theoretical curves are compared with those experimentally obtained in order to get structural and chemical information. The procedure was first applied to vacancies in InP (Alatalo et al. 1995). The isolated phosphorus and indium vacancies were distinguished from each other by the magnitude of the core-electron annihilation. Furthermore, the zinc decoration of phosphorus vacancies in InP:Zn was identified. First-principles calculations have been presented to calculate the momentum distribution of the annihilating electron–positron pair (e.g. Gilgien et al. 1994; Hakala et al. 1998; Tang et al. 1997). Complete angular correlation of annihilation radiation spectra can be obtained and from them, by integration and convolution, whole Doppler-broadening spectra.

2.3.3 Angular Correlation of Annihilation Radiation

The basic principle of angular correlation of annihilation radiation has already been illustrated in Fig. 2.1. The momentum components $p_{x,y}$ perpendicular to the propagation direction are related to the deviations $\Theta_{x,y}$ of the collinearity of the annihilation γ-rays according to (2.9). Since the annihilation γ-quanta are emitted simultaneously, $\Theta_{x,y}$ can be measured in a coincidence arrangement by position-sensitive detectors. The coincidence counting rate N_c is given as the integral of the momentum distribution of the annihilating positron–electron pair, $\sigma = \sigma(p_x, p_y, p_z)$, versus the momentum component p_z in the propagation direction of the annihilation radiation,

$$N_{\mathrm{c}}(\Theta_x,\Theta_y)=A_{\mathrm{c}}\int_{-\infty}^{\infty}\sigma(\Theta_x m_0 c,\Theta_y m_0 c,p_z)\,\mathrm{d}p_z\,. \tag{2.13}$$

A_c is a constant. As described above, the electron–positron momentum distribution in z direction gives the Doppler shift of the annihilation γ-quanta. The integration in (2.13) is due to the lack of energy-dispersive position-sensitive detectors, i.e. every measured point represents all events of the corresponding Doppler spectrum. As the momentum distribution is recorded in two dimensions, this technique is referred to as 2D-ACAR.

The first ACAR measurements in one dimension were realized with Geiger counters by Behringer and Montgomery (1942). A position-sensitive detection can be realized in the simplest way in one dimension (1D-ACAR) by the mechanical movement of a long scintillation detector (Hautojärvi and Vehanen 1979; Mijnarends 1979). The integration in one more dimension compared with (2.13) results in a counting rate of

$$N_c(\Theta_x) = A_c \int_{-\infty}^{\infty} \int_{-\infty}^{\infty} \sigma(\Theta_x m_0 c, p_y, p_z) \mathrm{d}p_y \mathrm{d}p_z \, . \qquad (2.14)$$

The angular resolution is realized by lead slits (Fig. 2.13). It can be adjusted in the range 0.2 to 5 mrad. The energy resolution of a corresponding Doppler-broadening experiment would range from 0.05 to 1.3 keV. Thus, ACAR has a much better momentum resolution than the Doppler-broadening technique.

The source is placed outside the visual field of the detectors. The sample–detector distance amounts to several meters. Because of this large distance, annihilation quanta from only a small solid angle are detected. Thus, much stronger sources compared with conventional positron lifetime and Doppler-broadening measurements are required (Sect. 2.1). In order to minimize the reduction of the counting rate due to the distance of several mm between sample and source, a strong magnetic field of about 1 T is usually applied to guide the positrons to the sample.

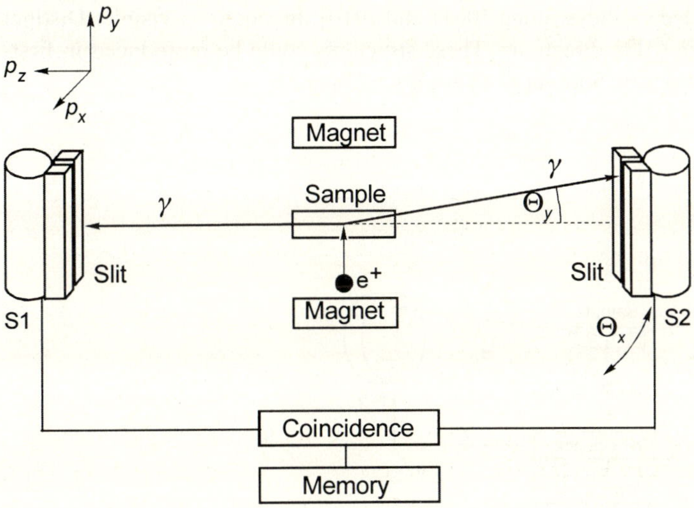

Fig. 2.13. Experimental setup for the measurement of one-dimensional angular correlation of annihilation radiation with a long-slit geometry. Sodium iodide crystals are used as scintillator rods (S1 and S2) in the y direction. The left arm with the detector and the lead collimators is fixed, while the right arm can be turned by the angle Θ_x.

The main application of angular correlation of annihilation radiation is the study of the electron structure of the bulk and of defects. The momentum resolution is significantly higher than for Doppler-broadening spectroscopy. A two-dimensional record of the momentum distribution is favorable for comparing with theoretical calculations. 1D-ACAR machines were hardly used for semiconductors, since the intensive study of semiconductors set in when 2D-ACAR machines became available. This two-dimensional detection (Fig. 2.14) can be carried out with position-sensitive detectors, such as multi-wire proportional chambers or Anger cameras. The electronics of the 2D-ACAR apparatus filters the coincident events of the two detectors and stores the angular deviation from collinearity in a two-dimensional memory. No movable parts are required. The typical resolution is in the range of 0.2×0.2 mrad^2. The time of the measurement to collect a single 2D-ACAR spectrum amounts to several days.

In principle, the line shape parameters S and W as described in Sect. 2.3.2.2 can also be used for angular correlation of annihilation radiation. In addition, the so-called peak-height parameter H has often been used for 1D-ACAR. The measurement of this parameter can be carried out very rapidly by fixing the movable detector arm in the collinear position. However, such a parameterization as broadly applied for Doppler-broadening spectroscopy is unusual for two-dimensional angular correlation of annihilation radiation as the details of the electron structure are almost lost. The two-dimensional electron momentum distribution is contained in the whole 2D plots of the coincidence events.

As an example, the perspective and contour plots of the 2D-ACAR distribution of gallium arsenide are shown in Fig. 2.15 (Tanigawa et al. 1995). A four-fold symmetry with deep valleys along [001] and [010] directions is visible. Distinct maxima appear in ⟨011⟩ directions. These structures could be reproduced in first-principle calculations by Saito et al. (1991); see Fig. 2.16.

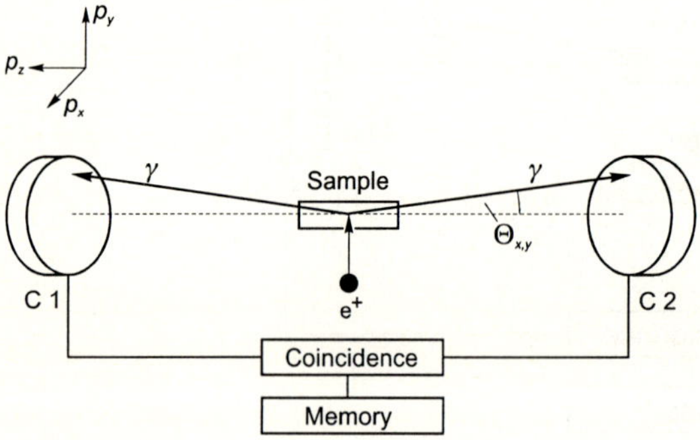

Fig. 2.14. Scheme of the apparatus for two-dimensional angular correlation of annihilation radiation. The angular deviation of the two γ-quanta, $\Theta_{x,y}$, is recorded by position-sensitive detectors (C1and C2) in a coincidence setup and stored in a two-dimensional memory array.

The electronic structure of defects can be studied in a similar way. However, in order to obtain the defect-specific plots, a normalization for saturated positron trapping, i.e. for capture of all positrons in defects, is necessary. This can be done by correlated positron lifetime measurements, which provide the fraction of positrons trapped in the defect. Such contour plots can also be directly compared with theoretical calculations to improve the understanding of the electronic structure of the defect.

The first depth-resolved measurements of the two-dimensional angular correlation of annihilation radiation were reported by Peng et al. (1996). The defect structure of the SiO_2/Si interface was studied by the combination of slow-positron-beam measurements and 2D-ACAR. The positron beam system used had a solid krypton moderator and an initial positron beam flux of about $6\times10^7\ s^{-1}$. The depth resolution was obtained by the variation of the incident positron energy in the

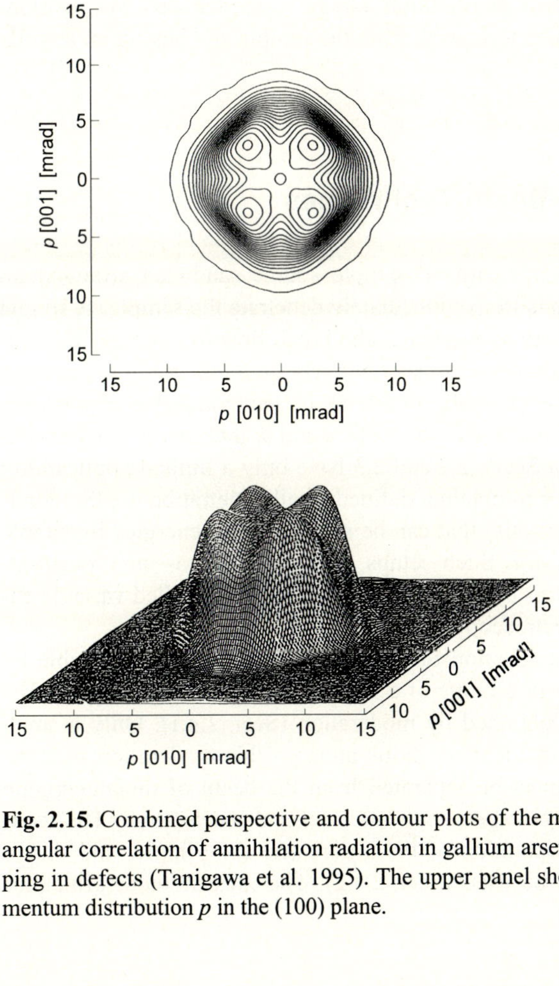

Fig. 2.15. Combined perspective and contour plots of the measurement of two-dimensional angular correlation of annihilation radiation in gallium arsenide exhibiting no positron trapping in defects (Tanigawa et al. 1995). The upper panel shows the contour map of the momentum distribution *p* in the (100) plane.

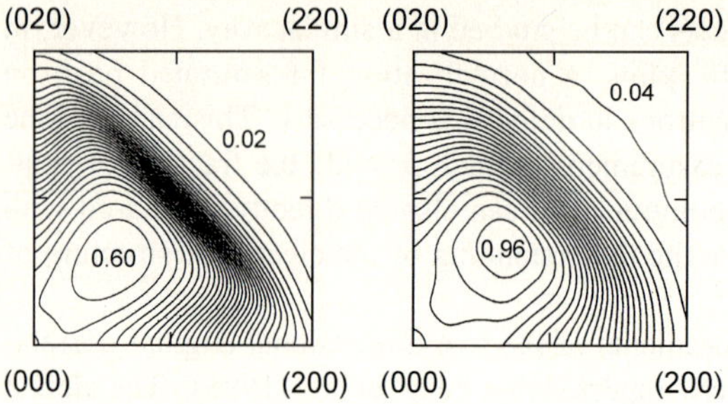

Fig. 2.16. Comparison of the two-dimensional angular correlation of annihilation radiation contour plot of the momentum distribution in the (001) plane of gallium arsenide obtained by ab-initio calculations (*left panel*) by Saito et al. (1991) with an experimental spectrum (*right panel*). The right panel represents a magnified part of the contour plot of Fig. 2.15.

range 0.2 to 20 keV. The two annihilation γ-rays were detected by position-sensitive Anger cameras located 8 m away from the sample and having an angular resolution of 1.4×1.4 mrad^2.

2.4 Slow-Positron-Beam Techniques

The main advantage of the conventional sample–source sandwich arrangement (Fig. 2.3) is that the emitted positrons immediately penetrate the sample. A simple measurement arrangement in air is possible. The broad positron energy distribution up to 540 keV for ^{22}Na sources causes a relatively high maximum penetration depth (see Sect. 3.1.3). However, many problems in semiconductor physics are related to thin layers and to defects near the surface and at interfaces. The conventional techniques described in Sects. 2.2 and 2.3 have only a limited applicability to such investigations. In order to obtain a defined small penetration depth, monoenergetic positrons (slow positrons) that can be set to defined energies by simple linear accelerators are necessary. Such setups are referred to as slow-positron-beam techniques throughout this book. They are elsewhere also called variable energy positron annihilation spectroscopy (VEPAS).

Monitoring of defects as a function of depth (defect profiling) is possible by adjusting the positron energy in a range of a few eV to several tens of keV. The monoenergetic positrons are obtained by moderation (Sect. 2.4.1). Only a small fraction of less than 1 % of incident positrons undergo this moderation process. The unmoderated positrons must be separated from the beam of monoenergetic positrons that is used for defect experiments after defined acceleration. The moderation requires the spatial separation of the source and the sample, and thus a beam guidance system must be used (Sect. 2.4.2). The measurement principles and the application for defect profiling of the different slow-positron-beam techniques are explained in Sects. 2.4.3 and 2.4.4.

2.4.1 Positron Source and Moderation

The energy emission spectrum of radioactive ^{22}Na sources is shown in Fig. 2.17. The moderation is based on the fact that a negative positron work function Φ_+ exists for many solids. In most cases, a transmission geometry with a thin moderator foil placed directly on top of the source capsule is used. The thickness of the foil is much smaller than the mean penetration depth and, therefore, only a small fraction of positrons thermalizes and starts to diffuse there. If the surface is reached during the diffusion, the positrons are spontaneously emitted from the moderator foil (Fig. 2.18) with a kinetic energy equal to the thermally broadened work function Φ_+. Materials with high atomic numbers are favorable for moderation, because the ratio of the mean diffusion length to the thermalization distance is larger. A suitable material is a single-crystal tungsten foil in a (100) orientation with a thickness of a few μm or a (110) tungsten single crystal for application in backscattering geometry. Since the positrons may be trapped in defects during their diffusion to the surface, a foil containing only a small number of positron traps must be prepared by annealing. The work function of a (110)-oriented tungsten single crystal was measured to be $\Phi_+ = -3.0$ eV and a moderation efficiency of 3×10^{-3} could be achieved (Vehanen et al. 1983). The moderation efficiency is given as the ratio of the number of moderated slow positrons to the total number of incident positrons. Routinely, efficiencies in the order of 10^{-4} can be obtained. Polycrystalline tungsten foils having an only slightly lower moderation efficiency have been also suc-

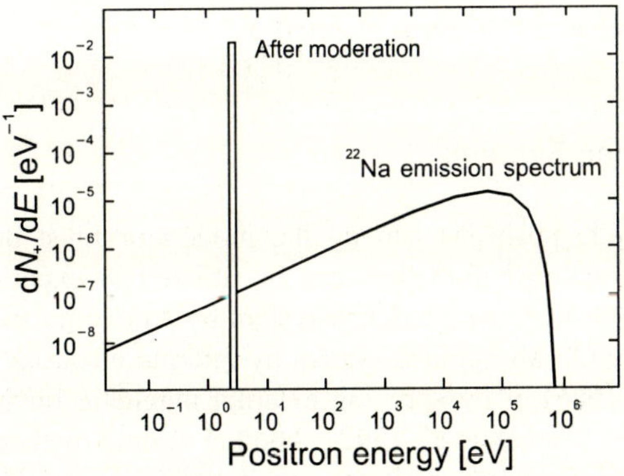

Fig. 2.17. Scheme of the positron emission spectrum of a ^{22}Na source. dN_+/dE is the number of positrons per energy channel E. The narrow curve centered at 3 eV illustrates the energy distribution after moderation in tungsten.

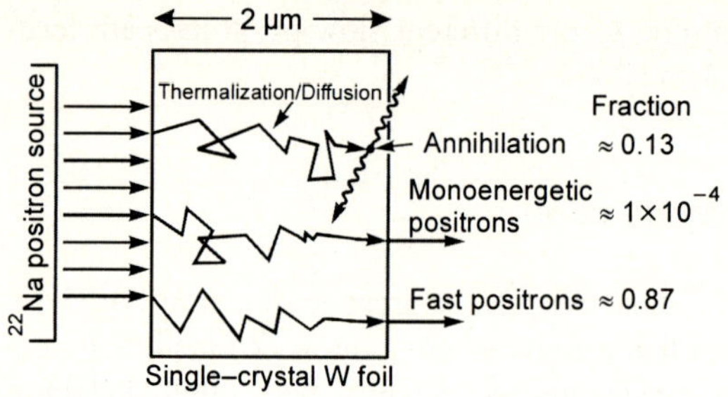

Fig. 2.18. Scheme of positron moderation in transmission geometry by a (100) tungsten foil. Most positrons leave the moderator foil with a high residual energy. A smaller fraction stops in the foil and annihilates there. When the surface is reached during diffusion, the positrons may be spontaneously emitted due to the negative work function of tungsten. The moderation efficiency amounts to about 10^{-4}.

cessfully tested (Brusa et al. 1992).

Recently, solid-state rare-gas moderators, which are obtained by depositing a neon or krypton layer on a carrier foil at low temperatures, have been described (Khatri et al. 1990; Mills and Gullikson 1986). They have an extremely high efficiency, up to 10^{-2}, which cannot yet be conclusively explained. It is supposed that a drift due to an internal electric field superimposing the diffusive motion of the positrons plays an important role. Another future development may be the use of SiC as a field-assisted positron moderator (Bauer-Kugelmann et al. 1997; Beling et al. 1987; Brauer et al. 1997; Störmer et al. 1996a). This material is the only semiconductor known so far with a negative positron work function.

The low moderation efficiency requires much stronger positron sources for positron beam techniques than those used in conventional positron lifetime and Doppler-broadening spectroscopy. Intensive radiation protection is needed because the source activities are as high as 5×10^9 Bq (135 mCi).

2.4.2 Positron Beam Guidance Systems

The small fraction of monoenergetic positrons leaving the moderator must be separated from the unmoderated fraction before they can be utilized in experiments. This separation takes place in the beam guidance system by an energy filter, which may be realized in a magnetically guided system by internal electrodes in an **E**×**B** filter (Hutchings et al. 1986) or by applying external magnetic fields perpendicular to the beam direction (Liszkay et al. 1992). Another simple method is the use of bent solenoids. The unmoderated positrons are stopped in a shield. High vacuum conditions of 10^{-5} Pa are sufficient for the guidance system and positron studies near the sample surface. Ultra-high vacuum is only required in connection with surface studies. In this case, the specimen chamber should be

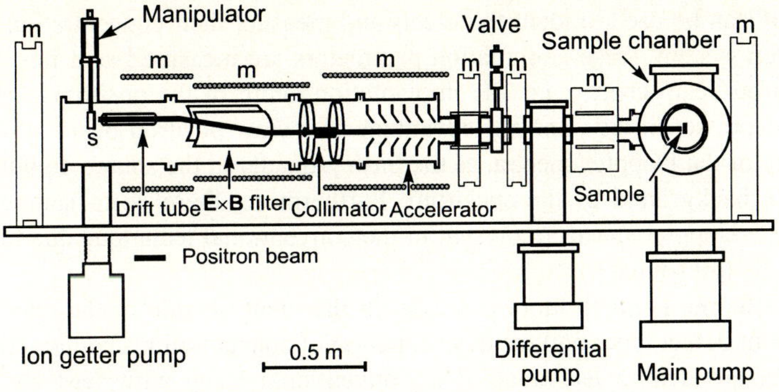

Fig. 2.19. Section view of the slow-positron-beam system POSSY at Martin Luther University Halle–Wittenberg. Positrons emitted from the source–moderator arrangement (s) are fed through the collimator by the **E**×**B** filter and are accelerated to the sample. The overall longitudinal magnetic field is generated by the guidance coils (m). The vacuum at the source side is kept in the 10^{-6} Pa range, while a pressure of 10^{-9} Pa may be achieved in the specimen chamber by means of the differential pumping stage.

separated by a differential pumping station. As an example, the magnetically guided positron beam system which is realized at Martin Luther University Halle–Wittenberg is shown in the section view of Fig. 2.19. The source–moderator arrangement is placed in front of a drift tube, at the end of which the positrons are entering the **E**×**B** filter. The monoenergetic positron beam is guided into the system axis and through a linear accelerator supplying a maximum energy of 50 keV. The fast positrons are stopped in the collimator.

The positron emission angle to the normal of the moderator foil amounts to a few degrees (Fischer et al. 1986). A longitudinal magnetic field, which is spread over the whole beam system, forces the positrons onto a helical trace. This ensures that all emitted slow positrons reach the target. A system of guidance coils (Fig. 2.19) generates this longitudinal field.

The beam guiding can also be done by a system of electrostatic lenses (Rosenberg et al. 1980). The main advantage is the possibility of focusing the beam. The design of such a lens system is, however, rather complicated due to the broad energy variation of the positron beam.

2.4.3 Measurement Principles

The Doppler-broadening technique can be applied for slow positrons similar to the conventional procedure with foil sources (Sect. 2.3.2). The Ge detector is mounted close to the sample outside the specimen chamber. This is possible because the energy of the annihilation radiation is high enough to allow the transmission of the γ quanta through a thin stainless-steel wall. As explained earlier, the line shape parameters S and W can be used to identify defects and measure their concentration. In a slow-positron system, these annihilation parameters are measured as a function of the positron beam energy, i.e. the implantation depth of the positrons. In addition to the depth sensitivity, another advantage of a slow-positron beam system is the quality of the Doppler spectra, as the birth γ-quanta in the source do not contribute to the background of the spectrum. Furthermore, there is no source contribution to the Doppler spectrum present in the conventional technique due to annihilations in the foil source.

The positron lifetime is much more powerful in the identification of the open volume of different defect types and the determination of concentrations compared with the Doppler-broadening technique. The conventional setup with start and stop detectors is not applicable in a slow-positron-beam system due to the strength of the source, since the start γ-quanta in the source (compare Sect. 2.2) cannot be correlated with annihilation events in the sample any more. Furthermore, the time of flight is much longer than the lifetime in the specimen. Thus, the measurement of the positron lifetime is only possible by pulsed beams supplying the start pulse by a specially designed bunching system. Such a technique is electronically expensive and has been realized up to now in only a few laboratories, e.g. in Munich (Kögel et al. 1988) and Tsukuba (Akahane et al. 1990). The performance limits of pulsed positron beams were discussed by Sperr and Kögel (1997).

In addition to lifetime and Doppler-broadening measurements, two-dimensional angular correlation of annihilation radiation can also be performed in a slow-positron-beam setup. The advantage is the analysis of the electronic structure at the surface, in thin epitaxial layers, or at interfaces (e.g. Howell et al. 1985; Peng et al. 1996).

The defect densities can be determined in a slow-positron system by a back-diffusion experiment in addition to the measuring principles of the momentum distribution and the positron lifetime. The fraction of positrons diffusing back to the surface, f_s, can be determined via the annihilation parameters at the surface or the fraction of positronium formed at the surface. This is possible because the surface annihilation parameters usually differ from the values in the interior of the sample, and positronium can usually only be formed at the semiconductor surface. The back-diffusing fraction of positrons is not only a function of the positron implantation depth and the diffusion constant, but also of the defect concentration. This is due to the fact that the trapped positrons cannot reach the surface. f_s is measured as a function of the incident positron energy. The corresponding fitting routines provide the trapping rate κ as a function of the depth. The main disadvantage of this procedure is that no information is available on the nature of the positron trap. Correlated positron lifetime measurements and Doppler-broadening measurements are recommended. On the other hand, total trapping rates for all positron traps are obtained by back-diffusion experiments. Hence, defect concentrations can also be determined in the case of saturated positron trapping, i.e. at very high concentrations, as was demonstrated for the case of ion-implanted silicon e.g. by Eichler et al. (1997a).

2.4.4 Defect Profiling

In order to obtain the defect depth profile from the measured variation of annihilation parameters as a function of the incident positron energy, knowledge of the positron implantation profile is required. The profiles resulting from Monte-Carlo simulations and experimental results are presented in Sect. 2.4.4.1. The computation of the defect depth profiles is nowadays carried out with computer programs. These procedures are described in Sect. 2.4.4.2.

2.4.4.1 Positron Implantation Profile

In contrast to the techniques described in Sects. 2.2 and 2.3, the variation of the positron energy allows the detection of defects as a function of the penetration depth z, i.e. defect depth profiling. The implantation or penetration profile $P(z,E)$ of monoenergetic positrons having the energy E is given by

$$P(z,E) = \frac{mz^{m-1}}{z_0^m} \exp\left[-\left(\frac{z}{z_0}\right)^m\right] \quad \text{with} \quad z_0 = \frac{AE^r}{\rho\Gamma\left(1+\frac{1}{m}\right)}. \tag{2.15}$$

m, r, and A are empirical parameters. ρ is the mass density of the sample and Γ the gamma function. Widely used empirical values are: $A = 4.0\ \mu\mathrm{g\,cm^{-2}\,keV^{-r}}$, $m = 2$, and $r = 1.6$ (Vehanen et al. 1987).

$$\bar{z} = AE^r/\rho \tag{2.16}$$

represents the mean penetration depth. The positron implantation profile is called a Makhov profile, named after Makhov's original electron implantation experiments (Makhov 1961). The parameters of this profile can be obtained theoretically from Monte-Carlo simulations (Valkealahti and Nieminen 1983, 1984). Ghosh (1995) showed by several Monte-Carlo calculations that the parameters A and r are material dependent. The positrons after thermalization, i.e. at the beginning of diffusion, exhibit a depth distribution as given by (2.15). Examples of such Makhov profiles are shown in Fig. 2.20. The limitations of defect profiling for high positron energies become visible, as sharp defect structures lying relatively deep in the sample are smeared out when folded with the broad positron implantation profile. Ghosh (1995) obtained different A and r values depending on the Monte-Carlo schemes used. Although the main features of the positron implantation profile can be described by the analytical Makhov profile, small deviations are found. These deviations were the reason to approximate the profile obtained by Monte-Carlo simulations by a parameterized function (Ghosh and Aers 1995).

The empirical parameters of the Makhov profile (2.15) were also determined experimentally (Gebauer et al. 1997b; Leung et al. 1995). The results of the former study are shown in Fig. 2.21. The parameters of the implantation profile could be determined from S versus E plots of amorphous silicon layers, since the thickness was determined independently by cross-section scanning electron microscopy (SEM). The a-Si layers of a thickness in the range 120 to 1000 nm were grown by molecular beam epitaxy on a SiO_2 buffer layer, which was obtained by thermal oxidation at 1000 °C. The S parameter curves were fitted using the VEPFIT program (see Sect. 2.4.4.2) using systematically varying A and r parameters of the Makhov profile. For every (r, A) pair, the mean quadratic deviation ξ was calculated from the deviation of the layer thickness obtained from the fit of the positron results, z_i^{fit}, and the layer thickness measured by SEM, z_i^{SEM},

$$\xi = \frac{1}{5}\sum_{i=1}^{5}\left(\frac{z_i^{\text{SEM}} - z_i^{\text{fit}}(A,r)}{z_i^{\text{SEM}}}\right)^2 . \tag{2.17}$$

The deviation ξ is averaged in (2.17) over all five measured $S(E)$ profiles. Gebauer et al. (1997b) found a minimum deviation for the Makhov profile parameters of $A = (2.75 \pm 0.25)\ \mu\text{g}\,\text{cm}^{-2}\,\text{keV}^{-r}$ and $r = 1.7 \pm 0.05$, as can be seen in the contour plot of Fig. 2.21b. These parameters are in a good agreement with the results of the Monte-Carlo simulation of Ghosh (1995).

2.4.4.2 Computation of Defect Depth Profiles

The positron trapping into defects can occur after thermalization during diffusion. The positron diffusion can be described as usual by the Fick's second law (compare Sect. 3.1.4). For the determination of depth profiles, only the one-dimensional diffusion equation has to be solved.

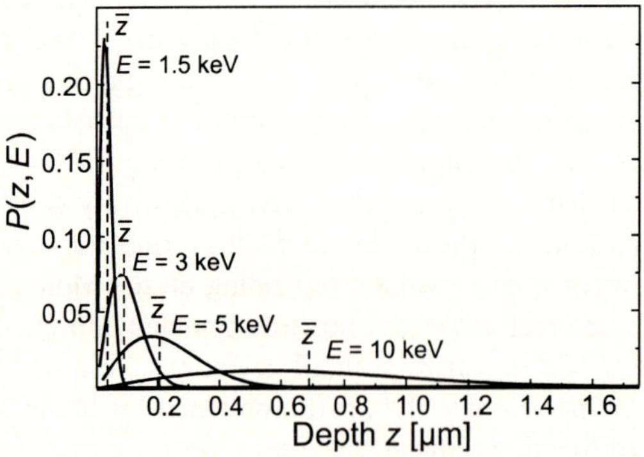

Fig. 2.20. Makhov profiles $P(z, E)$ in silicon calculated for four incident positron energies according to (2.15) with the parameters $A = 4.0\ \mu\text{g}\,\text{cm}^{-2}\,\text{keV}^{-r}$, $m = 2$, and $r = 1.6$. The dashed lines correspond to the mean penetration depth $\bar{z}$.

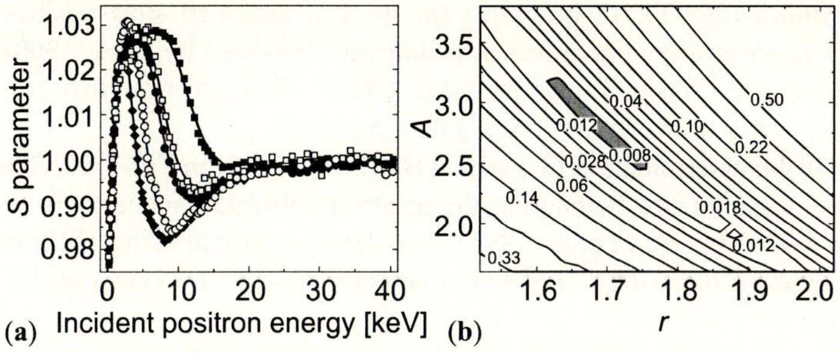

Fig. 2.21. Experimental determination of the parameters A and r of the Makhov profile (2.15). (**a**) S parameter as a function of the positron energy for amorphous silicon layers (■ 1000 nm, □ 485 nm, ● 350 nm, ○ 200 nm, ◆ 120 nm thickness) grown on 600 nm thick silicon dioxide. The oxide layer was obtained by 1000 °C annealing of a Czochralski-grown silicon wafer. The solid lines are numerical fits with the VEPFIT (variable energy positron fit) program. The A and r parameters were varied in order to reproduce the thickness of the Si layer determined independently by cross-section scanning electron microscopy (SEM). (**b**) Mean quadratic deviation ξ of the thickness determined from the positron experiment by VEPFIT and SEM shown as contour lines in an A versus r plot. The parameter A is given on the axis in units of $\mu g\,cm^{-2}\,keV^{-r}$. All five curves were taken into account for the calculation of ξ. The minimum deviation is indicated by the shaded area at $A = (2.75 \pm 0.25)\ \mu g\,cm^{-2}\,keV^{-r}$ and $r = 1.7 \pm 0.05$ (Gebauer et al. 1997b).

Possible quantities of the measurement are the line shape parameters (S or W parameter), the positron lifetime components τ_i and their intensities I_i, or the so-called positronium fraction F (see Sect. 3.1.6). The profile of these annihilation parameters in terms of the depth is the defect profile in the sample convolved by the penetration profile $P(z, E)$ and influenced by the positron diffusion. The diffusion equation must be solved numerically. However, the defect depth profile can hardly be identified directly. For the numerical procedure, the sample is usually divided into slices which are taken thin enough to assume a constant defect concentration and a constant positron density there. The annihilation parameters are fitted as a function of the energy in a non-linear procedure. In this way, the defect distribution perpendicular to the surface is obtained.

The VEPFIT[1] program is a package for the evaluation of slow-positron beam data (van Veen et al. 1990, 1995). A Gaussian curve as an analytical function of the defect profile or a simple slice structure with a constant defect concentration in the particular layer must be taken as a program input. Both a Gaussian and a step function of the defect concentration may reflect the experimental data appropriately. Often, it cannot be decided which function is the better choice to mimic the real defect profile. This is due to the broad implantation profile of the positrons and the positron diffusion, which is itself a function of the defect concentration.

[1] Variable energy positron fit.

Another algorithm (POSTRAP program) for the evaluation of slow-positron depth profiles in order to carry out defect profiling was presented by Aers (1990). The program includes defects and the effect of electric field on positron diffusion. It allows arbitrary forms of the positron implantation profile.

An example of the determination of a defect profile is shown in Fig. 2.22. The *S* parameter was measured as a function of the positron implantation energy in arsenic-implanted silicon (Gebauer et al. 1995). The data were fitted using VEPFIT assuming a structure of four slices of different defect densities or a Gaussian-like defect distribution.

The computer programs mentioned above also provide the fraction of positrons diffusing back to the surface, f_s. Such back-diffusion experiments have the advantage that the positron trapping rate can also be determined in the case of saturated

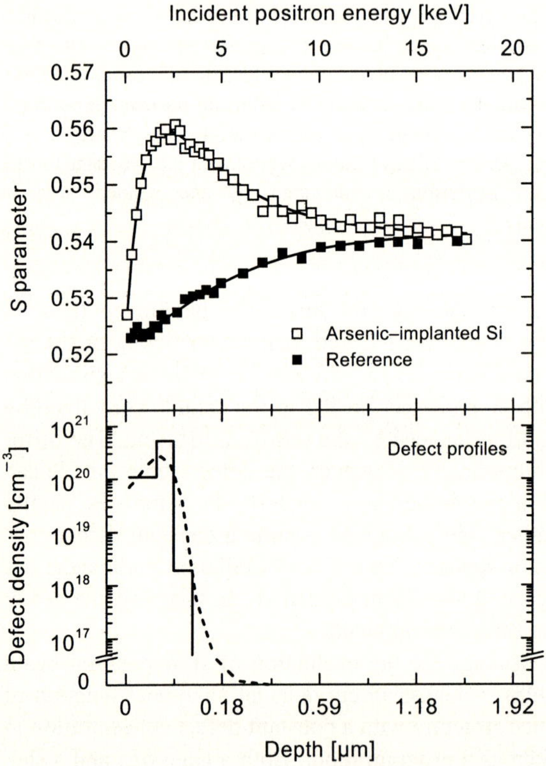

Fig. 2.22 *S* parameter as a function of the incident positron energy for a silicon sample implanted with 5×10^{13} cm^{-2} arsenic (60 keV). The curve through the data points was obtained by VEPFIT (variable energy positron fit). The data and the fit for as-grown Si free of positron traps are shown as reference. Four slices of different defect densities or a Gaussian-like defect distribution were assumed for the fit. The layer structure providing the best fit is shown in the lower panel (Gebauer et al. 1995).

positron trapping in defects, in contrast to the conventional techniques described in Sects. 2.2 and 2.3. This means that there is in principle no upper sensitivity limit. However, the lower sensitivity limit is comparable to the conventional techniques due to the requirement of a minimum fraction of positrons annihilating from defects. The annihilation fraction η can be calculated from f_s as a function of incident positron energy E, i.e. as a depth profile,

$$\eta(E) = 1 - \frac{f_s(E)}{f_s^{\mathrm{ref}}(E)}, \tag{2.18}$$

where $f_s(E)$ are values in the sample under consideration and $f_s^{\mathrm{ref}}(E)$ the corresponding back-diffusion fractions in a defect-free reference sample (Mäkinen et al. 1986).

2.5 Other Positron Techniques

In addition to the experimental setups mentioned above, new positron techniques have become available in recent years. They are, however, partly still under development or have a limited applicability to defect issues in semiconductors. They are therefore only briefly introduced here for the sake of completeness.

The slow-positron-beam techniques described in Sect. 2.4 have only a limited spatial resolution. The development of highly focused brightness-enhanced positron beams allows the investigation of matter with the positron microscope or microprobe (Sect. 2.5.1). The combination of positron lifetime and Doppler-broadening spectroscopy is called age–momentum correlation (AMOC). It is described in Sect. 2.5.2. The chemically sensitive Auger electron process has been applied in a modified manner by positron-annihilation-induced Auger spectroscopy (PAES). This method is extremely surface-sensitive and is outlined in Sect. 2.5.3. Another new interesting application is low-energy positron diffraction (LEPD), which is presented in Sect. 2.5.4.

2.5.1 The Positron Microscope and Microprobe

The first transmission positron microscope (TPM) was realized by van House and Rich (1988). They achieved a magnification of 55 in transmitting positrons through a polymer foil. TPM may have a variety of new applications compared with transmission electron microscopy (TEM). The comparison of the contrast formation in TEM and TPM may provide valuable information on particular terms in the scattering cross-sections. The reduced small-angle scattering of positrons is governed by the much more effective screening of the nuclei. A strongly Z-dependent (Z atomic number) difference in the amplitude contrast exists between electron and positron microscopy. Thus, the comparison of the contrast mechanisms of both techniques could provide sensitive chemical information.

The positron re-emission microscope (PRM) utilizes the effect of negative positron work functions at certain surfaces (for the data see Schultz and Lynn 1988). The surface of a sample is exposed to a positron beam with an energy of several keV. The penetrating positrons thermalize and diffuse, as a result of which a large fraction reaches the surface. The lateral distribution of re-emitted positrons is used to form an image by a position-sensitive detector, such as a channel plate. A dark contrast is obtained in this image, when near-surface defects prevent positrons from emission. The PRM represents thus an extremely surface- and near-surface-sensitive tool for the detection of positron-active defects. Such a microscope was realized in a reflection geometry by van House and Rich (1988). They achieved a spatial resolution of 2.3 μm.

Positron microscopy is still in an early stage of development (for a review see Waeber 1990). The main drawback is the lack of a positron beam with sufficient brightness. The spatial resolution is limited because positron point sources are not available.

The reflection geometry of the positron re-emission microscope allows in principle operation in a microprobe mode (Hulett 1995). This is due to the fact that the chemical composition of the surface considerably influences the work function of the positrons used for the imaging (Wissman et al. 1992). The realization of a positron microprobe requires sufficiently intense positron microbeams.

In addition to the re-emission, other annihilation parameters, e.g. the Doppler-

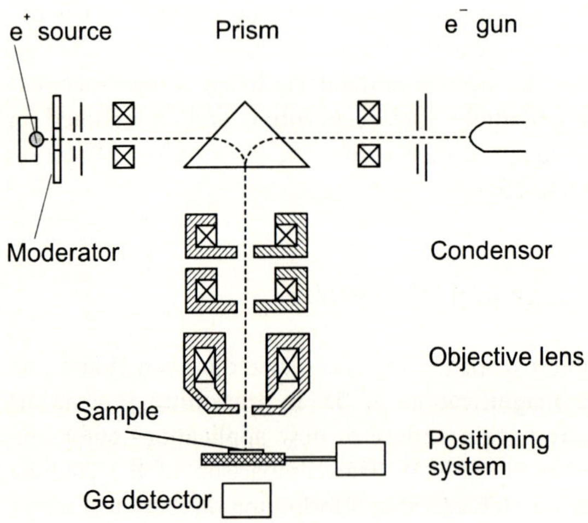

Fig. 2.23. Combination of a scanning electron microscope (Zeiss/Leo) and a positron microprobe as realized at Bonn University (Greif et al. 1997). The positron beam is guided through the optical axis of the microscope. The electron beam is used to select the area of interest for Doppler-broadening measurements.

broadening S parameter or the positron lifetime, can be detected in a positron microscope. A promising development at Bonn University represents the combination of a scanning electron microscope with a positron microprobe. The SEM image is used to select areas of interest on the sample for subsequent positron measurements with high lateral resolution. For this purpose, a monoenergetic positron source was integrated in the optical system of a commercially available scanning electron microscope (Greif et al. 1997). The electron gun is replaced by a magnetic prism for energy filtering of the positrons. The electron and the positron beams are guided through the prism with the same magnetic field in the optical axis of the microscope (Fig. 2.23).

As a first application, the fatigue damage profile along the cross-section of a bent copper plate was investigated (Greif et al. 1997). The observed S parameter profile agreed well with the fatigue damage distribution. A lateral resolution of 30 μm was achieved. A counting rate of 35 s^{-1} in the germanium detector was obtained. The annihilation events were accumulated over several hours per beam position. Such low counting rates hardly allow the formation of images and thus mainly lines scans are possible.

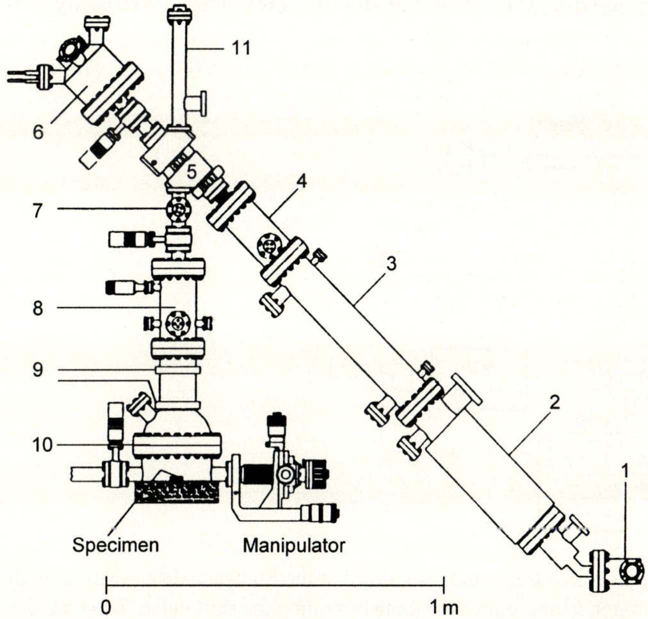

Fig. 2.24. Scheme of the Munich scanning positron microscope (David et al. 1997). (1) Positron source and moderator, (2) drift tube for pulse forming, (3) first buncher, (4) accelerator, (5) beam switch, (6) re-moderator, (7) second buncher, (8) main accelerator, (9) scanning coils, (10) sample chamber, (11) electron gun. Some components are rotated into the plane of drawing.

A very ambitious project for a scanning positron microscope for defect analysis is pursued in Munich (David et al. 1997; Triftshäuser et al. 1997). The scheme of the microscope is shown in Fig. 2.24. It represents the combination of a scanning electron microscope with a pulsed variable-energy positron beam allowing positron lifetime measurements. The positron beam, which is formed during a double-stage moderation, has a spot diameter of 1 μm or below. The positron beam, with an energy of 0.5 to 30 keV, can be used for defect imaging over sample areas of 0.6×0.6 mm^2. An overall counting rate of $2\times10^4\ s^{-1}$ has been achieved at the current stage of development.

2.5.2 Age–Momentum Correlation

The age–momentum correlation (AMOC) technique combines positron lifetime spectroscopy with Doppler-broadening spectroscopy using the same annihilation event. Such a measurement is based on the registration of the collinear γ-quanta for the positron lifetime as the stop signal and simultaneously for the measurement of the Doppler broadening of the annihilation line. The start signal may be provided from a scintillator in the conventional sandwich arrangement (Fig. 2.3), which must not be arranged in a line with the stop detector. This is necessary, be-

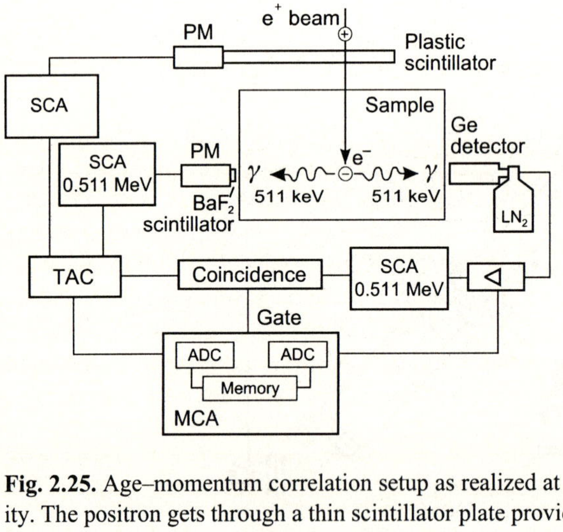

Fig. 2.25. Age–momentum correlation setup as realized at the Stuttgart MeV positron facility. The positron gets through a thin scintillator plate providing the start pulse. The time difference between this pulse and the stop signal of the second scintillator is registered as the positron lifetime. The collinearly emitted second annihilation γ-quantum is used for the Doppler-broadening measurement and is stored in coincidence with the lifetime signal in a two-dimensional memory (PM—photomultiplier, SCA—single-channel analyzer, TAC—time-to-amplitude converter, MCA—multi-channel analyzer, ADC—analog–digital converter).

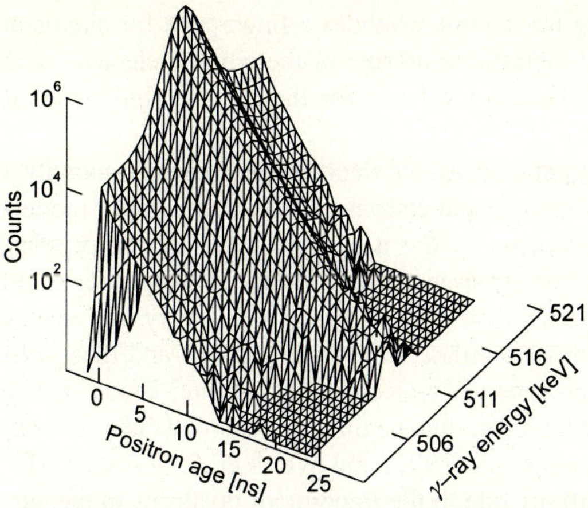

Fig. 2.26. Age–momentum profile of positron annihilation in quartz (Siegle 1997).

cause the Ge detector for the coincident detection of the Doppler signal must be arranged opposite the scintillator for the stop signal. A special setup using a positron beam is realized at the Stuttgart MeV positron facility (Stoll et al. 1991). The start signal is generated by the transition of a fast positron with an energy of several MeV through a thin scintillator plate (Fig. 2.25). The stop pulse is registered conventionally by a BaF_2 scintillator. In this way, a β^+–γ coincidence is realized. The positron lifetime measured in a fast–fast spectrometer (Fig. 2.4) is registered in coincidence with the Doppler signal and stored in a two-dimensional memory array.

A typical spectrum is shown in Fig. 2.26. The profile is composed of individual Doppler curves, which are measured as a function of the positron lifetime (age). The technique has the advantage of measuring the dependence of the electron momentum distributions on the positron lifetime. In this way, Doppler curves for ortho-positronium events with a very long lifetime can be filtered out of the annihilation spectrum. The main applications are found in positronium chemistry. The only study known in the field of semiconductors was carried out for diamond (Stoll et al. 1994). It was shown that a large Doppler broadening in natural and synthetic diamonds is correlated with a very short lifetime component.

2.5.3 Positron-Annihilation-Induced Auger Electron Spectroscopy

The Auger process is a non-radiative transition of electrons, where a punched-out inner-shell electron is replaced by an electron from an outer shell. The released energy is transferred simultaneously to another electron, which is called an Auger electron. The energy levels of elements are known, and thus the analysis of the energy of the Auger electron leaving the sample provides a fingerprint for chemical analysis. The mean free path until inelastic scattering of the emitted electron is in the range of several nanometers. This is the basis for the surface sensitivity of Auger electron spectroscopy (AES).

The main disadvantages of conventional Auger electron spectroscopy, namely a high background level caused by secondary electrons and the comparably moderate surface sensitivity, can be overcome by the use of positrons as the primary particles. This method is called positron-annihilation-induced Auger electron spectroscopy (PAES). Positrons are incident on the surface with a very low energy of several eV. Some of the positrons annihilate with core electrons and thus initiate the Auger process. Virtually no secondary-electron background occurs in the energy range of the Auger lines, which can thus be directly recorded. The first experiments were reported by Soininen et al. (1991) and Weiss et al. (1989). PAES has greatly increased surface sensitivity due to the trapping of positrons in the surface state potential prior to annihilation (Weiss 1995). This means that the Auger electron originates practically from the outermost monolayer. However, the intensity is rather low compared with conventional Auger spectroscopy, which can be partly overcome by single-electron counting.

A scheme of the positron Auger electron spectrometer realized at the University of Texas at Arlington is shown in Fig. 2.27 (Weiss 1995). The energy of the Auger electron is analyzed by an **E**×**B** filter. This electron is detected in a micro-channel plate in coincidence with the registration of one of the annihilation γ-quanta in an NaI scintillator. The spread of the Auger electrons along the vertical axis of the micro-channel plate is due to their different energies and allows the simultaneous detection of the complete Auger spectrum.

Yang et al. (1996) combined conventional Auger electron spectroscopy with PAES in order to study the structure of annealed Au/Si (100). An example of a

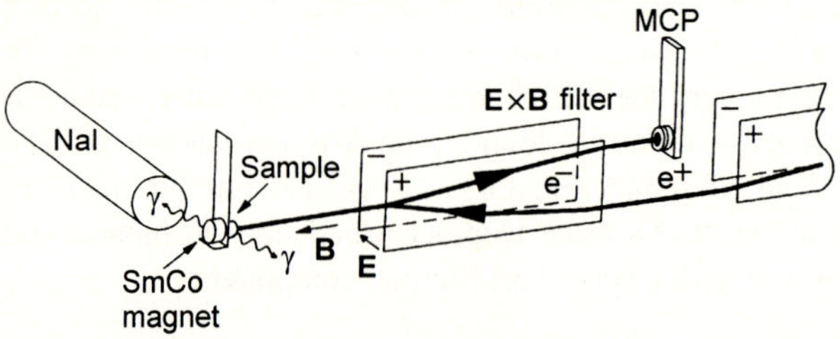

Fig. 2.27. Schematic view of the energy analyzer system of the positron-annihilation-induced Auger electron spectrometer at the University of Texas at Arlington (Weiss 1995). The incoming positron beam hits the sample and invokes the Auger process. One of the emitted annihilation γ-quanta is detected in a sodium iodide scintillator counter. The emitted Auger electron passes the **E**×**B** filter and is registered in a micro-channel plate (MCP). The electron is counted in coincidence with the detected γ-quantum in order to reduce the background. The samarium cobalt permanent magnet placed behind the sample is used to improve the positron beam guidance.

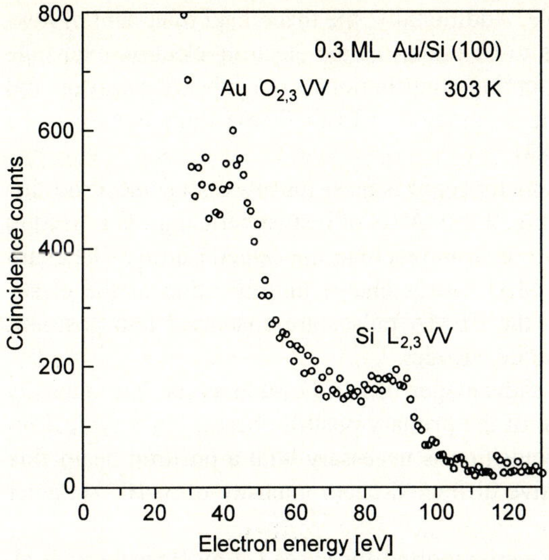

Fig. 2.28. A typical positron-annihilation-induced Auger electron spectrum obtained from a sub-monolayer of gold (0.3 monolayers) deposited on (100) silicon at room temperature (Yang et al. 1996). The peaks at 43 and 88 eV correspond to the Au OVV and the Si LVV transitions, respectively.

positron-annihilation-induced Auger electron spectrum is shown in Fig. 2.28. Two Auger peaks at 43 and 88 eV corresponding to the Au OVV and Si LVV transitions, respectively, are visible. The sharp increase at low electron energies is due to secondary electrons exited by the 30-eV positron beam. From a series of deposition experiments combined with annealing treatments, it was concluded that the gold concentration amounts to about 100 % to a depth of 1 monolayer and then decreases continuously. This conclusion can hardly be drawn from conventional AES alone.

2.5.4 Low-Energy Positron Diffraction

Low-energy electron diffraction (LEED) has become a powerful tool for the investigation of surfaces. The diffraction of electrons having energies up to 200 eV is used for the identification of the crystalline structure of surfaces, such as relaxation, reconstruction, and adsorbed layers. It is believed that the use of positrons instead of electrons (low-energy positron diffraction—LEPD) could provide additional information about complex surfaces. Particularly in compound semiconductors, there are large differences between the calculated and measured diffracted-beam intensities in LEED. The uncertainties of the analyses may be less using a combination of low-energy electron and positron diffraction, as two independent data sets become available. Additionally, the theoretical treatment of low-energy positron diffraction seems to be easier, as the electron–electron exchange interaction is absent. Multiple scattering contributions from subsurface layers and many-body effects can restrict the accuracy of LEED. These limitations may be less for positrons (Weiss et al. 1983).

In principle, the same diffractometer setup is used for low-energy positron diffraction and electron diffraction (Fig. 2.29). A set of hemispherical grids is used to retard inelastically scattered positrons from reaching the detector array. The channel electron multiplier array (CEMA) causes charge amplification of the elastically scattered positrons. Finally, the CEMA pulses are converted into position-sensitive signals in the resistive anode encoder.

There are, however, distinct disadvantages of LEPD, such as the low intensity and the relatively high divergence of the primary positron beam. For a typical intensity profile, a week of data acquisition is necessary with a positron beam flux of $5\times10^3\ s^{-1}$ and an average relative diffracted-beam intensity of 5×10^{-5} (Canter 1995).

The first LEPD measurements were reported for (111) Cu by Rosenberg et al. (1980). The positrons were electrostatically focused onto the sample. The scattered positrons were detected versus the scattering angle with a single-particle detector. The first measurements were reported on semiconductor surfaces, such as GaAs (Lessor et al. 1992), CdSe (Horsky et al. 1989), and CdTe (Duke et al. 1997) cleavage planes. The CdSe investigations were carried out in conjunction with LEED at $(11\bar{2}0)$ and $(10\bar{1}0)$ cleavage planes. The aim was to determine the relaxed atomic positions of the cleavage planes. The positron diffraction analyses yielded optimal fits at smaller relaxations perpendicular to the top layer than low-

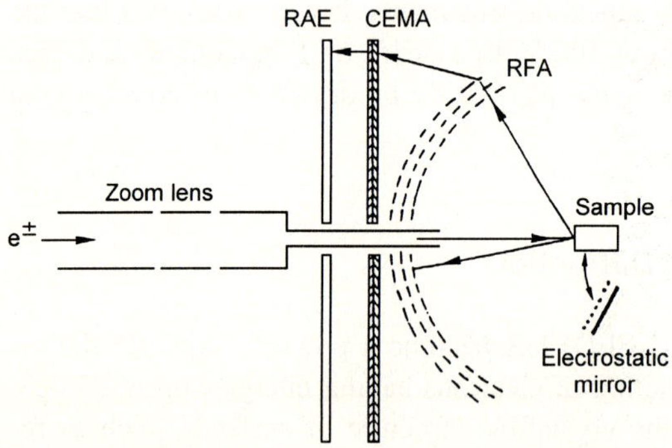

Fig. 2.29. Diffractometer used for low-energy electron diffraction (LEED) and low-energy positron diffraction (Canter 1995). The energy of the diffracted particles is analyzed by the retarding field analyzer (RFA).The positrons are detected in a channel electron multiplier array (CEMA) and the resistive anode decoder (RAE). The sample is mounted together with an electrostatic mirror on a manipulator to enable examination by a reverse view LEED display (not shown).

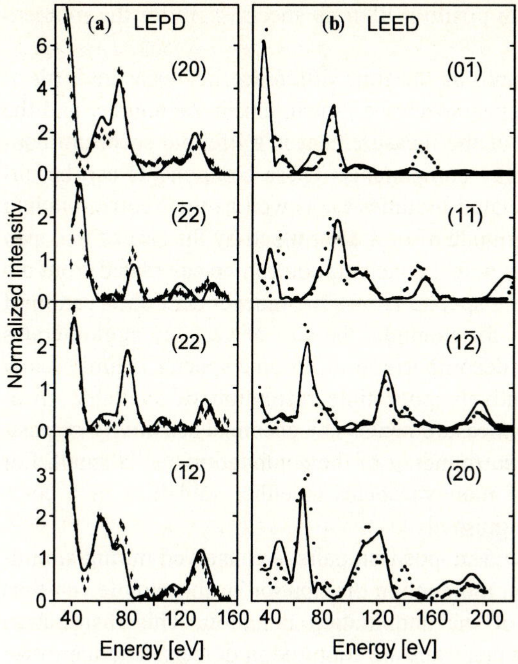

Fig. 2.30. Comparison of low-energy positron (**a**) and electron (**b**) diffraction (LEPD and LEED) profiles of cadmium selenide $(11\bar{2}0)$ cleavage planes (Horsky et al. 1989). Solid lines are the result of theoretical calculations utilizing the best-fit analysis of the structures. The positron intensities are given in absolute reflectivity units of 10^4 and in arbitrary units for low-energy electron diffraction.

energy electron diffraction for both cleavage faces. A better agreement between theoretical and experimental intensity profiles was obtained for LEPD (Fig. 2.30).

2.6 Summary

Various positron annihilation techniques are suitable for the characterization of defects in semiconductors. The special strength of positron annihilation is the sensitivity to open-volume defects. The positrons are usually obtained from isotope sources such as ^{22}Na, which can be easily handled in the laboratory. The positron experiment is arranged in such a way that as many positrons as possible penetrate the sample. They thermalize, and diffuse, and may finally be trapped in defects. In the latter case, characteristic parameters of the annihilation radiation change. These parameters can be used to identify defects and to determine their concentrations. They are mainly obtained in positron lifetime spectroscopy or the momentum distribution techniques.

The positron lifetime is measured as the time difference between the birth of the positron, monitored by the appearance of a γ-quantum in the source, and the annihilation γ-quanta. The result of the measurement is a lifetime spectrum consisting of several exponential decay components. These components can be numerically separated, and defect-related lifetimes τ_d, as well as their corresponding intensities, are obtained. The magnitude of τ_d is determined by the size of the open volume of the detected defect. The defect concentration can be calculated from the intensity. The decomposition of the spectra allows the independent observation of various defect types. In this way, for example, the size of vacancy agglomerates can be determined. However, the decomposition of lifetime spectra is limited to a few components due to the relatively poor time resolution of available spectrometers and the statistics of the measurements. The chemical sensitivity of positron lifetime spectroscopy to the surrounding of the annihilation site is small. For example, the positron lifetime of monovacancies in either sublattice of a compound can frequently not be distinguished.

The total momentum of the electron–positron pairs is conserved during annihilation. Information on the electron momentum distribution in the sample can thus be obtained from the analysis of the annihilation radiation. This distribution changes in a characteristic way if positrons are captured in defects. There are two basic techniques, which are distinguished by the observed momentum component. The angular correlation of annihilation radiation measures two momentum components perpendicular to the propagation direction of the γ-quanta. A two-dimensional spectrum is obtained with a high momentum resolution. The main application is the study of the electron structure in the bulk and in defects. The drawback of low counting rates and relatively large dimensions of the ACAR machines prevents the routine use of this method for defect characterization.

The momentum component in propagation direction leads to a Doppler broadening of the annihilation line. Doppler-broadening spectroscopy has the advantage of a compact setup with a counting rate and allows the simultaneous measurement of the positron lifetime. Despite the lower momentum resolution compared with angular correlation of annihilation radiation, the Doppler technique permits the registration of many data points in a rather short time and is thus a standard tool in a positron laboratory. The Doppler-broadened line is frequently characterized by line-shape parameters. The valence annihilation parameter (S parameter) is more sensitive for the structure of the defect such as the size of the open volume, while the core annihilation parameter (W parameter) provides information on the chemical surrounding of the annihilation site.

The recent methodical development of the Doppler-broadening coincidence technique is the basis of the measurement of background-reduced spectra with improved momentum resolution. Such high-quality Doppler spectra can be compared with theoretical calculations carried out for various defect configurations. The annihilation site, i.e. the chemical nature of neighbor atoms, may be identified more reliably. Momentum distribution techniques cannot be decomposed in several defect components as carried out in positron lifetime spectroscopy. If more than one defect type contributes significantly to positron trapping, the line-shape parameters cannot be easily interpreted. It is thus essential to combine lifetime and momentum techniques for the same set of samples in order to obtain the optimum information gain from positron experiments.

Monoenergetic positrons can be used for the study of thin layers. Defect profiling is possible by the variation of the energy of the incident positron beam (slow-positron-beam technique). In principle, all three positron techniques mentioned above can be applied also in combination with a positron beam. The realization of slow-positron-beam lifetime measurements is rather complicated and has been realized only in a few laboratories. The main measurement principle is thus the Doppler-broadening technique. In contrast to conventional positron lifetime and Doppler techniques, where an upper sensitivity limit of the defect density of about 10^{19} cm^{-3} exists, higher defect concentrations, as obtained e.g. after ion implantation, may be determined by slow-positron-beam measurements. For this purpose, the back-diffusion of positrons towards the surface is observed as a function of the incident positron energy.

A more recent progress is the increase in the lateral resolution in positron beam experiments. A spot diameter of 1 μm is predicted for positron microscopes currently under development. In addition to positron lifetime spectroscopy and the momentum techniques, there are other positron methods, e.g. low-energy positron diffraction and positron-annihilation-induced Auger spectroscopy, which are realized only in a few laboratories and which have a limited applicability to defect issues in semiconductors.

3 Basics of Positron Annihilation in Semiconductors

This chapter is dedicated to the fundamentals of the interaction of the positron with the semiconductor from first encounter to the annihilation event. No comprehensive treatment of the theory of positrons in solids is intended (for a review see Puska and Nieminen 1994). The interaction processes of positrons with solids comprise backscattering, channeling, thermalization, diffusion, possible trapping in lattice defects, and interaction with surfaces. The treatment of these effects is general to all solids, but the peculiarities of positron behavior in semiconductors are emphasized (Sect. 3.1). Section 3.2 describes how absolute defect concentrations can be determined. This is always achieved by the determination of the positron trapping rate κ which is proportional to the defect concentration. Principally, there are two different approaches to obtain the quantity κ: (i) the measurement of the annihilation parameters, viz. the positron lifetime τ or the line shape parameter S of Doppler-broadening spectroscopy, or (ii) positron diffusion experiments, i.e. the determination of the positron diffusion length. Theoretical calculations of the positron lifetime and the annihilation line shape in the defect-free bulk of the sample and in open-volume defects are described in Sect. 3.3.

3.1 Interaction of Positrons with Solids

A positron approaching a solid surface may be scattered back or may penetrate the sample. In the latter case, the positron rapidly loses its high kinetic energy, i.e. it thermalizes. The thermalized positron diffuses through the sample until annihilation. It may be localized during diffusion by positron traps, such as lattice defects, and then it annihilates there. The processes of backscattering, channeling, thermalization, diffusion, and trapping are described in Sects. 3.1.1 to 3.1.5. The special case of positronium formation is the subject of Sect. 3.1.6.

3.1.1 Backscattering

Depending on the incident energy of positrons and on the sample material, there is a certain probability for the backscattering of highly energetic positrons from the sample. The interaction of the positron with the nucleus is, similar to α-particles or electrons, an elastic scattering process. The differential elastic-scattering cross-section was calculated by Valkealahti and Nieminen (1984). The important difference between electrons and positrons is that the cross-section for electrons is much larger than for positrons and the probability of large-angle scattering is higher for electrons. These differences are due to the fact that the interaction of electrons with the nucleus is attractive, whereas it is repulsive for the positrons. The back-scattering probability was treated theoretically by Monte-Carlo simulations taking into account both elastic and inelastic scattering. It was compared with experimental results (Mäkinen et al. 1992b) for Si, Ge, and Au (Fig. 3.1). Experimental investigations of Al and Au were compared with Monte-Carlo simulations by Massoumi et al. (1994).

It is obvious that a high atomic number of the target leads to a significantly larger fraction of backscattered positrons. By using conventional foil sources, these positrons give rise to a larger source component, since the positrons are penetrating the source more than once (see discussion of source correction in Sect. 2.1). Special care must be taken in slow-positron-beam experiments to avoid the recording of annihilation events originating from backscattered positrons. This is necessary because backscattered positrons annihilate outside the sample, leading to distortion of the spectra. This effect can be avoided by sufficiently large specimen chambers and a careful shielding of the annihilation radiation detector. It can be further improved by measurements detecting the second collinear γ-quantum in coincidence. A further problem occurs in the lifetime measurements using a bunched positron beam. Some of the backscattered positrons may reach the accelerator stage and may be re-accelerated to the sample. Hence, an additional small but distinct peak may appear in the lifetime spectrum. This effect can be diminished by a large spatial separation of the specimen and the accelerator.

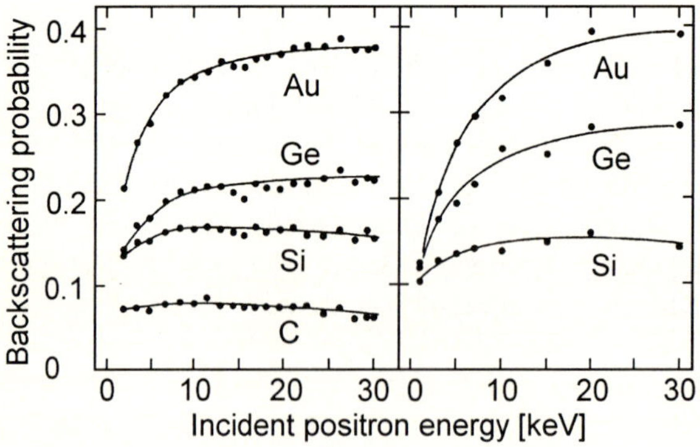

Fig. 3.1. Backscattering probability of positrons entering the surface of the indicated solids as a function of the incident positron energy. The measured results are given in the left panel, the Monte-Carlo simulations in the right panel (Mäkinen et al. 1992b).

3.1.2 Channeling

A charged particle penetrating a single-crystalline target may travel extraordinarily long distances through the crystal in low-indexed directions. This effect is called channeling and was investigated for electrons, protons, and ions. Rutherford backscattering utilizes the channeling of ions for the investigation of crystal defects. In contrast, the channeling of light particles is usually detected as channeling pattern or spectra in transmission. Positron channeling experiments using a 1-MeV positron beam were performed with gold and silicon thin crystals (Andersen et al. 1971; Pedersen et al. 1972).

A number of interesting phenomena are related to channeling of electrons and positrons. The particles undergo an undulatory motion during the transition through a channel. This gives rise to the emission of the so-called channeling radiation (Sørensen and Uggerhøj 1987), the intensity of which is considerably higher than the bremsstrahlung background. There are promising features in this radiation. It is highly monochromatic with energies in the range of hard X-rays. If the particles are channeled over appreciable distances for incident energies in the MeV or GeV range, the intensity of the channeling radiation could compete with that of synchrotron radiation.

The channeling of 50-keV positrons through a thin silicon foil was investigated by Schultz et al. (1988a). The positron yield as a function of the angular deviation from the direction of the incident positron beam is shown in Fig. 3.2 with two different detector positions. The yield increases significantly in low-indexed crystallographic directions for the detector on the axis of the incident beam, and con-

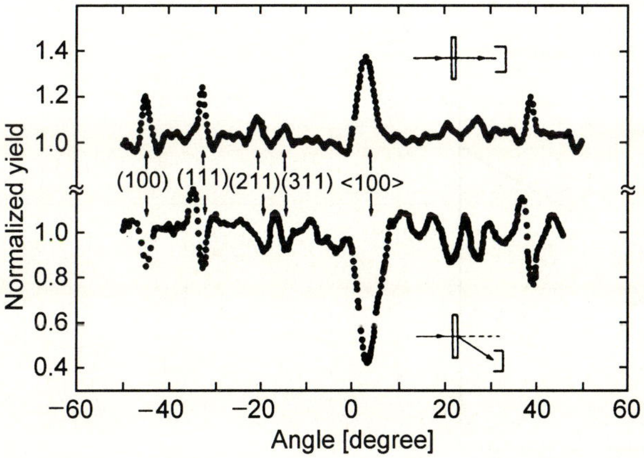

Fig. 3.2. Channeling effect of 50-keV positrons transmitted through a 0.26 μm single-crystalline silicon foil (Schultz et al. 1988a). Angular scans of the normalized positron yield measured with a surface barrier detector on the axis of the incident positron beam (*upper curve*) and off axis (*lower curve*). Low-indexed crystallographic directions are indicated.

versely, it decreases for such orientations for the detector position off axis. The prerequisite for positron channeling is a beam with very low divergence. This is not given for magnetically guided positron beams, which are frequently used for defect studies (Sect. 2.4.2). The positrons on a helical trajectory hit the sample under different incident angles. Thus, channeling effects are of minor importance for defect profiling with such beam systems.

3.1.3 Thermalization

The kinetic energy of positrons entering the solid is much larger than the thermal energy of diffusion, independent of the origin of the positron (isotope foil source or monoenergetic positron beam). The difference between these energies is dissipated in the crystal within the first few picoseconds. This process is called thermalization and its duration is much shorter than the total positron lifetime. The energy transfer rate is very high at large energies and proceeds by core ionization processes (Perkins and Carbotte 1970). This means that the positron energy is decreased rapidly. It is illustrated by the energy-loss rate plotted against the energy in Fig. 3.3. Additionally, plasmon excitations can occur (Oliva 1980). The positron energy is not much higher than the thermal energy for most of the thermalization time. In this energy range the excitation of conduction electrons dominates for metals. The dominant process in semiconductors is the excitation of electron–hole pairs for energies higher than the bandgap width. Scattering of phonons dominates for positron energies reaching the sub-electronvolt level (Perkins and

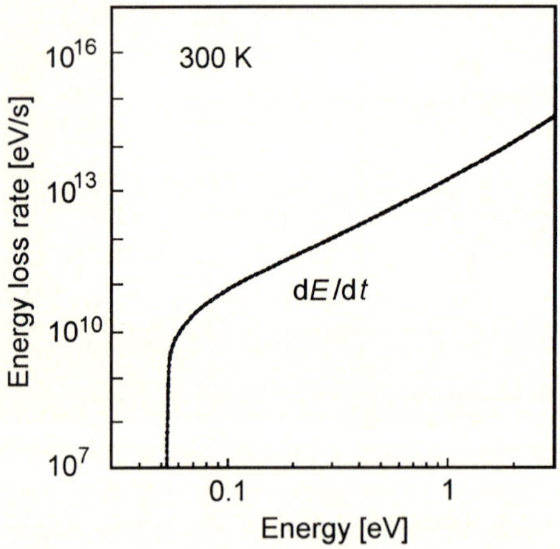

Fig. 3.3. Energy-loss rate d*E*/d*t* of positrons in aluminum as a function of the kinetic energy *E* of the positron at a temperature of 300 K (Jensen et al. 1990).

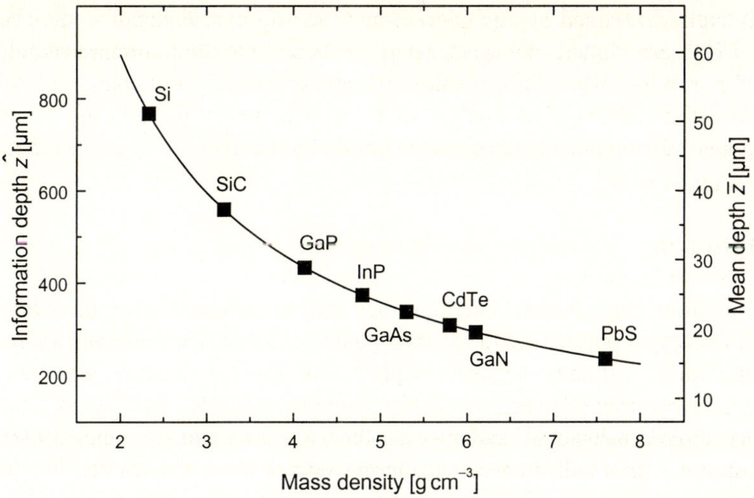

Fig. 3.4. Positron penetration depth of semiconductors as a function of their mass density. In the calculation of the mean penetration depth $\bar{z}$ (defined by $P(\bar{z}) = 1/e$) and the information depth $\hat{z}$ (defined by $P(\hat{z}) = 0.999$), positrons of a ^{22}Na source with a maximum energy of $E_{max} = 540$ keV are assumed. The values of important semiconductors are indicated as squares.

Carbotte 1970).

The penetration depth profile $P(z, E)$ is obtained as the result of the thermalization process of positrons of the energy E. Usually, it can be simulated by Monte-Carlo procedures; see discussion in Sect. 2.4.4.1. The fitting parameters depend on the particular Monte-Carlo scheme used (for a review see Ghosh and Aers 1995).

The implantation profile for monoenergetic positrons was experimentally determined for amorphous Si layers by slow-positron-beam measurements (Gebauer et al. 1997b) and a reasonable agreement with the parameters of Ghosh (1995) applying the scheme of Jensen et al. (1990) was found.

While these profiles are the relevant stopping profiles for monoenergetic beam experiments (see Sect. 2.4.4.1), the positron implantation profile resulting from the application of conventional foil sources, P, must take into account the broad energy distribution of the β^+-decay that is shown in Fig. 2.17. An implantation profile, which was given as the empirical equation (Brandt and Paulin 1977)

$$\overline{P}(z) = \exp(-\alpha z) \quad \text{with} \quad \alpha[\text{cm}^{-1}] = 17 \frac{\rho[\text{gcm}^{-3}]}{E_{max}^{1.43}[\text{MeV}]}, \tag{3.1}$$

is obtained. α is the positron absorption coefficient, E_{max} the maximum positron energy, and ρ the mass density of the solid. In order to illustrate the function $P(z)$ characterizing the information depth of the method, Fig. 3.4 illustrates the dependence of the penetration depth on the mass density of semiconductors. The mean penetration depth $\bar{z}$, defined by $P(\bar{z}) = 1/\mathrm{e}$, and the information depth $\hat{z}$, defined by $P(\hat{z}) = 0.999$, are plotted. $\hat{z}$ represents at the same time the minimum sample thickness required for conventional positron experiments using foil sources in order to minimize perturbation of data by positrons penetrating through the whole sample and annihilating outside the sample volume of interest.

3.1.4 Diffusion

When the positrons reach thermal energies, they start to diffuse through the lattice and behave as charged particles. Due to their positive charge, the positrons are repelled by the nuclei and have their largest position probability density in interstitial regions. The positron diffusion can be described similarly to electrons with the semiclassical three-dimensional random-walk theory (Puska and Nieminen 1994). The positron mean free path between scattering events $\langle l \rangle$ is determined by the microscopic scattering mechanisms and is a function of temperature,

$$\langle l \rangle = \frac{3D_+}{\sqrt{\langle v^2 \rangle}} \quad \text{with} \quad \langle v^2 \rangle = \frac{3k_\mathrm{B}T}{m^*}. \tag{3.2}$$

v is the thermal positron velocity, k_B the Boltzmann constant, T the temperature, and m^* the effective positron mass. The most important process in semiconductors is the scattering by longitudinal acoustic phonons, leading to a temperature dependence of the positron diffusion constant according to $D_+ \propto T^{-1/2}$. Such a functional dependence was theoretically predicted using the deformation-potential approximation (Bardeen and Shockley 1950) and was experimentally confirmed for several metals by slow-positron-beam back-diffusion experiments (Soininen et al. 1990).

Especially at higher temperatures, the scattering by optical phonons may have a non-negligible contribution according to $D_+ \propto T^{-3/2}$ (Soininen et al. 1992). The diffusion is strongly impeded by the presence of a high concentration of ionized impurities. This was shown in the case of 10^{20} cm^{-3} phosphorus donors in silicon. The diffusion length was significantly reduced (Soininen et al. 1992), in accordance with the theoretical expectation of $T^{3/2}$ (Fig. 3.5). Additionally, the contributions of different phonon scattering mechanisms are visible for undoped silicon (open symbols in Fig. 3.5). Scattering at electrons in the conduction band does not play any significant role in semiconductors.

The positron diffusion length L_+ is limited due to the finite lifetime of positrons in the defect-free bulk, τ_b,

$$L_+ = \sqrt{\tau_\mathrm{b} D_+} \quad \text{with} \quad D_+ = \tau_\mathrm{r} \frac{k_\mathrm{B}T}{m^*}. \tag{3.3}$$

τ_r is the relaxation time for the dominant scattering mechanism. The mean free path $\langle l \rangle$ and the positron diffusion length[1] L_+ of some representative semiconductors at room temperature are presented in Table 3.1. The effective positron mass m* is larger than the rest mass of the free positron m_0, $m^* = 1.3$ to $1.7\,m_0$, due to phonon scattering, screening of positrons by electrons, and the effect of the periodic lattice (Schultz and Lynn 1988).

Positron diffusion can be generally described by the time-dependent diffusion equation

$$\frac{\partial}{\partial t} n_+(\mathbf{r},t) = D_+ \nabla^2 n_+(\mathbf{r},t) - \nabla\left[v_\mathrm{d} n_+(\mathbf{r},t)\right] - \lambda_\mathrm{eff} n_+(\mathbf{r},t). \tag{3.4}$$

$n_+(\mathbf{r},t)$ is the positron density at the position $\mathbf{r}$ and time t, and v_d is the positron drift velocity, which depends on the strength of the electric field. In contrast to (3.3), the positron trapping in defects was taken into account, resulting in the effective annihilation rate

$$\lambda_\mathrm{eff} = 1/\tau_\mathrm{b} + \kappa(\mathbf{r}), \tag{3.5}$$

and consequently in the effective positron diffusion constant

$$D_+^\mathrm{eff} = L_+^2 \lambda_\mathrm{eff}\,. \tag{3.6}$$

$\kappa(\mathbf{r})$ is the positron trapping rate, which has to be taken in general as a function of position.

Positron experiments performed with slow-positron beams (see Sect. 2.4) are only sensitive for the direction perpendicular to the surface. Hence, (3.4) is reduced to a one-dimensional problem. Despite this reduction, the analysis of depth-dependent lifetime spectra is a non-trivial problem. The time-dependent diffusion/trapping equations were analyzed by Britton and Störmer (1995) and Kögel (1995, 1996). No standard computer programs have been available up to now, in contrast with conventional lifetime measurements using foil sources (e.g. by Kirkegaard et al. 1989).

A further simplification is obtained when depth-dependent Doppler experiments are performed. In this special case only the time-independent one-dimensional diffusion equation has to be solved with appropriate starting and boundary conditions,

$$0 = D_+ \frac{\mathrm{d}^2}{\mathrm{d}z^2} n_+(z) - \frac{\mathrm{d}}{\mathrm{d}z}\left[v_\mathrm{d} n_+(z)\right] - \lambda_\mathrm{eff} n_+(z) + P(z,E). \tag{3.7}$$

[1] The diffusion length L_+ according to (3.3) is a result of the solution of the diffusion equation and has to be distinguished from the microscopic quantity of the root-mean-square diffusion distance $\overline{X}$, which is given as $\overline{X} = \langle X^2 \rangle^{1/2} = (2dD_+\tau_\mathrm{b})^{1/2}$, where d is the dimension of the diffusion.

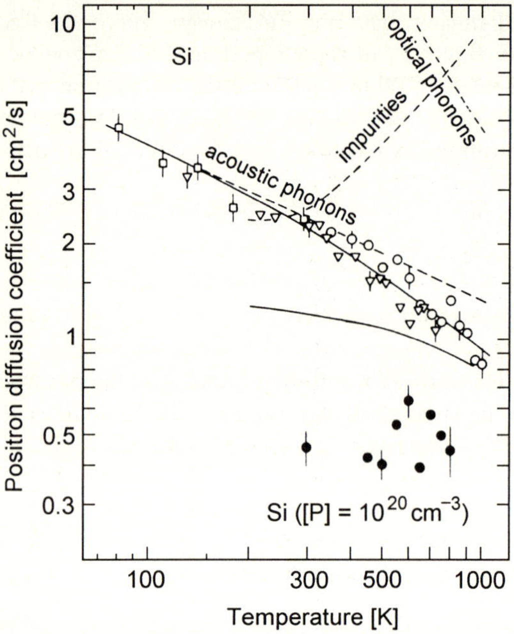

Fig. 3.5. Measured positron diffusion coefficient in undoped silicon (open symbols; different symbols stand for different samples) and in Si doped with 10^{20} cm^{-3} phosphorus (closed circles) obtained by slow-positron-beam experiments (Soininen et al. 1992). The contributions of scattering at phonons and ionized impurities are indicated by dashed lines. The upper solid curve is obtained by combining the contributions from acoustic and optical phonons. The lower solid curve additionally includes impurity scattering.

$P(z, E)$ is the positron implantation profile as a function of incident energy (2.15). Equations (2.15) and (3.7) are the basis of defect depth-profile analysis (see Sect. 2.4.4).

The back-diffusion of positrons to the surface is an important tool for defect

Table 3.1. The mean free path $\langle l \rangle$ and the positron diffusion length L_+ at 300 K for some semiconductors containing no positron traps. The values were calculated from the measured diffusion constants given in the references according to (3.2) and (3.3) using an effective positron mass of $m^* = 1.5m_0$.

Material	$\langle l \rangle$ [nm]	L_+ [nm]
Si	6.9[a], 6.6[b], 8.5[b,c]	219[a], 214[b], 243[b,c]
GaAs	5.3[a]	198[a]
Ge	5.3[a]	200[a]

[a]Soininen et al. (1992), [b]Schultz et al. (1988b), [c]Nielsen et al. (1985).

profiling, because there is practically no upper sensitivity limit for the determination of the defect concentration any more. The back-diffusion probability is calculated as the Laplace transformation of the implantation profile,

$$J(E) = \int_0^\infty \exp\left[-\frac{z}{(D_+ / \lambda_{\text{eff}})^{1/2}}\right] P(z,E)\,\mathrm{d}z\,. \tag{3.8}$$

The trapping rate as a measure of the defect concentration can be calculated via λ_{eff} in (3.5) and (3.8). The back-diffusion can also be evaluated using the VEPFIT routine (see Sect. 2.4.4.2). A simpler method had been applied before VEPFIT became available. The depth profiles were characterized by the so-called E_0 parameter, defined by $L_+ = AE_0^r$ (Vehanen 1989). The empirical parameter A and the exponent r are determined by the positron implantation profile (see Sect. 2.4.4.1). E_0 corresponds to that positron energy where $J(E)$ is reduced to 0.5.

The positron diffusion can be significantly influenced by the presence of electric fields. This was experimentally verified in the positron diffusion study by Corbel et al. (1989) at a gold/silicon Schottky barrier (see Fig. 4.21). It was shown that the fraction of positrons diffusing back to the Au/Si interface becomes smaller or larger according to the polarity of the applied bias. A positron drift over distances as long as 2 μm could be achieved with a field strength of about 10^6 V/m. The positron diffusion in the system Au/GaAs was extensively studied by Shan et al. (1994, 1996, 1997b). A positron diffusion coefficient of 1.8 $\text{cm}^2\,\text{s}^{-1}$ and a mobility of 70 $\text{cm}^2\,\text{V}^{-1}\text{s}^{-1}$ at room temperatures was determined by back-diffusion measurements. It was concluded from the temperature dependence of these quantities that the dominant scattering process in GaAs was, in contrast to Si, scattering at optical phonons.

Another experiment demonstrating the influence of internal electric fields on positron diffusion was carried out by Britton et al. (1994). Both a thin intrinsic layer grown on a p-type substrate and highly n-doped δ-layer buried in intrinsic Si formed effective barriers to positron diffusion, although no positron traps could be detected.

The presence of crystal defects and electric fields influences the effective diffusion length L_{eff},

$$L_{\text{eff}} = \frac{1}{\sqrt{\dfrac{\lambda_{\text{eff}}}{D_+} + \left(\dfrac{eE_{\text{drift}}}{2k_B T}\right)^2} - \dfrac{e|E_{\text{drift}}|}{2k_B T}}\,. \tag{3.9}$$

Taking into account the Einstein relation, $\mu_+ = eD_+/(k_B T)$, this equation is similar to that given by Shan et al. (1996). μ_+ is the positron mobility and E_{drift} the electric field strength. The effective diffusion length can be measured by positron back-diffusion experiments (see Sect. 3.2.3). The presence of positron traps is taken into account via λ_{eff} (3.5). Both the influence of an electric field and the variation of the

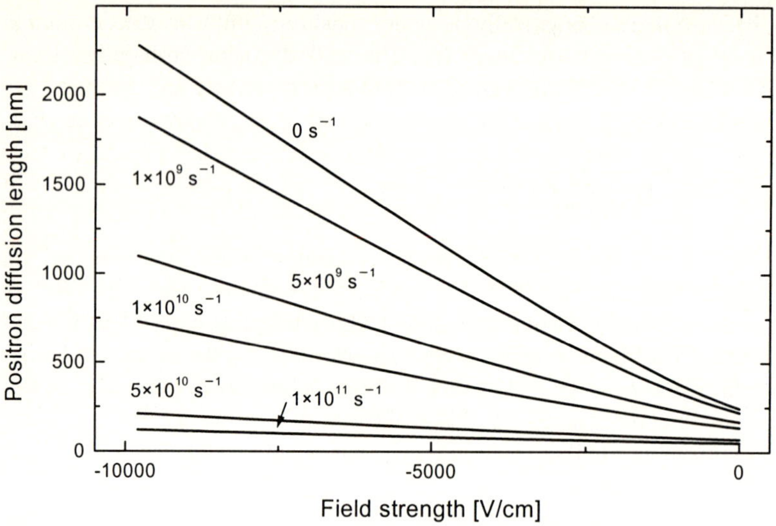

Fig. 3.6. Effective positron diffusion length in silicon calculated according to (3.9) as a function of the electric field strength for various defect concentrations expressed by the trapping rate κ. The calculation was carried out for 300 K using the positron diffusion coefficient of $D_+ = 2.7$ cm^2/s and the positron bulk lifetime of 218 ps.

trapping rate (defect concentration) on the positron diffusion length are shown in Fig. 3.6. The effective diffusion length increases with the electric field strength E_{drift} in (3.9). A similar result was obtained by Monte-Carlo simulations of the positron diffusion under the influence of an electric field (Eichler et al. 1997b). However, the number of positrons diffusing back to the surface in the positron experiment is higher or lower as a function of the sign of the applied bias. The increase in L_{eff} for a negative bias is more distinct for low trapping rates (Fig. 3.6).

It is obvious that the determination of trapping rates (defect concentrations) from the measurement of the diffusion coefficient may be incorrect if electric fields are not taken into account. On the other hand, it should be noted that L_{eff} is determined from back-diffusion experiments with the slow-positron-beam technique. If the electric field is directed in such a way as to enhance the positron diffusion away from the surface, it results in steeper S versus E curves, giving an apparent reduction in the diffusion length, although L_{eff} is actually increased by the influence of the field.

The electric field and the extension of the depletion region of a Schottky contact are influenced by the number of charged centers in this layer. Positrons may be used to monitor defect charge transients by their sensitivity to the electric field. For this purpose, the back-diffusion of positrons to the metal–semiconductor interface can be measured. The charge state transients depend on the frequency of the applied bias. The reason is the frequency dependence of the electron or hole emission rate. The effect was demonstrated for the Au/GaAs Schottky contact and the EL2 0/+ transition (Au et al. 1994; Shan et al. 1997a). The back-diffusion of positrons to the contact was detected by annihilation in voids at the interface, exhibiting a defect-related lifetime of 410 ps. The positron lifetime spectra were measured as a function of the frequency for reverse and forward bias half cycles separately by blocking the spectrometer during the unwanted half cycle. A peak in the intensity of the void-related positron lifetime occurred for reverse bias. The peak position was related to two competing processes, i.e. the increase of the mean electric field with time and the decrease of the width of the depletion region. The numerical fit of the positron diffusion and the charge-density profile allows the electron emission rates and the energy level of the trap in the bandgap to be determined. In this way, the 0/+ ionization level of the EL2 defect was obtained at $E_{\mathrm{cond}} - E_{\mathrm{d}} = (0.8 \pm 0.05)$ eV, in accordance with earlier results of deep level transient spectroscopy (Lagowski et al. 1982).

3.1.5 Trapping

The different positron techniques for the detection of defects are based on the fact that positrons may be trapped during the diffusion by lattice defects. The positron trapping during the thermalization was also discussed as non-thermal or epithermal trapping. Jensen and Walker (1990), however, showed by solving the Boltzmann equation for the positron momentum distribution in a homogeneous medium taking into account the positron scattering at electrons and phonons that non-thermal trapping was not important for positron studies of vacancies in Al. It is commonly accepted that this is also valid for semiconductors.

The trapping rate is obtained from Fermi's golden rule as

$$\kappa = \frac{2\pi}{\hbar} \sum_{i,f} P_i M_{if}^2 \delta(E_i - E_f), \tag{3.10}$$

where P_i is the occupation probability of the initial state i, M_{if} the transition matrix element between initial and final states i and f, and E_i and E_f the respective energies (Hodges 1970). The initial state corresponds to the delocalized Bloch state of the positron and the final state to the positron trapped in the defect.

In most cases, the dominant trapping process is determined either by the diffusion of positrons to the defects or by the transition of positrons to deep bound states related to the defects. These two processes are termed diffusion-limited and transition-limited trapping. The characteristics of positron capture to the relevant defect types are presented in the following and the influences of the defect charge state on positron trapping and of the temperature are taken into account. The positron trapping rate κ is proportional to the defect concentration C in both cases (Brandt and Paulin 1972), although the equations are different for transition-limited trapping,

$$\kappa_{\mathrm{tl}} = \mu C, \tag{3.11}$$

and for diffusion-limited trapping for a spherical defect,

$$\kappa_{\mathrm{dl}} = 4\pi r_{\mathrm{d}} D_{+} C. \tag{3.12}$$

κ_{tl} and κ_{dl} are the trapping rates for transition-limited and diffusion-limited trapping, respectively. C is the defect concentration and μ the positron trapping coefficient. μ is a constant for a given defect and must be determined by a correlation between the positron experiment and an independent method providing both κ_{tl} and C in (3.11). It can be understood as the product of the thermal velocity v and the positron trapping cross-section σ, $\mu = v\sigma$. The cross-section σ is not identical to the mere extension of the vacancy, but also takes into account the limitation of the trapping process by the transition into the bound state. The results of the experimental determination of the trapping coefficient in semiconductors are presented in Sects. 3.2.4 and 3.2.5.

The prefactor $4\pi r_{\mathrm{d}}$ in (3.12) is related to an extended defect of spherical shape with radius r_{d}. For cylindrical defects (Brandt 1974), such as pipes or dislocations, the trapping rate was given in the framework of diffusion-limited trapping as

$$\kappa_{\mathrm{dl}} = \frac{4\pi D_{+} \rho_{\mathrm{dis}}}{\ln\left(\pi \rho_{\mathrm{dis}} r_{\mathrm{d}}^{2}\right)^{-1}}. \tag{3.13}$$

ρ_{dis} is the dislocation density.

The temperature dependence of positron diffusion is determined by the scattering mechanisms which were described in Sect. 3.1.4. Since the other quantities in (3.12) are supposed to be temperature independent, the same temperature dependence of D_{+} is also expected for the trapping rate. For the most frequent case of positron scattering at acoustic phonons, a dependence according to $D_{+} \propto T^{-1/2}$ and consequently $\kappa \propto T^{-1/2}$ occurs.

In general, the positron trapping rate κ is given as the superposition of transition- and diffusion-limited trapping,

$$\frac{1}{\kappa} = \frac{1}{\kappa_{\mathrm{tl}}} + \frac{1}{\kappa_{\mathrm{dl}}}. \tag{3.14}$$

If it is clear from the context what type of trapping is dealt with, the indices dl and tl are omitted in the following. In the next sections, the positron interactions with various important defect types are treated in detail.

3.1.5.1 Vacancies

Vacancies are the most important positron traps in semiconductors. Their concentration often lies in the sensitivity range of the different positron techniques. Positron annihilation parameters, e.g. the positron lifetime, change in a defined manner when positrons are trapped in a vacancy. The trapping process is based on an attractive potential for the positrons which originates basically from the missing repulsing force of the absent nucleus.

Positron trapping in vacancies and its temperature dependence were theoretically considered by Puska et al. (1990). The model potentials for positron capture in differently charged and neutral vacancies, which were used for the calculation of the temperature dependence of the trapping coefficient, are shown in Fig. 3.7. A simple square-well potential is supposed to describe the positron attraction for the neutral vacancy V^0. This leads to a temperature-independent trapping coefficient (Fig. 3.8).

A charge of the vacancy is included by superimposing the square-well potential on a long-range Coulomb potential. This results in an additional repulsion or attraction in case of a positive or negative charge, respectively. The delocalization of the charge is taken into account by a Coulomb-tail cut-off radius (Fig. 3.7). In addition to the deep bound states of the square-well potential, a series of Rydberg states is induced. For positively charged vacancies, the repulsion at low temperatures is extremely strong so that the positrons are repelled and trapping is practically prohibited. Higher temperatures may induce a thermal excitation over the potential barrier, so that the trapping coefficient of positively charged vacancies

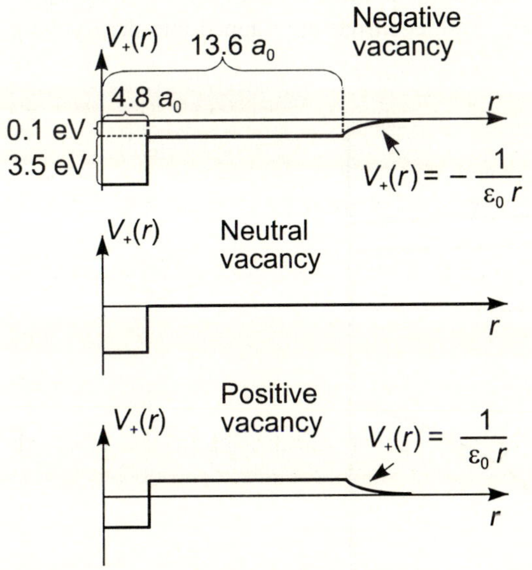

Fig. 3.7. Scheme of the positron potential V_+ for negative, neutral, and positive vacancies in silicon according to Puska et al. (1990). a_0 is the Bohr radius. The deep potential of the vacancy (3.5 eV) is superimposed on the Coulomb potential, $V_+(r) = \pm 1/(\varepsilon_0 r)$, when the vacancy is carrying a charge (ε_0—dielectric constant). The Coulomb tail is cut off at a distance $r = 13.6\,a_0$ due to the delocalization of the charge. 4.8 a_0 corresponds to the radius of the uncharged vacancy.

V^+ may be only one order of magnitude smaller than for neutral vacancies. Thus, according to the theoretical calculations, a distinct positron trapping in positively charged vacancies should be possible already at room temperature, but has not been experimentally observed (compare discussion in Sect. 3.2.4).

The additional attractive Coulomb potential of the negative vacancy V^- is especially effective at low temperatures, as can be seen in Fig. 3.8, where the increase in the trapping coefficient in this temperature range is shown as a result of theoretical calculation. The coefficient μ is expected to be about one order of magnitude larger than that for the neutral vacancy at room temperature. The physical reason for this sensitivity enhancement is the occurrence of extended Rydberg states induced by the long-range Coulomb potential. These states "collect" the positron in a more extended volume compared with a neutral vacancy. Positrons may be trapped in the Rydberg states with a high transition rate from the free delocalized state. The final trapping state at the bottom of the vacancy square-well potential is reached very fast. As a result of this two-stage process, the trapping becomes more efficient and the total trapping coefficient is larger than for a neutral vacancy. The positron binding energy at the Rydberg states amounts to some 10 meV. Thus, the trapping coefficient decreases with increasing temperature for negatively charged vacancies.

In order to describe the positron trapping in vacancies more quantitatively (Puska et al. 1990), a simple two-state model, which replaces the Rydberg series by one single level, was chosen (Fig. 3.9). Positrons are trapped into the Rydberg

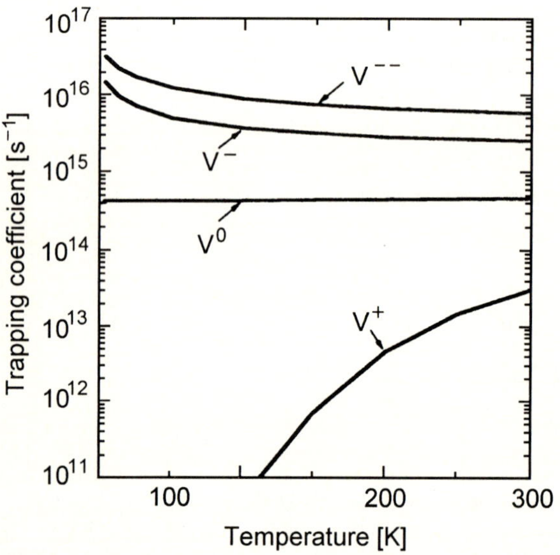

Fig. 3.8. Temperature dependence of the positron trapping coefficient of negatively and positively charged and neutral monovacancies in silicon according to Puska et al. (1990) and Puska and Nieminen (1994).

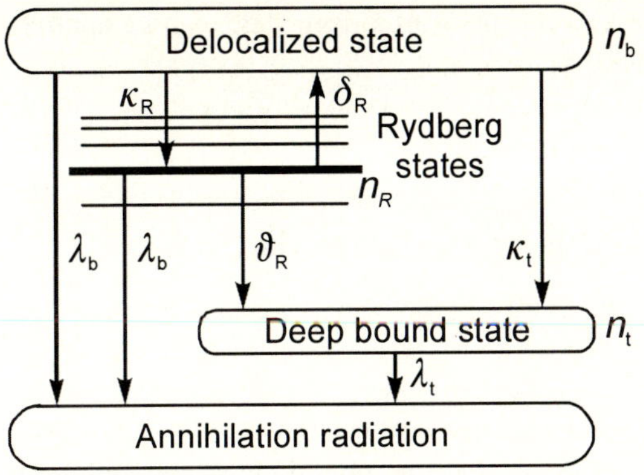

Fig. 3.9. Scheme of the two-stage positron trapping into a negative vacancy. The Rydberg series (*thin lines*) is replaced by a single level for simplicity. The positron can be trapped from the delocalized state in the bulk (with the number of positrons n_b) in the shallow Rydberg level with the trapping rate κ_R. Alternatively, it can directly pass over to the deep ground state of the positron trap with the rate κ_t or it can annihilate from the bulk with the annihilation rate λ_b. Once trapped in the Rydberg state (number of positrons n_R), the positron may annihilate with an annihilation rate supposed to be equal to λ_b. Furthermore, it may re-escape to the bulk state with the detrapping rate δ_R or it may be further localized by a transition to the ground state with the transition rate ϑ_R. The annihilation rate from the ground state (number of positrons n_t) is λ_t.

state with the trapping rate κ_R or directly into the deep ground state of the vacancy with the trapping rate κ_t, respectively. In contrast to this ground state, where no thermal escape is assumed, positrons trapped in the Rydberg state may escape thermally stimulated with the detrapping rate δ_R or they can pass over to the deep level with the transition rate ϑ_R. λ_b and λ_t are the annihilation rates (reciprocal lifetimes) in the delocalized state of the bulk and of the trap (vacancy), respectively. Positrons annihilating from spatially extended Rydberg states are expected to feel the electron density of the bulk. Thus, nearly the same annihilation rate λ_b is supposed there.

The thermally stimulated re-escape of positrons from shallow Rydberg states is characterized by the detrapping rate δ_R, calculated by Manninen and Nieminen (1981) as

$$\delta_R = \frac{\kappa_R}{\rho_v}\left(\frac{m^* k_B T}{2\pi\hbar^2}\right)^{3/2} \exp\left(-\frac{E_R}{k_B T}\right). \tag{3.15}$$

ρ_v is the vacancy density, E_R the positron binding energy to the Rydberg state.

The term scheme of Fig. 3.9 is used with the additional assumptions of $\kappa_t \ll \lambda_b$ and $\vartheta_R \gg \lambda_b$. The series of Rydberg states is combined in a single energy level. A net trapping rate of such a two-step trapping, i.e. positrons are first trapped in a shallow Rydberg state as a precursor of a deep state, can be approximated according to Puska et al. (1990) as

$$\kappa = \frac{\kappa_R \vartheta_R}{\vartheta_R + \delta_R} . \tag{3.16}$$

The trapping in negatively charged vacancies is then given with (3.15) as

$$\kappa = \frac{\vartheta_R \rho_v \kappa_R}{\vartheta_R \rho_v + \kappa_R \left(\frac{m^* k_B T}{2\pi \hbar^2} \right)^{3/2} \exp\left(-\frac{E_R}{k_B T} \right)} . \tag{3.17}$$

Dannefaer and Kerr (1993) studied the positron trapping at high temperature in as-grown GaAs and InSb. They interpreted the decreasing annihilation fraction above 450 K in terms of thermally stimulated re-escape of positrons from the deep ground state of monovacancies. Supposing that the trapping rate is temperature independent, they could determine the positron binding energy to be 0.3 ± 0.02 eV for GaAs. In the case of InSb the binding energy is constant below 540 K (0.3 ± 0.02 eV), but increases to 0.6 ± 0.05 eV close to the melting point. This increase in the binding energy was theoretically predicted by Manninen and Nieminen (1981) and was related to a strong influence of lattice vibrations and lattice expansion on the binding energy.

A similar decrease in the trapping rate with increasing temperature was observed in undoped or slightly n-doped GaAs (Corbel et al. 1988; Krause et al. 1989a). However, the temperature-dependent measurements of heavily silicon-doped GaAs of Gebauer et al. (1997c) provided, up to a temperature of 600 K, no significant decrease in the trapping rate. Since monovacancies were detected both in weakly and in heavily doped GaAs, it has to be concluded that the reason for the decrease of the trapping rate with increasing T is not the increasing fraction of detrapped positrons from the ground state of the vacancy. A comparable temperature dependence was found by Krause-Rehberg et al. (1998) in CdTe for the cadmium monovacancy as part of the A center. The positron binding energy to the ground state of vacancies has thus to be larger than estimated by Dannefaer and Kerr (1993).

In the case of negative vacancies, the positron trapping occurs at first to extended Rydberg states with low binding energies. Hence, the positron trapping in these charged vacancies via the Rydberg levels may not be treated as transition limited any more, but as diffusion limited, because of the large spatial extension of the Rydberg states. This leads to an additional term as a function of temperature for the trapping rate. The temperature dependence of the diffusion to the defect is mostly determined by scattering at acoustic phonons, leading to a dependence of the diffusion constant according to $D_+ \propto T^{-1/2}$ (see Sect. 3.1.4). Consequently, the trapping rate κ_R should have the same temperature dependence, $\kappa_R = \kappa_{R0} T^{-1/2}$. κ_{R0} is the trapping rate at a certain low temperature, e.g. at 20 K. The overall trapping rate is then obtained as

$$\kappa = \frac{\vartheta_R \rho_v \kappa_{R0} T^{-1/2}}{\vartheta_R \rho_v + \kappa_{R0} \left(\frac{m^* k_B}{2\pi\hbar^2} \right)^{3/2} T \exp\left(-\frac{E_R}{k_B T} \right)} . \tag{3.18}$$

Indeed, such a temperature dependence has been verified for vacancies in silicon irradiated at low temperatures with high-energy electrons (Mäkinen et al. 1992a) and protons (Mäkinen et al. 1990b) and for semi-insulating GaAs (Le Berre et al. 1995). The positron trapping rate in gallium vacancies is shown in Fig. 3.10. The temperature dependence below 130 K corresponds to the $T^{-1/2}$ power law caused by the diffusion limitation of positron trapping in the Rydberg states. The detrapping from these precursor states becomes important above that temperature, leading to a stronger temperature dependence according to (3.18). This is approximated by a $T^{-1.2}$ power law (Le Berre et al. 1995). It should be noted that in the case of GaAs, the assumption of negligible direct trapping in the ground state, $\kappa_t \ll \lambda_b$, is not fulfilled and (3.18) has to be modified accordingly.

The influence of the negative charge of the vacancy on the positron trapping decreases due to the screening effect of the charge induced by high concentrations of free carriers. This case is important for very high doping concentrations or in the case of intrinsic conductivity, when a high number of carriers are thermally excited in small-bandgap material. The screening effect is characterized by the

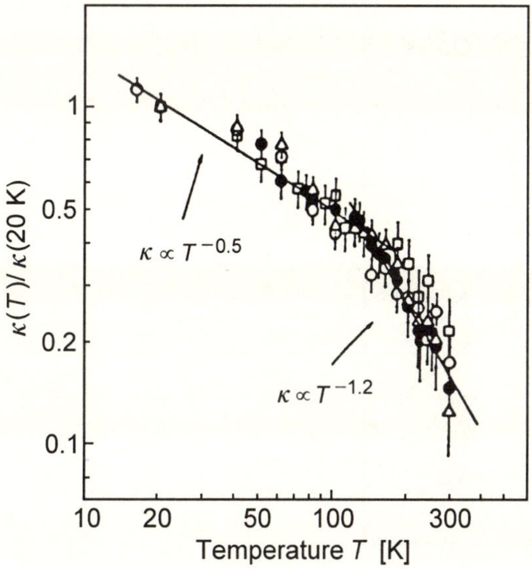

Fig. 3.10. Positron trapping rate κ in negatively charged gallium vacancies determined in semi-insulating gallium arsenide as a function of temperature T (Le Berre et al. 1995). The trapping rate is normalized to the value measured at 20 K. Different symbols stand for different samples.

Debye length,

$$L_{\mathrm{D}} = \sqrt{\frac{\varepsilon_0 \varepsilon_{\mathrm{r}} k_{\mathrm{B}} T}{e^2 n}}. \tag{3.19}$$

n is the carrier density, e the elementary charge, ε_0 the dielectric constant, and ε_{r} the relative permitivity. A characteristic capture radius r_{c}, where positron trapping in the vacancy may occur, may be defined by the condition that the electrostatic interaction energy equals the thermal energy $3/2\, k_{\mathrm{B}}T$ (Krause-Rehberg and Leipner 1997),

$$\frac{Ze^2}{4\pi\varepsilon_0\varepsilon_{\mathrm{r}} r_0} \exp\left(-\frac{r_{\mathrm{c}}}{L_{\mathrm{D}}}\right) = \frac{3}{2} k_{\mathrm{B}} T. \tag{3.20}$$

The parameter r_{c} can be numerically calculated and is plotted in Fig. 3.11 against the free-carrier density for GaAs as an example. It is obvious that the capture radius decreases with increasing carrier concentration or temperature. The formation of Rydberg states of a negatively charged vacancy is constrained to the sphere with the radius r_{c}. The promotion of positron trapping by the negative charge is hence reduced with the reduction of r_{c}. In consequence, the formation of Rydberg states may be suppressed by a sufficiently small capture radius. This case

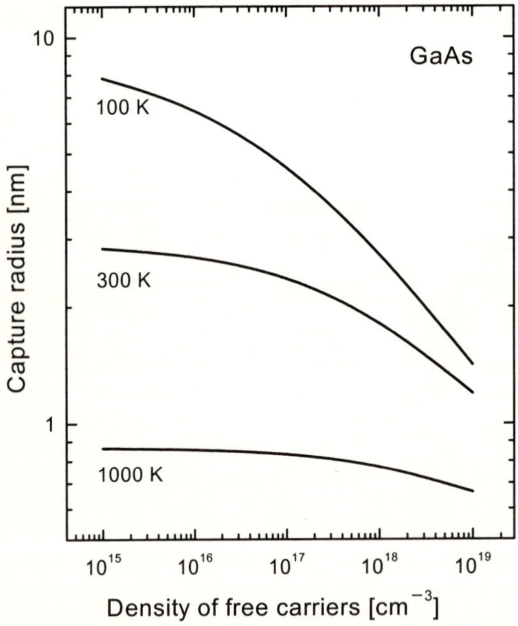

Fig. 3.11. Capture radius defined by (3.20) versus the free-carrier density in gallium arsenide for three different temperatures (Krause-Rehberg and Leipner 1997).

is known as the screening effect in metals, where the capture radius is of the order of the lattice constant due to the high number of free electrons. The influence of the reduction of the capture radius with increasing carrier concentration can also be found in semiconductors. The dependence of the trapping coefficient on the screening, i.e. on the capture radius, was observed in the small-bandgap material $Hg_{0.78}Cd_{0.22}Te$ (Krause-Rehberg et al. 1995a). When intrinsic conduction was obtained at elevated temperatures, the trapping coefficient of the negatively charged Hg vacancy was significantly reduced to values usually obtained for vacancies in metals (see Sect. 3.2.4.5).

3.1.5.2 Shallow Positron Traps

The existence of shallow positron traps in semiconductors was first clearly shown by Saarinen et al. (1989) by slow-positron-beam experiments on undoped GaAs. The term shallow trap does not refer to the energetic position of the defect in the bandgap, but to the weak positron binding energy. It was found that the positron diffusion constant decreases with decreasing temperature in contrast to the expected $T^{-1/2}$ dependence. It was concluded that negatively charged, non-open volume defects are responsible for positron trapping at lower temperatures. Examples of such defects are acceptor-type impurities and intrinsic defects like antisites, interstitials, or defect complexes, respectively.

The appearance of shallow positron traps in semiconductors is not only restricted to the case of negatively charged non-open-volume defects as described so far. Shallow trapping of positrons may also occur in the case of a very small open volume related to a small binding energy of the positron to the defect. Such a case has been discussed, e.g. for the *A* center in silicon (see Sect. 4.1.3.1) and for undisturbed dislocations (see Sect. 3.1.5.3).

In the case of negatively charged defects without open volume acting as shallow positron traps, the trapping is based on the capture of positrons in shallow Rydberg states caused by the long-range Coulomb potential (Fig. 3.12). The attractive potential of the negative charge is superimposed on the strong repulsive potential of the nucleus. The weak binding energy of the Rydberg states results in a pronounced temperature dependence of the positron trapping by the occurrence of thermally induced detrapping. This positron detrapping can be described (Manninen and Nieminen 1981) by the detrapping rate δ,

$$\delta = \frac{\kappa}{\rho_{st}} \left(\frac{m^* k_B T}{2\pi\hbar^2} \right)^{3/2} \exp\left(-\frac{E_{st}}{k_B T} \right). \tag{3.21}$$

κ is in this case the trapping rate of shallow positron traps, ρ_{st} their density, and E_{st} the binding energy of the positron to the shallow trap. This equation is basically identical to (3.15). The observable annihilation parameters, such as the positron lifetime or the Doppler *S* parameter, are very close to their corresponding bulk

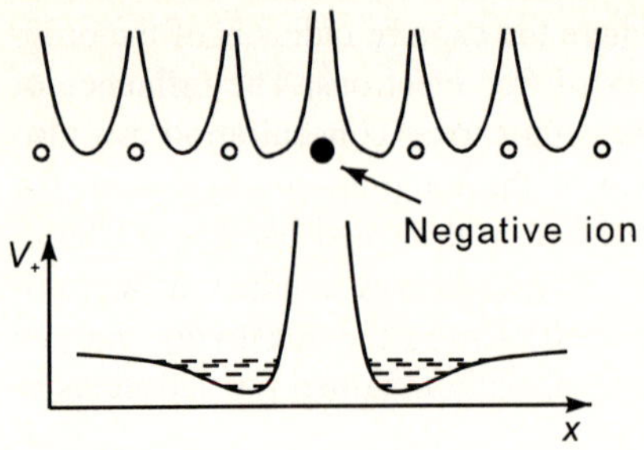

Fig. 3.12. Schematic presentation of the periodic Bloch potential for positrons disturbed by a negatively charged non-open volume defect (*upper part*). In the lower part the attractive positron potential $V_+(x)$ of the shallow trap is shown. The positron bound states are formed by a series of Rydberg states (*dashed lines*).

values, because the position probability density of the positron is extended into the bulk surrounding the shallow trap and is strongly reduced at the site of the nucleus (Fig. 3.12). In order to detect a defect, the defect-characteristic annihilation parameters have to differ from the bulk values. Hence, the shallow trap cannot be directly detected in most cases.

The pronounced temperature dependence of the trapping in shallow traps can, however, be used for this purpose. One has to detect the shallow trap as a competing trapping center to an open-volume defect exhibiting annihilation parameters significantly different from the bulk. In this case, typical temperature dependences, which are characterized by the approach of the annihilation parameter to the bulk value with decreasing temperature, are obtained (Fig. 3.13). For the sample calculation, competitive trapping in negative or neutral vacancies and in shallow positron traps was assumed. The shallow traps were negatively charged defects without open volume. Positron lifetimes of $\tau_v = 270$ ps and $\tau_{st} = \tau_b = 218$ ps were taken. Typical values were chosen for the trapping coefficients. The calculations were carried out with trapping rates low enough to avoid saturated trapping in one of the defects. The positron binding energies in the shallow positron traps were between 30 and 40 meV. The average positron lifetime was calculated using the equations in Sects. 3.1.5.1, 3.1.5.2, and 3.2.1.2. A temperature-independent course of $\bar{\tau}$ is obtained for the neutral vacancy as the only defect, according to the case of transition-limited trapping (compare Fig. 3.8).

The typical decrease in $\bar{\tau}$ with lower temperatures is obtained if, in addition to the neutral vacancy, a shallow positron trap is taken into account. The competitive trapping in a negatively charged vacancy and a shallow positron trap gives a more complex temperature dependence of the average positron lifetime (Fig. 3.13). Such temperature dependences in the case of the presence of vacancies and shallow positron traps can indeed be experimentally observed; see e.g. Fig. 4.11 or Fig. 6.2. The case of negative vacancies is demonstrated by the measurements of Fig. 4.3. A temperature-independent course of the average lifetime due to neutral

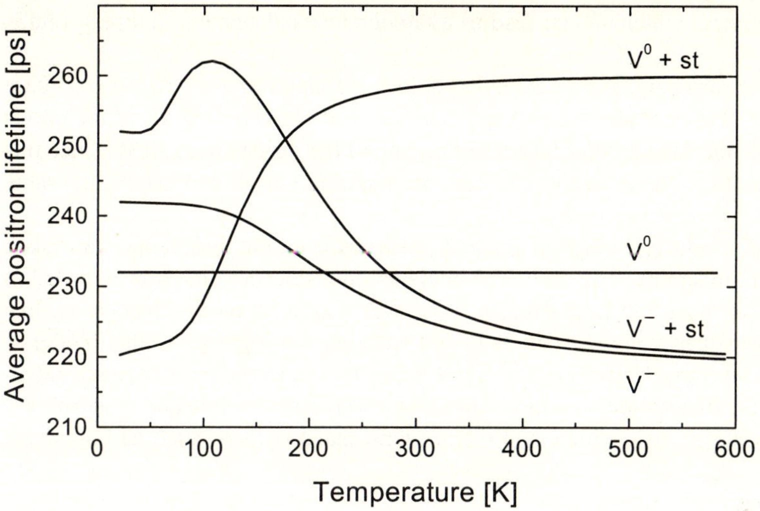

Fig. 3.13. Characteristic course of the average positron lifetime in the presence of vacancies and shallow positron traps. The temperature dependence of the average positron lifetime was calculated for vacancies V in different charge stages and shallow positron traps st in silicon (Eichler 1998).

vacancies has not been reported for semiconductors, but is frequently found in metals.

There is one way for more direct detection of shallow traps: the positron diffusion length measured by back-diffusion experiments is affected by the presence of any positron trap. The decrease in L_+ with decreasing temperature may indicate the presence of shallow traps. This was demonstrated in undoped GaAs by Saarinen et al. (1989) by means of a slow-positron-beam back-diffusion measurement (Fig. 3.14). Although the average positron lifetime and the S parameter in the bulk did not indicate any open-volume defect, the diffusion length L_+ decreased at low temperatures. It was concluded that this behavior was due to the existence of shallow positron traps.

3.1.5.3 Dislocations

There are several approaches to the treatment of the positron interactions with dislocations. For the analysis of positron data after plastic deformation in metals, a simple model of the positron trapping at the dislocation line was applied. The dislocation is treated as an open-volume defect, similar to a row of vacancies. Hence, the trapping rate can be taken proportional to the dislocation density $\kappa = \mu\rho_{\mathrm{dis}}$ (Hashimoto et al. 1992; McKee et al. 1974). This means that a transition-limited trapping process which should lead to a temperature-independent trapping rate is supposed.

Since the dislocation is an extended defect, diffusion-limited trapping (Brandt and Paulin 1972) was also considered. A cylindrical shape (radius r_d) of the trap is assumed for the calculation of the trapping rate of the dislocation (3.13). Since this equation contains the diffusion constant, the trapping rate should exhibit the same temperature dependence as D_+.

Thermally induced escape of trapped positrons from the dislocation was taken into account in another approach. The undisturbed dislocation line was regarded as a shallow positron trap (Smedskjaer et al. 1980). Defects on the dislocation line with a larger open volume, such as bound vacancies or jogs, act in this model as deep positron traps. Häkkinen et al. (1990) carried out molecular dynamics calculations of the dislocation structure. These data were used to calculate positron annihilation data in dislocations and dislocation-related defects, such as bound va-

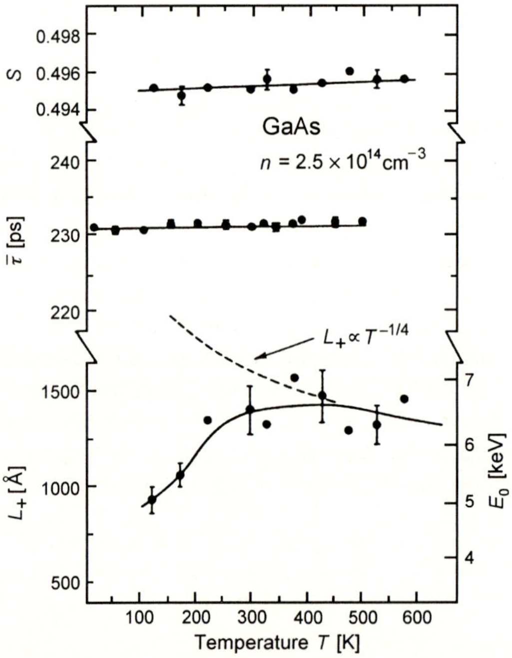

Fig. 3.14. Results of positron experiments in undoped gallium arsenide (Saarinen et al. 1989). The positron diffusion length L_+ was determined as a function of the temperature (*lower curve*) using a slow-positron-beam back-diffusion experiment. The dashed line represents the $T^{-1/4}$ dependence expected for L_+ when no shallow positron traps are presents. On the right scale the E_0 parameter as introduced in Sect. 3.1.4 is shown. For reference, the S parameter and the average positron lifetime $\bar{\tau}$ are presented in the upper part as a function of the temperature.

Fig. 3.15. Atomic configuration of a screw dislocation in gallium phosphide calculated with linear elasticity theory (Lazar and Wagner 1997). The screw, running perpendicular to the plane of drawing, is dissociated into two 30° partial dislocations, indicated as A and B, which are separated by a stacking fault. The gallium and the phosphorus atoms are represented by small and large circles, respectively.

cancies and jogs. They obtained for undisturbed edge dislocations in Al and Cu a lifetime very close to τ_b. A similar result was obtained by Kamimura et al. (1995) for dislocations in iron. The atomic configuration of dislocations in semiconductors was calculated as well. It was found that the open volume is rather small for most dislocation types. It can be assumed therefore that the positron lifetime in dislocations in semiconductors should be close to the bulk value. An example is shown in Fig. 3.15 with a screw dislocation in gallium phosphide dissociated into two 30° partial dislocations. The configuration was calculated by linear elasticity theory (Lazar and Wagner 1997).

In contrast to the positron lifetime at the undisturbed dislocation, vacancies or jogs at dislocations yielded lifetimes between the bulk value and the positron lifetime typical for monovacancies (Häkkinen et al. 1990). These theoretical calculations have been corroborated by experimental results. For example, a positron lifetime of about 225 ps was calculated for a dislocation-bound vacancy as well as for a jog in Al. This corresponds well to the experimental values of 215 (Hidalgo et al. 1989) and 230 ps (Hautojärvi et al. 1970). As a consequence, the model of Smedskjaer et al. (1980) of dislocations as shallow positron traps and deep positron trapping centers associated with the dislocations seems to be very likely. The positron binding energy was theoretically calculated to be of the order of 0.1 eV (Häkkinen et al. 1990). A strong temperature dependence of the trapping rate below room temperature should be expected, related to the small binding energy of the positron to this shallow state.

The dislocation line itself enhances the trapping probability of these deep traps and appears as a precursor state (Hashimoto and Kino 1991; Smedskjaer et al.

1980; Xiong and Lung 1989). The trapping scheme is very similar to the two-stage trapping process described for positron trapping in negative vacancies (Fig. 3.9). Since direct trapping in the deep trap is neglected, the assumptions of (3.17) are also fulfilled. The trapping scheme is analogous to that described for the negatively charged vacancy (Fig. 3.9), where the Rydberg state corresponds to the shallow state of the undisturbed dislocation line. Manninen and Nieminen (1981) calculated with a thermodynamic approach the ratio of detrapping and trapping rate for a dislocation acting as a shallow positron trap,

$$\frac{\delta}{\kappa}=\frac{m^* k_{\mathrm{B}}T}{2\rho_{\mathrm{dis}}\hbar^2}\mathrm{erf}^{-1}\left(\sqrt{\frac{E_{\mathrm{dis}}}{k_{\mathrm{B}}T}}\right)\exp\left(-\frac{E_{\mathrm{dis}}}{k_{\mathrm{B}}T}\right). \tag{3.22}$$

This equation differs from (3.21) because of the different geometry and the different origin of the shallow state. E_{dis} is the positron binding energy to the undisturbed dislocation line. The net trapping rate of a dislocation containing additional vacancy-like defects can be calculated similarly to (3.14).

Dislocations in semiconductors have been treated so far in a similar way to metals (Kawasuso et al. 1995d; Krause-Rehberg et al. 1993a). However, as a further complication, they may be charged, leading to additional attractive or repulsive terms of the trapping potentials.

No conclusive decision between the three models, viz. the transition-limited trapping in dislocations, the diffusion-limited trapping, and the shallow-trap model, presented can be made with the current experimental data in semiconductors. The temperature dependence of the positron trapping to dislocations should be studied to distinguish between them. However, such experiments are not easy to perform, since other defects introduced in addition to dislocations during plastic deformation may influence the analysis (Krause-Rehberg et al. 1994b). A first attempt made by Kawasuso et al. (1995d) seems to confirm the validity of the shallow-trap model of dislocations in semiconductors, but a further verification is necessary.

3.1.5.4 Voids

According to the theoretical calculations of Nieminen and Laakkonen (1979), the agglomeration of vacancies leads to an increase in the trapping coefficient. At least for a small number of vacancies i ($i \leq 5$) forming the vacancy cluster, the trapping coefficient of the vacancy cluster $\mu_{i\mathrm{v}}$ is proportional to i,

$$\mu_{i\mathrm{v}} = i\mu_{\mathrm{v}}, \tag{3.23}$$

where μ_{v} is the trapping coefficient of a monovacancy. If the agglomerate grows further, (3.23) is not valid any more and the trapping coefficient saturates at a high value (Fig. 3.16). This is connected to the fact that the positron mobility starts to limit the trapping rate.

A distinct change in the annihilation parameters is observed during agglomeration of vacancies into clusters. The positron lifetime as well as the *S* parameter increase significantly. The lifetime was calculated for different cluster sizes in Si. The results are presented later in Fig. 4.4.

Voids in semiconductors have been studied by positron annihilation in only a few cases (e.g. Dannefaer et al. 1991; Halec et al. 1994; Krause-Rehberg et al. 1993a; Perovic et al. 1991b). However, a lot of experimental data on void formation is available for plastically deformed or irradiated metals. A theoretical treatment of positron trapping in voids was given by Nieminen et al. (1979). The positron trapping was treated as a diffusion problem. The intensity of the void component of the positron lifetime spectra was calculated as a function of the sample temperature. The experimentally observed dependence of the intensity I_2 for voids in Al (Fig. 3.17) is in a good agreement with these semiclassical calculations. The quantum reflection at the potential well of the void may play an important role at lower temperatures, especially for large voids. Additionally, at low temperatures, the positron mobility is large and the capture rate is small. The trapping process is transition-limited there. A crossover to diffusion-limited trapping occurs as the mobility decreases and the capture rate increases. This effect is called self-shielding of the voids from the positron flux (Nieminen et al. 1979).

Dannefaer et al. (1991) analyzed the positron trapping in voids observed after plastic deformation of gallium arsenide in the framework of the model of Nieminen et al. (1979). The positron trapping coefficient as a function of temperature was treated with more general model assumptions, leading to a modified temperature dependence (Fig. 3.18).

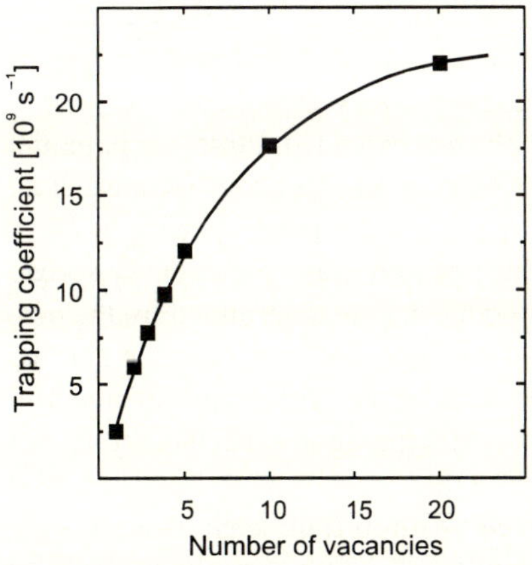

Fig. 3.16. Trapping coefficient versus the number of vacancies in a void in aluminum. The numerical date were taken from Nieminen and Laakkonen (1979). The line is drawn to guide the eye.

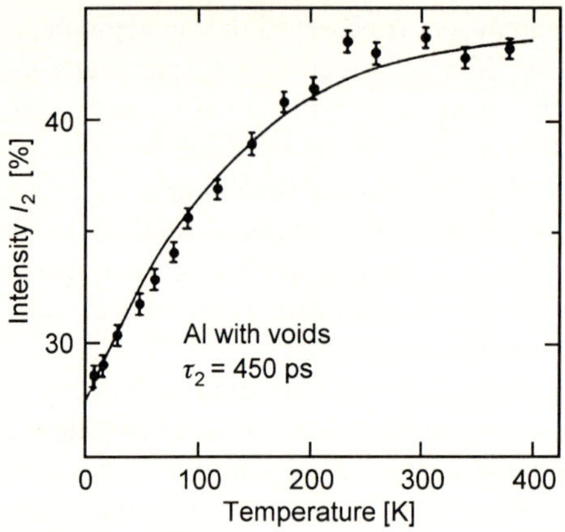

Fig. 3.17. Intensity I_2 of the void-related lifetime component as a function of temperature for positrons annihilating in aluminum irradiated with fast neutrons. The average radius of the voids was 25 nm, the void density 2×10^{14} cm^{-3}. The solid curve is the result of the theoretical calculations (Nieminen et al. 1979).

3.1.5.5 Precipitates

Positron trapping in precipitates was extensively studied in aluminum alloys (Dlubek 1986). It was found that positrons may be trapped by coherent precipitates, i.e. by particles completely free of defects in the interior and at the interface. The reason for positron trapping lies mainly in the different positron affinities for different chemical elements.

A table of positron affinities for most elements can be found in the paper of Puska et al. (1989a). The values are −6.95 and −6.69 eV for silicon and germanium, respectively. Fig. 3.19a shows the weak, but extended attractive potential for the example of Guinier–Preston zones in Al–Zn or Al–Ag alloys (Krause et al. 1986a, 1986b). Precipitates with a diameter of 2 nm and a mean spacing of 10 to 20 nm cause saturated positron trapping. The positron wave function is spread over the whole precipitate. The annihilation parameters are determined by the corresponding bulk parameters of the precipitated material. It is thus possible that the measured lifetime lies below the value of the host lattice. In the case of positron trapping in coherent precipitates, diffusion-limited trapping is expected and (3.12) must be applied to calculate the positron trapping rate.

The coherent precipitate may contain a deep positron trap, such as a vacancy (Fig. 3.19b). For positrons trapped in the precipitate in a first step, the probability to be captured in this additional trap is very high. The reason is the strong confinement of the positron wave function to the volume of the precipitate. Thus, a

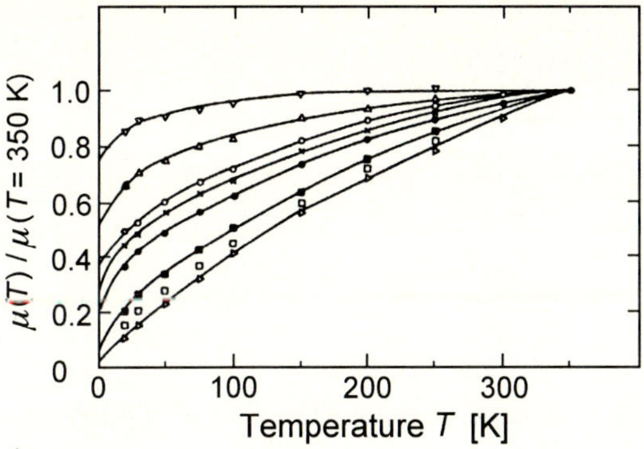

Fig. 3.18. Theoretically calculated trapping coefficient μ and their temperature dependence for different void radii indicated by different symbols given in Table 3.2. The calculations performed by Dannefaer et al. (1991) were done using a modified model of Nieminen et al. (1979). The plotted trapping coefficients were normalized to their respective values at $T = 350$ K, which are also given in Table 3.2.

Table 3.2. Values and symbols used in Fig. 3.18 for the calculation of the trapping coefficient of voids of different sizes (Dannefaer et al. 1991).

Symbol	▽	△	×	○	●	■	□	▸
μ(350 K) $[10^{-7}\ \mathrm{cm^3 s^{-1}}]$	4	6	12	8	17	70	237	1371
r_d [nm]	5	8	15	15	20	50	100	250
m^*/m_0	1	1	1	1.5	1	1	1	1
i	25	102	675	675	1600	2.5×10^4	2×10^5	3.1×10^6

μ is the trapping coefficient, r_d the radius of the defects (voids), m^* the effective positron mass, m_0 the positron rest mass, i the number of vacancies per void.

single vacancy is sufficient to determine the trapping in the precipitate. In this case, the annihilation parameters provide a signal of this inner defect (Dlubek et al. 1983).

The interface of incoherent or semicoherent precipitates always contains misfit defects, which can act as positron traps (Fig. 3.19c). The positron wave function is strongly localized in the interface region and the annihilation parameters exhibit a characteristic defect signal. In contrast to the former two cases, no positron capture to the precipitate must precede the trapping in these defects. This case was observed in incoherent Si precipitates in Al–Si alloys (Krause et al. 1985a) and in semicoherent Guinier–Preston zones in Al–Zn (Krause et al. 1985b).

The examples for metals given above are relatively deep positron traps. It is likely that for a smaller difference in the positron affinity, the attractive potential of the precipitate may be weak, so that it acts as a shallow positron trap. The de-

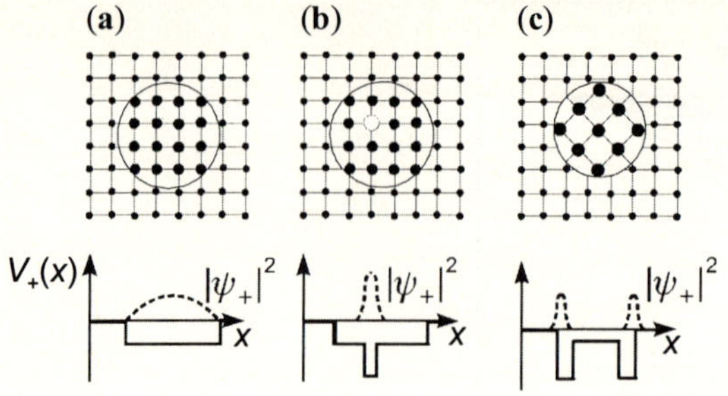

Fig. 3.19. Positron potentials $V_+(x)$ and wave functions ψ_+ of different types of precipitates as experimentally found e.g. in different aluminum alloys. (**a**) Fully coherent precipitate, (**b**) fully coherent precipitate containing a vacancy, and (**c**) semicoherent or incoherent precipitate.

trapping behavior of positrons from three-dimensional shallow traps was discussed by Manninen and Nieminen (1981). The temperature dependence of the detrapping rate was obtained as

$$\frac{\delta}{\kappa} = \frac{1}{V_t \rho_t} \left[\frac{\sqrt{\pi}}{2} \operatorname{erf}\left(\sqrt{\frac{E_t}{k_B T}} \right) - \sqrt{\frac{E_t}{k_B T}} \exp\left(-\frac{E_t}{k_B T} \right) \right] \exp\left(-\frac{E_t}{k_B T} \right), \tag{3.24}$$

where E_t, V_t, and ρ_t are the positron binding energy, the volume, and the density of three-dimensional traps (precipitates).

Precipitates may be also important in semiconductors. However, distinct positron trapping is only expected in the rare case, when the mean distance between the precipitates is of the order of the positron diffusion length. For instance, the density of arsenic precipitates may be as high as 10^{17} cm^{-3} in annealed low-temperature-grown gallium arsenide layers. Distinct positron trapping was found in such layers and attributed to arsenic precipitates (see Sect. 5.2.4).

3.1.5.6 Surfaces

Positrons implanted into a solid can be backscattered into the vacuum or can thermalize and start to diffuse. When the implantation energy amounts to only a few keV, positrons have a high probability of reaching the surface during diffusion. In contrast to electrons, the positron work function of surfaces may have negative values, i.e. positrons may be spontaneously emitted. In the following, the processes occurring at a surface with a negative work function (Chu et al. 1981) are described.

Positrons can be re-emitted into the vacuum (Fig. 3.20) and they have a kinetic energy corresponding to the thermally spread work function (Fischer et al. 1986).

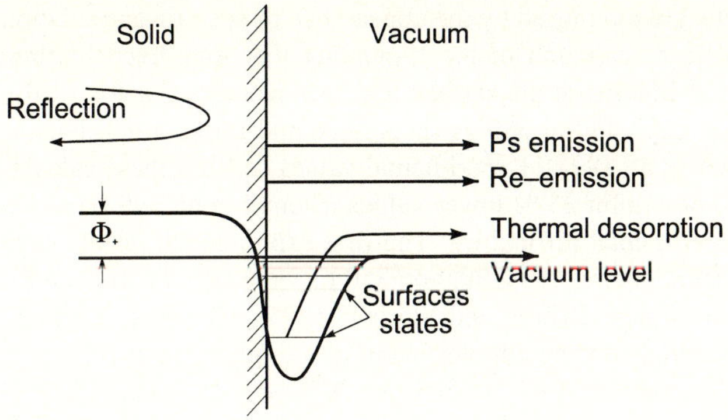

Fig. 3.20. Interaction mechanisms of thermalized positrons diffusing back to a surface with a negative work function.

This effect is used for the moderation of positrons in order to get a monoenergetic positron beam (see Sect. 2.4.1). As an alternative to re-emission, positrons can be trapped in the surface image potential or in surface defects. This process is more likely at low temperatures. The increase in the temperature results in thermally stimulated desorption into the vacuum. Another possible process is the formation of positronium followed by its escape from the solid.

With a low probability, the annihilation of positrons in the surface potential can also take place with a core electron. This leads to an Auger process, which is used in positron-annihilation-induced Auger emission spectroscopy, already described in Sect. 2.5.3. Finally, positrons may be reflected from the surface potential step (Britton et al. 1989). This quantum mechanical process is especially important at low temperatures.

The positron work function has mainly been found with positive values in semiconductors. In this case, the scheme of positron behavior at surfaces remains valid, but the re-emission channel does not occur. As one of the rare exeptions, SiC was reported to have a negative positron work function (Brauer et al. 1996b). This material is the subject of intensive investigations, because it opens up the possibility of the construction of field-assisted positron moderators. The positron can reach the surface not only by diffusion in such a moderator, but the electric field gives an additional drift contribution to increase the number of positrons reaching the surface, i.e. to increase the moderator efficiency.

The various processes described above can be distinguished experimentally. This is due to the variation of the annihilation parameters in the different positron states. The re-emitted positrons can be separated by their kinetic energy from the backscattered ones with a retarding field experiment. For semiconductors and metals, positronium is only formed outside the crystal and must originate from one of the channels mentioned. The detection of positronium is based on the three-γ-quanta annihilation of ortho-positronium (see Sect. 3.1.6). The positrons captured in the surface-image state are monitored by the appearance of specific annihilation parameters, e.g. a significant variation of the S parameter or long-lived lifetime components. The positron lifetime at the surface has been calculated theoretically for several metals in the range 450 to 660 ps (Nieminen and Puska 1983; Nieminen et al. 1984; Rubaszek et al. 1993). Experimental values confirm these calculations (Lynn et al. 1984) or exhibit 25 % lower values (Steindl et al. 1992), which have been related to experimental difficulties. The first experimental results have been published for semiconductors. 410, 330, and 350 ps were found at n-Si, p-Si, and p-InP surfaces, respectively (Britton and Störmer 1995). However, it should be noted that the surfaces studied were not cleaned in ultra-high vacuum.

3.1.5.7 Interfaces

The strong image potential as described for surfaces does not exist at interfaces. No positron trapping occurs at coherent interface structures. However, a lattice mismatch is connected to interface defects, such as misfit dislocations. Such interfaces are potential positron traps. In respect to the interaction with positrons, grain boundaries also act as incoherent interfaces. The situation is similar to incoherent or semicoherent precipitates (Sect. 3.1.5.5), where misfit defects occur as well as positron traps. The annihilation parameters may change and specific interface values are measured.

Similar to dislocations, it is conceivable that the open volume at an interface is small. The interface may thus act as a shallow positron trap. This case represents a two-dimensional shallow trap. The temperature dependence of the detrapping rate was calculated by Manninen and Nieminen (1981). The ratio of the detrapping and trapping rates was obtained as

$$\frac{\delta}{\kappa} = \frac{1}{\rho_\mathrm{t}} \sqrt{\frac{m^* k_\mathrm{B} T}{2\pi\hbar^2}} \, \frac{\exp\left(-\frac{E_\mathrm{t}}{k_\mathrm{B}T}\right)}{1-\exp\left(-\frac{E_\mathrm{t}}{k_\mathrm{B}T}\right)}, \tag{3.25}$$

where E_t and ρ_t are in this case the positron binding energy and the surface to volume ratio of planar traps, respectively.

In a real case, the interface defects may be spread over a certain thickness. For semiconductors, several investigations exist on the technologically important SiO_2/Si interface (see Sect. 4.1.5). The existence of vacancy-like defects and voids in the interface region was reported by Peng et al. (1996). A specific interface state having a smaller S parameter than the bulk of Si and SiO_2 layers, respectively, has been found (Asoka-Kumar et al. 1994). This finding can hardly be due to the occurrence of simple open-volume-type defects. In order to interpret the results, the trapping center was supposed to be associated with dangling-bond sites and oxygen.

Interface structures have been extensively studied by various surface-sensitive methods for several epitaxial metal layers. In contrast to these established techniques, slow-positron-beam experiments are promising for the study of buried defect structures. Therefore, such interfaces were also studied by slow positrons. The first results were published by Huttunen et al. (1990) for the systems Cu/Ag(111) and Ag/Cu(111). Short positron diffusion lengths were found by back-diffusion experiments. They were interpreted by misfit dislocations (density of ~ 10^{10} cm^{-2}) or alternatively vacancies (~ 10^{18} cm^{-3}) in the interface region. The density was higher by a factor of three in the Cu/Ag system, where open-volume defects are expected for the reduction of the interface strain.

Another study was carried out on Cu/Ru(0001) layers (Hahn and Krause-Rehberg 1997). Scanning tunneling microscopy provided detailed information on the layer growth and a model of the interface could be given (Günther et al. 1995). Similar to the results of the Cu/Ag system, the positron diffusion length was very short for thin Cu layers, resulting in unrealistically high defect concentrations (Hahn and Krause-Rehberg 1997). Also, no indications for positron trapping in interface defects, detected by other methods, could be found in the $S(E)$ plots. It was concluded that misfit defects in a layer with a thickness of a few atomic distances can hardly be detected by positrons.

An interesting quantum mechanical phenomenon at the interface is the appearance of a potential step ΔV_+ for positrons. This potential step equals the difference in the positron affinity of the substrate and the layer. Thermal positrons penetrating the interface gain additional kinetic energy. For layers with a thickness in the same range as the positron mean free path, the probability is high that the positrons reach the surface without further energy losses. The positrons re-emitted from this surface have an energy that is the sum of the work function Φ_+ and ΔV_+. This effect was found by Gidley and Frieze (1988) by studying the energy distribution of positrons re-emitted from Cu layers, which were epitaxially grown with different thicknesses on top of a Ni(110) surface. In the opposite case for Ni on Cu, the escape of thermal positrons is prevented by the interface potential.

Most of the semiconductors have a positive work function and, therefore, re-emission experiments to study the positron energy distribution in thin layers are hardly possible. However, a similar interface potential step is expected.

3.1.5.8 Fine-Grained Material and Powders

The ability of grain boundaries to trap positrons has already been described in the last section. It is important to know in fine-grained materials and powders what fraction of positrons is able to reach grain boundaries. The size of the crystallites obviously plays an important role. The dimensions of the grains are the same as the extension of interface regions in the extreme case of nanocrystalline materials (Gleiter and Marquardt 1984). This means that the grain diameter is smaller than the positron diffusion length, and the positron may penetrate several grains on its

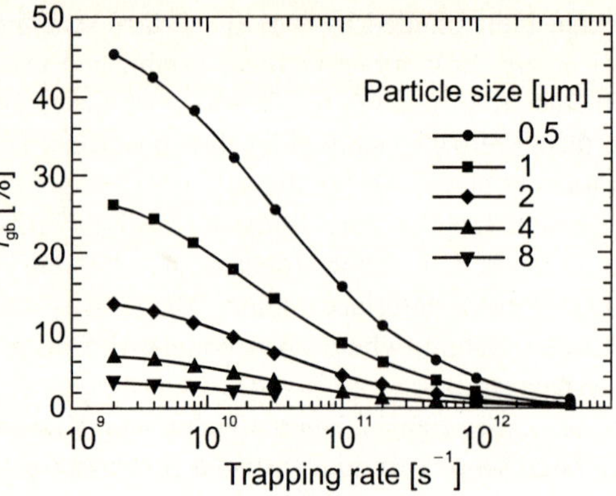

Fig. 3.21. Monte-Carlo simulation of positron diffusion in a fine-grained solid (Hübner et al. 1994). The fraction of positrons f_{gb} reaching the surface of the grain during the diffusion is plotted versus the trapping rate of open-volume defects present in the interior of the grain. The parameter of the curves is the diameter of the grains, supposed to have a spherical shape. The calculation was carried out for copper.

diffusion path. Indeed, positron lifetime measurements of nanocrystalline metals and silicon proved the effect of grain boundaries as positron traps (Würschum et al. 1987).

Grains with diameters of the order of 1 μm occur frequently in metals and some semiconductors. The possible influence of grain boundaries on positron trapping was studied for this case. A Monte-Carlo simulation of the positron diffusion in spherical and ellipsoidal grains was carried out by Hübner et al. (1994). The fraction of positrons reaching the grain surface, f_{gb}, was investigated as a function of the grain dimensions and the concentration of open-volume defects in the interior of the grain. f_{gb} becomes very small for grain sizes larger than 5 μm (Fig. 3.21). On the other hand, approximately 50 % of all positrons reach the grain surface for a particle size of about 500 nm. This holds for negligible trapping in open-volume defects in the grain interior. f_{gb} reduces significantly with increasing trapping rate of such defects. This fraction may be additionally influenced in polycrystalline semiconductors by the existence of internal fields due to band bending. Geometry effects were considered in the simulation of Hübner et al. (1994) by variation of the ratio of the half axes of ellipsoidal particles. The smaller half axis determines f_{gb} for a ratio larger than 1 : 2.

The diffusion-controlled trapping in grain boundaries was theoretically described by Dupasquier et al. (1993) and Seeger (1992). The positron lifetime measurements obtained in fine-grained ZnAl alloys were analyzed with this model. The grains were assumed to be spherical and the grain boundaries as zones rich in open-volume defects. Trapping and detrapping at the grains and no trapping within the grains were considered. The diffusion equation was solved and an analytical solution given.

3.1.6 Positronium Formation

Positronium (Ps) represents a bound state between a positron and an electron, which can be treated formally as an hydrogen atom. The Schrödinger equation for positronium is identical to that for hydrogen, where the reduced mass of the hydrogen atom is replaced by one half of the electron mass (Akhiezer and Berestetskii 1965). The eigenvalue equation for the center of mass of Ps can be solved. The eigenvalues are very close to half of those of hydrogen. The ionization energy of positronium is 6.8 eV.

There are several models of positronium formation, which were reviewed in a monograph by Mogensen (1995). The first general theory was published by Ore (1949). He assumed that positronium is formed during slowing down of positrons at energies E_+ in the "Ore gap", $E_{\text{ion}} - 6.8\text{ eV} < E_+ < E_{\text{ion}}$. E_{ion} is the ionization energy of molecules or atoms in the crystal, i.e. the energy necessary to release the electron for the Ps formation. However, this model can only be applied for solids. The bubble model of the Ps states in liquids was introduced by Ferrell (1958). The most accepted model nowadays is the "spur model" by Mogensen (1974). Positronium is formed mainly by the reaction between the positron and an excess electron in the spur. The positron spur is the terminal track of the positron, formed when it loses the last part of its kinetic energy. The positronium yield can be quantitatively treated with reaction kinetics of positron spur reactions (Mogensen 1995).

Positronium is basically formed in two states, 75 % as ortho-positronium with parallel spins (triplet state) and 25 % as para-positronium with antiparallel spins (singlet state). The energy difference between these spin states (hyperfine splitting) is only 8.4×10^{-4} eV (Akhiezer and Berestetskii 1965). The electron density in matter must be sufficiently low in order to allow the formation of positronium. The lifetime of para-Ps was theoretically calculated to be 123.91 ps (Rich 1981). The best theoretical value for ortho-Ps is 142.08 ns. Experimentally, it was found to be 141.88 ns (Nico et al. 1990).

Para-positronium annihilates intrinsically (i.e. annihilation between the particles forming the Ps) mainly into two γ-rays of 511 keV. However, ortho-positronium annihilates in vacuum into three γ-rays in order to conserve spin. In matter, the positron wave function overlaps with electrons outside the positronium. The annihilation with such electrons having an antiparallel spin decreases the lifetime by about two orders of magnitude and amounts thus to about 1 to 5 ns. This process is called pick-off annihilation and results in two γ-rays.

The occurrence of positronium has been extensively studied in ice (Mogensen 1995). This material is one of the rare cases, where Ps is delocalized in a Bloch state leading to a peak in angular correlation of annihilation radiation with a negligible width. A corresponding experiment is shown in Fig. 3.22 (Mogensen et al.

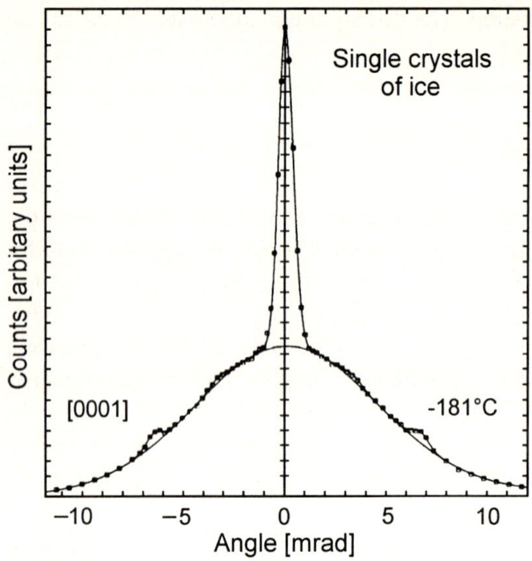

Fig. 3.22. Angular correlation of annihilation radiation curves of [0001]-oriented single crystals of H_2O ice measured at −181 °C (Mogensen 1995). The sharp peak in the center is caused by intrinsic para-positronium annihilation.

1971; Mogensen 1995). Both the central and the side peaks are ascribed to intrinsic para-positronium formation from Bloch states. The rest of the curve is due to pick-off annihilation of ortho-positronium and annihilation of positrons not forming Ps. Such annihilation from Ps Bloch states can thus be used to determine the angular resolution function of a angular correlation of annihilation radiation machine, as was demonstrated for quartz by Bisson et al. (1982).

Positronium is normally found in a defect-free solid in a self-trapped state, i.e. the positronium creates by itself a "cage" by pushing away the surrounding atoms. The exception of the delocalization of Ps, as in the case of ice (Fig. 3.22), was only found in very rigid solids. In addition to self-trapping, positronium can also be captured in structural defects, such as vacancies and vacancy clusters, as observed e.g. in plastic crystals (Eldrup et al. 1979). In its time of existence, positronium can react with atoms and molecules to form bound states.

Under the influence of a static magnetic field, new positronium eigenstates are formed as mixed states of ortho-Ps and para-Ps (Hyodo 1995). The decay rates are linear combinations of the decay rates of ortho- and para-positronium, respectively, with field-dependent coefficients. Experimentally, a modification of the positronium-related lifetimes and the reduction of the ortho-Ps fraction are to be observed. This effect is called magnetic quenching of ortho-positronium. The quenching ratio was defined as the ratio of annihilation events of ortho-Ps with and without magnetic field (Bisi et al. 1987). In the normal lifetime spectrum, it

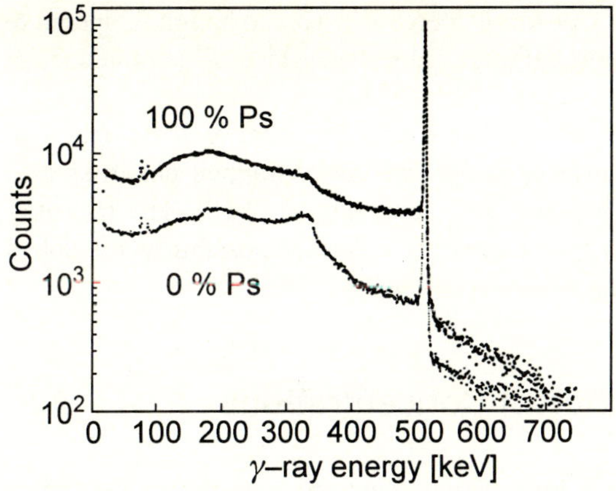

Fig. 3.23. Annihilation γ-spectra measured at an Al(110) surface for sample conditions, where no positronium is formed (0 % Ps) and where 100 % Ps emission occurs (Lahtinen et al. 1989). The first condition is realized by a high incident positron energy of 15 keV, the second one by a low incident energy of 40 eV at a sample temperature of 400 °C. Both curves were normalized to equal height of the 511-keV line.

can be determined by using a time window containing only ortho-positronium contributions in a good approximation, i.e. at large output pulses of the time-to-amplitude converter.

In special cases, positronium formation was also reported for semiconductors (Dannefaer et al. 1996b; Itoh et al. 1994), e.g. for porous Si (see Sect. 4.1.10). Provided that no such large open-volume defects are present in the bulk of semiconductors, positronium may exist only at the semiconductor surface. Hence, the occurrence of positronium may be taken as a measure of the fraction of positrons diffusing back to the surface, f_s. Back-diffusion measurements may be used in order to obtain the positron diffusion constant D_+ and the effective annihilation rate λ_{eff} (see Sect. 3.2.3). The latter procedure allows the determination of the defect concentration practically without an upper sensitivity limit. It is based on the quantification of the positronium formation at the surface. The respective measure is called positronium fraction F. It is calculated (Lynn and Welch 1980; Marder et al. 1956; Mills 1978) as

$$F = \left(1 + \frac{A_{\mathrm{P}}^{1}}{A_{\mathrm{P}}^{0}} \frac{R^{1} - R^{\mathrm{f}}}{R^{\mathrm{f}} - R^{0}}\right)^{-1}, \tag{3.26}$$

where $R^{i} = (A_0^{i} - A_{\mathrm{P}}^{i}) / A_{\mathrm{P}}^{i}$, with A_0^{i} and A_{P}^{i} as the integrated counts in the peak and total area of the annihilation spectrum (Fig. 3.23). The superscripts i = 1, 0 correspond to the cases of 100 % and 0 % positronium formation, respectively. Practically no positrons are diffusing back to the surface at a high positron implantation energy, and no positronium can be formed. Maximum positronium formation can be obtained when a low positron energy is used and a high temperature is applied to ensure the positronium emission from the surface. R^{f} is obtained for the sample under consideration.

As the positronium fraction is assumed to be proportional to the fraction of positrons diffusing back to the surface, f_{s}, the fraction of trapped positrons as a function of incident energy can be calculated according to (2.18). The measurement of the positronium fraction thus represents a further possibility for defect depth profiling in semiconductors.

3.2 Determination of Defect Concentrations

Defect spectroscopy should provide information about the kind and concentration of the defects under consideration. This section is dedicated to the question of how absolute values of the defect concentration can be determined by the different methods of positron annihilation. Positron capture can be phenomenologically described by a set of differential rate equations (Sect. 3.2.1). The positron capture in defects is characterized by the trapping rate κ being proportional to the defect concentration, (3.11) and (3.12). All positron techniques provide this trapping rate as an experimental quantity. The determination of the trapping rate from positron lifetime measurements is dealt with in Sect. 3.2.1, while the quantification of Doppler-broadening and slow-positron-beam back-diffusion measurements are treated in Sects. 3.2.2 and 3.2.3, respectively. The proportionality constant between trapping rate and defect concentration, which is called the trapping coefficient μ, was introduced in (3.11). Its determination requires a reference method which yields the defect concentration independently. Results for semiconductors are summarized in Sects. 3.2.4 and 3.2.5. If the trapping coefficients for different defect types are known, the sensitivity range of the positron methods can be determined.

3.2.1 Trapping Models and Interpretation of Positron Lifetime Spectra

A phenomenological description of positron trapping was given by Bertolaccini and Dupasquier (1970) and by Connors et al. (1971) and was later generalized (e.g. Brandt and Paulin 1972; Frank and Seeger 1974). This model is referred to as the trapping model. The aim is a more quantitative analysis of the influence of positron trapping in defects on the annihilation spectra. Information on the dominant positron trapping centers and their concentrations may be obtained if the analysis of the spectra with the trapping model is feasible. The prerequisites of a description of positron trapping in the simple rate equations, given in the following, are that positrons are not significantly trapped before thermalization and that the defects are homogeneously distributed in the sample volume under consideration and do not interact with each other. The wave function of the trapped positron is localized at lattice defect sites. The rate equations can be derived under these assumptions from the time-dependent positron diffusion equation (3.4).

The general task of data analysis on the basis of the trapping model is the determination of the trapping rates and thus the defect concentrations of the dominant positron trapping centers. Two kinds of computer programs are available for the analysis of the experimental spectra. The first one is based on Gauss–Newton fitting procedures, and up to four lifetime components can be separated (LIFSPECFIT program by Puska 1978; POSITRONFIT program by Kirkegaard et al. 1989). The special problems of data treatment using these programs, such as source corrections and the determination of the number of components, were treated by Staab et al. (1996) and Somieski et al. (1996). The second kind of computer programs gives distributions of lifetimes (CONTIN program by Provencher 1982; MELT program by Shukla et al. 1993). The latter type of programs has the advantage that the number of components of the spectra is not an input parameter but is provided in contrast to the former type of programs by the fit itself. Nevertheless, a source correction has to be carried out as well. Both kinds of evaluation procedures are restricted to a small number of components in the spectra, at most five.

The simplest case of the trapping model, including only one open-volume defect type, is presented in Sect. 3.2.1.1 before a more general case of independent defects is described (Sect. 3.2.1.2). Section 3.2.1.3 deals with the occurrence of combined traps, e.g. deep positron traps on a dislocation line, which acts as a shallow precursor state. A collection of important equations for different trapping models is given in Appendix A2.

3.2.1.1 Positron Capture by One Defect Type

Positron capture in a single open-volume defect type, such as a vacancy, is used to demonstrate the simplest trapping model. The scheme of this trapping model is shown in Fig. 3.24. After thermalization, the positrons may annihilate from the defect-free bulk of the sample with the annihilation rate λ_b. Provided that the concentration of the open-volume defect d is sufficiently high, a certain fraction of positrons diffusing through the sample may be trapped there with the trapping rate κ_d. A thermally induced escape from the bound state in the attractive potential of the defect is excluded in this model. This assumption is usually fulfilled for vacancies. Since the electron density in such defects is reduced, the corresponding annihilation rate λ_d is expected to be smaller than λ_b. The rate equations are

$$\begin{aligned}\frac{\mathrm{d}n_{\mathrm{b}}(t)}{\mathrm{d}t} &= -(\lambda_{\mathrm{b}} + \kappa_{\mathrm{d}})n_{\mathrm{b}}(t),\\ \frac{\mathrm{d}n_{\mathrm{d}}(t)}{\mathrm{d}t} &= -\lambda_{\mathrm{d}} n_{\mathrm{d}}(t) + \kappa_{\mathrm{d}} n_{\mathrm{b}}(t).\end{aligned} \tag{3.27}$$

n_{b} and n_{d} are the number of positrons in the bulk and in the defect at time t. λ_{b} and λ_{d} are the annihilation rates in the bulk and in the defect, and κ_{d} is the positron trapping rate of the defect. The starting conditions are $n_{\mathrm{b}}(0) = N_0$, where N_0 is the total number of positrons at $t = 0$, and $n_{\mathrm{d}}(0) = 0$. This corresponds to the condition that no trapping occurs before thermal energies are reached. The solution of (3.27) is the decay spectrum of positrons,

$$D(t) = I_1 \exp\left(-\frac{t}{\tau_1}\right) + I_2 \exp\left(-\frac{t}{\tau_2}\right), \tag{3.28}$$

with the abbreviations

$$\begin{aligned} \tau_1 &= \frac{1}{\lambda_{\mathrm{b}} + \kappa_{\mathrm{d}}}, & \tau_2 &= \frac{1}{\lambda_{\mathrm{d}}},\\ I_1 &= 1 - I_2, & I_2 &= \frac{\kappa_{\mathrm{d}}}{\lambda_{\mathrm{b}} - \lambda_{\mathrm{d}} + \kappa_{\mathrm{d}}}. \end{aligned} \tag{3.29}$$

Equation (3.28) is identical to (2.2) with the number of defects $k = 1$. The lifetime spectrum is obtained as the number of annihilating positrons per time channel in the multi-channel analyzer and, therefore, is written as the absolute value of the time derivative of $D(t)$,

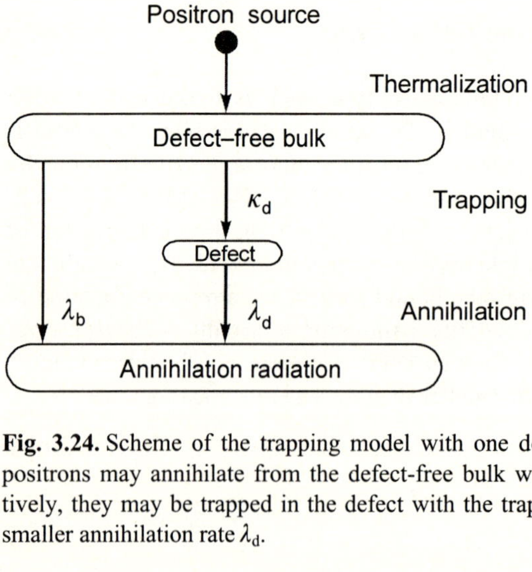

Fig. 3.24. Scheme of the trapping model with one defect type. After thermalization, the positrons may annihilate from the defect-free bulk with the annihilation rate λ_{b}. Alternatively, they may be trapped in the defect with the trapping rate κ_{d} and annihilate with the smaller annihilation rate λ_{d}.

$$N(t)=\left|\frac{\mathrm{d}D(t)}{\mathrm{d}t}\right|=\frac{I_1}{\tau_1}\exp\left(-\frac{t}{\tau_1}\right)+\frac{I_2}{\tau_2}\exp\left(-\frac{t}{\tau_2}\right). \tag{3.30}$$

This corresponds to (2.3) with number of defects $k = 1$. The spectrum experimentally obtained contains in addition to (3.30) an almost constant background contribution. Equation (3.30) is furthermore convolved by the experimental time resolution function, which can be approximated by a sum of Gaussians (Sect. 2.2.2).

It is more common to use the lifetimes $\tau_i = 1/\lambda_i$ instead of the annihilation rates λ_i. The individual terms of (3.30) are called lifetime components τ_i with the corresponding intensities I_i. According to (3.29), the values of τ_1, I_1, and I_2 are in contrast to τ_2 dependent on the trapping rate κ_d and thus on the defect concentration. The τ_i and I_i are obtained after background reduction and source correction by a non-linear fit from the experimental lifetime spectrum (Sect. 2.2.2). They can be used to calculate the trapping rate κ_d as the experimental result being proportional to the defect concentration,

$$\kappa_d=\mu C=I_2\left(\frac{1}{\tau_1}-\frac{1}{\tau_2}\right)=\frac{I_2}{I_1}\left(\frac{1}{\tau_b}-\frac{1}{\tau_d}\right). \tag{3.31}$$

τ_1 is sometimes also called reduced bulk lifetime. τ_2 determined experimentally from the two-component decomposition (3.28) and (3.29) is identical to τ_d. It is a measure of the reduced electron density in the open-volume defect and is thus larger than the bulk lifetime τ_b. Because τ_d is independent of the defect concentration, it can be taken as a characteristic value of the open volume of the defect. In this way the agglomeration of vacancies can be observed. For example, the ratio τ_d/τ_b amounts to 1.25 for monovacancies in Si (Fuhs et al. 1978) and may increase up to 2.6 for voids (Krause-Rehberg et al. 1993a). τ_d is called, in the following, the defect-related lifetime, and it can be determined in principle from a single lifetime measurement. However, in order to reduce the influence of statistical uncertainties, the average of several measurements is usually taken.

Three values are determined experimentally, viz. the two lifetimes and one intensity. τ_2 does not depend on the trapping rate, but from each of the other two observables the trapping rate can be calculated. The system of equations is overdetermined. The reliability of the chosen number of components used in the numerical fit can be checked by comparing τ_1 determined experimentally and τ_1 from the trapping model (3.29) obtained with κ calculated from I_2. If these two values differ significantly, more lifetime components should be taken into account for the decomposition of the spectra. A further hint for more components to be taken into account is a bad variance of the fit (Somieski et al. 1996).

The average positron lifetime $\bar{\tau}$ is given as

$$\bar{\tau}=\sum_{i=1}^{k+1} I_i\tau_i\,. \tag{3.32}$$

In most cases, this is rather insensitive to the numerical fitting procedure applied. Provided that the values of τ_b and τ_d are known, the trapping rate κ_d can also be determined from the average lifetime $\bar{\tau}$ with (3.29) and (3.32) as

$$\kappa_d = \frac{1}{\tau_b}\frac{\bar{\tau}-\tau_b}{\tau_d-\bar{\tau}} = \frac{\eta}{\tau_b(1-\eta)} . \tag{3.33}$$

The parameter η is the annihilation fraction (frequently also called the fraction of trapped positrons[1]), given by

$$\eta = \int_0^\infty n_d(t)\mathrm{d}t = \frac{\kappa_d}{\lambda_b+\kappa_d} . \tag{3.34}$$

For very high defect concentrations, i.e. the average spacing between defects is much smaller than the positron diffusion length in the bulk, all positrons are trapped and a one-component spectrum with $\bar{\tau} = \tau_d$ and $I_2 = 1$ is obtained. Thus, only a lower-limit estimate of the defect concentration can be given. This case is called saturated positron trapping.

3.2.1.2 Trapping in Several Independent Defect Types

When more than one defect type is involved in positron trapping, the model of Sect. 3.2.1.1 is not sufficient. Not all possible models are discussed here, but only one extended model that can be used for the interpretation of many experimental results to be described later. This model (Fig. 3.25) includes two different types of open-volume defects and one shallow positron trap exhibiting thermal detrapping with the detrapping rate δ. All three defects are considered to be non-interacting. The corresponding differential equations are

$$\begin{aligned}
\frac{\mathrm{d}n_b(t)}{\mathrm{d}t} &= -(\lambda_b+\kappa_{d1}+\kappa_{d2}+\kappa_{d3})n_b(t)+\delta n_{d1}(t),\\
\frac{\mathrm{d}n_{d1}(t)}{\mathrm{d}t} &= -(\lambda_{d1}+\delta)n_{d1}(t)+\kappa_{d1}n_b(t),\\
\frac{\mathrm{d}n_{d2}(t)}{\mathrm{d}t} &= -\lambda_{d2}n_{d2}(t)+\kappa_{d2}n_b(t),\\
\frac{\mathrm{d}n_{d3}(t)}{\mathrm{d}t} &= -\lambda_{d3}n_{d3}(t)+\kappa_{d3}n_b(t).
\end{aligned} \tag{3.35}$$

The solution of (3.35) corresponds to (2.3) with the total number of independent defect types $k = 3$. Defect d1 is a shallow positron trap, and d2 and d3 are open-volume defects, such as vacancies and vacancy agglomerates. The starting conditions are $n_b(0) = N_0$ and $n_{di}(0) = 0$ ($i = 1$ to 3). With k defect types, $k + 1$ lifetime

[1] Both fractions are identical only when no detrapping occurs.

components are obtained. From (3.35), the intensities I_1 to I_4 and the lifetimes τ_1 to τ_4 are obtained as

$$\tau_1 = \frac{2}{\Lambda + \Xi}, \quad \tau_2 = \frac{2}{\Lambda - \Xi}, \quad \tau_3 = \frac{1}{\lambda_{d2}}, \quad \tau_4 = \frac{1}{\lambda_{d3}},$$

$$\begin{aligned}
I_1 &= 1 - (I_2 + I_3 + I_4), \\
I_2 &= \frac{\delta + \lambda_{d1} - \frac{1}{2}(\Lambda - \Xi)}{\Xi} \\
&\times \left[1 + \frac{\kappa_{d1}}{\delta + \lambda_{d1} - \frac{1}{2}(\Lambda - \Xi)} + \frac{\kappa_{d2}}{\lambda_{d2} - \frac{1}{2}(\Lambda - \Xi)} + \frac{\kappa_{d3}}{\lambda_{d3} - \frac{1}{2}(\Lambda - \Xi)} \right], \\
I_3 &= \frac{\kappa_{d2}(\delta + \lambda_{d1} - \lambda_{d2})}{\left[\lambda_{d2} - \frac{1}{2}(\Lambda + \Xi)\right]\left[\lambda_{d2} - \frac{1}{2}(\Lambda - \Xi)\right]}, \\
I_4 &= \frac{\kappa_{d3}(\delta + \lambda_{d1} - \lambda_{d3})}{\left[\lambda_{d3} - \frac{1}{2}(\Lambda + \Xi)\right]\left[\lambda_{d3} - \frac{1}{2}(\Lambda - \Xi)\right]}.
\end{aligned} \tag{3.36}$$

The abbreviations in (3.36) are

$$\begin{aligned}
\Lambda &= \lambda_b + \kappa_{d1} + \kappa_{d2} + \kappa_{d3} + \lambda_{d1} + \delta, \\
\Xi &= \sqrt{(\lambda_b + \kappa_{d1} + \kappa_{d2} + \kappa_{d3} - \lambda_{d1} - \delta)^2 + 4\delta\kappa_{d1}}\,.
\end{aligned} \tag{3.37}$$

The two long-lived lifetimes are equal to the defect-related lifetimes, i.e. $\tau_3 = \tau_{d2}$ and $\tau_4 = \tau_{d3}$. The average positron lifetime $\bar{\tau}$ is given for four components by (3.32) with $k = 3$.

Cases where the number of independent defects $k < 3$ can easily be obtained from (3.35) by setting the appropriate trapping rates equal to zero. In this way, the case for one defect type (3.27) is obtained.

3.2.1.3 Combined Positron Traps

In certain defects, positron capture into deep states occurs via precursor states. A point-like trap is located in a more extended positron trap. The precursor state promotes the trapping and acts as a "funnel" for the final localization of the positron. Such a promotion of the trapping, called the "funnel effect" by Maier (1997), is discussed for dislocations (Sect. 3.1.5.3). The undisturbed dislocation is assumed to be a shallow trap guiding the positrons to deep traps, such as vacancies located on the dislocation line (Hashimoto and Kino 1991; Smedskjaer et al. 1980). Another case is the occurrence of deep positron traps in coherent precipitates (Sect. 3.1.5.5).

Positron capture in such a defect can phenomenologically be described according to Fig. 3.26 with a set of rate equations,

$$\begin{aligned}
\frac{dn_b(t)}{dt} &= -(\lambda_b + \kappa_t + \kappa_{st} + \kappa_v)n_b(t) + \delta n_{st}(t), \\
\frac{dn_{st}(t)}{dt} &= -(\lambda_{st} + \delta + \vartheta)n_{st}(t) + \kappa_{st}n_b(t), \\
\frac{dn_t(t)}{dt} &= -\lambda_t n_t(t) + \kappa_t n_b(t) + \vartheta n_{st}(t), \\
\frac{dn_v(t)}{dt} &= -\lambda_v n_v(t) + \kappa_v n_b(t).
\end{aligned} \tag{3.38}$$

The quantities are basically the same as in (3.27) and (3.35). The indices st and t stand in this case for the shallow precursor and the final deep trap, respectively. The shallow level is characterized by the detrapping rate δ and the transition rate to the deep state ϑ. The annihilation rate of the shallow trap, λ_{st}, is expected to be very close to the bulk annihilation rate, at least for a negatively charged defect without open volume (see Sect. 3.1.5.2). In addition to the combined shallow/deep trap, another deep positron trap is included. This corresponds to the usual situation after plastic deformation, where not only dislocations are present but also point defects (vacancies).

The lifetime components and their corresponding intensities are obtained (Staab et al. 1998) as

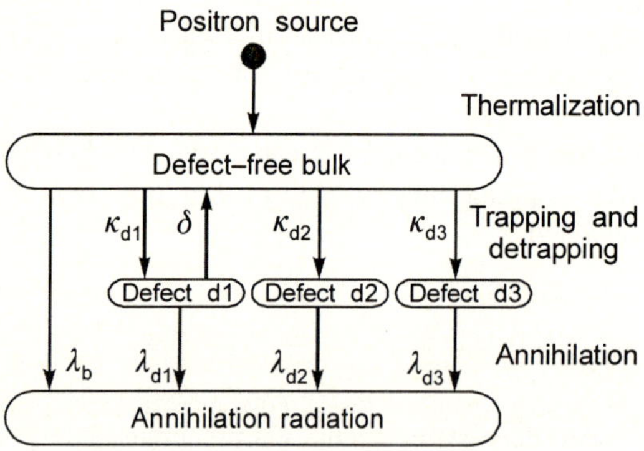

Fig. 3.25. Scheme of the extended trapping model including two types of open-volume defects, d2 and d3, and one shallow positron trap, d1. The latter exhibits thermally induced detrapping with the temperature-dependent detrapping rate δ. The individual trapping rates κ_{di} and the corresponding annihilation rates λ_{di} are indicated ($i = 1$ to 3). λ_b is the bulk annihilation rate.

$$\tau_1 = \frac{2}{\Lambda + \Xi}, \quad \tau_2 = \frac{2}{\Lambda - \Xi}, \quad \tau_3 = \frac{1}{\lambda_t}, \quad \tau_4 = \frac{1}{\lambda_v},$$

$$I_1 = 1 - (I_2 + I_3 + I_4),$$

$$I_2 = \frac{\lambda_{st} + \delta + \vartheta - \frac{1}{2}(\Lambda - \Xi)}{\Xi} \left\{ 1 + \frac{\kappa_{st}}{\lambda_{st} + \delta + \vartheta - \frac{1}{2}(\Lambda - \Xi)} \right. \tag{3.39}$$

$$\left. + \frac{\kappa_t \left[\lambda_{st} + \delta + \vartheta - \frac{1}{2}(\Lambda - \Xi) \right] + \kappa_{st}\vartheta}{\left[\lambda_t - \frac{1}{2}(\Lambda - \Xi) \right] \left[\lambda_{st} + \delta + \vartheta - \frac{1}{2}(\Lambda - \Xi) \right]} + \frac{\kappa_v}{\lambda_v - \frac{1}{2}(\Lambda - \Xi)} \right\},$$

$$I_3 = \frac{\kappa_t(\lambda_{st} + \delta + \vartheta - \lambda_t) + \kappa_{st}\vartheta}{\left[\lambda_t - \frac{1}{2}(\Lambda + \Xi) \right] \left[\lambda_t - \frac{1}{2}(\Lambda - \Xi) \right]},$$

$$I_4 = \frac{\kappa_v(\lambda_{st} + \delta + \vartheta - \lambda_v)}{\left[\lambda_v - \frac{1}{2}(\Lambda + \Xi) \right] \left[\lambda_v - \frac{1}{2}(\Lambda - \Xi) \right]}.$$

The abbreviations Λ and Ξ are in this case

$$\Lambda = \lambda_b + \kappa_{st} + \kappa_t + \kappa_v + \lambda_{st} + \delta + \vartheta, \tag{3.40}$$

$$\Xi = \sqrt{(\lambda_b + \kappa_{st} + \kappa_t + \kappa_v - \lambda_{st} - \delta - \vartheta)^2 + 4\delta\kappa_{st}}.$$

A four-component decomposition of the experimental positron lifetime spectrum is necessary for the determination of the trapping rates for the trapping schemes (3.35) and (3.38). However, such a decomposition is often not possible and only the average positron lifetime can be determined.

3.2.2 Evaluation of Momentum Distribution Measurements

It was demonstrated in Sect. 2.3 that information about defects can also be obtained by Doppler-broadening spectroscopy. The momentum distribution of the annihilating electrons significantly changes in a defect compared with the bulk. This results in different spectra of the angular correlation of annihilation radiation technique or different shapes of the Doppler-broadened annihilation lines. The latter effect has been characterized by the line shape parameters introduced in Sect. 2.3.2.2. The appearance of more than one annihilation site makes the data analysis complicated if only the line shape parameters S and W are used. Recently, more sophisticated methods, such as the S versus W plots (Sect. 3.2.2.2) and fin-

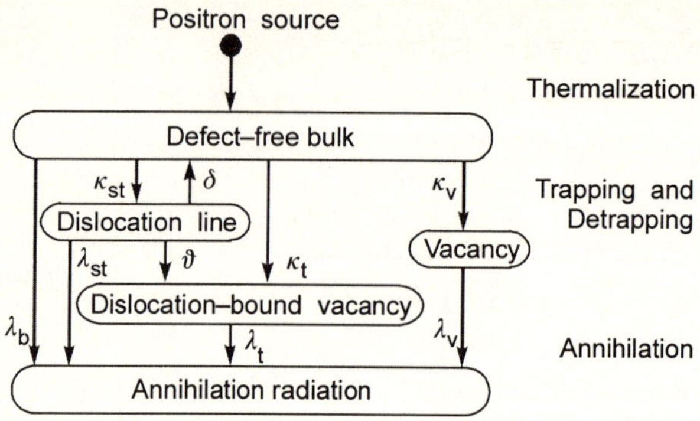

Fig. 3.26. Scheme of positron trapping in dislocations. A deep positron trap (dislocation-bound vacancy) is connected to a shallow precursor state (undisturbed dislocation line). The trapping rate of the shallow trap is κ_{st}. Thermally induced detrapping occurs from the shallow state with the detrapping rate δ. The transition rate from the shallow state to the deep ground state is indicated as ϑ, and the direct trapping into the deep state as κ_t. λ_b and λ_t are the annihilation rates of the bulk and the deep ground state, respectively. The shallow level has the annihilation rate λ_{st}. In addition to the combined positron trap, another deep positron trap (vacancy) is included in the scheme with the trapping rate κ_v and the annihilation rate λ_v.

gerprint methods including the use of difference (Sect. 3.2.2.3) or ratio curves (Sect. 3.2.2.4), have been developed.

3.2.2.1 Line Shape Parameters

The annihilation of positrons from a localized defect state is related to a significant change in the annihilation line shape. Usually, for open-volume defects, e.g. vacancies, a narrowing of the line width is observed (Fig. 2.7). The reason is the increased contribution of annihilations with valence electrons having smaller momentuma compared with core electrons and leading thus to a smaller Doppler broadening of the annihilation line.

In the presence of one defect type, the apparent S parameter, defined in Sect. 2.3.2.2, is the weighted sum of the bulk and defect S parameters, S_b and S_d, respectively,

$$S = (1 - \eta) S_b + \eta S_d . \tag{3.41}$$

The weighting factor η is the fraction of positrons annihilating in the defect and was defined with the integral of (3.34). It can be determined from a single Doppler measurement when S_b and S_d are known from reference experiments.

The trapping rate κ is calculated similarly to (3.33),

$$\kappa_{\mathrm{d}} = \frac{1}{\tau_{\mathrm{b}}} \frac{S - S_{\mathrm{b}}}{S_{\mathrm{d}} - S}. \tag{3.42}$$

The concentration of the defect can then be obtained from (3.11) or (3.12).

As soon as more than one defect type ($k > 1$) has a significant contribution to positron trapping, (3.41) must be extended by further terms,

$$S = \left(1 - \sum_{i=1}^{k} \eta_i\right) S_{\mathrm{b}} + \sum_{i=1}^{k} \eta_i S_{\mathrm{d}i}. \tag{3.43}$$

In this case, several weighting factors η_i need to be determined from the measured S value, which is impossible without further assumptions. Thus, a multi-defect analysis of Doppler data can hardly be carried out from a single S parameter measurement. In this case, only the determination of an effective trapping rate averaging over the various defect types according to (3.41) and (3.42) is possible.

3.2.2.2 *S versus W Plots*

As described in Sect. 3.2.1.2, positron lifetime experiments can provide the trapping rates independently for more than one defect type. Combining lifetime and Doppler measurements, the defect-related $S_{\mathrm{d}i}$ parameters in (3.43) can be determined, since the η_i can be calculated from the trapping rates ($i = 1$ to k). The $S_{\mathrm{d}i}$ are specific for the defect types and independent of the defect concentrations. In addition to the defect-related positron lifetimes $\tau_{\mathrm{d}i}$, these quantities provide further characteristic information about the nature of the defects.

Equation (3.41) can be given in a similar way for the W parameter defined in (2.11),

$$W = (1 - \eta) W_{\mathrm{b}} + \eta W_{\mathrm{d}}. \tag{3.44}$$

Consequently, (3.42) and (3.43) can also be written for the W parameter using W instead of S. The combination of (3.41) and (3.44) provides the R parameter for one defect type, which has already been introduced in Sect. 2.3.2.2. It follows immediately that the R parameter (2.12) is not a function of η, i.e. of the defect concentration. It thus depends only on the kind of the defect (Mantl and Triftshäuser 1978). Equation (2.12) may be used to define a straight line in the W–S plane,

$$S - S_{\mathrm{b}} = R(W - W_{\mathrm{b}}), \tag{3.45}$$

with the slope R. Such S versus W plots can be used for defect identification (Liszkay et al. 1994). Alternatively, the W versus S plots are also used for this purpose.

An S versus W plot is shown in Fig. 3.27b. The plot was obtained from slow-positron-beam measurements of silicon implanted with boron. The S parameter is

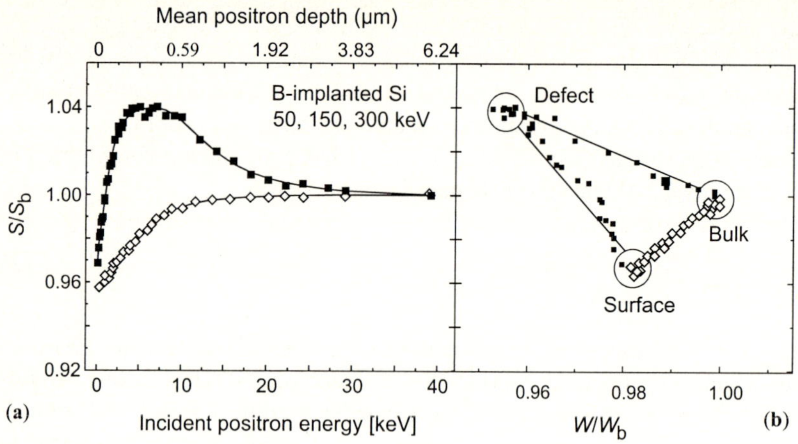

Fig. 3.27. Results of a slow-positron-beam experiment on silicon implanted at 300 K with 2×10^{14} cm^{-2} boron (Eichler et al. 1997a). (**a**) *S* parameter versus the incident positron energy. The full line is the result of the VEPFIT (variable energy positron fit) routine. For comparison, the curve measured in defect-free Si (open symbols) is shown. (**b**) *S* versus *W* plot. The curve of the implanted sample runs with increasing positron energy clockwise from the surface state through the characteristic point of the defect to the bulk value. The values for the defect-free reference sample lie on a straight line connecting the surface and the bulk states. The straight lines are drawn to guide the eye. The line shape parameters are normalized to their bulk values, S_b and W_b, respectively.

shown as a function of the incident positron energy in Fig. 3.27a. The steep increase starting from the surface-related *S* parameter up to about 1.04 is due to the presence of implantation-induced open-volume defects, such as divacancies and larger agglomerates (for details see Sect. 4.1.8).

A straight line is obtained in the *S* versus *W* plot (Fig. 3.27b) between the characteristic points of the surface and the defect; a second straight line between the defect state and the bulk. The characteristic (W, S) pairs of values are indicated as circles. With increasing incident positron energy, i.e. increasing depth, the *S* versus *W* curve runs clockwise through different states. The typical (W, S) values for annihilation at the surface were obtained with the smallest positron energy. All positrons implanted into a medium depth are trapped in implantation defects. The bulk parameter is observed for the highest positron energies, because almost all positrons reach the defect-free region beneath the damaged layer. As there are no positron traps in the reference sample of Fig. 3.27b, all values lie on a straight line connecting the surface and the bulk states.

Where the measured points lie on a straight line connecting two characteristic states, these states must be considered as the dominant annihilation sites. This is observed for the points between the characteristic values for the bulk and the implantation defects. However, a distinct deviation from the straight line would be a clear sign of the simultaneous positron trapping in more than two annihilation sites (Clement et al. 1996), as observed for the data between the characteristic points of the surface and the defect. In this case, a small fraction of positron annihilation in the bulk causes the deviation of the data from the straight line.

3.2.2.3 Fingerprint Difference Curves

The plots of two-dimensional angular correlation of annihilation radiation show clearly that the momentum distribution spectra contain detailed information about the electronic structure of the defect (Sect. 2.3.3). The S parameter may be determined from such a 2D-ACAR curve by a twofold integration. However, in this way, most of the defect-specific structures in the curves are lost. In the measurement of the momentum distributions by the Doppler technique, the integration of

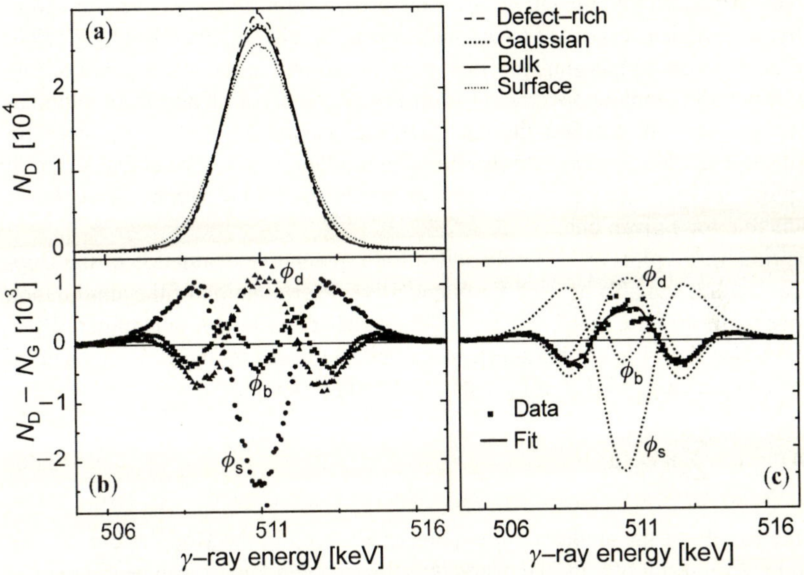

Fig. 3.28. Fingerprint curves in boron-implanted silicon (Krause-Rehberg and Eichler 1997). (**a**) Normalized Doppler spectra obtained for silicon with no positron traps, for annihilation in the surface state of silicon covered by a thin silicon dioxide layer, and for annihilation in implantation-induced defects in silicon. 1×10^7 counts were collected in each spectrum. (**b**) Difference curves calculated as the difference between the Doppler curve N_D and the Gaussian N_G indicated in (a). (**c**) Fit of the difference curve obtained from slow-positron measurements of boron-implanted silicon (1.5 keV positron energy). The fingerprint spectra for bulk silicon, ϕ_b, surface states, ϕ_s, and defect-rich Si, ϕ_d, from (b) are shown again for comparison as dotted lines. A numerical fit according to (3.46) provided the difference curve $\Delta N(E_\gamma) = N_D - N_G = 0.154\,\phi_s + 0.822\,\phi_d + 0.024\,\phi_b$.

the data in two directions is unavoidable. Since the detector resolution is poor, even in high-statistics Doppler-broadening spectra (10^7 counts) no fine structure becomes directly visible (Fig. 3.28a). However, difference curves exhibit pronounced structures (Fig. 3.28b). The origin lies in the anisotropy of the momentum distributions of the annihilating electrons, as described earlier for the results of two-dimensional angular correlation of annihilation (see Fig. 2.15).

An obvious approach for the calculation of difference curves is to take the Doppler spectrum of the defect-free bulk as the reference (Manuel 1995). However, all difference curves of defect-rich samples also exhibit in this case the structures of the defect-free sample. This effect is most pronounced for the case of saturated trapping. Hence, it is more favorable to subtract a smooth and structure-free curve. For Si a Gaussian N_G is suitable; for GaAs a sum of two Gaussians can be taken. Fig. 3.28a shows as an example the normalized Doppler spectra of defect-free bulk silicon, of the annihilation in the surface state of silicon covered by a thin SiO_2 layer, and of the annihilation in implantation-induced defects in silicon. All of these spectra were chosen to have positron annihilation from only one annihilation site, so that the curves are characteristic for the respective site. Additionally, a Gaussian was fitted to the bulk spectrum and was then used for calculating the difference plots shown. These plots are shown in the lower panel of Fig. 3.28b. Since the Doppler spectra N_D from Fig. 3.28a are obtained from annihilations of only one annihilation site, the difference curves $\Delta N = N_D - N_G$ represent the "fingerprint plots" of the corresponding annihilation sites. The actual shape of these plots in Fig. 3.28b obviously depends on the selected reference curve, but is reproducible for a given one.

The fingerprint plots are not only suited for a qualitative evaluation of the Doppler spectra, but can also be used for quantitative determination of the annihilation fractions η_i. In the case that $k + 1$ annihilation sites contribute to a Doppler difference spectrum $\Delta N = f(E_\gamma)$, this spectrum can be composed by the weighted sum of fingerprint spectra $\phi_i = f(E_\gamma)$ using a standard fitting routine,

$$\Delta N(E_\gamma) = \sum_{i=1}^{k+1} \eta_i \phi_i(E_\gamma) . \tag{3.46}$$

E_γ is the energy of the annihilation γ-quanta registered in the Doppler broadening spectrum. As a result of a fit, the η_i and thus the trapping rates κ_i can be obtained.

As an example, a fit taken from a depth profile analysis of a boron-implanted silicon sample is presented in Fig. 3.28c. It was shown that the annihilation fractions of three annihilation sites η_i could be determined from such a fit. This procedure represents a considerable improvement compared with the standard data analysis of the line shape with the parameters S or W. This is due to the fact that the S parameter not only depends on the character of the defects but also on their density, so that the determination of only one annihilation fraction is possible. By using the fingerprint curves, a "decomposition" of the Doppler broadening spectra may be possible. The case of depth profile measurements in silicon as used in Fig.

3.28c is especially suited for such a fingerprint evaluation, since the ϕ_s curve is, due to the SiO_2 overlayer, significantly different from the bulk curve ϕ_b. A high statistics measurement with more than 10^7 registered annihilation events is required in order to separate the individual annihilation sites in the fit.

It may be useful to weight the difference curves by the statistical error of the individual channel content as a function of the γ-ray energy in order to obtain more reliable information in the high-momentum region of the Doppler curve. An example is shown in Fig. 5.41. The advantage of difference curves in comparison to ratio curves is that the influence of positrons annihilating in the source in conventional measurements is reduced to a great extent. This holds also for slow-positron-beam measurements, where the fraction of positrons annihilating outside the sample after backscattering also does not contribute essentially to the difference curves.

The fingerprint spectra have to be determined at least once for complete positron annihilation from the corresponding state. To ensure this condition, positron lifetime measurements may be used. The theoretical calculation of complete angular correlation or Doppler-broadening curves may represent a promising development to obtain the fingerprint curves from first principles. The first such calculations for semiconductors have been presented recently (Hakala et al. 1998; Saito et al. 1997; Tang et al. 1997).

3.2.2.4 Ratio Curves

The interpretation of Doppler-broadening spectra may be improved by dividing the measured high-resolution spectra by a reference spectrum. A spectrum of defect-free samples is usually taken as such a standard spectrum. In the high-momentum part of the spectra, the annihilation with core electrons carrying chemical information is dominant. Characteristic structures in the ratio curves, which may be related to chemical elements located at the annihilation site, are revealed as far as 10 keV away from the annihilation peak at 511 keV. Background-reduced measurements with high statistics in a coincidence setup are the prerequisite to obtain suitable Doppler spectra.

An example for experimental ratio curves measured in arsenic-implanted silicon is shown in Fig. 3.29. The Doppler-broadening spectra were divided by the spectra of bulk Si. The curves measured after different annealing treatments are compared with the ratio curve obtained for pure arsenic (Myler et al. 1996). Two distinct structures appear in the implanted Si. The first peak, located at 514 keV, could be attributed to annihilation in the native oxide surface layer. Its intensity became smaller upon annealing. The second structure in the curves at 517 keV became dominant in the implanted Si layer after annealing. It was similar to the peak in the arsenic ratio curve, and it was thus concluded that arsenic–vacancy complexes were formed.

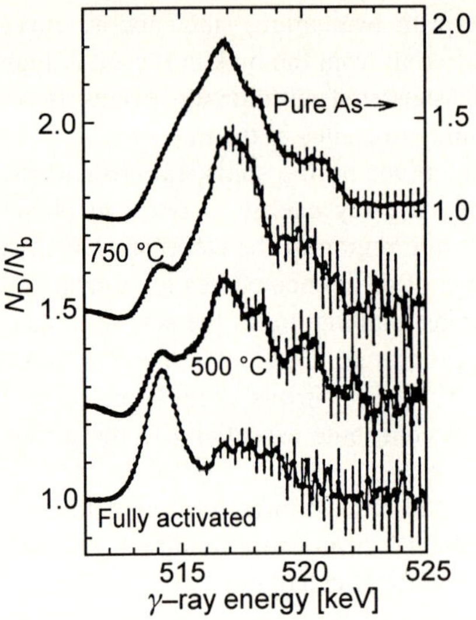

Fig. 3.29. Doppler-broadening ratio curves N_D/N_b of arsenic-implanted silicon (Myler et al. 1996). The spectra measured within the implanted layer with an incident positron energy of 12 keV were divided by the spectra of bulk silicon N_b after laser melt annealing (marked as fully activated), and after subsequent annealing at 500 and 750 °C. The curves are compared with that of pure arsenic, which was also divided by the spectrum of bulk silicon. The curves are offset by 0.25 each for clarity.

The concept of chemical sensitivity of Doppler-broadening spectroscopy was further pursued by Myler and Simpson (1997) by the measurement of a large number of element-specific ratio curves. The specific spectrum of a certain chemical element is conserved in different environments, as demonstrated for oxygen in SiO_2/Si and MgO/Mg. An attempt was made to identify various chemical environments of open-volume defects in Si by the measurement of ratio curves (Szpala et al. 1996). The experimental curves were described as linear combinations of known elements or defects.

Kruseman et al. (1997a, 1997b) used ratio curves for the identification of annihilation sites in oxygen-implanted Si. Large vacancy–oxygen complexes appear after annealing at 800 °C. The fraction of positrons annihilating in these defects, in the bulk, and at the surface could be determined by fitting fingerprint spectra to measured ratio curves, similarly to (3.46).

3.2.3 Back-Diffusion Measurements

The determination of the positron diffusion length L_+ can be utilized for the evaluation of the positron trapping rate, i.e. the defect concentration. L_+ is usually

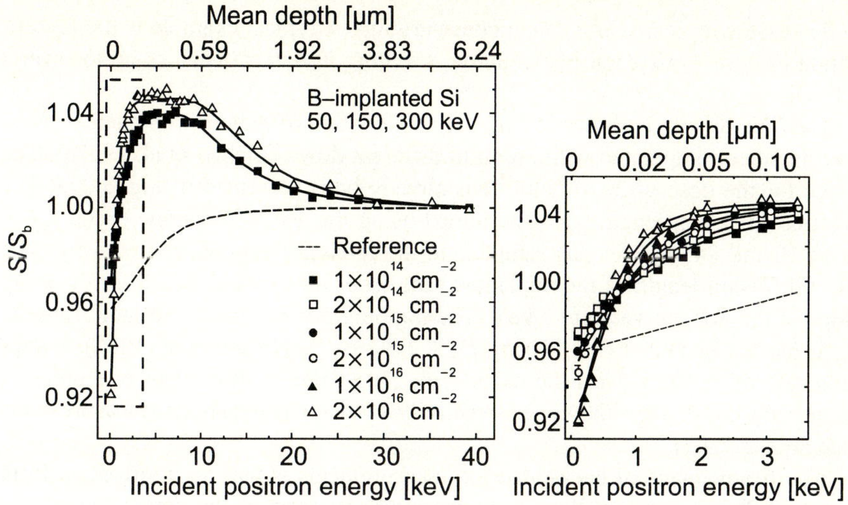

Fig. 3.30. Result of a slow-positron-beam experiment on a series of boron-implanted silicon wafers (Eichler et al. 1997a). In order to obtain a homogeneous defect distribution, a multiple implantation with ion energies of 50, 150, and 300 keV was applied. The dashed line is the typical S parameter curve as function of the incident positron energy of silicon free of positron traps. The low-energy part of the curves (*marked area*) is shown in the left panel at an enlarged scale to visualize the different slopes of the samples implanted with different doses ranging from 1×10^{14} to 2×10^{16} cm^{-2}. The lines are best fits to the positron diffusion equation by the variable energy positron fit procedure (VEPFIT).

measured with slow positrons diffusing back to the surface after their implantation. The method is called the back-diffusion technique. It is based on the dependence of L_+ on the effective annihilation rate λ_{eff} being a function of the trapping rate κ; see (3.5) and (3.6). The fraction of positrons diffusing back to the surface is measured in such an experiment as a function of the incident positron energy (Sect. 2.4.3). This measurement can be performed using the dependence of the annihilation parameters, such as the S parameter, on the positron implantation energy E. Another possibility is the registration of the positronium fraction as a function of E (Sect. 3.1.6).

The determination of the defect concentration requires a homogeneous defect distribution, i.e. a depth-independent trapping rate κ, and knowledge of the positron diffusion coefficient D_+, which can be determined in a defect-free reference sample. In comparison with conventional positron lifetime and Doppler-broadening measurements, higher defect concentrations can be determined by a back-diffusion experiment. This method remains sensitive to the defect concentration also in the case of saturation trapping, i.e. when all positrons are captured by defects. In this special case, the annihilation parameters do not change any more with the defect concentration. However, the positron diffusion length L_+ continues to decrease with increasing defect concentration. A typical example is the investigation of defects after ion implantation, resulting in defect densities up to several 10^{20} cm^{-3}.

The S parameter is shown in Fig. 3.30 as function of the positron energy measured in silicon implanted with boron to different doses (Eichler et al. 1997a). The curve for the dose of 2×10^{16} cm^{-2} has already been presented in Fig. 3.27a. The positron diffusion length was determined using the VEPFIT routine (Sect. 2.4.3) to be 10 and 245 nm for this sample and the reference sample, respectively. The short diffusion length of the implanted samples is already obvious from the steep slope of the $S(E)$ curve up to 2 keV. The trapping rate can be determined from the L_+ value via (3.5) and (3.6). Using $D_+ = 2.7$ cm s^{-1} (Nielsen et al. 1985), a trapping rate of 3×10^{12} s^{-1} can be calculated. This value is more than one order of magnitude higher than the upper sensitivity limit for conventional measurements (see Sect. 3.2.4.7).

The disadvantage of back-diffusion measurements is that no information from the annihilation site and, thus, from the microstructure of the defect can be obtained directly. This is the reason that in principle only an effective positron trapping rate for the various defect types contributing to positron trapping can be determined and is measured as a sum effect. A further difficulty of the back-diffusion technique is the sensitivity of the results to the appearance of electric fields. As was demonstrated in Fig. 3.6, the determination of the trapping rate is strongly influenced by the field strength.

3.2.4 Determination of the Trapping Coefficient of Vacancies

The positron trapping rate can be calculated from the vacancy concentration according to (3.11). The proportionality constant μ is called the trapping coefficient (also called the specific trapping rate). Its determination requires the application of an experimental technique which provides the defect concentration independent of the positron experiment. At the beginning of positron research on defects in semiconductors, numerous experimental data on the trapping coefficient in metals were available (e.g. Nieminen and Manninen 1979; West 1979). In comparison with metals, it is more difficult to determine absolute vacancy concentrations in semiconductors, especially in compounds. A reason is that the concentration of vacancies in thermal equilibrium is lower than in metals. Thus, experiments which combine precision lattice parameter and density measurements with positron annihilation at higher temperatures as performed e.g. for copper (Kluin and Hehenkamp 1991) are hardly possible in semiconductors. Hence, in the early studies of vacancies in semiconductors, the trapping coefficients of metals were used, leading to unreasonably high concentrations (Bourgoin et al. 1988; Dlubek et al. 1986a). A number of positron experiments correlated with other techniques have been carried out in recent years in order to determine precisely the trapping coefficients of defects in semiconductors. Contrary to the former assumptions, the trapping coefficient was found for negatively charged vacancies to be significantly higher, leading to smaller defect concentrations. The available data for vacancies are summarized in the following. The few data reported for the trapping coefficients of shallow positron traps and dislocations are given in Sect. 3.2.5.

In the case of neutral vacancies, positrons are localized at a deep state in a square-well potential. No temperature dependence of the trapping coefficient is expected (Fig. 3.8). This is the result of semiclassical calculations by Puska et al. (1990) in a treatment according to Fermi's golden rule, where the trapping coefficient is the transition rate between the delocalized and the localized state in the positron trap. The trapping process is expected to be transition limited, i.e. the capture is not limited by the diffusion of positrons to the trap, but by the necessary energy dissipation during the transition from the Bloch state to the bound state.

For negatively charged vacancies, the square-well potential is superimposed by a long-range Coulomb potential (Fig. 3.7) leading to shallow Rydberg states. They have a strong effect on positron trapping, especially at low temperatures. It was calculated (Puska et al. 1990) and experimentally shown (Mäkinen et al. 1992a) that the transition rate from the delocalized state to the Rydberg states may be rather high. Thus, the Rydberg states act as an intermediate state before trapping in the ground state (Fig. 3.9). This enhances the trapping probability ("funnel effect") and becomes especially important with decreasing temperatures, where thermally induced detrapping from the Rydberg states becomes inefficient as the competing process. Due to the high transition rate, the positron trapping in the spatially extended Rydberg states is expected to be limited by the diffusion of positrons to the defect, leading to a distinct temperature dependence of the positron trapping (3.1.5.1). The overall temperature behavior is thus rather complex.

Despite the fact that (3.11) only describes the case of transition-limited trapping, this equation will be extended to the more complex trapping in negatively charged vacancies, where diffusion-limited trapping in the Rydberg states, thermally induced detrapping, and transition-limited trapping in the ground state are considered simultaneously. The effective trapping coefficients obtained for such charged defects depend on temperature, i.e. they increase with decreasing temperature. For all vacancies for a given charge state, the proportionality between defect concentration C and trapping rate κ holds at least for a constant temperature, $\kappa = \mu(T)\, C$. Moreover, a very high concentration of free carriers leads to a Debye screening of the vacancy charge. In this rare case, μ may be reduced (Fig. 3.11).

Although the trapping coefficient is precisely defined by (3.11), complications arise in the case of compound semiconductors, where vacancies occur in both sublattices. The trapping coefficient should be defined with respect to the total number of available lattice sites. In an elementary semiconductor, this number is identical to the number of atoms. The unit of μ appears then as $[\mu] = \mathrm{s}^{-1}$, meaning that the corresponding defect concentration is given as the number of defects per atom. For a compound semiconductor, the number of available lattice sites for a vacancy of a certain sublattice is reduced to the number of atoms in the corresponding sublattice. Consequently, this number is better used when calculating the trapping coefficient from a reference experiment. In the early attempts to determine the trapping coefficient in compounds this fact was ignored.

This complication does not appear if the number of defects is not related to the number of lattice sites but given per volume (defect density). This leads according to (3.11) to a unit of the trapping coefficient of $[\mu] = \mathrm{cm}^3\,\mathrm{s}^{-1}$, meaning that the defect density is given per cm^3. If the trapping coefficients are given in units of s^{-1}, the existing convention of the literature is followed throughout the book, i.e. the defect concentration is related to the total number of lattice sites. The advantage of this unit is the comparability of different materials independent of their mass density. The conversion of the trapping coefficient given in the two units can be done using the number of atoms per unit volume, N_{at}, according to

$$\mu[\mathrm{cm}^3\mathrm{s}^{-1}] = \frac{\mu[\mathrm{s}^{-1}]}{N_{\mathrm{at}}}. \tag{3.47}$$

Table 3.3 summarizes available experimental data on the trapping coefficients for vacancy-type defects in different semiconductors (Krause-Rehberg and Leipner 1997). Only such values were included where a reference method for the independent determination of the density of the corresponding defect studied by positrons was applied to the same sample. The trapping coefficient is given both in units of s^{-1} and $\mathrm{cm}^3\,\mathrm{s}^{-1}$, respectively. The temperature of the positron experiments is given as well as the reference methods applied.

Only a few reliable correlated measurements on the same samples are known. Beside a couple of measurements on electron-irradiated Si, data on several III–V, II–VI, and IV–VI compound semiconductors exist. In the following sections, the determination of the trapping coefficients for the defects in Table 3.3 is discussed. A further description of positron investigations of these defects is presented in Chaps. 4 to 6.

3.2.4.1 Electron-Irradiated Silicon

Defects after electron irradiation were investigated in a study by Kawasuso et al. (1995c), who used Hall effect measurements and positron lifetime spectroscopy of Si doped with different phosphorus concentrations. For higher doping levels the so-called *E* center, consisting of a phosphorus donor and a vacancy, was studied. The trapping coefficients of this defect in the neutral and negative charge states were measured. A value of $\mu = 7\times10^{14}\ \mathrm{s}^{-1}$ was found at room temperature for the uncharged vacancy–phosphorus pair. This trapping coefficient was calculated as the ratio of the positron trapping rate and the defect concentration, which was determined independent of positron annihilation by Hall effect measurements.

The neutral vacancy–phosphorus pair was also investigated by Mäkinen et al. (1992a). The upper limit of the concentration of the complex was estimated from

Table 3.3. Compilation of positron trapping coefficients of vacancy-type defects experimentally determined in various semiconductors (Krause-Rehberg and Leipner 1997). Only such experiments where the independent reference method was applied to the same samples were taken into account.

Material	Defect	Trapping coefficient $[10^{15}\,s^{-1}]$	$[10^{-8}\,cm^3s^{-1}]$	T [K]	Reference method	Authors
Si:P	$(VP)^0$	0.68	1.4	300	Hall effect	a
		> 1.3	> 2.6	300	Resistivity	b
Si:P	$(VP)^-$	18	36	300	Hall effect	a
		> 2	> 4	300	Resistivity	b
Si	V_2^0	0.8 ± 0.40	1.5 ± 0.8	300	EPR	c
		0.8	1.6	300	Hall effect	a
Si	V_2^-	2.6 ± 1.3	5.2 ± 2.6	300	EPR	c
		10	20	300	Hall effect	a
Si	V_2^{2-}	5.2 ± 2.7	10.5 ± 5.3	300	EPR	c
		29	58	300	Hall effect	a
Si	V_2^+	< 0.1	< 0.2	300	Hall effect	a
GaP	V_P^0	0.8± 0.3	1.5 ± 0.6	473	Hall effect	d
GaP	V_P^-	1.9 ± 0.5	3.8 ± 1.0	473	Hall effect	d
GaP	V_P^+	< 0.1	< 0.2	473	Hall effect	d
GaAs:Te	$(V_{Ga}Te_{As})^-$	1.1 ± 0.2	2.5 ± 0.5	300	Hall effect	e
GaAs:Si	$(V_{Ga}Si_{Ga})^-$	0.7 ± 0.2	1.6 ± 0.5	300	STM	f
GaAs	V_{Ga}^- (EL2*)	> 3	> 7	25	IR absorption	g
		> 30	> 68	20	IR absorption	h
GaAlSb	V_{Ga}^- (DX)	1 ± 0.3	2.9 ± 1	300	DLTS	i
HgCdTe	V_{Hg}^{2-}	2.1 ± 0.3	7 ± 1	300	Hall effect	j
		0.1 ± 0.025	0.3 ± 0.1	800	Hall effect	j
PbSe	V_{Pb}^{2-}	0.1 ± 0.01	0.3 ± 0.03	300	Hall effect	k
CdTe	$V_{Cd}^{2-}Cl_{Te}^+$	1.7 ± 0.4	5.2 ± 1.2	300	PL	l

T temperature of the positron experiment; EPR—electron paramagnetic resonance; STM—scanning tunneling microscopy; IR—infrared; DLTS—deep level transient spectroscopy; PL—photoluminescence.

[a]Kawasuso et al. (1995c), [b]Mäkinen et al. (1992a), [c]Mascher et al. (1989b), [d]Krause-Rehberg et al. (1993c), [e]Krause-Rehberg et al. (1995b), [f]Gebauer et al. (1997c), [g]Krause et al. (1990b), [h]Le Berre et al. (1994), [i]Krause-Rehberg et al. (1993b), [j]Krause-Rehberg et al. (1995a), [k]Polity et al. (1993), [l]Krause-Rehberg et al. (1998).

resistivity measurements and the irradiation dose. In this way, the lower limit of the trapping coefficient was found to be $1.3\times10^{15}\,s^{-1}$ at room temperature.

The negatively charged vacancy–phosphorus complex provided in the investigations by Kawasuso et al. (1995c) a significantly higher trapping coefficient of $1.8\times10^{16}\,s^{-1}$ at room temperature. The corresponding lower limit, as estimated by

Mäkinen et al. (1992a), resulted in a value of $> 2\times10^{15}$ s^{-1}. In addition, the temperature dependence of the positron trapping was studied.

The pronounced increase found with decreasing temperature could be explained by the occurrence of Rydberg states as the result of the Coulomb potential of the negative charge. The trapping can thus be described as a two-stage process according to Fig. 3.9.

Mascher et al. (1989b) presented a correlated investigation of electron-irradiation-induced divacancies in silicon by positron annihilation, electron paramagnetic resonance (EPR), and infrared absorption. The defects were created by 2-MeV irradiation of Czochralski-grown Si at room temperature to a dose of 1×10^{18} cm^{-2}. Monovacancies as primary irradiation defects are expected to anneal at the irradiation temperature. Divacancies of different charge states with a density of about 10^{17} cm^{-3} were quantitatively detected by positron lifetime spectroscopy and electron paramagnetic resonance. The trapping coefficients were determined for different charge states of the divacancies, identified by EPR measurements. They were found to be 0.8×10^{15}, 2.6×10^{15}, and 5.2×10^{15} s^{-1} at room temperature for uncharged, single, and doubly negative divacancies, respectively. The temperature dependence of the trapping coefficient for the charged divacancies was found to be proportional to T^{-m}, with $2 < m < 3$.

In another positron study of divacancies in Si correlated to Hall effect measurements, Kawasuso et al. (1995c) reported the same trapping coefficient as Mascher et al. (1989b) for the neutral divacancy at room temperature. However, μ was by factors of about 4 and 6 higher for the single and doubly negative divacancies, respectively. The positron trapping in positively charged divacancies was below the detection limit in the study of Kawasuso et al. (1995c). The trapping coefficient was estimated to be $\mu < 1\times10^{14}$ s^{-1} at room temperature.

3.2.4.2 Phosphorus Vacancies in Gallium Phosphide

The phosphorus vacancy was identified in different charge states in as-grown and in electron-irradiated GaP by Krause-Rehberg et al. 1993c). The $-/0$ and $0/+$ ionization levels of V_P were found by temperature-dependent positron lifetime measurements in the upper half of the bandgap at $E_{cond} - E_d = (250 \pm 30)$ meV and > 400 meV, respectively, where E_{cond} is the lower edge of the conduction band and E_d the ionization level of the defect. The first of these levels was corroborated by temperature-dependent Hall effect measurements on the same sample in the same temperature range and it was concluded that one of the ionization levels of the phosphorus vacancy was detected (see detailed discussion in Sect. 5.1). The concentration of the V_P defect could thus be independently determined. The corresponding positron trapping rate obtained at 473 K is plotted in Fig. 3.31 versus the vacancy density for different samples in different annealing states. A straight line according to (3.11) is obtained. The slope represents the trapping coefficient for the negatively charged phosphorus vacancy, $\mu = 1.9\times10^{15}$ s^{-1}. In an annealing ex-

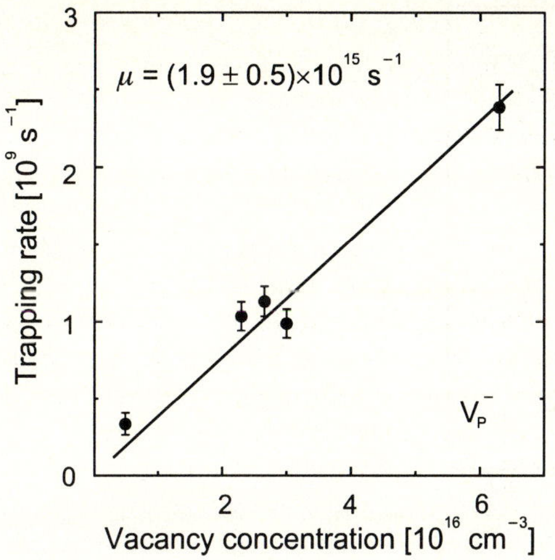

Fig. 3.31. Determination of the trapping coefficient of the negative phosphorus vacancy in gallium phosphide at 473 K. The vacancy density was measured independently of the positron experiment by the Hall effect (Krause-Rehberg et al. 1993c). The full line is the fit to the experimental data. The trapping coefficient μ was obtained from the slope of the line.

periment on an electron-irradiated sample, the Fermi level was shifted across the entire upper half of the bandgap, leading to a change in the vacancy charge from positive to neutral and negative. The trapping coefficient of the neutral phosphorus vacancy was obtained to be $\mu = 0.8\times10^{15}\ s^{-1}$. No positron trapping was observed for positive vacancies, leading to the upper-limit estimate of $\mu < 1\times10^{14}\ s^{-1}$.

3.2.4.3 Vacancy–Donor Complexes in Arsenide

Tellurium dopants on arsenic sites, Te_{As}, act as donors in GaAs. An autocompensation effect is observed, i.e. the concentration of acceptors C_{acc} increases simultaneously with the donor concentration C_{don}. The C_{acc}/C_{don} ratio is called the degree of compensation and is about 0.25 (Wolfe and Stillman 1975). It was suggested in a model by Hurle (1979b) that the compensating acceptor consists in GaAs:Te of the Te donor and a Ga vacancy, i.e. the negatively charged $V_{Ga}Te_{As}$ complex. The concentrations of both the donors and acceptors can be independently determined by measurement of the temperature-dependent Hall effect. The model mentioned above was supported by a combined positron and Hall effect study (see Sect. 5.2.3.2), showing that the compensating center contains a vacancy (Krause-Rehberg et al. 1995b). In a series of GaAs:Te samples with different dopant concentrations, both the number of acceptors and the vacancy concentration were de-

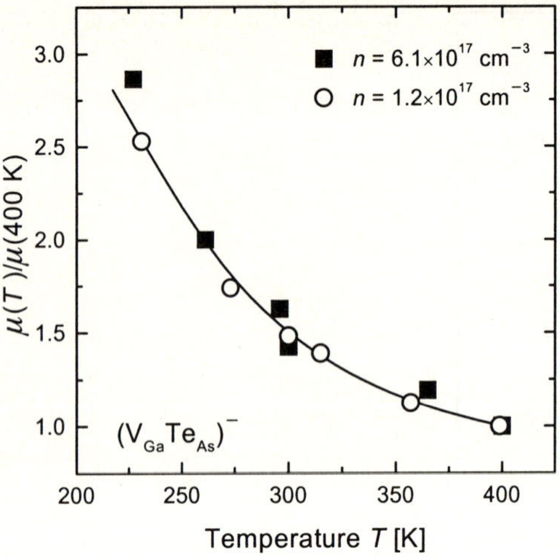

Fig. 3.32. Positron trapping coefficient μ of negatively charged compensating centers $V_{Ga}Te_{As}$ in tellurium-doped gallium arsenide versus the sample temperature T for two carrier densities n (Krause-Rehberg and Leipner 1997). The plotted values are normalized to the value measured at 400 K, $\mu(T)/\mu(400\text{ K})$. The solid line represents the fit according to (3.18).

termined and the trapping coefficient was obtained to be $\mu = 1.1\times10^{15}\text{ s}^{-1}$ at room temperature.

The temperature dependence of the positron trapping coefficient for these compensating centers is shown in Fig. 3.32 for temperatures above 200 K, where the influence of shallow positron traps (Sect. 3.1.5.2) can be neglected (Krause-Rehberg et al. 1995b). The normalized trapping coefficient $\mu(T)/\mu(400\text{ K})$ is plotted versus sample temperature T for two differently doped GaAs:Te samples (Krause-Rehberg et al. 1995b and unpublished data). The solid line is the fit according to (3.18). Obviously, the two-stage trapping model of negatively charged vacancies (Fig. 3.9) is well suited to describe the temperature dependence in the investigated range. The expected increase in μ with decreasing temperature is visible. With the temperature decreasing from 400 to 200 K, the trapping coefficient changes by a factor of three.

Si_{Ga} donors are increasingly compensated in highly doped GaAs:Si with increasing Si content (see Sect. 5.2.3.1). Recently, it has been quantitatively proven by scanning tunneling microscopy (STM) that, in addition to Si_{As} acceptors and Si clusters, negatively charged $V_{Ga}Si_{Ga}$ complexes are responsible for the compensation (Domke et al. 1996). It could be shown by the combination of STM with positron lifetime spectroscopy that the concentration of the defects detected at the surface of cleavage planes corresponds to the bulk concentration (Gebauer et al. 1997c). Thus, scanning tunneling microscopy represents an independent reference method for the determination of the trapping coefficient. The concentration of

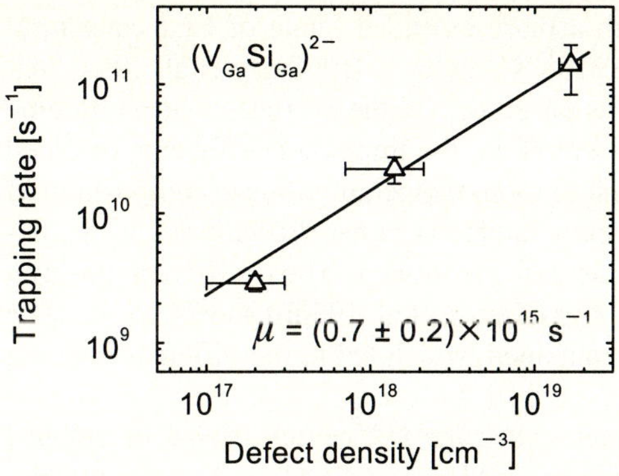

Fig. 3.33. Positron trapping rate measured by positron lifetime spectroscopy versus the density of $V_{Ga}Si_{Ga}$ complexes determined by scanning tunneling microscopy at (110) cleavage planes of highly silicon-doped gallium arsenide. The trapping coefficient μ was obtained from the fit through the data (Krause-Rehberg and Leipner 1997).

negatively charged $V_{Ga}Si_{Ga}$ complexes was measured at room temperature for three GaAs:Si samples with both positron lifetime spectroscopy and STM. The trapping coefficient was found as the slope of the straight line in Fig. 3.33 to be $0.7\times10^{15}\ s^{-1}$ (Krause-Rehberg and Leipner 1997).

3.2.4.4 Metastable Defects in III–V Compounds

The EL2 defect is supposed to be identical to the As_{Ga} antisite defect in GaAs (Chadi and Chang 1988; Dabrowski and Scheffler 1989). Its metastable state, EL2*, which is obtained by illumination at low temperatures, should contain a Ga vacancy according to this model. The occurrence of this vacancy could be proven by positron lifetime spectroscopy (see Sect. 5.2.1.2). The EL2 concentration can be independently determined by infrared absorption measurements as a reference method. With the trapping rates and the EL2 concentration given by Krause et al. (1990b), the trapping coefficient can be calculated as $\mu = 3\times10^{15}\ s^{-1}$ at 25 K. However, shallow positron traps as competing trapping centers frequently occur in GaAs at low temperatures. They can be usually identified by temperature-dependent positron lifetime measurements. Such measurements are not possible for the study of EL2*, since the stable state is recovered for temperatures higher than 110 K. At temperatures below 110 K, trapping to EL2* exhibits a strong temperature dependence. Thus, the concentration of shallow traps cannot be determined and the given value for the trapping coefficient can only be taken as the lower limit. Furthermore, the trapping coefficients of EL2* exhibit a strong temperature dependence.

A subsequent study on GaAs with a more extended range of EL2 concentrations gave a trapping coefficient of 3×10^{16} s^{-1} at 20 K (Le Berre et al. 1994). Although the same trapping behavior was observed, i.e. the average positron lifetime was identical for the same EL2* concentration, the trapping coefficients obtained were significantly different. In contrast to room temperature, the decomposition of the positron lifetime spectra at low temperatures is rather difficult due to the occurrence of shallow positron traps, as just mentioned. Thus, different positron lifetimes in the EL2* defect of 255 ± 8 ps (Krause et al. 1990b) and 245 ± 3 ps (Le Berre et al. 1994), respectively, were obtained, which led to the different trapping coefficients of both studies.

Another important metastable defect, called the DX center, occurs in n-doped pseudo-ternary III–V compounds, such as GaAlAs and GaAlSb. Also for this defect, a vacancy plays an important role (Sect. 5.3). However, in contrast to EL2, the vacancy is expected in the stable ground state at high temperature (Chadi and Chang 1989; Dabrowski and Scheffler 1992). The vacancy disappears in the metastable state under illumination at low temperature. This theoretical prediction was confirmed by a combined study of positron lifetime spectroscopy and deep level transient spectroscopy (DLTS) in GaAlSb (Krause-Rehberg et al. 1993b). The decomposition of the lifetime spectra was carried out for spectra measured at room temperature, i.e. independent of any influence of shallow positron traps. The obtained trapping rate was 7.2×10^{9} s^{-1} and the corresponding DLTS measurement resulted in a DX density of 2.5×10^{17} cm^{-3}. This gives a trapping coefficient of 1×10^{15} s^{-1}.

3.2.4.5 Cation Vacancies in Cadmium Mercury Telluride and Lead Selenide

The concentration of cation vacancies in II–VI and IV–VI compounds can be adjusted by post-growth annealing under defined vapor pressure conditions. Furthermore, they are electrically active as ionized acceptors. Their concentrations can thus be obtained by Hall effect measurements, when the concentration level of residual impurities is much lower. Annealing experiments in the temperature range from 20 to 500 °C were performed for $Hg_{1-x}Cd_xTe$ of different compositions and the formation of mercury vacancies was observed by He et al. (1989). The number of V_{Hg} was calculated according to thermodynamic calculations (Vydyanath 1981). The corresponding model is described in Sect. 6.1.2.1. The trapping coefficient of Hg vacancies was determined to be 5×10^{14} s^{-1}.

A series of annealing experiments was performed in p-type $Hg_{1-x}Cd_xTe$ with the composition $x = 0.22$ (Krause-Rehberg et al. 1995a). The same set of samples was used for both positron lifetime and Hall effect measurements. The result of this study is shown in Fig. 3.34, where the trapping rate determined by positron lifetime measurement is plotted against the density of free carriers measured at 77 K. It could be shown that this number equals the concentration of doubly negative Hg vacancies (see the more detailed discussion in Sect. 6.1.2.1). The trapping coefficient was found as the slope of the straight line in Fig. 3.34 to be $\mu = 2.1\times10^{15}\ \mathrm{s}^{-1}$ at room temperature.

In-situ positron measurements carried out at elevated temperatures (500 to 800 K) during annealing experiments resulted in a drastic decrease in the sensitivity of the positrons to the detection of Hg vacancies. The decrease in the trapping coefficient was interpreted as the reduction of the capture radius r_c (defined in Sect. 3.1.5.1) as the result of intrinsic conductivity at higher temperatures in this small-bandgap material (Krause-Rehberg et al. 1995a). With decreasing r_c, i.e. the more effective screening of the defect charge by the free carriers, the influence of the charge of the vacancy on positron trapping is increasingly lost. This effect becomes visible in Fig. 3.35, where the trapping coefficient of the negatively charged Hg vacancy is plotted versus the capture radius. r_c was varied in the experiment by changing the concentration of intrinsic carriers with the temperature. The upper saturation value of the trapping coefficient corresponds to the case where the effect of screening of the negative charge can be ignored. With increasing carrier concentration, i.e. decreasing capture radius, the trapping coefficient reaches a value typical for metals.

Post-growth annealing experiments could be carried out in order to change the vacancy concentrations in most compound semiconductors. An example was demonstrated by Polity et al. (1993) for PbSe. They obtained, with a combination of Hall effect measurements and positron lifetime spectroscopy, a trapping coefficient of $1\times10^{14}\ \mathrm{s}^{-1}$ at room temperature. This value is rather low for negatively charged lead vacancies at this temperatures. The screening effect of the negative

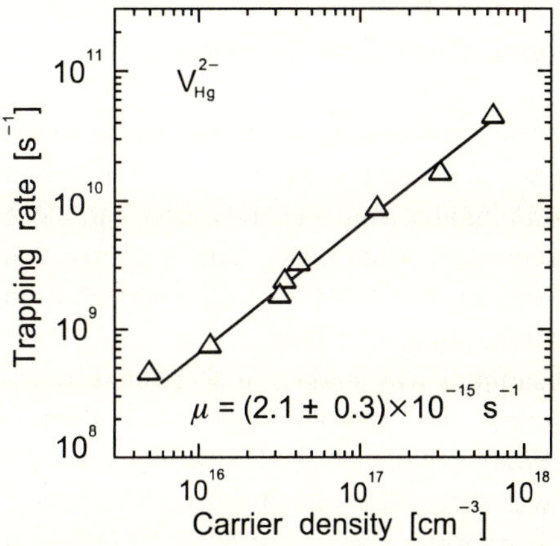

Fig. 3.34. Positron trapping rate at room temperature versus the density of free holes at 77 K measured by the Hall effect in a series of $Hg_{0.78}Cd_{0.22}Te$ samples (Krause-Rehberg et al. 1995a). The density of free carriers corresponds to the density of doubly negative mercury vacancies. The trapping coefficient μ was obtained from the fit to the data (full line).

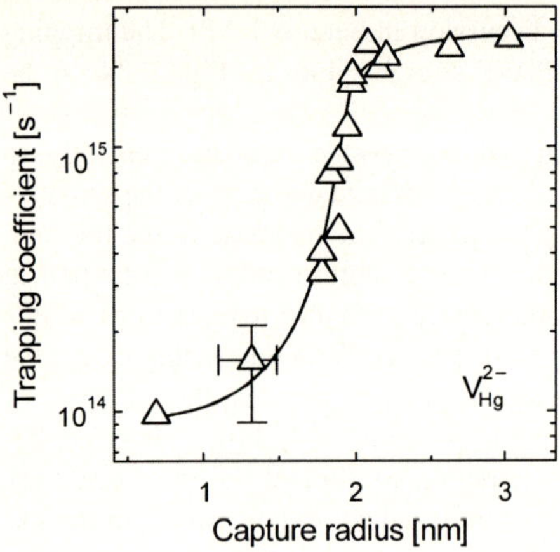

Fig. 3.35. Trapping coefficient versus the capture radius in $Hg_{0.78}Cd_{0.22}Te$ (Krause-Rehberg and Leipner 1997). r_c was numerically calculated from (3.20). The trapping coefficient is reduced by more than one order of magnitude with a decreasing capture radius as a result of effective screening caused by a high concentration of free carriers, which is obtained by intrinsic conduction at elevated temperatures in this small-bandgap material. The data for the trapping coefficients of the doubly negative mercury vacancy were taken from Krause-Rehberg et al. (1995a). The solid line is drawn to guide the eye.

vacancy charge is probably responsible for this small value, because the density of holes is of the order of magnitude of 10^{19} cm^{-3}.

3.2.4.6 The A Center in Chlorine-Doped Cadmium Telluride

The *A* center in CdTe:Cl was identified by electron paramagnetic resonance as the Cd vacancy paired off with a chlorine donor at a nearest-neighbor site (Hofmann et al. 1992). The donor–acceptor pair recombination from a shallow donor to the *A* center can be identified in photoluminescence measurements. The *A* center line shifts to higher energies with increasing doping level and thus the concentration can be quantitatively determined (Krause-Rehberg et al. 1998).

A series of differently doped CdTe:Cl samples was studied in correlated investigations by positron lifetime spectroscopy and photoluminescence (Krause-Rehberg et al. 1998; Polity et al. 1995b; Stadler et al. 1995). The vacancy as part of the acceptor-type *A* center, $V_{Cd}^{2-}Cl_{Te}^{+}$, was detected by positrons and the corresponding trapping rate was determined. A trapping coefficient at room temperature can be calculated to be 1.7×10^{15} s^{-1} as the average value of three samples.

3.2.4.7 Summary and Comparison with Theory

The experimentally obtained trapping coefficients are discussed in the following and compared with theoretical calculations of differently charged monovacancies. Such calculations have only been carried out for silicon. First, the room-temperature values for the different charge states are considered.

Most of the experimental data are available for single negatively charged monovacancies or monovacancy–impurity complexes. The trapping coefficients at room temperature scatter around a value of $\mu = 1\times10^{15}\ \mathrm{s}^{-1}$ within a factor of 2. Only the value of Kawasuso et al. (1995c) for the vacancy–phosphorus pair in Si is significantly higher. It should be noted that with this one exception the scattering of the data is rather low, despite the fact that different vacancy-type defects in elemental and compound semiconductors were studied, and different reference methods for the independent determination of the vacancy concentration were applied. Puska et al. (1990) calculated the trapping coefficients for monovacancies of different charge states in Si (Fig. 3.8). According to this, a value of $2.4\times10^{15}\ \mathrm{s}^{-1}$ is expected.

Only a few experiments, in which the ratio of the trapping coefficients for negative and neutral monovacancies was determined at room temperature, are known (Kawasuso et al. 1995c; Mäkinen et al. 1992a). For the vacancy–phosphorus pair in Si a ratio of about 2 was obtained. The ratio of trapping coefficients of the negative and neutral phosphorus vacancy in GaP measured at 473 K gave the same value (Krause-Rehberg et al. 1993c). Both the ratio for the vacancy–phosphorus pair in Si and the ratio for V_P in GaP are smaller than the theoretical prediction by a factor of 6, calculated for the Si monovacancy (Puska et al. 1990). Hence, the semiclassical theory is able to provide only an estimate of the order of magnitude of the trapping coefficient. The initial and final positron states before and after trapping were taken in the theoretical approach as products of independent electron and positron wave functions. Furthermore, the positron wave functions were approximated, e.g. as a spherically symmetric Gaussian function for the trapped state and a plane wave for the delocalized state (Nieminen and Laakkonen 1979). The attractive potential of the vacancy was assumed to be a square-well potential. Despite the simplifications in the theory, the agreement with the experimental values is rather good.

The Hg vacancy in HgCdTe was studied at room temperature (Krause-Rehberg et al. 1995a), where this defect acts as a doubly negative acceptor. μ was obtained to be $2.1\times10^{15}\ \mathrm{s}^{-1}$ and is thus twice as high as the trapping coefficient found in other semiconductors for single negatively charged monovacancies, as is expected from Fig. 3.8.

The negative charge of vacancies may be screened by a high concentration of free carriers (Sect. 3.1.5.1). The promotion of positron trapping due to the existence of extended shallow Rydberg states is thus reduced, leading to a smaller trapping coefficient (Fig. 3.35). The proportionality between the trapping rate and the vacancy concentration C (3.11) may be disturbed in the special case, when the

Table 3.4. Trapping coefficients for different charge states of monovacancies in semiconductors averaged from positron lifetime experiments at room temperature (Krause-Rehberg and Leipner 1997). The data are compared with the theoretical calculations by Puska et al. (1990) for monovacancies in silicon. The sensitivity range of the positron lifetime spectroscopy (POLIS) for the vacancy detection in semiconductors is taken from Fig. 3.36.

Charge state	Trapping coefficient [10^{15} s^{-1}]		Sensitivity range of POLIS	
	Experiment	Theory	Lower limit	Upper limit
doubly negative	2	6	3.1×10^{-8}	7.5×10^{-5}
negative	1	2.4	6.2×10^{-8}	1.5×10^{-4}
neutral	0.5	0.5	1.5×10^{-7}	3.0×10^{-4}
positive	< 0.1	0.03	2.1×10^{-6}	3.3×10^{-3}

concentration of free carriers, and thus the capture radius, is determined by a high concentration of vacancies as ionized donors or acceptors. This requires at least vacancy densities of the order of 10^{19} cm^{-3}. In consequence, the trapping coefficient becomes a function of C, which is a rather rare case.

The interaction of positrons with positively charged vacancies has not been studied extensively. Up to now, no such trapping has been observed. One experiment on the positively charged phosphorus vacancy in GaP (Krause-Rehberg et al. 1993c) provided the data for an estimate of the upper limit of the trapping coefficient of 1×10^{14} s^{-1}. The theoretical calculation for the positive monovacancy in Si gave 3×10^{13} s^{-1} (Puska et al. 1990).

According to the same calculations, a prediction for divacancies can be made on the basis of (3.23). Values of 9×10^{14}, 5×10^{15}, and 1.2×10^{16} s^{-1} are obtained for neutral, negative, and doubly negative divacancies, respectively. The value for neutral divacancies is in good agreement with the experimentally obtained ones. The theoretically calculated trapping coefficients for negatively charged divacancies lie in between the experimental values (see Table 3.3, p. 103) of Mascher et al. (1989b) and Kawasuso et al. (1995c). The experimental estimate for the upper limit of the trapping coefficient for positively charged divacancies (< 10^{14} s^{-1}, Kawasuso et al. 1995c) is again in agreement with the theoretical prediction of 6×10^{13} s^{-1}.

The trapping coefficients experimentally obtained at room temperature for monovacancies of different charge states give consistently similar values and are summarized and compared in Table 3.4 with the trapping coefficients theoretically calculated for Si. Only monovacancies have been taken into account for the table, since data on divacancies are rare. The trapping coefficient for negatively charged monovacancies was found to be close to 1×10^{15} s^{-1} for almost all studied semiconductors at room temperature. This result was obtained although different reference methods for the independent determination of vacancy concentrations were applied. The neutral monovacancy gave a value smaller by a factor of two, and the doubly negative monovacancy a value higher by a factor of two compared with the negative monovacancy.

No evidence was found for trapping in positive vacancies and the upper limit of the trapping coefficient can be given as $10^{14}\ s^{-1}$, as mentioned above. Similarly to negatively charged vacancies, the positive charge can be screened by very high carrier concentrations. It may be expected that the Coulomb barrier, impeding the trapping of positrons is reduced by this screening and results in the increase in the trapping coefficient. However, it should not exceed the value for neutral vacancies.

The increase in the trapping coefficient for negative vacancies at low temperatures can be satisfactorily described by the two-stage trapping model (Puska et al. 1990), where positron trapping is promoted by extended Rydberg states.

The sensitivity range of positron lifetime spectroscopy can be calculated with the values for the trapping coefficients at room temperature given above (Krause-Rehberg and Leipner 1997). The average positron lifetime $\bar{\tau}$ was calculated as a function of the vacancy concentration using the trapping coefficients experimentally determined for different charge states (Table 3.4). For the positive vacancy the theoretical estimate of $\mu = 3\times10^{13}\ s^{-1}$ was taken (dashed line). $\bar{\tau}$ is shown in Fig. 3.36 in a normalized form as the increase between the bulk lifetime τ_b and the defect-related lifetime τ_d. It was calculated together with the corresponding trapping rate according to the one-defect trapping model (Sect. 3.2.1.1). In the upper part of Fig. 3.36, showing the trapping rate as a function of the vacancy concentration, the sensitivity range of positron lifetime spectroscopy is indicated by two horizontal lines. The corresponding ranges of the vacancy concentrations are given in Table 3.4. The lower sensitivity limit was calculated supposing a measurable minimum deviation of 1 ps from the bulk lifetime τ_b and the upper limit with a chosen deviation of 2 ps from the defect-related lifetime τ_d. For the calculation of the sensitivity range, a ratio of $\tau_d/\tau_b = 1.24$ was assumed (i.e. the value of the monovacancy in Si, see Table 4.2, p. 182). If this ratio is significantly smaller, the sensitivity range becomes only slightly narrower, e.g. for mercury vacancies in HgCdTe with $\tau_d/\tau_b = 1.11$ (see Sect. 6.1.2).

The lower limit is related to the fact that the positrons must be trapped during their limited diffusion path through the sample. When the vacancy concentration is too low, this process becomes improbable, independent of the positron technique applied. In case, the vacancy concentration exceeds the upper limit, positron lifetime spectroscopy shows only the existence of vacancies but is no longer sensitive to their concentration, i.e. saturation trapping is reached. The limitation at very high vacancy concentrations can be overcome and the trapping rate is still measurable when back-diffusion experiments using slow-positron beams are performed (Sect. 3.2.3).

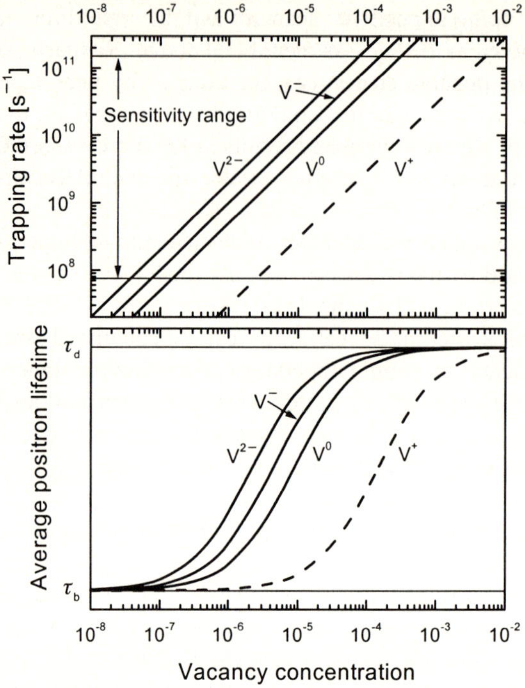

Fig. 3.36. Illustration of the sensitivity of positron lifetime spectroscopy for monovacancies of different charge states (Krause-Rehberg and Leipner 1997). The average positron lifetime was calculated as a function of the vacancy concentration by using the trapping coefficients at room temperature given in Table 3.4. For the positive charge state the theoretically estimated value of $3\times10^{13}\ \mathrm{s}^{-1}$ was taken. The lifetime is given in a normalized form between the bulk lifetime τ_b and the defect-related lifetime τ_d. In the upper part, the corresponding trapping rates are plotted. The upper and lower sensitivity limits of the positron lifetime technique are indicated as horizontal lines, determined by a measurable minimum deviation of 1 ps from the bulk value and of 2 ps from the defect value.

3.2.5 Trapping Coefficients of Shallow Positron Traps and Dislocations

Negatively charged non-open-volume defects may act as effective positron traps at low temperature. They are described as shallow positron traps (Sect. 3.1.5.2). The capture is based on the existence of extended Rydberg states and can be described as a diffusion-limited process. Trapping coefficients which are a function of the temperature have been given.

Equation (3.21) contains the defect concentration of shallow traps C_{st} and the trapping rate, which may be obtained in principle from the trapping model independently. Thus, the trapping coefficient can be obtained from a fit of the temperature-dependent annihilation parameters, e.g. from the average positron lifetime. Corbel et al. (1992b) simulated the temperature dependence of the average positron lifetime as a function of the positron binding energy and the concentration of shallow positron traps. They found a reasonable agreement with their experimental results on semi-insulating GaAs with a trapping coefficient in the range 10^{15} to 10^{16} s^{-1} at low temperatures.

Saarinen et al. (1995a) measured the average lifetime as a function of the temperature in electron-irradiated n-doped GaAs, fitted the curves according to the trapping model, and took into account the temperature dependence of the detrapping rate of the shallow traps, as given by (3.21). The corresponding trapping coefficient for shallow traps obtained from the fit was given as a function of the temperature T as $\mu = (5 \pm 2)\times 10^{16}\,(T/\mathrm{K})^{-1/2}\,s^{-1}$. The shallow traps were supposed to be negatively charged Ga_{As} antisite defects.

A similar determination of the trapping coefficient was performed by Polity and Engelbrecht (1997a) in electron-irradiated undoped InP. The fit was compatible with trapping coefficients for shallow traps between 7×10^{15} and 1×10^{18} s^{-1} at 20 K in differently annealed samples. The nature of the shallow traps in InP has not been identified.

The determination of the trapping coefficient has been based in the experiments described above, on the fit of the temperature dependence of positron trapping (3.21) without any independent reference method for determination of the shallow trap concentration. Due to statistical uncertainties, the fitting parameters exhibit mutual numerical dependences and the fits allow in principle only an estimate of a range comprising at least two orders of magnitude of the trapping coefficient. It has to be concluded that, similar to open-volume defects, another method must provide the concentration of the shallow positron traps independently in order to determine accurate trapping coefficients.

Such an experiment was carried out using scanning tunneling microscopy. Gebauer et al. (1997c) studied highly silicon-doped GaAs. Negatively charged Si_{Ga} centers were identified by STM as the only candidate for shallow positron traps. An increasing number of these centers were found with increasing Si content. The number of shallow positron traps increased as well. Provided that the same defect is detected by positron lifetime spectroscopy, the trapping coefficient can be estimated as 5×10^{16} s^{-1} at 20 K.

Positron trapping in dislocations was described in Sect. 3.1.5.3. It has been concluded that undisturbed dislocation lines act as shallow precursor states for deep positron traps, such as bound vacancies or jogs. There are first indications that the situation in semiconductors is quite similar (Hübner et al. 1997a; Kawasuso et al. 1995d). The two-stage trapping process has been treated as simple transition-limited trapping (Dannefaer et al. 1983a), leading to a trapping rate proportional to the dislocation density,

$$\kappa = \mu \rho_{\mathrm{dis}}. \tag{3.48}$$

Trapping coefficients of 4.5×10^{15} to $8\times10^{15}\ \mathrm{s}^{-1}$ were obtained in silicon deformed at different strains (Dannefaer et al. 1983a). This corresponds to 3.3 to 5.9 $\mathrm{cm}^2\,\mathrm{s}^{-1}$ if the dislocation density is given as usual in units of cm^{-2}. However, the dislocation density was not measured directly, but estimated from theoretical considerations. A direct determination of ρ_{dis} was carried out by transmission electron microscopy in the work of Krause-Rehberg et al. (1993a). They calculated a trapping coefficient of $1\times10^{15}\ \mathrm{s}^{-1}$ ($0.7\ \mathrm{cm}^2\,\mathrm{s}^{-1}$) and $4.8\times10^{15}\ \mathrm{s}^{-1}$ ($4\ \mathrm{cm}^2\,\mathrm{s}^{-1}$) for dislocations in n-type silicon and germanium, respectively. Kawasuso et al. (1995d) determined the dislocation density in plastically deformed silicon by measuring the etch pit density. A trapping coefficient of $5.3\times10^{15}\ \mathrm{s}^{-1}$ was given with the trapping rate calculated for dislocations. No dislocations were detected by positron lifetime spectroscopy in p-type Si deformed in a similar way to n-Si, indicating that dislocation-related positron traps must be positively charged. It could be shown by transmission electron microscopy that the dislocation density was in the same range for both the n-type and p-type deformed Si (Krause-Rehberg et al. 1993a; Leipner et al. 1993).

Brandt and Paulin (1972) treated positron trapping in dislocations as diffusion limited, where the dislocation is assumed to be a cylindrical trap. The trapping rate for this case was given in (3.13). Krause-Rehberg et al. (1993a) discussed their results in addition to transition-limited trapping with the diffusion limitation. The diffusion coefficient increases with decreasing temperature, which should lead to an increasing trapping rate (Sect. 3.1.4). Although such a dependence was observed, it cannot be taken as a proof for the diffusion limitation, because the trapping in the dislocation as a shallow trap should exhibit a similar temperature dependence. Additionally, the charge state of dislocations may influence the trapping similarly to vacancies.

In the framework of the model of the dislocation as a shallow positron trap as a precursor state for deep traps, the special case of high temperatures leads to the usual transition limitation of positron trapping. This is due to the fact that at high temperatures the shallow trap becomes ineffective, as thermally induced detrapping dominates. Thus, there is virtually no shallow precursor state and the positrons are directly trapped in the deep dislocation-related defects. In this case, the positron trapping rate of (3.48) becomes independent of the temperature. Obviously, the dislocation density does not directly determine the trapping rate, since the density of deep traps on the dislocation line, such as jogs or dislocation-bound vacancies, is the dominant factor. It may be varied by the deformation conditions.

Despite these uncertainties, an estimate of the sensitivity range for positron trapping in dislocations can be obtained. It can be concluded from the experimental results given above that a dislocation density of at least $10^8\ \mathrm{cm}^{-2}$ is necessary to be detectable by positron annihilation. Such high densities are usually found in bulk semiconductors only after plastic deformation.

3.3 Theoretical Calculation of Annihilation Parameters

The localization of the positron in defects results in annihilation parameters which can be taken as fingerprints for the defect identification. Such quantities are the details of the annihilation line shape and the positron lifetime. The lifetime of a certain defect has a characteristic value which is determined by the electron density felt by the trapped positron. As an example, the lifetime increases significantly when vacancies cluster to larger vacancy agglomerates. The line shape parameters are influenced by the variation of the electron momentum distribution within or near the defect capturing the positron. Thus, the line shape parameters are more sensitive to the chemical environment of the annihilation site compared with the positron lifetime.

In the case of positron lifetime measurements, it is possible to determine experimentally the defect-related lifetimes by the decomposition of the spectra in lifetime components, i.e. saturated positron trapping is not required. The determination of the open volume of different defects may be successful when the decomposition of the lifetime spectra can be reliably carried out (Sect. 2.2.2).

Annihilation of trapped positrons with core electrons of the surrounding atoms may reveal chemical information when measurements based on the momentum distribution of electrons, i.e. Doppler-broadening spectroscopy or angular correlation of annihilation radiation, are compared with theoretically simulated curves. The high-momentum part of the annihilation line, which is used for the analysis, mainly contains the contributions from annihilations with core electrons. The decomposition of these spectra for various types of positron traps is not possible self-consistently for DOBS or ACAR measurements. The comparison of theoretically simulated and experimentally obtained annihilation line shapes requires that the fraction of positrons annihilating in different traps is known. This fraction can be determined independently by positron lifetime spectroscopy. Such correlated measurements are not necessary when saturated trapping in only one defect type occurs. However, such a case is rather rare.

In the following, calculation methods for the positron lifetime are compiled (Sect. 3.3.1). Thereafter, the calculation scheme of the high-momentum part of the annihilation line is presented (Sect. 3.3.2).

3.3.1 Positron Lifetimes in the Semiconductor Bulk and in Defects

The theoretical calculation of positron annihilation parameters is becoming even more important in the identification of defects. In the simplest case, the Schrödinger equation is solved in the background potential built up by the atoms in the crystal. Hence, the crystal structure must be known in advance.

The ground state energy E of a system of electrons and a positron in an external potential V_{ext} can be written in two-dimensional density functional theory as a functional of the electron density n_- and the positron density n_+ (Puska 1987; Puska and Nieminen 1994),

$$E[n_-,n_+] = F[n_-] + F[n_+] + \int \mathrm{d}\mathbf{r} V_{\mathrm{ext}}(\mathbf{r})[n_-(\mathbf{r}) - n_+(\mathbf{r})] \\ - \int \mathrm{d}\mathbf{r} \int \mathrm{d}\mathbf{r}' \frac{n_-(\mathbf{r}) n_+(\mathbf{r})}{|\mathbf{r} - \mathbf{r}'|} + E_{\mathrm{c}}^{\mathrm{e-p}}[n_-,n_+], \tag{3.49}$$

where $F[n]$ is the one-component density functional for electrons and positrons,

$$F[n] = T[n] + \frac{1}{2} \int \mathrm{d}\mathbf{r} \int \mathrm{d}\mathbf{r}' \frac{n(\mathbf{r}) n(\mathbf{r}')}{|\mathbf{r} - \mathbf{r}'|} + E_{\mathrm{xc}}[n]. \tag{3.50}$$

$T[n]$ is the kinetic energy of non-interacting electrons and positrons, while $E_{\mathrm{xc}}[n]$ is the exchange correlation energy between indistinguishable particles. Finally, $E_{\mathrm{c}}^{\mathrm{e-p}}$ is the functional of the correlation energy of the electron positron pair.

The ground state of electron and positron densities which minimizes $E[n_-,n_+]$ is calculated with a generalized Kohn–Sham method (Boronski and Nieminen 1986). Therefore, it is necessary to solve the following system of one-particle Schrödinger equations for electrons and positrons,

$$-\frac{1}{2}\nabla^2 \psi_i^-(\mathbf{r}) + \left[\frac{\delta E_{\mathrm{xc}}[n_-]}{\delta n_-[\mathbf{r}]} - \phi_{\mathrm{c}}(\mathbf{r}) + \frac{\delta E_{\mathrm{c}}^{\mathrm{e-p}}[n_+,n_-]}{\delta n_-[\mathbf{r}]} \right] \psi_i^-(\mathbf{r}) = \varepsilon_i^- \psi_i^-(\mathbf{r}), \\ -\frac{1}{2}\nabla^2 \psi_i^+(\mathbf{r}) + \left[\frac{\delta E_{\mathrm{xc}}[n_+]}{\delta n_+[\mathbf{r}]} - \phi_{\mathrm{c}}(\mathbf{r}) + \frac{\delta E_{\mathrm{c}}^{\mathrm{e-p}}[n_+,n_-]}{\delta n_+[\mathbf{r}]} \right] \psi_i^+(\mathbf{r}) = \varepsilon_i^+ \psi_i^+(\mathbf{r}), \tag{3.51}$$

where

$$\phi_{\mathrm{c}} = \int \mathrm{d}\mathbf{r}' \frac{n_0(\mathbf{r}') - n_-(\mathbf{r}') + n_+(\mathbf{r}')}{|\mathbf{r} - \mathbf{r}'|} \tag{3.52}$$

is the total Coulomb potential and $n_0(\mathbf{r}')$ the (positive) charge density which arises from the external potential V_{ext}. The electron and positron densities are obtained by summing over all occupied states (E_{F} is the Fermi energy of the electrons and N_0 the total number of positrons),

$$n_-(\mathbf{r}) = \sum_{\varepsilon_i \le E_{\mathrm{F}}} |\psi_i^-(\mathbf{r})|^2, \quad n_+(\mathbf{r}) = \sum_{i=1}^{N_0} |\psi_i^+(\mathbf{r})|^2. \tag{3.53}$$

ψ_i^+ and ψ_i^- are the positron and electron wave functions, ε_i^+ and ε_i^- the energy eigenfunctions of the positron and the electron, respectively. Equations (3.51) to (3.53) must be solved self-consistently and simultaneously for electron and positron states, using an iterative method. An example is shown in Fig. 3.37 with the positron wave function in silicon. The highest position probability density of the positron is found in the perfect solid in interstitial regions (Fig. 3.37a). Localized bound states for the positron occur at open-volume defects. The degree of the localization and the isotropy vary for different defects (Fig. 3.37b, c).

The methods to solve (3.51) to (3.53) would be exact if the energy functional of the exchange correlation energy, $E_{xc}[n]$, and the functional of the correlation energy for electrons and positrons, E_c^{e-p}, were known. Most often the local-density approximation (LDA) is used in electron structure calculations, since the exchange and correlation effects can be handled simply and the accuracy is often sufficient. The LDA method is based on the accurate calculation of the homogeneous electron gas. The exchange correlation energy is approximated by

$$E_{xc}[n] = \int n[\mathbf{r}]\varepsilon_{xc}[n(\mathbf{r})]\mathrm{d}r, \tag{3.54}$$

where $\varepsilon_{xc}(n)$ is the well known exchange correlation energy for a homogeneous electron gas (Ceperley and Adler 1980).

The LDA is used for an approximate determination of the positron–electron correlation energy and the enhancement factor. The enhancement factor γ describes the increase in the electron density Δn_- as a result of the positron–electron interaction,

$$\gamma = \gamma[n(\mathbf{r})] = 1 + \frac{\Delta n_-}{n_-}. \tag{3.55}$$

This method does not describe properly the positron annihilation with core electrons in semiconductors and insulators. The calculated lifetimes are too short in comparison with experiments due to the overestimation of the annihilation with core electrons. The problem can be solved by reducing the fraction of annihilation with core electrons. The strength of the correlation energy and the enhancement

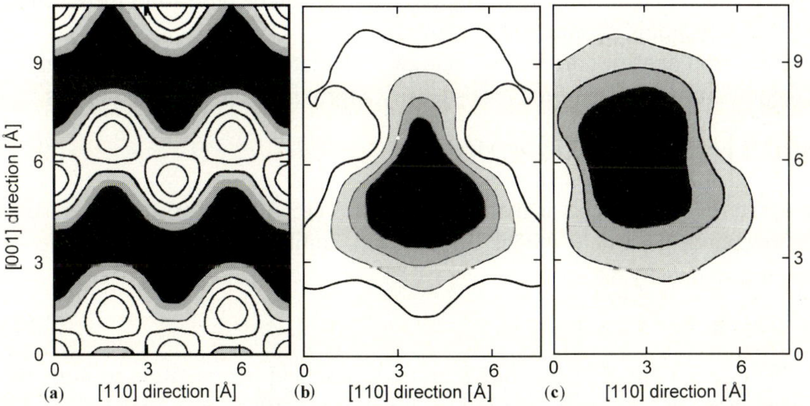

Fig. 3.37. Position probability density of the positron in bulk silicon (**a**), for a monovacancy (**b**), and a divacancy in Si (**c**) calculated by the superimposed-atom method (Puska and Nieminen 1983). The darker shading corresponds to larger values of the wave function.

factor are corrected in two ways: firstly, a semi-empirical approach is used, i.e. the screening is reduced by a parameter based on the high-frequency dielectric constant (Puska et al. 1989b). Secondly, the core enhancement factor is calculated by taking a constant enhancement factor.

Barbiellini et al. (1995, 1996) formulated a generalized gradient approximation (GGA) of the positron correlation energy and the enhancement factor. The GGA considers the gradient of the electron density, while only the electron density itself appears in LDA. The enhancement factor is improved to correct the LDA overestimation of the annihilation with core electrons. There is only one fit parameter in the calculations by the generalized gradient approximation. It is determined by the comparison between experimentally and theoretically determined lifetimes, i.e. it is a linear fit for the whole bulk lifetime range starting from body-centered cubic and transition metals (100 to 130 ps), followed by Zn, Al, Pb, and Mg (140 to 220 ps), through the semiconductors (220 to 300 ps), and finally for simple metals (330 to 400 ps). The most accurate lifetime values are obtained using the generalized gradient approximation with self-consistent electron densities (Barbiellini et al. 1996).

Calculating the electron density without the effect of localized positrons is justified because the positron and its screening cloud form a neutral quasi-particle which does not affect the remaining system of electrons. Hence, the total electron density is a superposition of the "clean" system density and the positron-induced screening cloud. Some terms cancel out in the self-interaction correction for small positron densities, and one obtains for the potential felt by the positron

$$V_+(\mathbf{r}) = -V_{\text{ext}}(\mathbf{r}) - \int \frac{n_-(\mathbf{r}')}{|\mathbf{r}-\mathbf{r}'|}\mathrm{d}\mathbf{r}' + \frac{\delta E_{\text{c}}^{\text{e-p}}[n_+,n_-]}{\delta n_+}. \tag{3.56}$$

This is used as the "conventional method" for localized states, i.e. it comprises positrons trapped in defects as well. Hence, a crystal can be built from the electron densities of free atoms (atomic superposition).

If the electron densities are known, one can calculate the annihilation rate in the two-component formalism of the local-density approximation, and one obtains

$$\lambda = \pi r_0^2 c \int n_+(\mathbf{r}) n_-(\mathbf{r}) g(0;n_+,n_-)\mathrm{d}\mathbf{r}, \tag{3.57}$$

where $r_0 = e^2/m_0c^2$ is the classical electron radius. The prefactor arises from results of the independent-particle model for positron annihilation in a homogeneous electron gas of density n_-,

$$\lambda_0 = \pi r_0^2 c n_- . \tag{3.58}$$

λ_0 is called the Sommerfeld annihilation rate. A practical interpolation formula for the calculation of the pair-correlation functional $g(0;n_+,n_-)$ was given by Boronski and Nieminen (1986).

If the positron and electron densities are known, the annihilation rate can easily be calculated according to (2.1). In the case that the positron is delocalized in a perfect crystal, one can consider the limiting case of a vanishing positron density of the pair-correlation functional $g(0; n_+, n_-)$ at the position of the positron. This limit is the enhancement factor γ of the electron density at the position of the positron introduced above. One obtains

$$\begin{aligned}\lambda &= \pi r_0^2 c \int n_+(\mathbf{r}) n_-(\mathbf{r}) \gamma(n_-) \mathrm{d}\mathbf{r} \\ &= \int n_+(\mathbf{r}) \Gamma[n_-(\mathbf{r})] \mathrm{d}\mathbf{r},\end{aligned} \tag{3.59}$$

where $\Gamma[n_-(\mathbf{r})]$ is the positron annihilation rate in a homogeneous electron gas of density n_-.

The annihilation rate (the reciprocal of the positron lifetime) is obtained from the overlap of the electron and positron wave functions taking into account the enhancement according to Boronski and Nieminen (1986). The method of conjugate gradients is used to perform the calculation. In this way, defects can easily be incorporated into the given lattice structure. Using the procedure of atomic superposition, i.e. no relaxation around defects, a three-dimensional potential for the positron is constructed and the Schrödinger equation is discretized. One obtains the energy eigenvalues and the wave function of the positron. The energy eigenvalues give the binding energy of the positron to traps (lattice defects), while the positron wave function is essential for the calculation of the annihilation rate.

Some results of positron lifetime calculations using the different methods described above are presented in Table 3.5, which gives bulk lifetimes for elemental and compound semiconductors. The results of non-self-consistent calculations according to the semiconductor model (positron lifetime $\tau_{\mathrm{th}}^{\mathrm{sc}}$) are compared with the

Table 3.5. Positron bulk lifetime values calculated theoretically according to the "conventional scheme" using the high-frequency dielectric constant (Puska et al. 1989b).

Material	$\tau_{\mathrm{th}}^{\mathrm{sc}}$ [ps]	$\tau_{\mathrm{th}}^{\mathrm{in}}$ [ps]	$\tau_{\mathrm{b}}^{\mathrm{exp}}$ [ps]	Material	$\tau_{\mathrm{th}}^{\mathrm{sc}}$ [ps]	$\tau_{\mathrm{th}}^{\mathrm{in}}$ [ps]	$\tau_{\mathrm{b}}^{\mathrm{exp}}$ [ps]
C	83	114	105	InP	246	237	241
Si	219	221	218	InAs	254	231	246
Ge	226	212	228	InSb	280	233	—
AlP	228	244	—	GaN	141	173	166
AlAs	235	242	—	SiC	134	163	142
AlSb	269	251	—	CdTe	292[a]	—	280
GaSb	257	228	253		286[b]		
GaP	218	224	223	HgTe	275[a]	—	274
GaAs	229	223	229		263[b]		

$\tau_{\mathrm{th}}^{\mathrm{sc}}$ is the positron lifetime calculated according to the semiconductor model (many-body theory for the screening of the positron in an electron gas), $\tau_{\mathrm{th}}^{\mathrm{in}}$ is the positron lifetime calculated according to the insulator model (starting point is the atomic polarizability), and $\tau_{\mathrm{b}}^{\mathrm{exp}}$ is the most reliable experimental bulk lifetime.

[a]calculated with tightly bound core electrons, [b]calculated with d electrons.

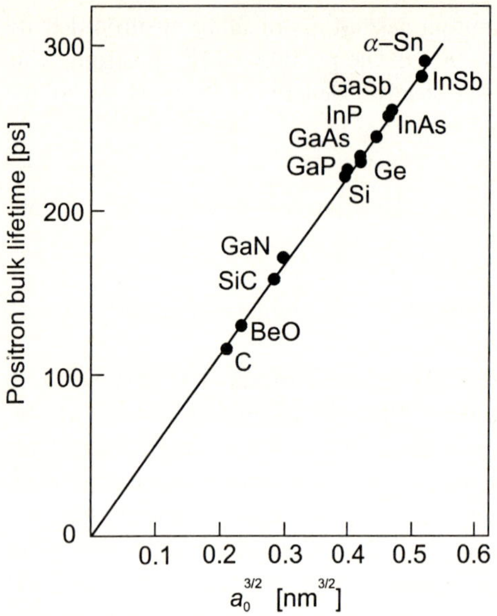

Fig. 3.38. Positron bulk lifetime τ_b versus the square root of the volume of the unit cell of various semiconductors (Siethoff 1998). τ_b was taken from the theoretical calculations of Puska et al. (1989b), except beryllium oxide (Brauer et al. 1984), gallium nitride (Dannefaer et al. 1977), and tin (Seeger et al. 1989), where experimental values were taken.

insulator model (positron lifetime τ_{th}^{in}) in the conventional scheme (Puska et al. 1989b). The high-frequency dielectric constant was used in both cases as a parameter to reduce the screening and to correct the overestimation of the annihilation arising from the pure local-density approximation. The comparison with the experimentally determined lifetimes shows that it is not possible to prefer one of these models. The experimental positron bulk lifetimes can be found in Chaps. 4 to 7. The most reliable values have been selected for Table 3.5.

A direct dependence exists between the unit cell volume and the positron bulk lifetime. A linear dependence was empirically found according to $\tau_b = C_1 a_0^{3/2}$, where the fit of Fig. 3.38 yielded the constant $C_1 = 543.8 \text{ ps nm}^{-1.5}$ (Siethoff 1998). This correlation holds for various semiconductors, despite different band-gaps, ionicities, and dielectric properties. It turned out that the positron bulk lifetime is inversely proportional to the square root of the valence electron density.

Table 3.6 gives defect-related lifetimes calculated according to the conventional scheme using the high-frequency dielectric constant. For comparison with experimentally obtained lifetimes, one should not compare the absolute defect-related lifetimes given in Table 3.6 but the ratios to the corresponding bulk lifetimes, τ_d/τ_b. Non-self-consistent (NSC) values of the positron bulk lifetime have to

Table 3.6. Defect-related positron lifetimes τ_d calculated theoretically according to the "conventional scheme" using the high-frequency dielectric constant (Puska et al. 1989b). The calculations were performed for carbon and gallium phosphide according to the insulator model and in the other cases according to the semiconductor model. The values for silicon carbide were taken from Brauer et al. (1996a).

	Defect	τ_d [ps]		Defect	τ_d [ps]		Defect	τ_d [ps]
C	V	146	Si	V	256	Ge	V	263
	V_2	206		V_2	309		V_2	316
AlP	V_{Al}	265	GaP	V_{Ga}	264	InP	V_{In}	295
	V_P	261		V_P			V_P	275
	$V_{Al}V_P$	319		$V_{Ga}V_P$	316		$V_{In}V_P$	340
AlAs	V_{Al}	271	GaAs	V_{Ga}	265	InAs	V_{In}	299
	V_{As}	274		V_{As}	268		V_{As}	285
	$V_{Al}V_{As}$	439		$V_{Ga}V_{As}$	321		$V_{In}V_{As}$	347
AlSb	V_{Al}	298	GaSb	V_{Ga}	287	InSb	V_{In}	315
	V_{Sb}	319		V_{Sb}	307		V_{Sb}	322
	$V_{Al}V_{Sb}$	455		$V_{Ga}V_{Sb}$	350		$V_{In}V_{Sb}$	369
CdTe	V_{Cd}	321	HgTe	V_{Hg}	304			
	V_{Te}	339		V_{Te}	315			
	$V_{Cd}V_{Te}$	384		$V_{Hg}V_{Te}$	362			
GaN	V_{Ga}	273						
	V_N	-						
	$V_{Ga}V_N$	348						
SiC	V_{Si}	196						
	V_C	153						
	2V	214						

τ_d is the defect-related lifetime.

be taken, since reliable self-consistent defect-related lifetimes are available only in very few cases.

Table 3.7 summarizes positron lifetimes, which were non-self-consistently calculated according to the conventional scheme. Plazaola et al. (1994) also calculated the positron bulk lifetimes in II–VI compound semiconductors self-consistently using thc lincar muffin tin orbital in the atomic sphere approximation (LMTO-ASA) within the local-density approximation. These bulk lifetimes are, however, too low compared with experimental values. Table 3.8 presents bulk lifetime calculations according to the generalized gradient approximation. Barbiellini et al. (1995, 1996) used linear muffin tin orbitals in the atomic sphere approximation, while Panda et al. (1997) used pseudo-potentials. One can see that for both cases the pure LDA results give bulk lifetimes too small in comparison with the experimental values, while the GGA calculations give nearly the right

Table 3.7. Defect-related positron lifetimes in II–VI compound semiconductors theoretically calculated according to the "conventional scheme" using the high-frequency dielectric constant (Plazaola et al. 1994).

	Defect	τ_b [ps]	τ_v [ps]	τ_v/τ_b	Defect	τ_{2v} [ps]	τ_{2v}/τ_b
ZnS	V_{Zn}	219	233	1.06	$V_{Zn}V_S$	292	1.33
	V_S		230	1.05			
ZnSe	V_{Zn}	234	245	1.05	$V_{Zn}V_{Se}$	310	1.32
	V_{Se}		252	1.08			
ZnTe	V_{Zn}	255	261	1.02	$V_{Zn}V_{Te}$	235	1.31
	V_{Te}		291	1.14			
CdS	V_{Cd}	242	265	1.10	$V_{Cd}V_S$	317	1.31
	V_S		249	1.03			
CdSe	V_{Cd}	254	272	1.07	$V_{Cd}V_{Se}$	331	1.30
	V_{Se}		265	1.04			
CdTe	V_{Cd}	276	288	1.04	$V_{Zn}V_{Te}$	355	1.29
	V_{Te}		301	1.09			
HgS	V_{Hg}	229	248	1.08	$V_{Hg}V_S$	297	1.30
	V_S		235	1.03			
HgSe	V_{Hg}	239	253	1.06	$V_{Zn}V_S$	306	1.28
	V_{Se}		249	1.04			
HgTe	V_{Hg}	257	267	1.04	$V_{Hg}V_{Te}$	329	1.28
	V_{Te}		281	1.09			

τ_b, τ_v, and τ_{2v} are the positron bulk lifetime, the vacancy-related lifetime, and the divacancy-related lifetime non-self-consistently calculated. The calculations were performed according to the semiconductor model.

ones. Using pseudo-potentials gives in the case of C, Si, and SiC significantly longer lifetimes than using LMTO-ASA. The difference between the LMTO-ASA calculations within the generalized gradient approximation in comparison with the atomic superposition method is that the GGA scheme is sensitive to self-consistency and shape of the electron density.

In addition to these non-self-consistent calculations of defect-related lifetimes, there are self-consistent calculations of the Helsinki group and others, which include relaxation effects due to atomic displacement near defects and due to positron capture in these defects (Gilgien et al. 1994; Laasonen et al. 1991; Puska et al. 1986; Saito and Oshiyama 1996). Disregarding the influence of the positron, usually a lattice relaxation inwards in a vacancy-type defect occurs. The effect is partly cancelled out under the influence of the positron and the atoms may relax outwards as well.

Table 3.8. Positron bulk lifetimes (in ps) calculated according to the generalized gradient approximation (GGA) and compared with results from the local density approximation (LDA). The pseudo-potential calculations were carried out by Panda et al. (1997), and the linear muffin tin orbital in the atomic sphere approximation (LMTO-ASA) and the atomic superposition calculations by Barbiellini et al. (1995, 1996).

	Pseudo-potential		LMTO-ASA		Atomic superposition		
	$\tau_{\rm th}^{\rm LDA}$	$\tau_{\rm th}^{\rm GGA}$	$\tau_{\rm th}^{\rm LDA}$	$\tau_{\rm th}^{\rm GGA}$	$\tau_{\rm th}^{\rm LDA}$	$\tau_{\rm th}^{\rm GGA}$	$\tau_{\rm b}^{\rm exp}$
C	87	100	86	96	—	—	105
Si	190	216	186	210	184	207	218
Ge	198	228	191	228	190	229	228
SiC	130	145	124	139	121	134	142
GaAs	197	232	190	231	190	232	229
InP	213	246	201	248	200	247	241
ZnS	—	—	179	223	179	232	230
CdTe	245	292	228	290	228	310	280
HgTe	—	—	222	285	222	310	274

$\tau_{\rm th}^{\rm LDA}$ and $\tau_{\rm th}^{\rm GGA}$ are the bulk lifetimes calculated for the local density approximation (LDA) and the generalized gradient approximation (GGA), respectively. $\tau_{\rm b}^{\rm exp}$ is the most reliable experimental bulk lifetime.

3.3.2 Electron Momentum Distributions

One obtains for the momentum distribution of the electron–positron pair with the state-dependent enhancement $u_j^2(0) = \lambda_j / \lambda_j^{\rm IPM}$, which has to be calculated in advance,

$$\sigma_j = \pi r_0^2 c u_j^2(0) \left| \int \mathrm{d}\mathbf{r} \exp(-i\mathbf{p}\cdot\mathbf{r}) \psi^+(\mathbf{r}) \psi_j^-(\mathbf{r}) \right|^2 . \tag{3.60}$$

λ_j is the annihilation rate of the state j and $\lambda_j^{\rm IPM}$ its approximation in the independent-particle model (Alatalo et al. 1996). This gives the momentum distribution σ_j for the annihilation with the electron state j. The total momentum distribution of the electron–positron pair is obtained by the summation over the different electron states.

Using wave functions for free atoms with the quantum numbers n and l and the isotropic positron wave function in (3.60), onc obtains a spherical momentum distribution $\sigma_{I,nl}(p)$ belonging to the ion I. The momentum distribution, which can be compared with the experimental Doppler spectra, evolves by summing up the contributions of the core states of all atoms of the system and then integrating along the high momentum part (Alatalo et al. 1996),

$$\sigma(p_z) = \sum_{I,nl} \int_{p_z}^{\infty} \mathrm{d}p\, p\, \sigma_{I,nl}(p) . \tag{3.61}$$

For comparison with experimental data in the high-momentum part, the momentum distribution (3.60) has to be normalized to the core-electron contribution of the total annihilation rate, since the experimental momentum distributions, which include contributions from both valence and core electrons, are usually normalized to unity. Furthermore, the theoretical momentum distribution should be convolved with a Gaussian with the experimentally determined line width.

Only a qualitative comparison was feasible in the work of Alatalo et al. (1995). A direct comparison with experimental data is only possible when the state-dependent enhancement is used in the calculations (Alatalo et al. 1996). In this way, theoretical and experimental data can be directly compared in one and the same figure.

The relative values of the W parameter, W_d/W_b are taken for the comparison between experiment and theory. The momentum window $[p_1, p_2]$ for the experimental and theoretical determination of the W parameter has to be the same. It is chosen in such a way that the W parameter reflects mostly core-electron annihilation of the system under investigation.

The W parameter is calculated from the Doppler curve according to (2.11). It is obtained from theory as a first approximation directly from core and total annihilation rates λ_c and λ at the defect d and in the bulk b,

$$\frac{W_d}{W_b} \approx \frac{(\lambda_c/\lambda)_d}{(\lambda_c/\lambda)_b}, \tag{3.62}$$

without the necessity to calculate the momentum distribution (Alatalo et al. 1996). Here, it is

$$\lambda_c/\lambda = \frac{\sum_{I,nl}^{\text{core}} \lambda_{I,nl}}{\lambda}, \tag{3.63}$$

where the sum runs over all core contributions.

Recently, there have been some attempts to calculate whole Doppler curves, i.e. including the contribution of the valence electrons. They can be obtained from the calculation of the theoretical calculation of the two-dimensional angular correlation of annihilation radiation (Hakala et al. 1998; Saito et al. 1997) by integration in one space direction.

4 Defect Characterization in Elemental Semiconductors

Chapter 3 was dedicated to the interaction of positrons with matter. It was shown that positrons may be trapped in defects, leading to a change in the annihilation parameters in a significant way. This effect can be utilized to study a variety of defect types present in semiconductors. Their concentration may be determined by evaluating the experimental data using the trapping model. The experimental results obtained by means of positron annihilation on silicon as the most important semiconductor are described in Sect. 4.1. Experiments on germanium, SiGe mixed crystals, and diamond are summarized in Sect. 4.2.

4.1 Silicon

Silicon single crystals are the most perfect crystals available nowadays. The concentration of in-grown vacancies is, in many cases, below the lower sensitivity limit of positron annihilation (Itoh et al. 1997). This is the reason that, in contrast to other experimental techniques, there are hardly any positron studies on crystals in the as-grown state. A weak increase in the average positron lifetime at low temperatures was found for a few float-zone silicon samples by Gebauer et al (1998a). This effect only became visible after repeated positron lifetime measurements with high statistics (Fig. 4.1). Provided that this increase is related to positron trapping in monovacancies, a defect density of 3×10^{14} cm^{-3} can be calculated with a trapping coefficient of 1×10^{16} s^{-1} at 20 K. Heat treatment at 500 °C did not alter the positron trapping. It was therefore concluded that rather stable agglomerates are present.

Extended studies exist on defects in silicon after electron irradiation. The occurrence of vacancies and their agglomerations as irradiation-induced defects was investigated in detail. Papers where the structure of such defects was dealt with are treated in Sects. 4.1.1 and 4.1.3. A similar attempt in comparison to earlier studies in metals was made to determine the formation enthalpy of vacancies at higher temperatures (Sect. 4.1.2).

The behavior of oxygen in Si, especially the formation of oxygen clusters, was of special interest (Sect. 4.1.3). It is demonstrated in Sects. 4.1.4 to 4.1.6 that positrons can be used to profile structural defects in silicon layers and at interfaces. Peculiarities specific to irradiation experiments, e.g. dose dependence, de-

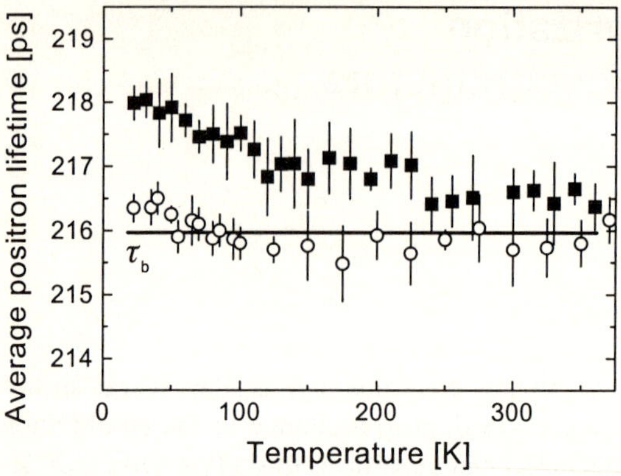

Fig. 4.1. Increase in the average positron lifetime with falling sample temperature in as-grown float-zone silicon (■). The results of a p-type boron-doped Si sample are shown for comparison (○). τ_b is the positron bulk lifetime. The data points were obtained from six measurements with 5×10^6 counts each (Gebauer et al. 1998a). The error bars correspond to the standard deviation of these measurements.

fect reactions, and annealing behavior, as well as implantation experiments, are treated in Sects. 4.1.7 and 4.1.8. The identification and annealing behavior of defects generated during plastic deformation are described in Sect. 4.1.9. Since the experimental results on plastically deformed germanium are very similar to those for Si, this section also includes data on Ge.

For technological relevance, most studies were performed in monocrystalline Si. Important results on porous, polycrystalline, and amorphous silicon are reviewed in Sect. 4.1.10.

4.1.1 Vacancy-Type Defects After Electron Irradiation

The appearance of vacancies, their complexes with impurities, and vacancy clusters has been extensively studied after electron irradiation, mostly performed at low temperatures. In contrast to compound semiconductors, defects in silicon and germanium can easily be studied by electron paramagnetic resonance as a reference method. Information about the structure and the charge state of vacancy-type defects obtained by EPR has been useful for the development of the understanding of the interaction of positrons with these defects.

The knowledge of the charge state of the defect is an important requirement for the interpretation of positron data, as already explained in Sect. 3.1.5.1. Before experimental studies are discussed, the ionization levels of different vacancy-type defects are given. Such data are shown in Fig. 4.2 for monovacancies and divacancies, as well as for the *A* center and the vacancy–phosphorus pair as an important *E* center (Mascher et al. 1989a). The monovacancy in Si exhibits a negative-U

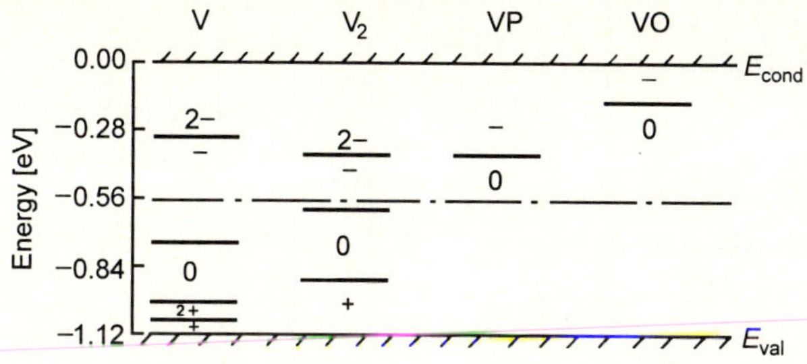

Fig. 4.2. Energy-level scheme in the bandgap of silicon for various charge states of monovacancies and divacancies, vacancy–phosphorus pairs (*E* centers), and vacancy–oxygen pairs (*A* centers) according to Mascher et al. (1989b). E_{val} and E_{cond} are the edges of the valence and conduction bands, respectively. The dash–dot line indicates the midgap position.

behavior. Five different charge states (+, 2+, 0, −, and 2−) were discussed (van Vechten 1986; Watkins and Troxell 1980). The 0/+ ionization level lies 0.05 eV above the top of the valence band, E_{val}, and the 2+/+ level at E_{val} + 0.13 eV (Watkins and Troxell 1980). V^+ is assumed to be, in contrast to earlier ideas, a metastable state, which decays either to V^0 or V^{2+} depending on the Fermi level position (Baraff et al. 1979).

It was proposed by van Vechten (1986) that the 0/− level lies 0.73 eV below the edge of the conduction band E_{cond} and the −/2− level at E_{cond} − 0.27 eV. Since positrons are very sensitive to the charge state of the vacancy (compare Fig. 3.7), the positron interaction with the monovacancy depends strongly on the position of the Fermi level. This remains valid also for divacancies occurring in the +, 0, −, 2− charge states. The corresponding levels were given by Evwaraye and Sun (1976) at E_{val} + 0.25 eV (+/0), E_{cond} − 0.41 eV (0/−), and E_{cond} − 0.23 eV (−/2−) based on DLTS measurements. An earlier paper of Watkins and Corbett (1965) attributed the level at E_{cond} − 0.41 eV to the 2−/− transition.

Defect complexes consisting of a vacancy and a group V impurity, such as VP pairs, exhibit a 0/− level found around E_{cond} − 0.45 eV by electron paramagnetic resonance and deep level transient spectroscopy (Chantre et al. 1983; Elkin and Watkins 1968; Hirata et al. 1966). Vacancy–oxygen pairs show similar charge states. The 0/− level is located at about E_{cond} − 0.2 eV (Kimerling 1977; Watkins and Troxell 1980). As a consequence of the dependence on the Fermi-level position, all defect types mentioned in Fig. 4.2 should be detectable by positrons in undoped and n-type doped silicon. However, no positron trapping is expected in heavily p-doped Si for isolated monovacancies and divacancies because of their positive charge, whereas the vacancy–impurity pairs should remain visible.

Isolated monovacancies are difficult to study in silicon due to their insufficient thermal stability. Electron paramagnetic resonance (Watkins 1964) and deep level

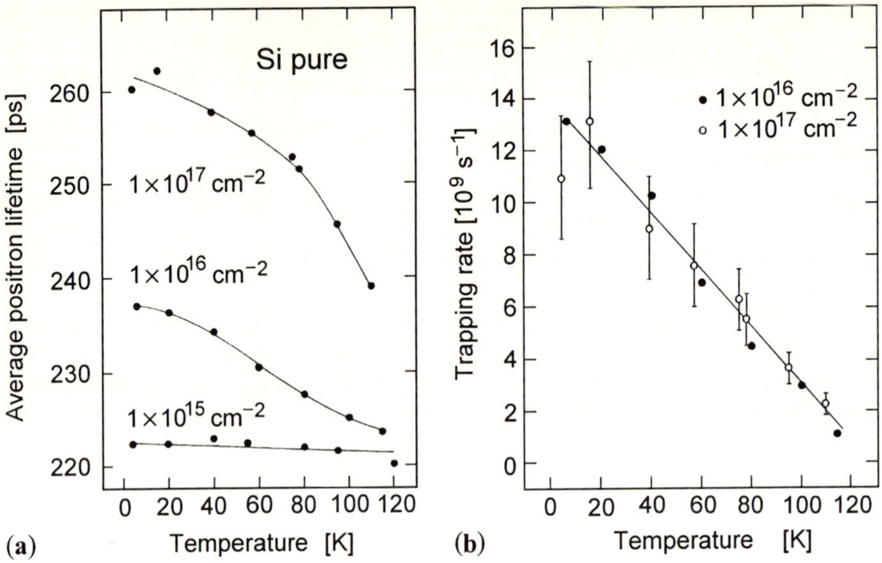

Fig. 4.3. Positron lifetime measurements of undoped float-zone silicon after 3-MeV electron irradiation at 20 K with different doses (Mäkinen et al. 1989). (**a**) Average positron lifetime as a function of the sample temperature. (**b**) Positron trapping rate calculated from the data presented in (a). The trapping rate for the 1×10^{16} cm^{-2} dose was scaled by a factor of 6.0 to compare its variation with the temperature to the 1×10^{17} cm^{-2} irradiated sample. The lines are drawn to guide the eye.

transient spectroscopy studies (Kimerling 1977) imply that these defects anneal below room temperature, although the annealing temperatures differ for both methods. The now mostly accepted value of 170 K is that of Kimerling (1977). They expected the vacancy–interstitial recombination to be the dominant annealing mechanism. Vacancies are generated by high-energy electron irradiation carried out at low temperatures. The sample must be kept at low temperatures (usually at 77 K) to prevent annealing until the experimental study of the monovacancy. In addition to extended studies by electron paramagnetic resonance, a couple of positron experiments of this type were performed and results on the defect structure, the annealing behavior, and the subsequent defect reactions were obtained (see below).

In order to characterize isolated monovacancies, further prerequisites are necessary. The Si crystal should be undoped and of low oxygen content to prevent the formation of vacancy–impurity pairs (*E* or *A* centers) or larger oxygen-related clusters, which may act as competitive positron traps. These conditions were fulfilled in a study of undoped float-zone Si by Mäkinen et al. (1989). After irradiation with 3-MeV electrons at 4 K, a clear increase in the average positron lifetime with irradiation dose was found (Fig. 4.3a). Furthermore, a distinct temperature dependence of $\bar{\tau}$ and of the calculated trapping rate (Fig. 4.3b) was observed. The temperature dependence was found to be completely reversible in the range between 4 and 120 K. This temperature behavior was taken as a clear indication for the negative charge of the vacancy defect under investigation. The model of the influence of the vacancy charge on the positron trapping as a function of temperature has already been discussed in Sect. 3.1.5.1. The increase in the trapping rate with decreasing temperature is a result of the promotion of positron trapping due to the extended shallow Rydberg states of the negatively charged vacancy. A further contribution results from the temperature dependence of the positron diffusion. Thus, the overall temperature dependence is rather complex. The straight line in Fig. 4.3b is not a fit according to (3.18) but only to guide the eye.

The decomposition of the positron lifetime spectra (the average lifetime is shown in Fig. 4.3a) yielded in the experimental study of Mäkinen et al. (1989) a defect-related lifetime component of $\tau_2 = (273 \pm 3)$ ps. Other authors also found lifetimes around 270 ps. Fuhs et al. (1978) reported a lifetime of 266 ± 10 ps after 1-MeV electron irradiation at 20 K. Würschum et al. (1989a) performed in undoped FZ Si the irradiation experiment with 1.5-MeV electrons at 90 K and found a 272 ps component, and Mäkinen et al. (1990a) reported after 12-MeV proton irradiation at 15 K a lifetime of 275 ± 5 ps. A systematic positron study after electron irradiation at 4 K with different doses was carried out by Polity (1997). The results obtained earlier were confirmed and a defect-related lifetime of 270 ps was given for the isolated monovacancy.

In order to attribute the lifetime values obtained experimentally to a certain vacancy configuration, a comparison with theoretically calculated values is necessary. The defect-related lifetime component was calculated for vacancy clusters of a size of one to ten vacancies (Hakala et al. 1998; Puska and Corbel 1988; Saito and Oshiyama 1996). The results of the calculations are summarized in Fig. 4.4. The values obtained for monovacancies are in the range 256 to 279 ps depending on the particular model used for the theoretical calculation (Sect. 3.3.1). It can be concluded that the experimental values mentioned above are in the same range and can thus be attributed to monovacancies. The best agreement is obtained for the calculations of Saito and Oshiyama (1996) taking into account the relaxation of the atomic position around the vacancy. Since the impurity content is less than 10^{16} cm^{-3} in the investigated undoped FZ Si samples, isolated monovacancies, but no vacancy–impurity complexes (*A* and *E* centers), are likely to be observed.

A distinct annealing stage of this monovacancy defect was reported for the temperature range around 150 K by Watkins (1986) using electron paramagnetic resonance measurements and by Mäkinen et al. (1989) using positron lifetime spectroscopy. The vacancies do not anneal completely, but form divacancies being stable up to 450 to 500 K (Kimerling 1977). These divacancies can also be obtained by electron irradiation at room temperature. The defect-related positron lifetime of V_2 was found at room temperature to be in the range of 318 to 327 ps (Dannefaer and Kerr 1986a; Fuhs et al. 1978; Kwete et al. 1990; Shimotomai et al. 1981; Würschum et al. 1989a). This is in accordance with the calculation of the

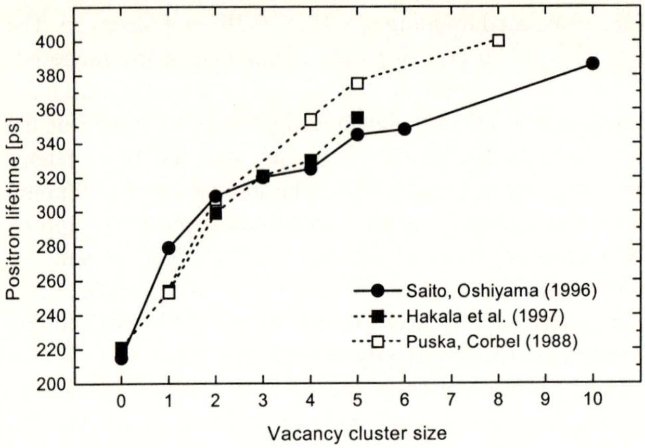

Fig. 4.4. Defect-related positron lifetime versus the vacancy size of neutral vacancy clusters in silicon. 0 corresponds to the positron bulk lifetime. Dashed and solid lines indicate the lifetime calculated for ideal and relaxed geometries, respectively.

fully relaxed neutral divacancy (Fig. 4.4). However, no detailed characterization of the charge state was performed.

Kawasuso et al. (1995a) combined positron lifetime spectroscopy with infrared absorption as a reference method to monitor the divacancy concentration during isochronal annealing. A good agreement was found for the formation and dissolution of divacancies indicated by infrared absorption at 0.343 eV with positron trapping rate that corresponds to the defect-related lifetime component of 320 ps (Fig. 4.5). It was concluded in other papers by the same authors from the investigation of differently doped silicon after 15-MeV electron irradiation at room temperature that the positron lifetime of the divacancy strongly depends on its charge state and on the sample temperature (Hasegawa et al. 1995b; Kawasuso et al. 1994). The position of the Fermi level was carefully checked by Hall effect measurements and the charge state of the divacancy could be determined. The defect-related positron lifetimes of divacancies are shown in Table 4.1. The characteristic change in τ_d for divacancies with the charge state and the temperature was attrib-

Table 4.1. Defect-related positron lifetimes for silicon divacancies measured in the doubly negative, single negative, and neutral charge state at two different temperatures (Hasegawa et al. 1995b; Kawasuso et al. 1994).

Defect	Defect-related lifetime [ps]	
	10 K	300 K
V_2^{2-}	260	320
V_2^{-}	278	320
V_2^{0}	290	295

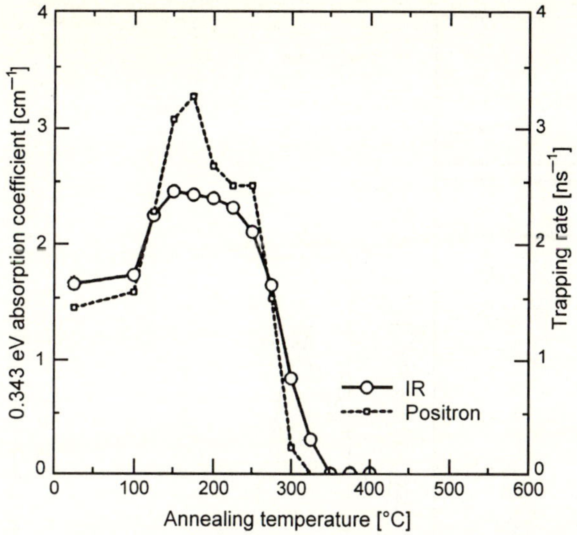

Fig. 4.5. Formation and dissolution of divacancies in electron-irradiated silicon during isochronal annealing. The positron trapping rate of the 320 ps lifetime component (*dashed line*) and the infrared (IR) absorption coefficient at 0.343 eV (*solid line*) associated with the divacancy are plotted (Kawasuso et al. 1995a).

uted to a temperature-dependent lattice relaxation around the divacancy with excess electrons. However, the divacancy density in these experiments was rather low (8×10^{14} to 3×10^{16} cm^{-3}, corresponding to a concentration of 1.6×10^{-8} to 6×10^{-7}), possibly leading to numerical problems in the decomposition of the lifetime spectra. The given concentrations are close to the lower sensitivity limit of the method (compare Fig. 3.36). Furthermore, the influence of vacancy–impurity pairs cannot be excluded, since doped Si samples were used. A further confirmation of the described charge effect is needed. Theoretical calculations of the defect-related lifetime as a function of the temperature and the divacancy charge taking into account lattice relaxation are also left to be done.

A distinct temperature dependence of the positron trapping is expected in the case of negatively charged vacancy-like defects (Sect. 3.1.5). This was confirmed by Kawasuso et al. (1994) for V_2^{2-} and V_2^-. Surprisingly, no difference in the trapping behavior was found for V_2^0, i.e. the trapping coefficient increased with decreasing temperature in the temperature range of 10 to 300 K in a similar way to the negatively charged divacancies (Fig. 4.6). The neutral charge state was verified in the work of Kawasuso and Okada (1997) by electron paramagnetic resonance, infrared absorption, and Hall effect measurements on the same samples used for the positron investigations. Provided that this determination of the charge state is correct, the conventional Rydberg state model for positron trapping in charged open-volume defects has to be modified. Kawasuso and Okada (1997)

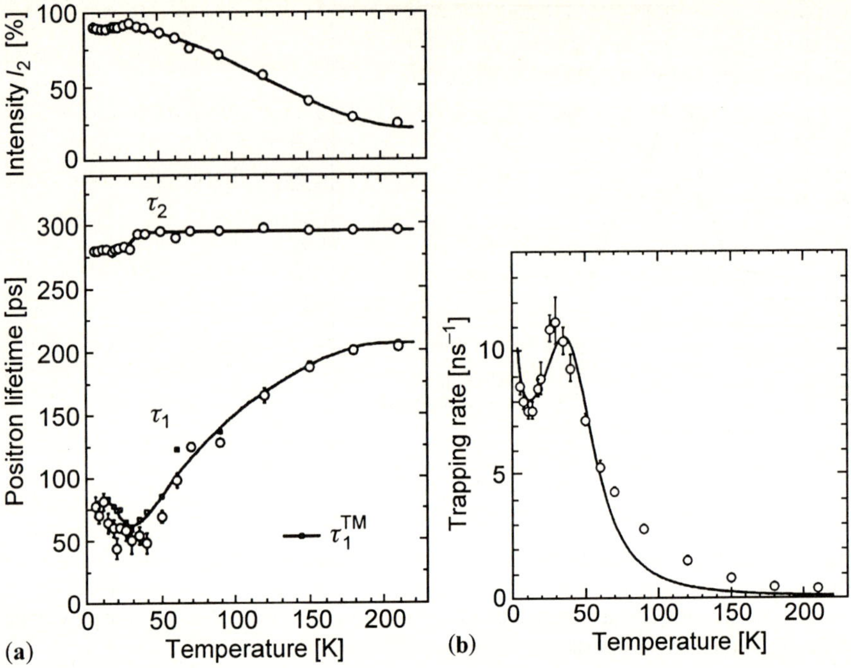

Fig. 4.6. Positron lifetime of electron-irradiated boron-doped (4×10^{14} cm^{-3}) float-zone silicon measured as a function of the sample temperature. The result of the two-component decomposition of the spectra with the lifetime τ_2 and the intensity I_2 attributed to neutral divacancies is shown in (**a**). τ_1^{TM} corresponds to the lifetime τ_1 calculated from the two-component trapping model (3.16). The positron trapping rate of neutral divacancies calculated from these data is presented in (**b**). The solid line corresponds to the fit according to the Breit–Wigner resonance formula (Kawasuso and Okada 1997).

interpreted the temperature dependence of positron trapping in terms of a Breit–Wigner-like resonance phenomenon (Mott and Massey 1965). This leads to the pronounced minimum and maximum of the trapping rate (Fig. 4.6b). In order to confirm such a model of resonance trapping, it is highly desirable to investigate the positron trapping as a function of temperature for other neutral vacancy-like defects.

The first attempt to calculate the influence of varying lattice relaxation on the positron lifetime was made by Mäkinen and Puska (1989). However, the lifetime was only calculated with fixed atom positions. More recent calculation on the vacancy relaxation were carried out by minimizing the total energy of the system, where the atom positions were free parameters (Saito and Oshiyama 1996). Charge-state-specific relaxation effects have not yet been taken into account for silicon.

Investigations using two-dimensional angular correlation of annihilation radiation were performed in order to study vacancy-type defects in irradiated silicon. First applications of the 2D-ACAR technique to semiconductor defects were reported by Tanigawa et al. (1990). A distinct difference in the spectra of 65-MeV proton-irradiated silicon compared with as-grown material was found. The observed effects were due to positron trapping in a variety of defects, which were not specifically identified.

More recent studies by two-dimensional angular correlation of annihilation radiation were carried out in order to study the anisotropy of the electron momentum distribution in differently charged divacancies (Chiba et al. 1995; Hasegawa et al. 1995a, 1995b, 1997). The spectra were obtained in a machine of the Anger camera type at the National Institute for Research in Inorganic Materials Tsukuba. The angular range was 36×36 mrad^2 with a bin size of 0.2×0.2 mrad^2. One spectrum was accumulated in about 4 days.

At first, electron-irradiated float-zone silicon samples were characterized by positron lifetime measurements in order to determine the annihilation fraction of divacancies, η_{2v}. The V_2 component could thus be calculated from the measured ACAR spectra. These spectra were decomposed into a centrally symmetric and an anisotropic part.

The results for neutral divacancies ($\eta_{2v} = 60$ %) are shown in Fig. 4.7. The 2D-ACAR spectra of non-irradiated Si show characteristic anisotropies (Fig. 4.7a, b). A twofold symmetry, which reflects the nature of the Si–Si bonding, is visible. The spectra of the divacancies in the different charge states are practically isotropic and very similar.

The isotropy is not due to a statistical distribution of the orientation of the divacancies, as was confirmed by an experiment with oriented divacancies. The orientation of divacancies was aligned by compressing the irradiated sample in the $[01\bar{1}]$ direction. According to Watkins and Corbett (1965), the orientation of the vacancy–vacancy axis could be achieved in this way.

Such 2D-ACAR experiments are a valuable source of experimental data for the electronic structure of defects in semiconductors. The first theoretical calculations of complete two-dimensional angular correlation spectra have been performed from first principles for the Si bulk, for the monovacancy, and the divacancy (Tang et al. 1997). It is hoped that the comparison of experimental results with such theoretical calculations of the electron momentum distribution will improve knowledge of the electronic structure of defects.

In the course of isochronal annealing of irradiated Si, further clustering of vacancies to quadrivacancies and larger clusters was observed (Dannefaer et al. 1980; Motoko-Kwete et al. 1989). A loose association of vacancies ("sponge defect") as a precursor of larger clusters, which was formed in electron-irradiated float-zone silicon after annealing, was reported to cause positron trapping (Avalos and Dannefaer 1996).

The appearance of vacancy–impurity pairs was discovered rather early by EPR studies of irradiated silicon. VO pairs are called *A* centers, while other vacancy–

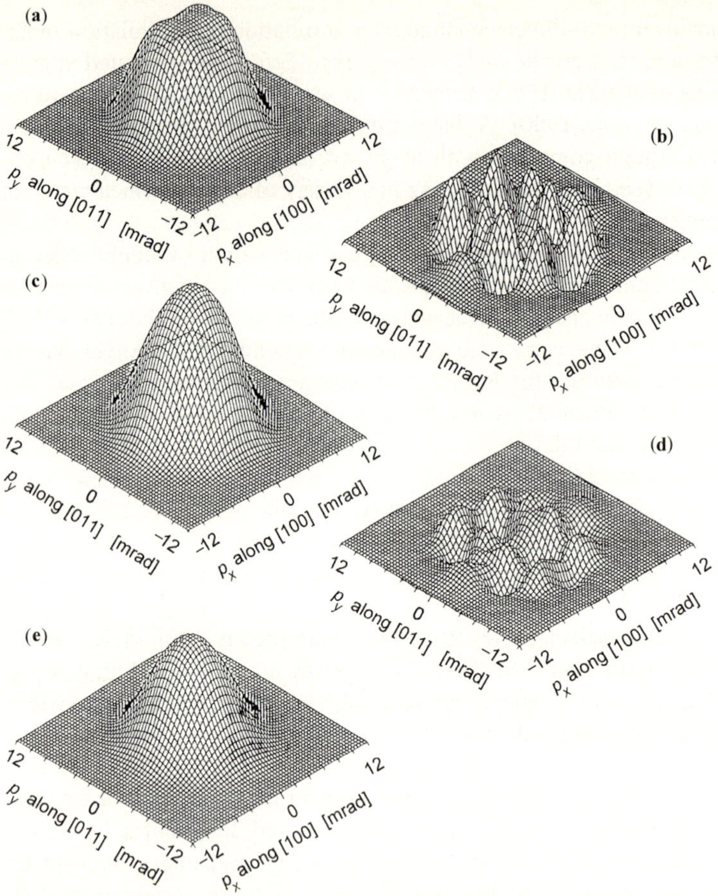

Fig. 4.7. Two-dimensional angular correlation of annihilation radiation spectra of silicon crystals in a $[01\bar{1}]$orientation before and after electron irradiation. (**a**) Unirradiated float-zone silicon, (**b**) the anisotropic part of this spectrum, (**c**) the spectra of the irradiated crystal, (**d**) anisotropic part of the spectrum in (c), and (**e**) its V_2 component (Hasegawa et al. 1995a).

impurity pairs are called *E* centers. Among the latter, the vacancy–dopant pairs, e.g. VP and VB pairs, are of particular importance. A review of ionization levels and annealing temperatures of the *A* and *E* centers studied by electron paramagnetic resonance and deep level transient spectroscopy was given by Bourgoin and Lannoo (1983).

A number of positron studies on *A* and *E* centers exist. The results of the *A* center are discussed in Sect. 4.1.3.1. The *E* center mostly investigated by positron annihilation is the vacancy–phosphorus pair. As early as in the 1970s, Arefiev et al. (1977, 1978) reported on angular correlation studies of *A* and *E* centers in irradiated Si. They concluded that these centers are attractive positron traps. Positron trapping in *E* centers was more conclusively confirmed by Mascher et al. (1987a, 1987b), who performed positron lifetime measurements on heavily phosphorus-doped float-zone and Czochralski-grown silicon. The high-temperature stability reported in this study could not be confirmed by a more recent systematic investigation of the annealing behavior. In accordance with DLTS and EPR measurements, Bourgoin and Lannoo (1983) and Kawasuso et al. (1995c) found the disappearance of VP pairs at around 150 °C.

In addition to the vacancy–phosphorus pair, further *E* centers have been investigated in irradiated Si by means of positron annihilation. Their characteristic annihilation parameters and annealing behavior are dealt with in Sect. 4.1.7.

4.1.2 Vacancies in Thermal Equilibrium

The formation enthalpy of vacancies has been extensively studied in metals, alloys, and intermetallic compounds. For these materials, its determination has become a standard application of the positron annihilation technique (see, e.g., Fluss et al. 1984; Hehenkamp 1985; Kluin and Hehenkamp 1991; Smedskjaer et al. 1985; Würschum et al. 1996). Attempts were made to perform similar experiments in elemental semiconductors. Dannefaer (1986) and Dannefaer et al. (1986) reported on high-temperature positron lifetime measurements in silicon, in order to determine the formation enthalpy of monovacancies. A conventional lifetime spectrometer as described in Sect. 2.2.1 was utilized. The usual foil sources (Sect. 2.1) cannot be used for high-temperature equilibrium measurements due to their insufficient thermal stability. The in-situ positron emitter was thus fabricated by irradiation and nuclear transmutation.

A distinct increase in the average positron lifetime was observed at temperatures exceeding 1320 K. This increase rose clearly above the almost linear slope of $\bar{\tau}$ due to the thermal lattice expansion. The additional increase was attributed to positron trapping in thermally generated vacancies. The vacancy concentration in thermodynamic equilibrium, C_v, is a function of the formation enthalpy H_f,

$$C_v = \exp\frac{S_f}{k_B}\exp\left(-\frac{H_f}{k_B T}\right), \qquad (4.1)$$

where S_f is the vacancy formation entropy. C_v was determined as a function of temperature from the positron lifetime data. H_f could be derived for neutral monovacancies from the Arrhenius plot of (4.1), $H_f = (3.6 \pm 0.2)$ eV. The positron data of Dannefaer (1986) and Dannefaer et al. (1986) were re-analyzed by van Vechten (1987) and a value of $H_f = 2.6$ eV was obtained.

In an earlier study of Tanigawa et al. (1979), the angular correlation line shape parameter H in Ge was measured as a function of temperature up to the melting point. No indication of vacancies in thermodynamic equilibrium was obtained. This was interpreted by the fact that the concentration of thermal vacancies is too low to be detected by positrons, even when the sample temperature is close to the melting point.

Similarly, Seeger and Swanson (1968) deduced theoretically a vacancy density in Si of 1×10^{13} to 9×10^{15} cm^{-3} near the melting temperature. This density is close to the lower sensitivity limit of the positron technique, as can be seen in Fig. 3.36 for the case of positron lifetime measurements.

Subsequent studies on vacancies in thermodynamic equilibrium in Si and Ge by Würschum et al. (1989a) and Throwe et al. (1989) also showed, in contrast to the results of Dannefaer (1986) and Dannefaer et al. (1986), no vacancy effect up to the melting temperatures. Again, the lack of a vacancy signal was attributed to the low level of the vacancy concentration in thermodynamic equilibrium. From the upper-limit estimate of the vacancy concentration at the melting point, a lower limit of the vacancy-formation enthalpy was estimated, $H_f > 3.6$ eV (Michel and Kimerling 1994) in Si. Throwe et al. (1989) supposed that in the experiments of Dannefaer (1986) and Dannefaer et al. (1986) problems with the thermal stability of the positron source, resulting in higher source contributions to the lifetime spectrum, led to an overestimation of the vacancy concentration.

In conclusion, it can be stated that vacancies in thermodynamic equilibrium cannot be detected in Si and Ge by positron techniques since their concentration is too low. Thus, for the vacancy-formation enthalpy in elemental semiconductors only the lower level can be estimated.

4.1.3 Oxygen-Related Defects

Oxygen is the most dominant impurity in silicon. In particular, Czochralski-grown Si contains up to 10^{18} cm^{-3} oxygen. Oxygen atoms are incorporated as electrically inactive interstitials. Various defect complexes containing oxygen, e.g. vacancy–oxygen complexes, thermal donors, and oxygen precipitates, occur during the processing and are of technological relevance (Fahey et al. 1989; Michel and Kimerling 1994; Shimura 1994). The dislocation mobility is reduced and the generation of dislocations during high-temperatures processes is impeded due to dislocation pinning (Yonenaga and Sumino 1985). Moreover, oxygen precipitates are gettering centers for impurities (Jastrzebski 1982). Oxygen-related defects have been investigated by all techniques of positron annihilation in as-grown, irradiated, and thermally treated silicon. The results are partially contradictory due to the lack of a precise defect identification by reference methods parallel to the positron experiment. Moreover, the defect concentrations are often close to the detection limit of the positron techniques. The decomposition of lifetime spectra and the interpretation of the obtained results is rather difficult in these cases.

In Sects. 4.1.3.1 and 4.1.3.2, the positron results on the vacancy–oxygen pair, i.e. the *A* center, and on the larger oxygen-related clusters, such as V_xO_y complexes and oxygen precipitates, are discussed.

4.1.3.1 The Monovacancy–Oxygen Complex

The formation of the vacancy–oxygen pair (*A* center) in silicon after electron irradiation was reported by Watkins and Corbett (1961). It was studied by electron paramagnetic resonance, infrared absorption (Bean and Newman 1971; Corbett et al. 1961), and deep level transient spectroscopy (Kimerling 1977). The detailed model of the *A* center was derived from these measurements. One Si–Si bond is bridged by the interstitial oxygen atom (Fig. 4.8) and thus the open volume is much smaller than that of the isolated monovacancy. There exists a $0/-$ ionization level (compare Fig. 4.2), which is located at $E_{\text{cond}} - 0.17$ eV (Hill 1959; Wertheim 1958).

No theoretical treatment of positron trapping in the *A* center has been known up to now. However, from the model of the *A* center it can be derived that the positron defect-related lifetime should be between the value for the monovacancy and the bulk lifetime due to the very small open volume of the defect. In contrast to the silicon monovacancy, the positron binding energy to the *A* center may be small enough that thermally stimulated positron detrapping may be observed below room temperature. This would lead to the typical behavior of a shallow positron trap already in the neutral charge state. It may, however, be possible that no bound positron states occur in the small open volume. In this case, only a negative charge state could lead to positron trapping (Sect. 3.1.5.2). For the prediction of the trapping behavior, a theoretical calculation of the defect-related lifetime and the positron binding energy is required.

The first positron study on *A* centers was reported by Arefiev et al. (1978). They concluded from their studies by angular correlation of annihilation radiation on electron-irradiated Si that these centers are most probably positron traps. The

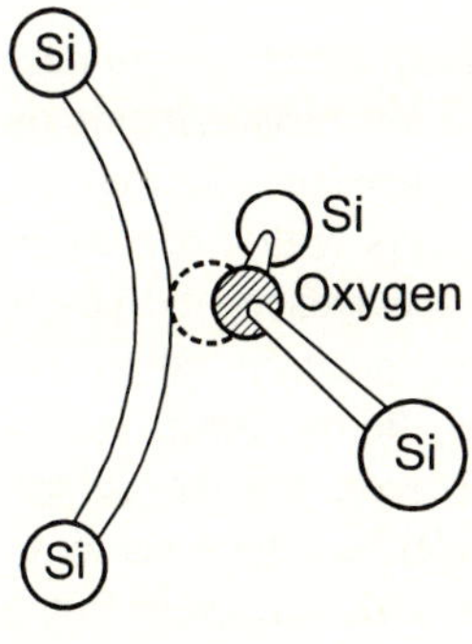

Fig. 4.8. Model of the *A* center (vacancy–oxygen pair) in silicon as derived from measurements by electron paramagnetic resonance (Watkins and Corbett 1961). The vacancy site is indicated by a dashed circle. The open volume of the vacancy is strongly reduced by the interstitial oxygen.

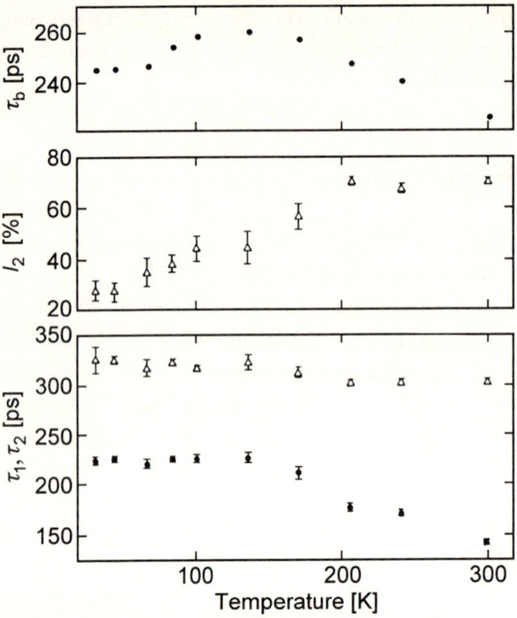

Fig. 4.9. Lifetime components τ_1 (•) and τ_2 (△), the intensity I_2, and the bulk lifetime τ_b, as calculated from the trapping model, versus the sample temperature in electron-irradiated n-type silicon (Mascher et al. 1989b). In addition to the 325-ps component due to trapping in divacancies, a 220-ps component was found in the low-temperature range. This lifetime close to the bulk lifetime was attributed to positron trapping in the *A* center.

observed changes in annihilation line shape parameters were attributed to these defects. However, in this and other early positron studies on oxygen-related defects in Si, no reference method was applied to determine the occurring defect types and their concentrations in an independent way. The defect assignment could not be carried out unambiguously, since a variety of possible positron traps may occur after electron irradiation. One has to distinguish between single and multiple vacancies, vacancy–dopant pairs (*E* centers), *A* centers, and different other vacancy–oxygen complexes.

Mascher et al. (1989b) reported on a systematic study using electron paramagnetic resonance, infrared absorption, and positron annihilation measurements on differently doped Si. The reference methods were applied to determine various defect types occurring after 2-MeV electron irradiation. *A* centers with a density of about 10^{17} cm^{-3} were detected after the irradiation. The occurrence of shallow positron traps could be observed in n-type samples. Complete positron trapping in defects was found at temperatures below 150 K. In this case, the two positron lifetimes of 320 and 225 ps decomposed from the lifetime spectra are the defect-related lifetimes (Fig. 4.9). They were attributed to divacancies and the *A* center in the negative charge state, respectively. The latter defect has a lifetime close to the

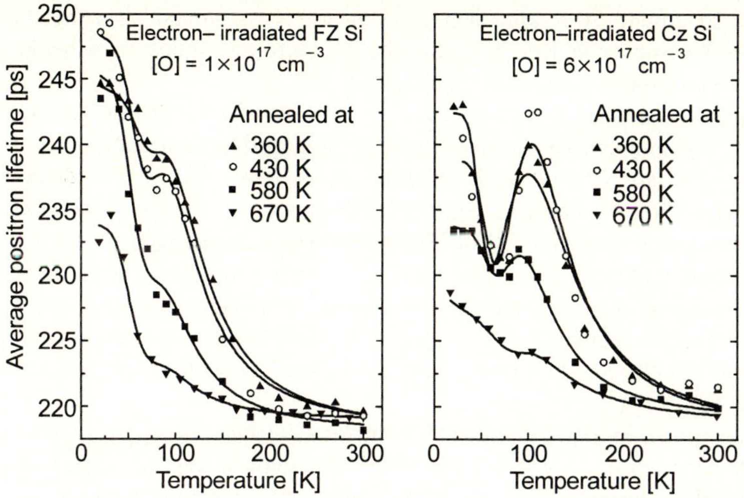

Fig. 4.10. Positron lifetime in float-zone (FZ) and Czochralski-grown (Cz) silicon irradiated with electrons at 4 K to a dose of 1×10^{18} cm^{-2} (Polity et al. 1998a). The average positron lifetime was measured as a function of the sample temperature after several annealing steps. The full lines are fits to the experimental data using a trapping model with a negatively charged divacancy and a shallow positron trap.

bulk lifetime and the temperature dependence of the positron trapping corresponds to the behavior of a shallow trap. At high temperatures, where the thermally induced detrapping from the shallow trap is very efficient, the trapping is controlled by a single defect, viz. the divacancy. Below 150 K, detrapping may be neglected and the *A* center behaves as a deep positron trap. After annealing at 670 K, no strong temperature dependence could be observed. This is likely due to the dissolution of the *A* centers in favor of the formation of larger oxygen clusters. Former annealing studies by Gregory and Barnes (1968) confirmed this interpretation.

No other defects with a structure creating a shallow-trap behavior for positrons could be found in comparable concentrations, so that the assignment of the *A* center to the observed shallow positron trap seems to be reasonable.

According to Fig. 4.2, the *A* center is expected to be neutral in highly resistive or p-doped Si. Although such Si samples were also investigated by Mascher et al. (1989b), no conclusions were drawn on trapping in neutral *A* centers.

The positron trapping behavior in undoped silicon with different oxygen content was studied by Polity et al. (1997a, 1998a). The samples were irradiated with electrons at 4 K to a dose of 1×10^{18} cm^{-2}. Divacancies with a defect-related lifetime of 310 ± 5 ps were observed after annealing at room temperature independent of the oxygen content. A pronounced increase in the average positron lifetime occurs with lowering the sample temperature (Fig. 4.10) in accordance with earlier results. This behavior can be understood by the presence of the divacancies, which are negatively charged. A distinct minimum in the average positron lifetime was found in Cz Si around 70 K after annealing at 360 K. A similar structure, but much weaker, may be recognized in FZ Si. This structure was almost invisible in a sample having an oxygen content of about 10^{15} cm^{-3}.

The temperature dependence of the average positron lifetime in Fig. 4.10 can be explained by competitive positron trapping in negative divacancies and in shallow positron traps (Polity et al. 1997a). The concentration of the shallow positron traps obviously increases with increasing oxygen content. It was concluded that oxygen is a constituent of this defect. The measurement of the temperature-dependent lifetime was carried out after several annealing steps. It becomes visible in Fig. 4.10 that the shallow positron traps anneal around 600 K. The defect was thus related to the *A* center, in accordance with the results of Mascher et al. (1989b) and to annealing studies of the VO center by infrared absorption (see Fig. 4.13).

The shallow positron trap was found by Polity et al. (1998a) in high-resistive Si in contrast to Mascher et al. (1989b), who detected a shallow trap in n-type material. Although the *A* center is expected in the neutral charge state in high-resistive Si, it was concluded that the *A* center is responsible for the occurrence of shallow positron traps (Polity et al. 1998a). The shallow-positron-trap behavior can only be explained by the small open volume of the defect, formed by the displacement of the oxygen (Fig. 4.8). In contrast to negative ions, the shallow-trap behavior is not related to positron capture in Rydberg states caused by a Coulomb potential, but to the smallness of the open volume of the defect. The positron binding energy to the shallow positron trap was determined from the fit to the experimental data (Fig. 4.10) to be 48 ± 10 meV.

In order to confirm the detection of the *A* center, Cz Si was investigated by correlated POLIS and infrared absorption measurements after electron irradiation at 4 and 300 K to different doses (Polity et al. 1998a). The result of the positron measurement (Fig. 4.11) indicates that the concentration of the shallow traps is a function of the dose for 4-K and 300-K irradiation. The effect of the shallow traps is weaker in the sample irradiated at 4 K. This finding is in accordance with infrared spectroscopy (Fig. 4.12). The absorbance at 835 cm^{-1} is a measure of the concentration of neutral *A* centers (Corbett et al. 1961).

Two-dimensional angular correlation of annihilation radiation measurements of the VO complex were reported by Hasegawa et al. (1995a, 1995b). The ACAR spectrum was found to be very isotropic, similar to the case of the divacancies as reported above (Sect. 4.1.1). The distribution of the VO component was similar to that of the V_2^0 component in the momentum region less than 7 mrad. It was enhanced in the higher momentum region, attributed to the broad momentum contribution of the oxygen electrons. However, it has not been clarified in the description of the experimental conditions how the separation of the three components (divacancies, *A* centers, and bulk contribution) was performed. No data have been provided on the shallow trap behavior and its temperature dependence. Further ACAR experiments correlated with detailed annealing studies and in conjunction

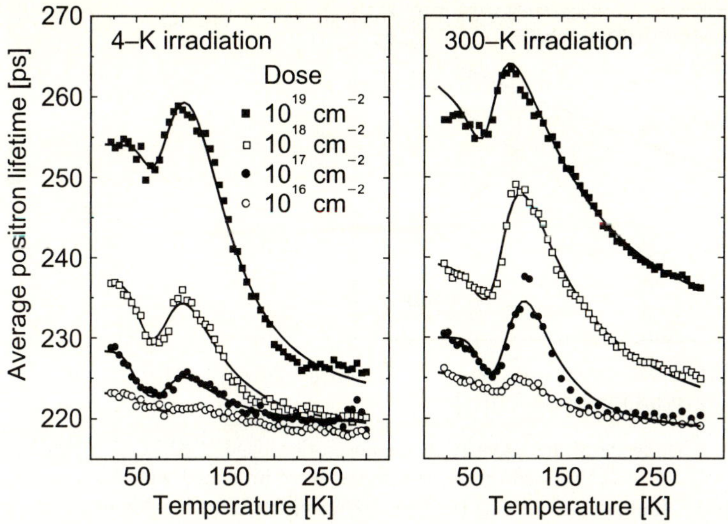

Fig. 4.11. Measurement of the average positron lifetime as a function of the sample temperature in electron-irradiated Czochralski-grown silicon. Samples were irradiated with 2-MeV electrons at 4 and 300 K to different doses (Polity et al. 1998a). The full lines are fits according to the trapping model with one negatively charged open-volume defect and a shallow positron trap.

with theoretical calculations are under progress by the same authors in order to enlighten the electronic structure of the *A* center.

4.1.3.2 Larger Oxygen-Related Clusters

Different V_xO_y complexes are known to be formed after heat treatment of electron-irradiated Czochralski-grown silicon (Lee and Corbett 1976). These vacancy–oxygen clusters are potential positron traps due to their open volume. Part of the sequence of formation and dissolution of various V_xO_y complexes becomes visible in the infrared absorption spectra (Fig. 4.13) measured after several isochronal annealing steps of electron-irradiated Cz Si (Ikari 1995).

A systematic study was presented by Kawasuso et al. (1995b), who performed IR absorption and positron lifetime measurements during isochronal and isothermal annealing between room temperature and 600 °C. They used Sb-doped Czochralski-grown Si with a carrier density of $n \approx 1\times10^{16}$ cm^{-3} after 3-MeV electron irradiation at room temperature to a dose of 10^{18} cm^{-2}. To check the Fermi level position, Hall effect measurements were carried out prior to all positron experiments. The trapping rates of the different defect complexes of the lifetime spectra are shown in Fig. 4.14.

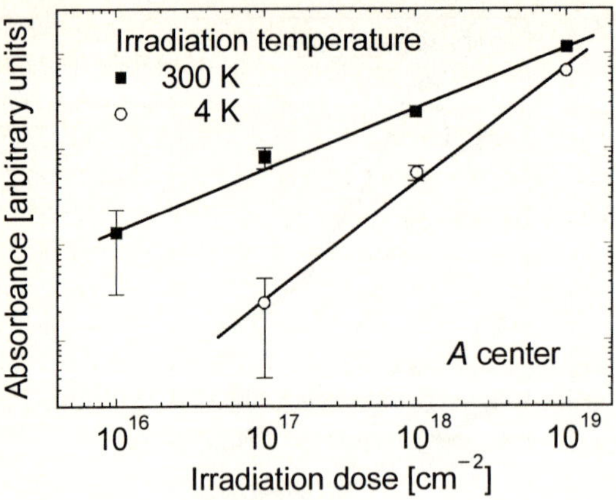

Fig. 4.12. Dose dependence of the intensity of the absorption band at 835 cm^{-1} related to the neutral *A* center. Czochralski-grown silicon samples were irradiated with 2-MeV electrons at 4 and 300 K (Polity et al. 1998a). The lines are linear fits in the log–log plot.

The two lifetime components of 320 and 240 ps observed at first disappear at an annealing temperature of 250 °C. They were attributed to the silicon divacancy V_2 and the vacancy–antimony complex VSb, respectively. The annealing of the latter defect at this temperature had been previously observed by Hall effect measurements (Hirata et al. 1969). The annealing of the divacancy component was related to the vanishing of the infrared absorption band at 2768 cm^{-1}. The V_2O complex is formed during the annealing of the divacancies via the reaction $V_2 + O \rightarrow V_2O$. The shortened lifetime of 295 ps is smaller than that of a divacancy and suggests a decrease in the open volume due to the incorporation of an oxygen atom. The V_2O complex disappears around 400 °C leading to an increase in the trapping rate related to lifetimes of 325 and 240 ps. This is likely due to a further clustering reaction of vacancy–oxygen complexes. Kawasuso et al. (1995b) interpreted these components as V_3O and V_2O_2 complexes, respectively. The further reduction of the defect-related lifetime of 295 ps (V_2O) to the rather short value of 240 ps can be understood in terms of the reduction of the open volume by building in an additional oxygen atom. Similarly, it can be explained that the lifetime of the V_3O complex is close to the silicon divacancy value. After annealing up to 600 °C, no positron trapping could be observed any more.

The activation energies for the annealing of the different vacancy–oxygen complexes were determined by isothermal annealing experiments (Kawasuso et al. 1995b). During the annealing of V_3O and V_2O_2, the formation of V_3O_2 was proposed. The latter complex anneals at 475 °C with an activation energy of 2.56 eV. The annealing of the VO and VO_2 complexes, respectively, was observed by infrared absorption, but no positron trapping in these centers could be detected. This result is not surprising since the positron lifetime experiments were performed at room temperature. As demonstrated by Mascher et al. (1989b) and Polity et al.

(1997a), the VO complex (*A* center) is very likely a shallow trap and thus distinct positron trapping can only be observed at lower temperatures. It could be expected that the VO_2 defect exhibits a similar shallow-trap behavior and could thus not be detected in the work of Kawasuso et al. (1995b).

Another study on oxygen microclusters in Czochralski Si was presented by Uedono et al. (1994c). Annealing experiments up to 1000 °C were carried out in as-grown and 3-MeV electron-irradiated Si material. Positrons were found to be trapped in oxygen microclusters. Positron trapping into complexes containing a monovacancy was observed with increasing annealing temperature. These defects were related to VO_y complexes ($y \geq 2$). It is, however, likely that these complexes with a small open volume exhibit a similar shallow-trap behavior to the VO complex.

The results of Uedono et al. (1994c) could, however, be understood with the presence of V_xO_y defects with $x \geq 2$. The vacancies of these clusters are probably not connected to the appearance of a large open volume, i.e. they do not form the usual voids, since the individual vacancies may be separated by oxygen atoms. Thus, the positron lifetimes of these clusters could be less than the value for sili-

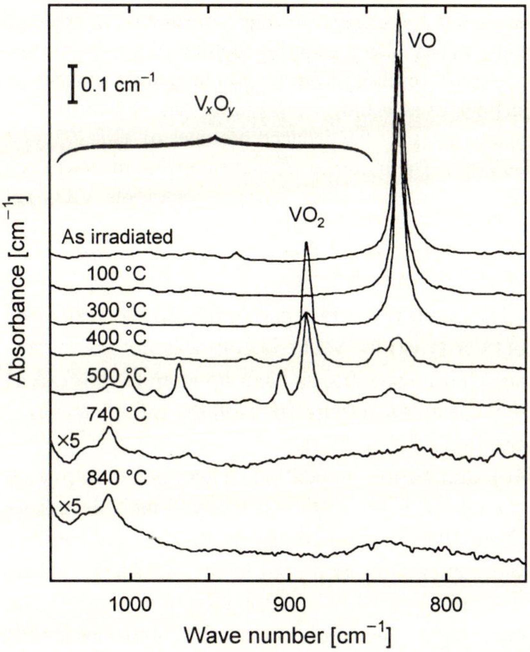

Fig. 4.13. Infrared absorption spectra of electron-irradiated Czochralski-grown silicon after isochronal annealing at the temperatures indicated. The disappearance of the VO center and the formation of VO_2 and higher V_xO_y complexes are monitored by the corresponding absorption bands (Ikari 1995).

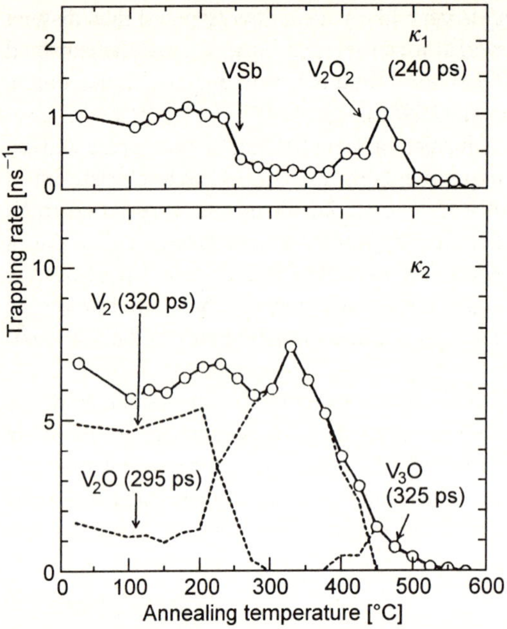

Fig. 4.14. Formation and dissolution of vacancy–oxygen clusters measured by positron lifetime spectroscopy. Positron trapping rates κ_1 for a positron lifetime of 240 ps (*upper figure*) and κ_2 (*solid line in the lower figure*) resulting from a two-component analysis of the lifetime spectra are shown as functions of annealing temperature. The broken lines in the lower figure correspond to the trapping rates for the deduced lifetimes of 320, 295, and 325 ps, respectively (Kawasuso et al. 1995b). The lifetimes were attributed to the vacancy–antimony complex, to divacancies, and to various vacancy–oxygen complexes V_xO_y, respectively.

con divacancies, $\tau_{2v} = 320$ ps, as described in the last paragraph on the results of Kawasuso et al. (1995b). This peculiarity of the V_xO_y defects was not taken into account in the decomposition of the lifetime spectra, as Uedono et al. (1994c) used fixed lifetimes of 270 and 320 ps being characteristic for isolated monovacancies and divacancies in silicon (compare Table 4.2, p. 182).

Ikari et al. (1992, 1994a, 1994b) studied the nucleation of oxygen precipitates in Czochralski Si crystals after quenching from 1390 °C. It was found by correlated infrared absorption and positron lifetime measurements that large vacancy–oxygen clusters are introduced by the quenching and lead to the enhanced formation of oxygen precipitates.

Interstitial oxygen clusters containing no vacancies were first measured with positron lifetime spectroscopy and Doppler-broadening spectroscopy by Dannefaer and Kerr (1986a). A short-lived lifetime component of about 100 ps, which did not appear in float-zone material, was observed in as-grown boron-doped Czochralski silicon wafers. This component was also found after heat treatment up to 650 °C. The authors interpreted the appearance of this lifetime component as positron trapping in interstitial oxygen clusters containing at least two oxygen atoms. These clusters represent silicon oxide precipitates, which are oherent due to the interstitial oxygen position. Trapping in coherent precipitates was reported earlier for metals by Dlubek et al. (1983) and Krause et al. (1986a, 1986b). In this case, positron trapping is not related to open-volume defects, but is due to a different positron affinity in the precipitate in comparison with the matrix (Sect. 3.1.5.5). It can be expected that the defect-related positron lifetime is strongly reduced, since the electron density in the precipitate increases in relation to the defect-free Si bulk due to the presence of oxygen in interstitial positions. The value of 100 ps attributed to the interstitial clusters is smaller than the bulk lifetime in Si, $\tau_b = 218$ ps. The results of the Doppler-broadening experiments of Dannefaer and Kerr (1986a) are in accordance with this interpretation. With the occurrence of these clusters, it was observed that the S parameter can reach values below the bulk parameter. For instance, for a Si sample taken from the center of a Czochralski-grown wafer, a S/S_b ratio of 0.979 was obtained. Usually for open-volume defects, a ratio greater than 1.0 is expected. Thus, it was concluded that the defect-related S parameter is determined by pronounced high-momentum components of a non-open volume defect containing different chemical species, viz. oxygen.

The growth of interstitial oxygen clusters proceeds under incorporation of vacancies between 450 and 650 °C. The dissolution of these clusters was observed after a 10 h annealing at 650 °C, where no positron traps were observed any more.

In the same work of Dannefaer and Kerr (1986a), the density of oxygen-related defects varied with position on the as-grown wafers with a diameter of 10 cm. Vacancy-related defects were predominantly found close to the edge of the wafers. The radial distribution of positron traps have been investigated in further studies. In 6" Si wafers, Puff et al. (1992a) found small interstitial oxygen clusters ($\tau_d = 113$ ps, $I_d = 6$ to 8 %), which showed virtually no distribution across the wafer. Vacancy-type defects were at or just below the detection limit. Sharma et al. (1992) measured lifetime spectra as the function of the radial displacement of the positron source from the center of Czochralski-grown 6" wafers. The decomposition of the spectra provided defect-related lifetimes of about 285 ps in the center and 256 ps at the edge of the wafer. The authors concluded that near the center of the wafer both monovacancies and divacancies were present. The intensity of the vacancy-related lifetime component increased towards the edge from 40 to 70 %. Thus, the vacancy concentration was higher at the edge of the wafer, which is in agreement with the results of Dannefaer and Kerr (1986a). No evidence of a short defect-related lifetime component, attributed to interstitial oxygen clusters, was found.

In a more recent study by Dannefaer and Bretagnon (1995), the association of vacancies with interstitial oxygen clusters was concluded from positron lifetime spectroscopy experiments in as-grown Czochralski silicon wafers. Monovacancies and divacancies supposed to be bound to the oxygen clusters had a nearly constant

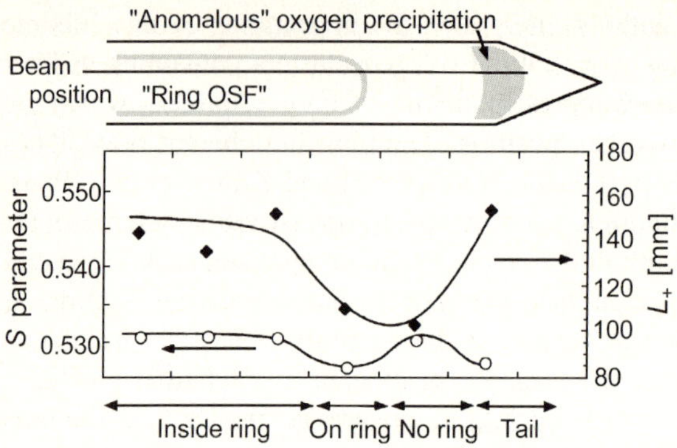

Fig. 4.15. Change in the S parameter in the bulk (*open circles*) and the positron diffusion length L_+ measured by the slow-positron-beam technique along a vertically sliced Czochralski silicon ingot. The U-shaped distribution of the oxygen-induced stacking faults (OSF) is indicated in the upper part (Ikari 1995). The lines are drawn to guide the eye.

total density of 3×10^{16} cm^{-3} across the wafer. Trapped monovacancies are stable up to 1000 °C, but divacancies are not.

Various defects have been found in Czochralski-grown silicon by chemical etching (Yamagishi et al. 1992), copper decoration (Yamauchi et al. 1992), and infrared tomography and transmission electron microscopy (Sadamitsu et al. 1993, 1995). These defects were proposed to be related to dislocation loops or small oxygen precipitates. Furthermore, the formation of oxygen-induced stacking faults (OSF) was reported (e.g. Sadamitsu et al. 1993). The OSF are observed in a ring-like region after thermal oxidation of the wafer.

The result of Doppler-broadening measurements using a slow-positron beam for a vertically sliced ingot containing oxygen-induced stacking faults is shown in Fig. 4.15 (Ikari 1995). A decrease in the S parameter, as well as in the positron diffusion length L_+ was observed in the ring. These effects were attributed to positron trapping in various oxygen clusters formed around the OSF ring. This interpretation was supported by the fact of the reduced positron diffusion length around the ring region compared with other parts of the sample, where a value of about 150 nm was measured. This is significantly smaller than the value for defect-free Si of 245 nm (Mäkinen et al. 1991a) due to the close vicinity of the measuring position to the ring of oxygen-induced stacking faults.

Defect annealing up to 1150 °C in Czochralski Si was studied by diffusion length measurements using a slow-positron beam (Kitano et al. 1992, 1994). They measured in as-grown Cz Si a diffusion length of about 125 nm. The increase in the diffusion length of up to 250 nm with increasing annealing temperatures was attributed to the outdiffusion of oxygen and the reduction of the corresponding concentration of positron traps. Surprisingly, the S parameter remained almost constant during the annealing experiment.

However, the effect of the variation of the surface *S* parameter during annealing due to the proceeding oxidation process and the influence of possible near-surface electrical fields has not been discussed by the authors. In order to support the conclusion of decreasing concentration of oxygen-related defects, a systematic study of these factors is required.

The formation of electrically active centers was observed during the annealing treatment of Czochralski silicon with a density of interstitial oxygen $[O_i] \sim 10^{18}$ cm^{-3} They act as donors and they are called "thermal donors" due to their generation under thermal treatment. They disappear above annealing temperatures of 550 °C (Michel and Kimerling 1994). These defects are discussed to be larger complexes containing oxygen but no vacancies. Due to their positive charge it is not expected that they act as positron traps.

The radial distribution of positron traps across device-quality 8" Czochralski-grown silicon wafers (manufacturer Wacker Chemitronics) was studied by Gebauer et al. (1998a). In contrast to former studies, no variation of the positron lifetime with the position was observed at room temperature, although the oxygen content was 1×10^{18} cm^{-3}. Despite the high statistics of 2×10^7 events, lifetime-spectra decomposition into more than one component was not possible. This lifetime was equivalent to the bulk lifetime of Si over the whole wafer. Obviously, the concentration of oxygen-related open-volume defects was, in these samples, below the sensitivity limit of positron annihilation.

The formation of agglomerates between the dopant and vacancy–oxygen complexes may be possible in highly doped Czochralski-grown Si. Dannefaer et al. (1989b) and Puff et al. (1989) studied by Doppler-broadening and positron lifetime measurements of differently doped as-grown Si the occurrence of positron traps and their thermal annealing behavior. No significant change in the *S* parameter was found for undoped and weakly doped (5×10^{16} cm^{-3} boron) Cz Si for isochronal annealing up to 1100 °C (Fig. 4.16). In contrast, the highly doped material (5×10^{18} cm^{-3} boron) exhibited a strong decrease in the *S* parameter around the annealing temperature of 800 °C. Provided that the observed *S* parameter of the as-grown undoped material corresponds to the Si bulk free of positron traps, the S/S_b ratio at the minimum of the annealing curve can be calculated to be 0.986. This value, which is significantly lower than defect-free Si, originated in high-momentum contributions of impurities. It was concluded that complexes formed during the annealing contain both boron and oxygen.

The results of positron lifetime measurements show that vacancies are generated in the same temperature range of 700 to 1000 °C. A lifetime component of 271 ps was found in addition to a short-lived component of 100 ps. The latter component may be due to small interstitial oxygen clusters, as explained above. The annihilation fraction of these clusters remained roughly constant over the whole temperature range. The lifetime component of 271 ps was close to the value for isolated monovacancies. However, from the fact that the vacancies are generated at rather high temperatures it must be concluded that they are part of a complex. Furthermore, these complexes were only found in highly boron-doped Czo-

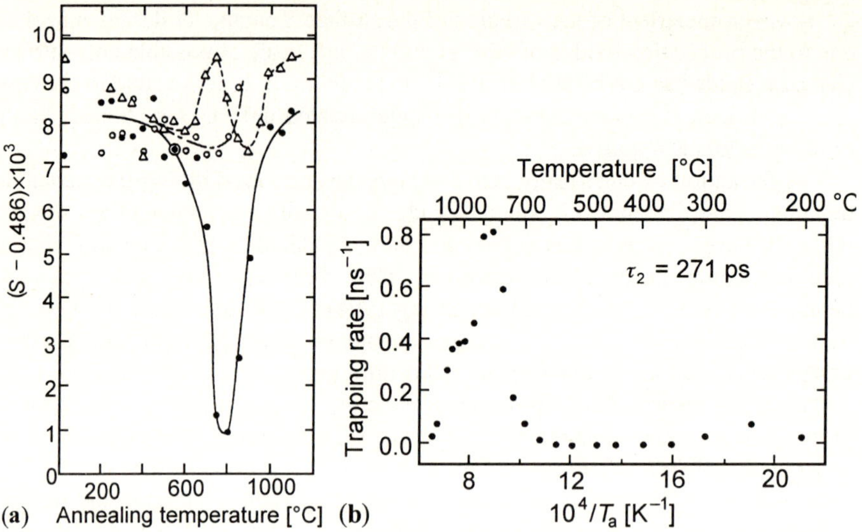

Fig. 4.16. Doppler-broadening S parameter for differently boron-doped (○ 5×10^{16} cm^{-3}, ● 5×10^{18} cm^{-3}) and undoped Czochralski silicon (△) as a function of the annealing temperature T_a. S was arbitrarily scaled. The lines are drawn to guide the eye. The horizontal line in panel (**a**) corresponds to the average value of undoped float-zone Si. The positron trapping rate attributed to a monovacancy-type lifetime component ($\tau_2 = 271$ ps) is shown in panel (**b**) as a function of the reciprocal annealing temperature (Dannefaer et al. 1989b).

chralski silicon and thus it was concluded that the vacancy complexes contain both oxygen and boron, respectively. The preferred model of the authors for the interpretation of these data is the reaction of substitutional boron with interstitial oxygen to form a new complex, $B_{Si} + O_i \leftrightarrows VOB$ (Dannefaer 1990; Dannefaer et al. 1989b). The specific atomic positions of the constituents of the complex formed could not be given.

The observed decrease in the S parameter below the bulk level and the simultaneous occurrence of a vacancy-type defect are not in contradiction. Normally, a distinct increase in the S parameter is expected in open-volume defects. This increase may be overcompensated in the case of the VOB complexes by the high-momentum contributions in the Doppler-broadening spectrum due to boron and oxygen. Thus, a smaller S parameter compared with the bulk may be obtained, despite the presence of a vacancy.

Similar studies on highly phosphorus-doped as-grown Si up to a P content of 1×10^{18} cm^{-3} did not result in any observable vacancy-type defect. This was related to a 30 % lower bond strength of P–O bonds compared with B–O bonds (Dannefaer et al. 1989b).

4.1.4 Epitaxial Silicon Layers

The first report on positron annihilation investigations of structural defects formed in epitaxially grown silicon was given by Druzhkov et al. (1980), but knowledge of the interaction of positrons with semiconductor defects was still poor at that time. The slow-positron-beam technique was applied to the study of defects in intrinsic Si layers grown by molecular beam epitaxy and at the interface with the n-type Si substrate (Tandberg et al. 1989). The *S* parameter curves as a function of the incident positron energy were modeled supposing a high density of interface defects ($\kappa = 10^{13}\ s^{-1}$) and a significantly lower defect concentration in the Si layer. Additionally, non-uniform electric fields located at the interface were necessary to describe the positron diffusion. MBE-grown layers were compared with Si layers grown by solid-phase epitaxy. The *S* parameter was found to be higher in the latter layers and attributed to a higher concentration of open-volume defects (Schut et al. 1991).

A large effort has been made to improve the understanding of low-temperature growth of Si layers. MBE at low temperatures is a way to reduce strain-induced intermixing, e.g. at the interfaces of SiGe superlattices. However, a transition from single crystalline to polycrystalline and finally amorphous growth occurs (Jorke et al. 1989) by lowering the growth temperatures. A high number of point defects, which were studied by slow-positron-beam techniques, occur in an intermediate temperature range. Perovic et al. (1991b) and Simpson et al. (1991a) studied the Doppler-broadening line shape parameters of differently doped Si layers grown below 260 °C.

The *W* parameter plotted versus the incident positron energy shows the usual behavior in the Czochralski-grown (001) Si reference sample (Fig. 4.17). The curve is determined by the superposition of the fractions of positrons annihilating in the thin oxide layer and in the defect-free bulk. The intermediate *W* parameter values are obtained by changing this fraction with increasing positron implantation depth. If the curve is slowly changing, as observed in the reference sample, it can be related to diffusion of positrons from large depths towards the surface. This is connected to a low concentration of lattice defects trapping positrons.

Implantation with 1-MeV Pt ions leads to the amorphization of a near-surface layer. This causes a steep decrease in the *W* parameter as a result of the high defect density and the reduction in the positron diffusion length. The Si epitaxial layer was measured in two states: with a B_2O_3 capping layer and after the removal of the top 1.2 μm layer. The *W* parameter is significantly lower than that after ion implantation. It was concluded that vacancy clusters with a large open volume were formed. Indeed, voids with diameters of 3 to 6 nm were found in transmission electron microscopy (Perovic et al. 1991a, 1991b). The voids are aligned in the [001] growth direction (Fig. 4.18), which was associated with the faceting of the (001) growth surface. The well-defined void formation was described as the precursor state of the breakdown of epitaxy via polycrystalline/amorphous transformation.

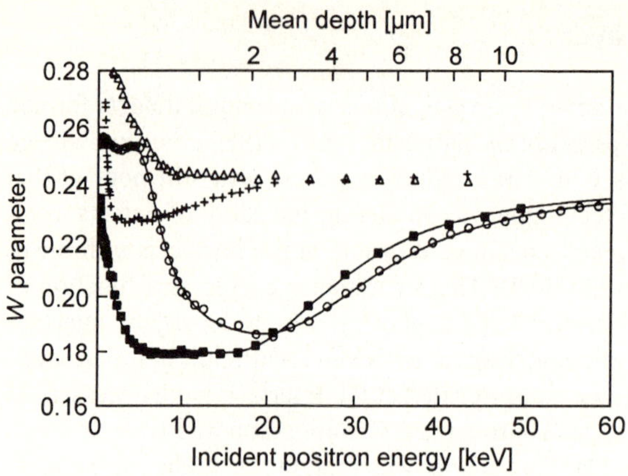

Fig. 4.17. Doppler-broadening *W* parameter measurements obtained in slow-positron-beam measurements of silicon layers grown by molecular beam epitaxy at about 260 °C (Perovic et al. 1991b). The following curves are shown: Czochralski-grown (001) silicon reference sample (△), Czochralski-grown Si wafer after amorphization by 1-MeV platinum ion implantation (+), low-temperature-grown epitaxial Si layer with a boron oxide capping layer (○), and Si epitaxial layer after chemical removal of 1.2 µm (■).

In order to confirm the assumption that the voids observed in TEM are responsible for positron trapping, slow-positron-beam lifetime measurements were carried out (Jackman et al. 1992; Störmer et al. 1995). The results are shown in Fig. 4.19. The void size was found to decrease with increasing growth temperature. The defect-related lifetime was determined as the average value for positron energies between 5 and 12 keV. Lifetimes of 538, 508, 355, and 380 ps were found for growth temperatures of 25, 400, 475, and 525 °C, respectively. No positron trapping occurred in samples grown at 575 and 700 °C. The layer thicknesses varied between 670 and 1130 nm. Consequently, the average lifetime approaches the bulk lifetime with increasing positron implantation energy, because the mean penetration depth is higher than the layer thickness.

The depth distribution of open-volume defects was studied in Si layers grown by molecular beam epitaxy at 300 °C (Szeles et al. 1995). With a combination of the slow-positron-beam technique and well-controlled chemical etching, a depth resolution of 10 nm, which is significantly better than with the positron beam alone, was obtained. The epitaxial layer was found to grow defect-free up to 80 nm thickness. Small vacancy clusters, however, appeared at the interface. The defect density in the layer increased with larger layer thicknesses. This result shows that the nucleation of small open-volume defects may be the precursor state to the breakdown of epitaxy and to the evolution of an amorphous film. The breakdown of epitaxy at low growth temperatures can be avoided by intermediate rapid thermal annealing up to 500 °C during the growth. The layers were shown to be free of positron traps even at growth temperatures as low as 220 °C (Asoka-Kumar et

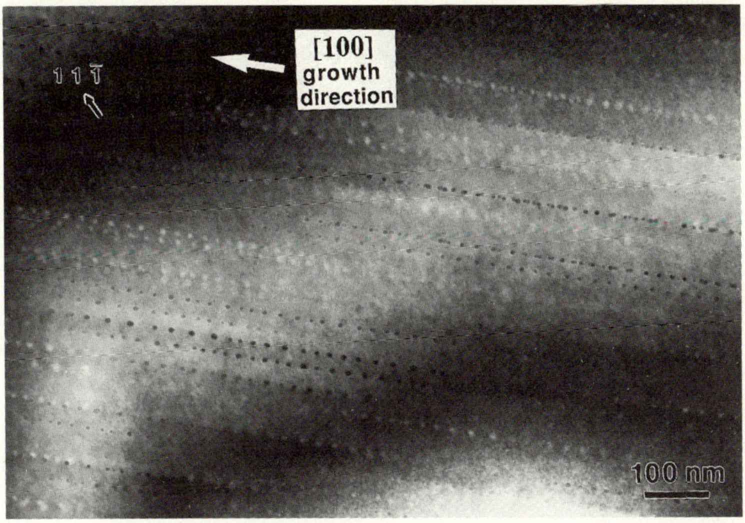

Fig. 4.18. Transmission electron microscopy dark-field image (diffraction vector $11\bar{1}$) of voids in a homoepitaxially grown silicon layer (Perovic et al. 1991a).

al. 1993). If the temperature of the rapid thermal annealing was lowered to 450 °C, vacancy-type defects with a density of 10^{18} cm^{-3} occurred in the layer.

Vacancy-type defects in heavily phosphorus-doped Si epitaxial films up to carrier densities of several 10^{21} cm^{-3} were studied by monoenergetic positron beams (Jia et al. 1993; Uedono et al. 1994e). The growth was carried out in the study of Jia et al. (1993) by plasma chemical vapor deposition at a very low temperature of

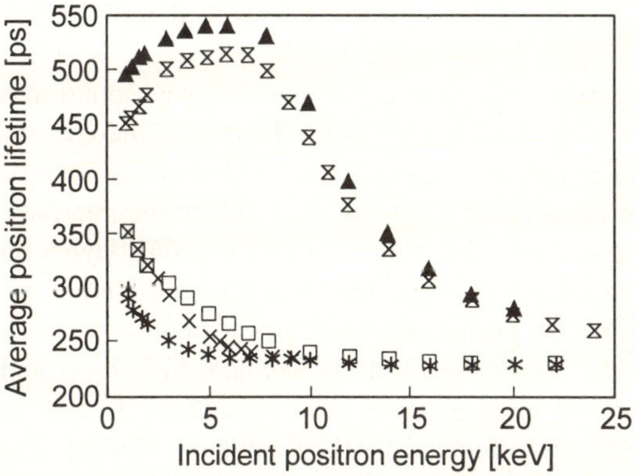

Fig. 4.19. Study of open-volume defects in homoepitaxially grown silicon layers by means of slow-positron-beam lifetime measurements (Störmer et al. 1995). The growth temperatures are: room temperature (▲), 400 °C (⧖), 475 °C (□), 525 °C (×), and 575 °C (∗).

250 °C. An electrical compensation was observed after annealing at 600 °C and related to the formation of a vacancy-type defect complex containing a vacancy and phosphorus atoms. This vacancy defect was detected in the slow-positron-beam experiment.

Divacancy–phosphorus complexes were detected in as-grown Si layers by Uedono et al. (1994e). The electrical compensation was related to these complexes. Vacancy clusters appeared at the interface to the substrate after annealing at 600 °C. They annealed out at 900 °C accompanied by an increase in the concentration of the divacancy–phosphorus complexes.

4.1.5 Silicon Dioxide Layers

The silicon dioxide is of high technological importance due to its application in modern metal-oxide semiconductor (MOS) device technology. The growth by thermal oxidation of very thin layers of the order of 10 nm is the key element for the success of this technology.

The first studies of SiO_2 layers and the interface with the Si substrate were published by Nielsen et al. (1987, 1988, 1989). The results obtained on a device-quality SiO_2 layer thermally grown on a Si layer are shown in Fig. 4.20. The existence of an interface state becomes evident from the shape of the curves. Altogether there are four states contributing to positron trapping: bulk Si, interface SiO_2/Si, the amorphous SiO_2 layer, and the free surface. The annihilation fractions in the former three states are shown in Fig. 4.20b. They were obtained by a numerical fit to the experimental S versus E curve for the sample measured at 470 K according to the procedures described in Sect. 2.4.4. Surprisingly, the S parameter in the interface is obviously lower than that for Si bulk and the SiO_2 layer. It was explained by a chemical decoration of the interface trapping sites by hydrogen. A similar decrease in the S parameter related to specific interface states was found by Asoka-Kumar et al. (1994). This finding was associated with dangling-bond sites and oxygen.

A distinct increase in the annihilation fraction at the interface was found at a higher measurement temperature. It was interpreted as the thermally induced increase in ortho-positronium formation in the interface region. The occurrence of Ps was taken as evidence for large open-volume defects, i.e. voids at the interface. Suzuki et al. (1991, 1992) investigated 47, 210, and 808 nm thick SiO_2 layers by pulsed-beam positron lifetime measurements. A positron lifetime component between 1500 and 2500 ps indicating the presence of ortho-positronium was obtained. At least for the thickest layer, the Ps occurred in the whole SiO_2 layer and not only at the interface. This result is in agreement with the detection of Ps in fused quartz (Dannefaer et al. 1993a) or vitreous SiO_2 (Uedono et al. 1993b). The formation of voids at the interface and in the SiO_2 layer was discussed by Rubloff et al. (1990) in conjunction with the microchemistry of the Si/SiO_2 system and the

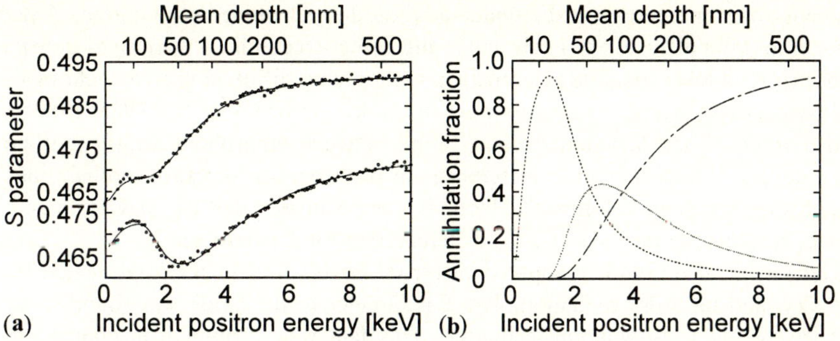

Fig. 4.20. Slow-positron-beam experiment of a 52 nm thick thermally grown silicon dioxide layer on silicon (Nielsen et al. 1989). (**a**) *S* parameter shown versus the incident positron energy measured at room temperature (upper curve) and at 470 K (lower curve). Both curves have been offset for clarity. The solid lines are the fits to the data. (**b**) The fraction of positrons annihilating in the oxide (dashed line), in the interface of the oxide layer to the Si substrate (dotted line), and in the silicon substrate (dashed–dotted line) as obtained from the fit to lower curve in (a) are shown versus the positron energy.

dielectric breakdown due to oxide decomposition. The void generation was seen as an inherent part of the thermal oxidation process.

Positronium formation was studied in an 166 nm thick SiO_2/Si layer by Doppler-broadening spectroscopy at temperatures between 50 and 300 K (Uedono et al. 1995b). Ps annihilating in localized states of voids was detected at high temperatures similar to the results mentioned above. No positronium was detected below 100 K. Competitive positron trapping in shallow traps in the SiO_2 film was supposed to be the reason for this effect.

Oxide layers were also investigated by Smith et al. (1991). The occurrence of interface states was concluded as well. The numerical evaluation may be influenced by an internal electrical field near the interface, which may alter the obtained positron diffusion length.

The passivation and activation of SiO_2/Si(100) interface traps by hydrogen was demonstrated in a comprehensive study (Asoka-Kumar et al. 1991; Lynn and Asoka-Kumar 1992). The reported reduction of the interface-related *S* parameter under hydrogen exposure was interpreted as the passivation of dangling bonds by hydrogen. The activation energy of the thermal dissociation of Si–H bonds was determined from *S* parameter measurements as a function of the annealing temperature to be about 2 eV.

Positron traps in silicon oxide layers were extensively studied by Fujinami and Chilton (1993a, 1995) and Fujinami et al. (1993a, 1993b). The treatment of Si with hydrofluoric acid and the subsequent rinsing in deionized water gave a reduction in the surface *S* parameter of up to 6 % (Fujinami and Chilton 1993a). The SiO_2 layer thickness determined by X-ray photoelectron spectroscopy was lower than 0.8 nm for the largest reduction in the *S* parameter. The *S* variation with the incident positron energy can be understood as the reduction in the surface *S* parameter and the diffusion of positrons in the defect-free bulk. A variety of defects occurring in thicker oxide layers obtained by applying different growth techniques and post-growth sample treatments were discussed (Fujinami et al. 1993a, 1993b). A difference in the *S* parameter was found between amorphous and crystalline oxide layers. It was related to differences in positronium formation probabilities due to differing open volume of defects. It was postulated that the Si–OH content may be responsible for the observed differences in the *S* parameter.

Uedono et al. (1994d) compared positron annihilation in SiO_2 fabricated by wet, dry, and ultra-dry oxidation. The *S* parameter in the oxide was found to be smaller for the wet-oxidation technique. This fact was attributed, in contrast to former interpretations, to the presence of oxygen interstitial clusters in the Si substrate. Wet-grown SiO_2 layers were also investigated by Anwand et al. (1997). The transition region between the oxide and the Si substrate was characterized by Auger electron spectroscopy and slow-positron-beam measurements. It extends beyond a few atomic monolayers and seems to increase with increasing SiO_2 layer thickness.

The two-dimensional angular correlation of annihilation radiation was combined with the slow-positron-beam technique by Peng et al. (1996) to study the defect structure of the SiO_2/Si interface. The formation and trapping of positronium in voids in the oxide and interface region was proven unambiguously. The sharp peak at small angular deviations typical for the occurrence of Ps (compare Fig. 4.36) was found in the ACAR spectra. Another test for the presence of positronium in SiO_2 is the magnetic quenching of ortho-positronium, which was carried out in fine grains of amorphous silica aerogel (Saito et al. 1995).

Positron diffusion in metal-oxide–silicon systems was investigated (Clement et al. 1997; Ohji et al. 1992; Uedono et al. 1992c). Positrons implanted into the Si substrate were accumulated at the SiO_2/Si interface by the application of an appropriately directed external electrical field. The data were analyzed by the one-dimensional diffusion model. Clement et al. (1997) used *S* versus *W* plots to identify the annihilation site of the positrons implanted into the MOS structure. By applying a field of 10^6 V/cm it was demonstrated that the positrons can be transported through the complete 100 nm thick oxide layer, although the defect concentration was high. The diffusion length is only approximately 10 nm in the case of no external field (Asoka-Kumar et al. 1994; Gebauer et al. 1997b).

Silicon-on-insulator (SOI) structures are very promising for devices in ultra-large scale integration. Such structures can be fabricated by separation of implanted oxygen (SIMOX). The implantation is carried out with doses as high as 10^{18} cm^{-2} at elevated sample temperatures. A layer structure of a homogeneous Si layer on top of an amorphous SiO_2 layer is formed during subsequent annealing at 1300 °C. SIMOX structures were investigated by slow-positron-beam Doppler-broadening and lifetime measurements (Kametani et al. 1992; Uedono et al. 1993a, 1994f). The dominant defects in the as-implanted wafer were identified as vacancy clusters with a defect-related lifetime of about 500 ps. A drastic increase in the formation probability of positronium was observed after high-temperature annealing, which was related to structural changes in the oxide layer.

4.1.6 Metal Films on Silicon

Metal films are of interest in Si technology for their use as Ohmic or Schottky contacts. Positron annihilation investigations on Au/Si Schottky barriers were carried out by Corbel et al. (1989). 10 nm gold was evaporated onto phosphorus-doped Si with a carrier density of 7×10^{14} cm^{-3}. The bias was varied from -5.5 to 0.45 V. The depletion region can be probed by slow-positron measurements. Obviously, the positron diffusion is strongly influenced by the external field (Fig. 4.21). The curve for zero bias is the result of the superposition of positron annihilation from four states: the gold surface, the interior of the Au layer, the Au/Si interface, and the Si bulk. A negative bias enhances the back-diffusion of positrons to the Au/Si interface. This becomes evident since all positrons implanted with 10 keV annihilate at the interface but not in the Si bulk. The bias of -4 V corresponds to an average electric field strength of 1.6×10^6 V/m. The positron drift can be over 2 µm during the average positron lifetime of 218 ps and is thus significantly larger than the positron diffusion length of 245 nm in defect-free Si without electric fields (Mäkinen et al. 1991a).

The random-walk model was used by Hübner et al. (1994) for a Monte-Carlo simulation of positron diffusion in systems with complex geometry (Sect. 3.1.5.8). The simulation procedure was extended by the consideration of electric fields (Eichler et al. 1997b). The results of Corbel et al. (1989) were simulated in this way (Fig. 4.21). The overall agreement is satisfactory. The steep increase in the S parameter for the lowest energies can only be reproduced by supposing a very high density of lattice defects in the gold layer.

The mechanism of silicide formation in the Ti/Si and Ni/Si systems was studied by Tanigawa et al. (1992a, 1992b) and Wei and Tanigawa (1994b). A reduction in the positron diffusion length, indicating a high number of positron traps in the silicides, were found. A very short L_{eff} was also found in the Si substrate for Ni/Si, leading to the conclusion that open-volume defects were present. It was concluded that in this case the Si atoms are the moving species. In contrast, open-volume defects were not detected in the substrate of Ti/Si and Ti was supposed to be the diffusing atom.

4.1.7 Annealing of Irradiation-Induced Defects

Defects were mainly found by positron annihilation in bulk silicon after sample treatments such as irradiation or plastic deformation. The identification of basic defect types occurring in silicon after irradiation has already been dealt with in Sects. 4.1.1 to 4.1.3. Only the peculiarities of the annealing of defects introduced

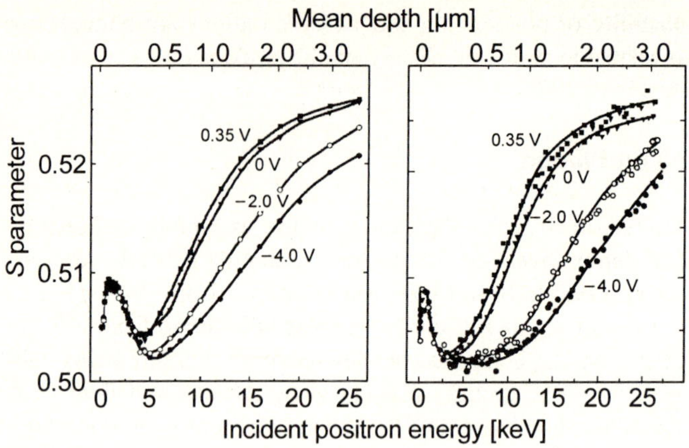

Fig. 4.21. Influence of an external electric field on positron diffusion at a gold–silicon Schottky contact. The measured annihilation line shape parameter S is shown versus the positron implantation energy for various bias voltages in the left panel (Corbel et al. 1989). The measured data (*right panel*) are compared with a Monte-Carlo simulation (*left panel*) of the positron diffusion considering the electric field (Eichler et al. 1997b). The full lines are to guide the eye in both panels.

by irradiation with protons, neutrons, and electrons are discussed in Sects. 4.1.7.1 and 4.1.7.2. The observed defect types depend on type, energy, and dose of the fast particles, as well as on the irradiation temperature. Defect reactions and the annealing behavior of irradiation-induced defects as far as observed by positron annihilation are described.

The radiation damage induced by ion implantation is of high technological importance, because of the wide application of ion implantation for doping in planar technology. Slow-positron-beam experiments on ion-implantation-induced defects are discussed in Sect. 4.1.8.

4.1.7.1 Proton and Neutron Irradiation

The first positron lifetime investigations on proton- and neutron-irradiated silicon were performed by Cheng and Yeh (1973) and Cheng et al. (1973). Dannefaer et al. (1976) studied p-type silicon after room-temperature irradiation with thermal and fast neutrons. Three annealing stages were separated. The stage between 30 and 150 °C was characterized by a defect-related lifetime of $\tau_2 = 325$ ps and a decrease in the corresponding intensity. It was attributed to the divacancy, which is the dominant defect type at room temperature. The stage at 150 to 250 °C was related to the aggregation of divacancies to quadrivacancies with $\tau_2 = 435$ ps. Comparing this lifetime with theoretically calculated values (Fig. 4.4), it seems that larger vacancy clusters were formed. The discrepancy between the vacancy cluster size of quadrivacancies supposed from experimental values and these theoretical lifetimes is also found in more recent papers to be discussed below. The decrease in τ_2 above 260 °C (between 250 and 420 °C) was interpreted as a break-up of vacancy clusters.

Snead and Lynn (1985) investigated Si after irradiation at 100 °C with a high neutron dose (2×10^{20} cm^2). They interpreted their positron results during isochronal annealing as a complex microstructure changing from the dominance of divacancies and quadrivacancies to either rod-like defects (Bartsch et al. 1984) or swirls (Föll et al. 1977). In agreement with earlier investigations, the divacancy was also identified to be the dominant defect formed after irradiation with neutrons of an energy of 14 MeV at temperatures between room temperature and 550 °C (Oshima et al. 1989). They found, in both float-zone and Czochralski silicon, that fast neutrons produce at least three times more lattice defects than fission neutrons. However, the annealing behavior detected by positrons was very similar. Transmission electron microscopy showed that vacancy-type defect clusters with a size of several nanometers were formed at irradiation temperatures below 200 °C. Dislocation loops and other microdefects were observed above the irradiation temperature of 450 °C. No defect clusters occurred in FZ samples irradiated at 550 °C.

Huang et al. (1992) detected a main annealing stage at 600 °C in Czochralski-grown Si single crystals irradiated with neutrons to a dose of 10^{18} cm^{-2}. After low-temperature irradiation, the dominant defects were attributed to divacancies. Positron trapping in monovacancies ($\tau_d = 270$ ps) and quadrivacancies was supposed to occur during further annealing. Only vacancy clusters (interpreted as V_4) with a specific lifetime of 392 to 460 ps and with an intensity of about 4 % were found above 650 °C. Surprisingly, these clusters were stable up to 1150 °C. The same main annealing stage also occurred in FZ silicon as well in the positron lifetime and in the Doppler-broadening S parameter (Fig. 4.22).

The clustering to stable vacancy agglomerates, regarded as V_4, during isothermal annealing after neutron irradiation of Cz Si was also reported (Meng 1994a; Meng et al. 1995a; Zhu et al. 1995). However, the vacancy clusters disappeared upon annealing at 600 °C. The occurrence of different types of V_xO_y complexes during the annealing was discussed in these papers. The clustering was monitored by perturbed angular correlation as a reference method (Zhu et al. 1995). In contrast, Meng and Puff (1994) did not find the long lifetime components attributed to vacancy clusters in silicon irradiated by different neutron doses.

The interaction of neutron-irradiation induced vacancy-type defects with hydrogen in FZ Si grown in a hydrogen atmosphere was discussed. A different annealing behavior after neutron irradiation due to the additional nucleation of hydrogen-related defects was seen (Meng 1994b, 1995; Meng et al. 1995b). The annealing stages found in the positron lifetime were compared with the annealing of different H-related absorption bands.

Two annealing stages were found after proton irradiation with 12 MeV at low temperature (15 K), the first starting at 100 K, the second starting at 400 K (Cho et

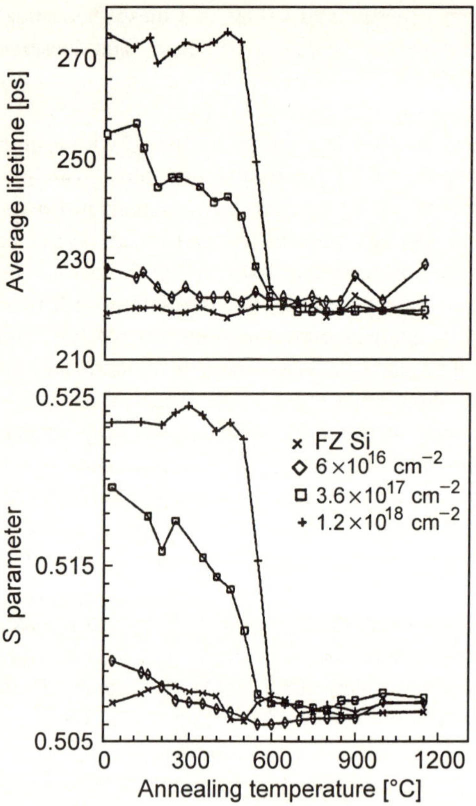

Fig. 4.22. Annealing experiment of neutron-transmutation-doped float-zone (FZ) silicon with different neutron doses (Meng and Puff 1994). The average positron lifetime and the Doppler-broadening *S* parameter are plotted versus the annealing temperature. Undoped float-zone (FZ) silicon was measured for reference.

al. 1992; Mäkinen et al. 1990b). The first stage was attributed to the isolated negatively charged monovacancy. The annealing stage at 400 K was connected to the annealing of a vacancy–impurity complex, for instance negative vacancy–oxygen pairs (Mäkinen et al. 1990b, 1991b). The measurement of the temperature dependence of the positron lifetime showed that the investigated defects were negatively charged.

The interaction of hydrogen with irradiation-induced defects was studied by positron lifetime spectroscopy after 6.95-MeV proton implantation (Mäkinen et al. 1991b). The Si–H bonds in hydrogen-containing monovacancies are suggested to break up above 550 K and also isolated divacancies are assumed to anneal out above 550 K. The annealing behavior of defects induced by hydrogen implantation was compared with that after He irradiation at low temperatures. In contrast, no annealing stage was found below 300 K, interpreted by the stabilization of monovacancies with He. Helium is suggested to be released above 600 K leading to the observed main annealing stage. The same findings after He irradiation were also reported by Motoko-Kwete et al. (1990).

The dominant defect after room-temperature proton irradiation is the divacancy (Dannefaer et al. 1993b; Uedono et al. 1994a). The annealing of the divacancy starts at 200 °C (Girka et al. 1992; Uedono et al. 1994a), and vacancy–oxygen complexes and clusters are formed up to 1000 °C (Dannefaer et al. 1993b). The annealing depends on the concentration of oxygen and dopants. The annealing of divacancies in undoped FZ Si with low oxygen concentration is completed at 700 °C (Dannefaer et al. 1993b). The vacancies in doped Si can be thermally stabilized by the formation of *E* centers (Sect. 4.1.1).

A special feature appeared after irradiation of elemental and compound semiconductors with supercurrent pulsed proton and electron beams (Pogrebnyak et al. 1982, 1984b; Pogrebnyak 1984). The measurement of the angular correlation of annihilation radiation showed a narrow peak at low angular deviations, typical of the presence of ortho-positronium. The existence of positronium was proven furthermore by magnetic quenching experiments. This unusual behavior, which has not been found in other irradiation experiments of semiconductors, may be explained by the introduction of large vacancy clusters. This is one of the rare cases, where Ps was observed in semiconductors. No evidence for positronium was found in plastically deformed semiconductors, where similar vacancy clusters were also detected (see Sect. 4.1.9).

4.1.7.2 Electron Irradiation

The usual method of introducing simple intrinsic defects in a controlled manner is irradiation with electrons at low sample temperatures. High-energy electrons lose their energy by inelastic interaction with bound electrons (ionization) and by elastic collisions with the nuclei (displacement). Electron irradiation creates mainly Frenkel defects. The displacement threshold energy was found to be between 4 and 30 eV for various semiconductors (Corbett and Bourgoin 1975). The transferred maximum energy is small because of the small electron mass. A high electron energy of at least several hundred keV is needed. Electron energies of 1 to 2 MeV are broadly used for irradiation experiments to have total defect introduction rates of the order of 1 cm^{-1}. It is generally accepted that at these energies primarily Frenkel pairs are formed.

Electron irradiation experiments had already been used for defect studies in semiconductors as early as in the 1960s (see e.g. the review of Corbett and Bourgoin 1975). In particular, electron paramagnetic resonance was successfully applied to characterize defects, such as monovacancies, divacancies, and the *A* center in silicon (Sects. 4.1.1 and 4.1.3). Arefiev et al. (1977, 1978) were the first to study electron-irradiated Si by means of positron annihilation. Positron trapping in *A* and *E* centers, as well as in divacancies was discussed.

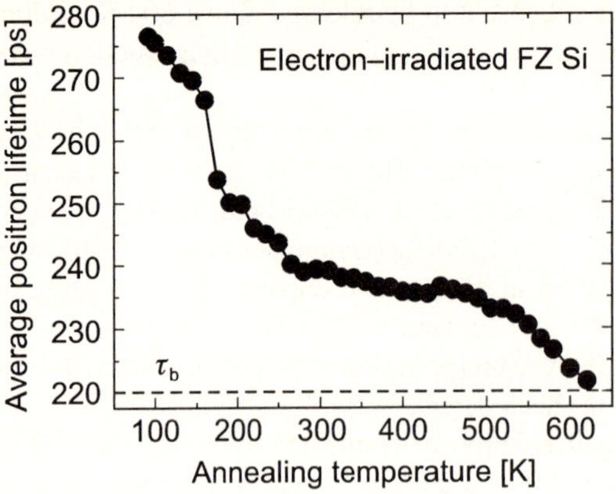

Fig. 4.23. Change in the average positron lifetime with annealing temperature in electron-irradiated undoped float-zone (FZ) silicon. The irradiation was carried out with 2-MeV electrons at 4 K (1×10^{18} cm^{-2} dose). The positron measurement was performed at 90 K (Polity et al. 1998a). τ_b is the bulk lifetime.

Fuhs et al. (1978) observed with positron lifetime spectroscopy the clustering of vacancy-type defects during isochronal annealing in the range from 12 to 600 K after 1-MeV electron irradiation of Si. They found defect-related lifetimes of 266, 318, and 420 ps during the annealing. The former two lifetimes were attributed to monovacancies and divacancies, respectively. The latter value was tentatively assigned to quadrivacancies, similar to the early studies of proton- and neutron-irradiated Si (Sect. 4.1.7.1). In the light of more recent theoretical calculations (Fig. 4.4), the attribution to monovacancies and divacancies is correct. However, the theoretically calculated lifetime for quadrivacancies is significantly smaller, so it is likely that larger clusters were formed during the annealing experiment. A similar lifetime of 420 to 450 ps, which was attributed in comparison with the theoretical calculation of Puska and Corbel (1988) to larger vacancy clusters, was reported by Motoko-Kwete et al. (1989).

Two annealing stages were found at 150 °C and at 230 °C in Cz silicon after 2-MeV electron irradiation at room temperature (Mascher et al. 1989b). They were attributed to the annealing of divacancies in the neutral and in the negative charge state, respectively. Kulkarni et al. (1992) studied the recovery after electron irradiation using rapid thermal annealing. The sample was kept for 30 s at the annealing temperature. A significant reduction in the defect concentration determined by positron lifetime spectroscopy sets in for temperatures higher than 700 °C.

Polity et al. (1997a, 1998a) investigated silicon of different oxygen contents and doping levels after electron irradiation to different doses at 4 K and 300 K. Fig. 4.23 shows the annealing curve of undoped float-zone Si. An average lifetime of 277 ps was measured at 90 K. A defect-related lifetime of 282 ps mainly due to

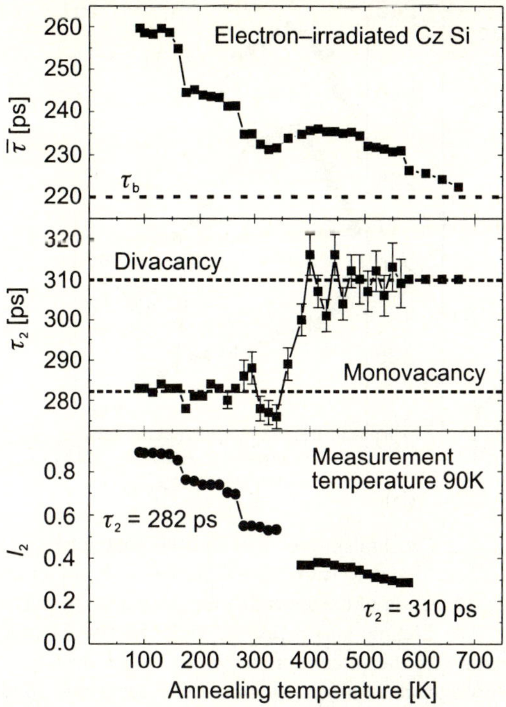

Fig. 4.24. Result of positron lifetime measurements in electron-irradiated Czochralski-grown (Cz) silicon (Polity et al. 1998a). The average positron lifetime $\bar{\tau}$ and the defect-related lifetime τ_2 together with its intensity I_2 are shown. The experimental conditions are the same as in Fig. 4.23. τ_b is the bulk lifetime.

monovacancies was separated. Using a trapping coefficient of $1\times10^{16}\ s^{-1}$ as an estimate for negative vacancies at 90 K, a vacancy density of $3\times10^{17}\ cm^{-3}$ may be obtained for the sample irradiated to a dose of $10^{18}\ cm^{-2}$.

Two steps, at 170 K and between 450 and 620 K, could be observed during the annealing of electron-irradiated undoped float-zone Si in accordance with former measurements of electron paramagnetic resonance (Watkins 1964) and deep level transient spectroscopy (Kimerling 1977). The first annealing stage was attributed to the annealing of negatively charged monovacancies and the second one to divacancies. The same experiment was repeated for undoped Czochralski-grown silicon. In addition to the annealing stages found in FZ Si, a further stage appears at 270 K (Fig. 4.24). This must be due to a defect with an open volume comparable to a monovacancy, as can be concluded from the defect-related lifetime τ_2. It is likely that oxygen is a constituent of the responsible defect. A divacancy signal is observed at annealing temperatures above 400 K.

The influence of doping on the annealing of irradiation-induced defects was also studied (Polity et al. 1997a, 1998a). The results obtained in phosphorus-doped

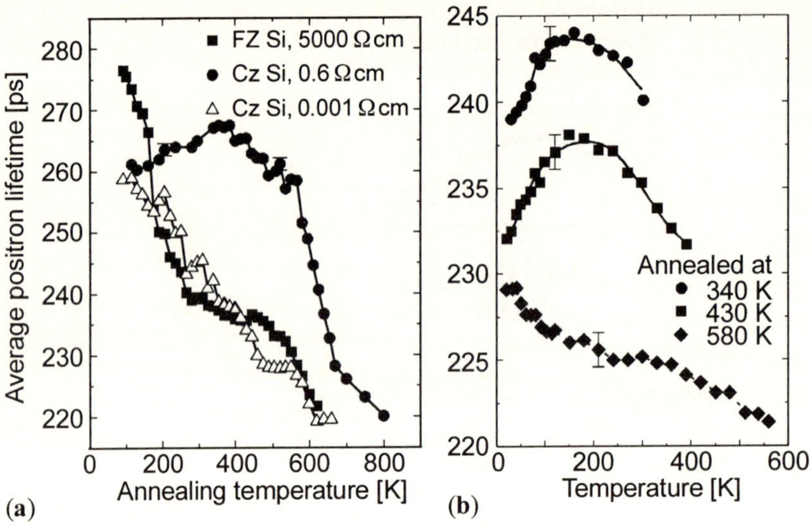

Fig. 4.25. Annealing of phosphorus-doped Czochralski-grown (Cz) and float-zone (FZ) silicon after irradiation to a dose of 10^{18} cm^{-2} at 4 K with 2-MeV electrons (Polity et al. 1998a). (**a**) Average positron lifetime as a function of the annealing temperature. The curve of Fig. 4.23 is included for reference. The lifetime was measured at 90 K. (**b**) Average positron lifetime versus sample temperature for the silicon sample with a carrier density of $n = 8 \times 10^{19}$ cm^{-3} (0.001 Ω cm resistivity) after three annealing steps. The upper two solid lines correspond to a fit according to the trapping model with a negative vacancy and a shallow positron trap.

Si are shown in Fig. 4.25. The average positron lifetime measured as a function of the sample temperature (right panel) is plotted for a highly n-doped Si sample after three annealing steps. The sample remained n-type after irradiation. The increase in $\overline{\tau}$ with lowering sample temperatures was attributed to the trapping in negatively charged phosphorus-related *E* centers (VP). The course of the lifetime at low temperatures after the first two annealing steps indicates the presence of a shallow positron trap. It is attributed to the *A* center. The shallow trap is missing after the annealing step at 580 K, in accordance to the annealing of the *A* center discussed earlier.

In contrast to the temperature dependence of $\overline{\tau}$ in undoped Cz Si, no increase occurs below 70 K (compare Fig. 4.24 and Fig. 4.25b). This difference may be understood by the negative charge of the *A* center in n-type Si. This leads to an additional temperature dependence of the trapping rate of the shallow traps, causing a higher sensitivity of positrons at very low temperatures. The defect-related lifetime is close to the bulk value, and thus the reduction in $\overline{\tau}$ is observed.

The reduction in $\overline{\tau}$, e.g. at a sample temperature of 200 K (Fig. 4.25b), is an indication of the annealing of the *E* center. This is more obvious in the annealing curve of Fig. 4.25a. In addition to the annealing stages of the undoped sample, the

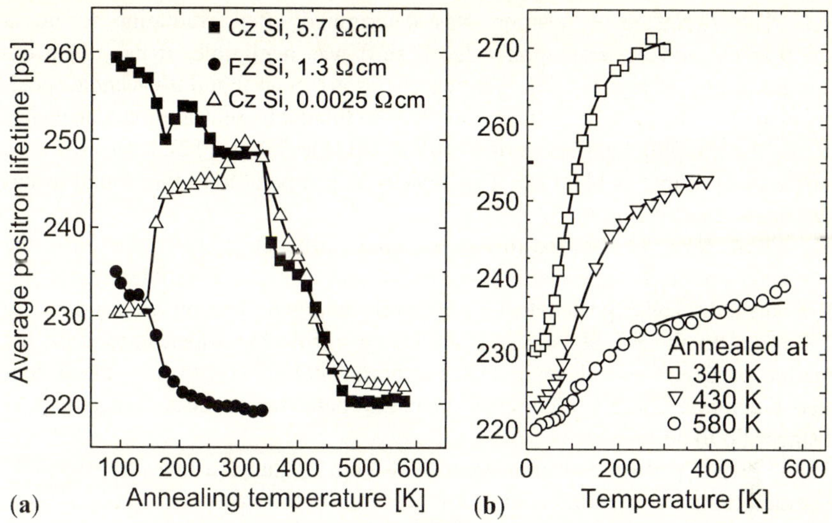

Fig. 4.26. Positron lifetime measurements in electron-irradiated (2 MeV, 4 K, 1×10^{18} cm^{-2}) boron-doped Czochralski-grown (Cz) and float-zone (FZ) silicon (Polity et al. 1997a, 1998a). (**a**) Average positron lifetime measured at 90 K as a function of the annealing temperature, (**b**) measurement of the temperature-dependent average positron lifetime in the highly doped Cz Si sample (0.0025 Ω cm) after three different annealing steps. The lines in (b) are the result of a fit with a neutral open-volume defect and a shallow positron trap.

stage at 400 K appears. It is attributed in accordance with Kimerling et al. (1975) to the annealing of VP complexes. Surprisingly, the annealing curves for the more weakly doped sample are different (Fig. 4.25a). A more stable monovacancy-type defect ($\tau_d = 280$ ps) is observed with an annealing stage at 600 K. It was attributed by Polity et al. (1998a) to more complex vacancy–oxygen clusters. Indeed, the annealing of V_2O was reported to appear in this temperature range (Kawasuso et al. 1995b; Lee and Corbett 1976). The annealing step due to such complexes is missing in highly phosphorus-doped Si, because the phosphorus content is at least two orders of magnitude higher and *E* centers are predominantly formed.

The result of positron lifetime measurements in electron-irradiated boron-doped Si is shown in Fig. 4.26. The measurement of the average lifetime as a function of the sample temperature (Fig. 4.26b) revealed distinct differences from n-type samples (Fig. 4.25b). The influence of shallow positron traps is much stronger, indicating a higher concentration. The responsible defects have an effect on the average positron lifetime up to the highest measurement temperatures applied and do not anneal at least up to 580 K. Therefore, the *A* center can be excluded as the responsible defect. The most likely candidate for the shallow positron trap is the boron acceptor itself.

The B-doped float-zone Si sample (carrier density $p = 8\times10^{15}$ cm^{-3}) shows only a slight increase in $\bar{\tau}$ after irradiation. One annealing stage appears at 170 K (Fig.

4.26a). The Fermi level position was determined after irradiation to be at E_{val} + 0.26 eV and its irradiation-induced shift was negligible. It has to be expected according to Fig. 4.2 that monovacancies are neutral and divacancies positively charged. The annealing stage may thus be related to neutral monovacancies. By using the trapping coefficient of 5×10^{14} s^{-1} (Table 3.4, p. 112), a vacancy density of 1.5×10^{17} cm^{-3} is obtained. This density is comparable to that found in undoped float-zone Si (Fig. 4.23).

The Fermi level was shifted to E_{val} + 0.63 eV after electron irradiation in the Czochralski-grown Si sample with $p = 2.5\times10^{15}$ cm^{-3} (Fig. 4.26a). Monovacancies and divacancies are expected to be negatively charged. The annealing stage at 170 K found in the FZ Si sample is allusively visible but superimposed on the formation of defect complexes containing boron and/or oxygen, e.g. the boron-related E centers. There is a complex annealing behavior of these defects in several stages up to 500 K.

The highly doped Cz Si sample ($p = 4\times10^{19}$ cm^{-3}) shows a low increase in $\bar{\tau}$ after irradiation. The measurement of the positron lifetime was carried out at 90 K and a distinct decrease in the average lifetime has to be expected due to the presence of shallow positron traps (Fig. 4.26b). Neither monovacancies nor divacancies should be visible for positrons due to their positive charge state during the whole annealing experiment. Thus, only vacancy-type defect complexes being negative or neutral in this degenerated p-type sample can be observed. Obviously, their number increases at 170 K when monovacancies become mobile. The annealing of the defect complexes has a similar course, as found for the weakly doped Cz sample.

4.1.8 Ion Implantation

Ion implantation is a well-established doping technology for the fabrication of planar devices. As a disadvantage, bombardment with high-energy ions leads to the formation of radiation damage, counteracting the desired doping effect. It is therefore of fundamental interest to understand the formation of defects, their interactions, and their annealing behavior. Since the development of the implantation technique, a great number of investigations of the induced defects have been carried out. However, the matter is very complex and many parameters, e.g. the implantation energy, dose, target temperature, atomic masses, impurities, exist. Therefore, several problems of ion implantation in semiconductors are still unsolved.

A widely used method for the characterization of layers after ion implantation is Rutherford backscattering, which is especially sensitive to interstitials and lattice distortions. The sensitivity to vacancy-type defects is, however, rather poor. Positron annihilation thus represents a suitable complementary method. The slow-positron beam allows the investigation of open-volume defects induced by ion implantation in the depth range of interest. The study of implanted layers has become one of the standard applications of the slow-positron-beam technique. Extensive studies were reported after implantation with a broad range of ions, starting from H^+ to heavy ions. The main emphasis is put in this section on studies with ions used for the doping of silicon and the ion energy range of interest for technological applications.

The application of the slow-positron-beam technique to defect issues in semiconductors began in the mid-1980s. One of the first studies of implantation-induced defects with a variable-energy positron beam was performed by Hautojärvi et al. (1986a). Implantation with 50 keV BF_2 was carried out, and vacancy-type defects were found up to 100 nm with a mean depth close to the maximum of the profile of implanted atoms.

A high number of positron annihilation studies of ion implantation in Si have been published since that time. The basic scheme is the measurement of the variation of the S parameter with incident positron energy for different implants, ion energies and doses. The main aim is to derive the depth profile of vacancy-type defects in comparison with the implantation profile. The depth distribution of the implanted species and the profile of the generated defects can be simulated in a Monte-Carlo scheme. The simulation is implemented in the TRIM[1] computer program (Ziegler et al. 1985). Usually, only the $S(E)$ curves are given as the result of the positron studies. The reason is that the real defect profile can hardly be separated from the measurement due to the broad depth distribution of the incident monoenergetic positrons, which is often approximated by a Makhov profile (2.15). It is, rather, necessary to assume a certain defect profile, e.g. a block profile or a Gaussian distribution, in order to fit the measured $S(E)$ plots and thus to determine the damage depth and the defect concentration.

Many investigations were carried out in the positron laboratory of Tsukuba University. The main emphasis was put on the effect of different implants with the variation of dose and energy (Lee et al. 1988b; Sugiura et al. 1992; Uedono et al. 1992d). The role of oxygen in the formation of defects during ion implantation was studied (Uedono et al. 1992a). Several types of vacancy–oxygen complexes were found. They show a complex annealing behavior with various defect reactions, similar to the findings in electron-irradiated Cz Si (Sect. 4.1.3). The influence of an SiO_2 cap layer on defect generation and distribution was also investigated. In addition to defects in the Si substrate, a large concentration of vacancy-type defects near the interface was derived from the measured $S(E)$ curves (Uedono et al. 1988). Amorphization and structural relaxation in Si implanted with Si^+, Ge^+, or P^+ were studied using slow positrons, electron paramagnetic resonance, and Raman spectroscopy (Hiroyama et al. 1993; Wei and Tanigawa 1994a). The ion-beam-induced recrystallization of Si layers pre-amorphized by Ge implantation was investigated by the combination of Rutherford backscattering and positron annihilation (Hayashi et al. 1993a, 1993b). The amorphization process led to a drastic increase in the positron lifetime. The maximum value of 400 ps

[1] Transport of ions in matter.

in the recrystallized silicon layer was attributed to the occurrence of small vacancy clusters.

Implantation with BF_2 is an important technique for the fabrication of shallow p-type Si layers. The behavior of boron and fluorine ions in Si and their interaction with intrinsic point defects is little known. A slow-positron-beam study of the implantation with B, F, and BF_2 was carried out by Uedono et al. (1997b, 1997c). The defects in the as-implanted samples below the amorphous region were identified as divacancies. The epitaxial regrowth of the amorphous region started upon rapid thermal annealing above 700 °C. Vacancy–fluorine complexes were introduced during this process. The mean size of the open volume of these complexes was estimated to be close to the size of monovacancies. The subsurface vacancy profile after 120-keV fluorine implantation in Czochralski-grown Si was studied by Fujinami and Chilton (1993b). It was found that the vacancy region reaches at least 1 μm deeper than expected from a TRIM simulation. In order to increase the depth resolution of the slow-positron-beam measurement, the sample was stepwise etched up to 1100 nm.

The characterization of the defect profile by slow-positron measurements (Jackman et al. 1989) after self-implantation was carried out in correlation with Rutherford backscattering, infrared absorption, and other techniques. The implantation with Si^+ was compared with He^+, and the formation and annealing of divacancies was studied (Simpson et al. 1991b). Furthermore, in addition to the divacancy signal, vacancy–impurity complexes were observed during the study of the dependence of the defect generation on the implantation dose, energy, and temperature (Goldberg et al. 1995). In the low-dose region, the concentration of divacancies increases linearly with the Si^+ dose (Nielsen et al. 1993). The clustering of the divacancies was observed in the range 200 to 300 °C (Fujinami et al. 1996; Nielsen et al. 1993). The clusters were removed during further heating up to 675 °C. The samples with low oxygen content returned to the state before irradiation, whereas oxygen–defect complexes were formed in oxygen-rich samples.

It was found in another study of self-implanted Si (Fujinami et al. 1996) that the recovery of defects strongly depends on the morphology of the implanted region controlled by the variation of the implantation dose below and above the critical value of amorphization. Three stages of annealing were found in completely amorphized Si. The amorphous layer was transformed into crystalline material by solid-phase epitaxy in the first stage at 600 °C. The recrystallized layer still contained a high number of voids, the further growth of which took place up to 700 °C. The annealing at 900 °C finally eliminated the vacancy clusters.

A silicon on insulator (SOI) structure was formed by implantation with 200-keV oxygen (dose about 2×10^{17} cm^{-2}) at elevated temperatures (Nielsen et al. 1991). Cavities were detected in the 80-nm Si top layer. Most of these defects annealed at 1300 °C. The dominant defects in the top layer after annealing were supposed to be oxygen inclusions and dislocations. The annealing behavior of defects generated by oxygen implantation (2×10^{15} cm^{-2}, 180 keV) at room temperature was studied in a non-amorphized sample by Fujinami (1996). The $S(E)$ profiles

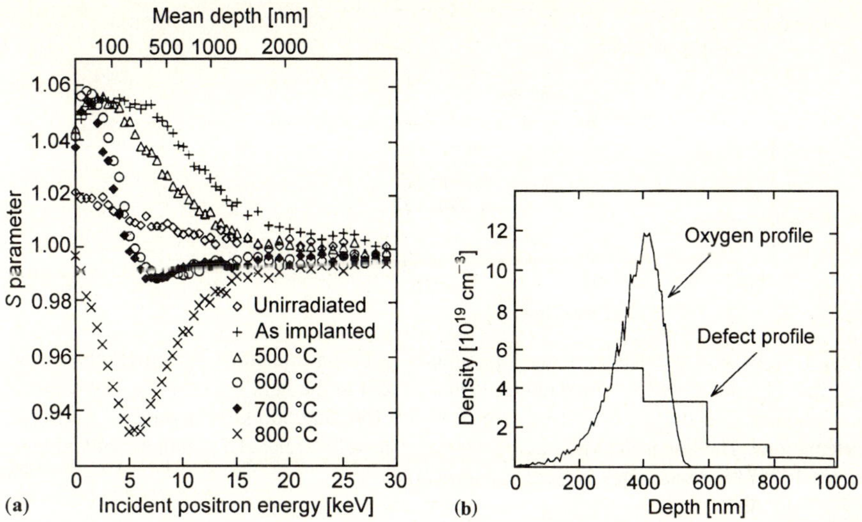

Fig. 4.27. Annealing experiment after oxygen implantation in silicon (Fujinami 1996). (**a**) *S* parameter versus incident positron energy for Si implanted with 2×10^{15} cm^{-2} oxygen (180 keV) after various annealing steps. (**b**) The oxygen profile as calculated by a TRIM (transport of ions in matter) simulation and the vacancy depth profile of the as-implanted sample in (a).

are shown in Fig. 4.27a. The profile of the implanted oxygen was obtained by a TRIM simulation (Fig. 4.27b). The profile of vacancy-type defects derived from the *S*(*E*) curves in the as-implanted Si sample is also given in Fig. 4.27b. Four strips of a constant defect concentration were assumed for the fit to the experimental data. The extension of the strips and the defect concentrations are fit parameters.

The first annealing range was characterized by the transformation of multivacancy–oxygen complexes into V_xO_y complexes in the ion-implanted region (see Sect. 4.1.3.2). A drastic change in the *S*(*E*) curve occurs above 800 °C annealing. The decrease in the *S* parameter to $S/S_b = 0.93$ cannot be understood by the vanishing of open-volume defects alone. It was attributed to a defect reaction of the implanted oxygen with open-volume defects. These defects were supposed to consist of several tens of oxygen atoms.

The defect evolution and annealing after ion implantation with B, As, P, and Si was extensively studied by Eichler et al. (1997a, 1998) and Gebauer et al. (1997a). After implantation with only one ion energy, a narrow peak in the *S*(*E*) curve is obtained, as is visible in Fig. 2.22. The disadvantage is that it can hardly be deduced whether saturated positron trapping occurs, because positron trapping in the bulk or in the surface states cannot be neglected for a narrow defect-rich layer. The defect-related *S* parameter can be determined if the layer of implantation-induced defects is broad. This was achieved in the study of Eichler et al. (1997a,

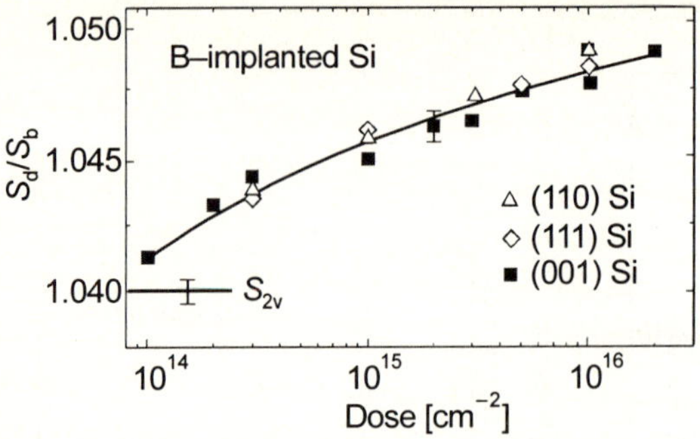

Fig. 4.28. Dose dependence of the *S* parameter in boron-implanted silicon (Eichler et al. 1997a). The defect-related *S* parameter is normalized to the bulk value, S_d/S_b. The multiple implantation was carried out with energies of 50, 150, and 300 keV in wafers of different orientations. The line is drawn to guide the eye. The defect-related *S* parameter of Goldberg et al. (1995) for divacancies is given for comparison.

1998) by multiple implantation with different energies. The necessary implantation energies and doses to obtain a flat defect profile up to a depth of 1 μm were determined by TRIM simulations. The defect-related *S* parameter can be recognized as the plateau in the *S*(*E*) curves (Fig. 4.28) provided that saturated trapping occurs. This was already the case for the lowest implantation dose of 1×10^{14} cm^{-2} applied.

The increase in the defect-related *S* parameter with a higher boron dose is shown in Fig. 4.28. It has been already known from Rutherford backscattering measurements that no amorphization can be achieved with B^+ implantation at 300 K. It is concluded that the increase in *S* is due to the increase in the concentration of open-volume defects with a size larger than divacancies. This conclusion is confirmed by infrared absorption measurements on the same samples, where it was found that the concentration of divacancies remained almost constant in the dose range of interest (Eichler et al. 1997a; Gebauer et al. 1997a). No dependence of the concentration of the vacancies on the implantation direction was found.

The defect concentration can be calculated from back-diffusion measurements (Sect. 3.2.3). The dependence of the vacancy density on the deposited energy is plotted in Fig. 4.29 for As, P, B, and Si implantation. The vacancy concentration for As^+, B^+, and P^+ implantation is proportional to $\Phi^{0.5}$, where Φ is the ion dose. This is in agreement with the model of homogeneous defect nucleation (Chadderton 1971). It is known, e.g. for B, that preferentially it occupies interstitial lattice sites after implantation. A remarkable difference occurs for the self-implantation. The concentration of Si interstitials is higher in the case of self-implantation. These interstitials can recombine with vacancies. A slight modification in the rate equations of the Chadderton model due to the vacancy–interstitial recombination has to be taken into account, and another slope results for the va-

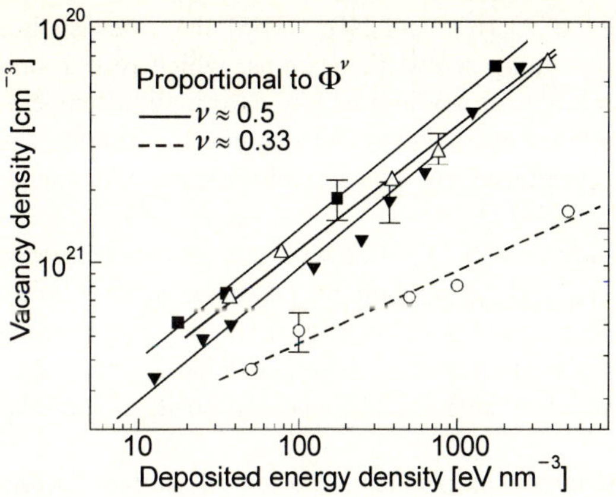

Fig. 4.29. Vacancy density determined from slow-positron-beam measurements as a function of the deposited energy for arsenic (■), phosphorus (△), boron (▼), and silicon (○) implantation (Eichler et al. 1998). Φ is the implantation dose.

cancy concentration versus the dose. A $\Phi^{0.33}$ dependence was found experimentally (Eichler et al. 1998).

The annealing of vacancy-type defects introduced by ion implantation was investigated in Cz and FZ Si (Fig. 4.30). The S parameter reduces in a sharp annealing stage between 700 and 800 K. This is an indication that divacancies are not responsible for positron trapping, since their annealing between 400 and 500 K is not observed in the implanted samples. The S parameter in Cz Si decreases significantly below the bulk value upon annealing above 800 K. Such behavior had already been reported by Fujinami (1996) after implantation with O^+. Obviously the oxygen present in Cz Si is sufficient to form vacancy–oxygen complexes dominating the positron trapping in this annealing range.

4.1.9 Plastic Deformation of Silicon and Germanium

The defect spectrum formed during plastic deformation of semiconductors is more complex than that after irradiation with high-energy particles. The movement and the interaction of dislocations as the primary defects are thought to be responsible for the generation of various types of point defects (Alexander 1986). Different mechanisms have been proposed, depending on parameters such as the deformation temperature, the stacking fault energy, and the self-diffusion coefficient. These mechanisms are related to different stages of dynamic recovery, such as cross-slip or dislocation climb (Siethoff 1992). The intersection of dislocations during the plastic deformation leads to the formation of jogs. These are segments of the dislocation line where the dislocation leaves its glide plane. They are in many cases not glissile, i.e. they can move only non-conservatively by the emission of point defects. The model of Barrett and Nix (1965) describes such a mechanism as the dragging of jogs along screw dislocations, which results in a high number of vacancies or interstitials at medium or low deformation temperatures. Several spectroscopic and microscopic methods were applied to characterize point defects and dislocations introduced by plastic deformation (Alexander 1986).

One of the first attempts to study defects in solids with positrons was carried out on plastically deformed silicon and germanium by Dekhtyar et al. (1969). However, the angular correlation curves did not alter after scratching and bending the samples. Obviously, the density of dislocations and other defects formed during plastic deformation was too small to have any influence on positron annihilation.

The first positron lifetime experiments on plastically deformed semiconductors

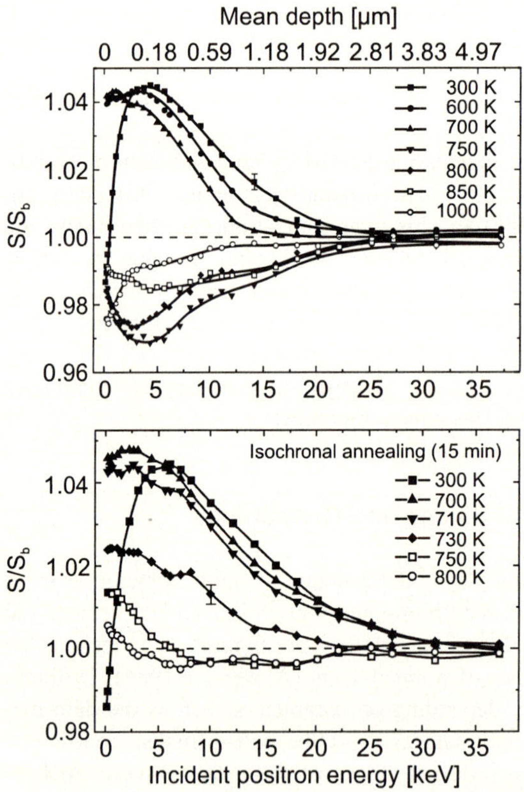

Fig. 4.30. Annealing behavior of boron-implanted Czochralski-grown (*upper panel*) and float-zone silicon (*lower panel*). The Doppler broadening S parameter is shown as function of the incident positron energy after implantation with energies of 30, 90, and 180 keV to a dose of 1×10^{14} cm^{-2} and after different isochronal annealing steps (Eichler et al. 1997a).

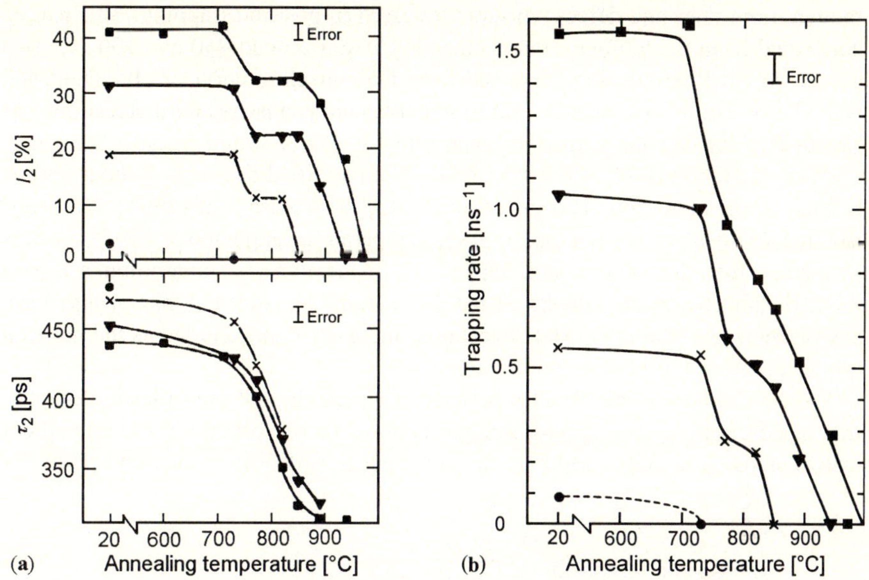

Fig. 4.31. Positron lifetime results of an isochronal annealing experiment of silicon (boron doped, $p = 4\times10^{14}$ cm^{-3}) deformed to (•) 2.5 %, (×) 5 %, (▼) 20 %, and (▪) 40 % (Dannefaer et al. 1983a). The intensity component I_2 of the two-component decomposition shows in (**a**) two distinct annealing stages of defects having defect-related lifetimes τ_2 of about 450 and 330 ps. The calculated trapping rate of this decomposition is shown in (**b**). The full and dashed lines are drawn to guide the eye.

were performed by Kuramoto et al. (1974). It was found in germanium that the positron trapping rate increased after uniaxial compression. This effect was attributed to positron trapping in dislocations, and it was found to be more pronounced in n-type than in p-type Ge. The results were discussed in terms of the charge distribution around dislocations as described by the Read model (Read 1954). In agreement with the results of Schröter and Labusch (1969) in both n- and p-Ge, the dislocation line was expected to be negatively charged, screened by a space-charge region (Read cylinder). Dislocations were thus regarded as positron traps in Ge of either conduction type. The defect-related positron lifetime of 310 to 350 ps obtained in a two-component decomposition was attributed to the charged dislocations. However, generated point defects were ignored. It must be concluded with the current knowledge that at least some of the positrons were trapped in secondary defects, such as vacancy clusters.

Dannefaer et al. (1983a) performed a more extended study on weakly boron-doped Si deformed up to the upper yield point, to the lower yield point, and to deformation degrees of 20 % and 40 %, respectively. The results are shown in Fig. 4.31. A two-component decomposition of the lifetime spectra was applied, although more than one defect type was involved in positron trapping. This can be concluded from the well-separated annealing stages around 750 and 900 °C. Two defect-related lifetimes were deduced from the annealing curves to be about 450 and 330 ps. These data were related to larger vacancy clusters and dislocations, respectively. The calculated trapping rate shown in Fig. 4.31b is virtually the superposition of all contributing positron traps. The identified lifetime of about 330 ps is close to the value found for divacancies (e.g. Kawasuso et al. 1994). However, isolated divacancies are not stable in the temperature range above 300 °C. It was concluded that divacancies are trapped by dislocations, so exhibiting a higher thermal stability, or that defects on the dislocation line such as kinks or jogs have a divacancy-like character. The dislocation line itself was excluded as the positron trap responsible for the 330 ps component.

Krause-Rehberg et al. (1993a) performed an investigation on plastic deformation of differently doped silicon and germanium. The deformation was carried out in two steps. In order to separate the contributions of dislocations and point de-

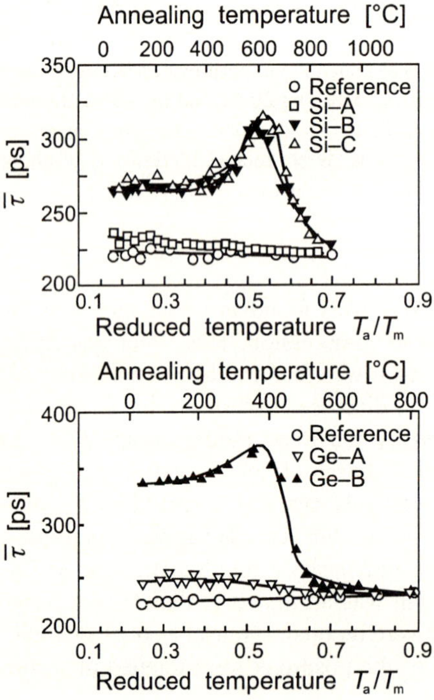

Fig. 4.32. Average positron lifetime $\bar{\tau}$ of plastically deformed n-type silicon and germanium versus the annealing temperature T_a (in K) reduced to the melting temperature T_m. Samples Si-A, Si-B, and Si-C differ mainly in the deformation degree (< 0.05, 0.6, and 1.4 %, respectively). Ge-A and Ge-B were deformed up to 0.05 % and 0.9 %, respectively (Krause-Rehberg et al. 1993a). The lines are drawn to guide the eye.

fects, an intermediate annealing at 800 °C was performed to reduce the density of residual defects measured by deep level transient spectroscopy and electron paramagnetic resonance after the pre-deformation at 800 °C and a resolved shear stress of 12 MPa. The main deformation was performed under high stress and low deformation temperature conditions (200 to 250 MPa and 390 to 420 °C). Defect structures after similar deformation procedures had been extensively studied by electron paramagnetic resonance (e.g. Kisielowski-Kemmerich 1990) and deep level transient spectroscopy (e.g. Omling et al. 1986)

The more heavily deformed samples (Si-A, Si-B, and Ge-A in Fig. 4.32) exhibit a strong increase in the average positron lifetime in the as-deformed stage. Two defect-related lifetime components could be separated, viz. 300 and 590 ps for Si, and 325 and 525 ps for Ge. Similar to Dannefaer et al. (1983a), they were attributed to positron trapping in dislocation-related defects (d1) and voids (d2), respectively.

The size of the voids was estimated to contain more than ten vacancies. This is in agreement with the calculations of Saito and Oshiyama (1996). As a special feature of the annealing curve (Fig. 4.33), a pronounced increase in the void-related trapping rate was observed both for Si and Ge at a reduced annealing temperature $T_a/T_m = 0.55$ (T_m melting temperature). This maximum was interpreted as an increase in the density of voids rather than the growth of void diameters. The main vacancy supply mechanism was attributed to non-conservative dislocation

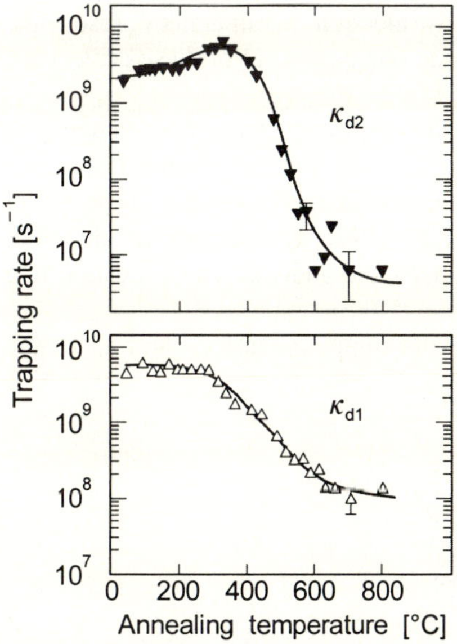

Fig. 4.33. Trapping rates κ_{d1} (dislocation-related defects) and κ_{d2} (voids) versus the annealing temperature in germanium plastically deformed up to 0.9 % (Krause-Rehberg et al. 1993a). The lines are drawn to guide the eye.

motion and relaxation during annealing. It is assumed as the dragging of jogs along screw dislocations (Barrett and Nix 1965). The edge-type jogs cannot follow the glide motion of the screw, and point defects, e.g. vacancies, are generated. A long row of vacancies is supposed to be unstable. The formation of vacancy clusters in the close vicinity of the jogs is sketched in Fig. 4.34.

The assignment of the shorter lifetime component to dislocation-related defects rather than to divacancies was deduced from the thermal stability observed during isochronal annealing, where these defects start to anneal at a temperature as high as 600 °C for Si and 400 °C for Ge. As an example, the calculated trapping rates for plastically deformed Ge are shown in Fig. 4.33. A distinct trapping in dislocation-related defects was found even at the highest annealing temperature of 825 °C, whereas the void component reached the lower detection limit.

The decrease in the trapping rate κ_{d1} (dislocation-related defects d1) in the temperature range 400 to 800 °C is not due to an overall decrease in the dislocation density, as was confirmed by transmission electron microscopy. It was concluded that the undisturbed dislocation line itself did not represent the deep positron trap detected. The perfectly reconstructed dislocation is connected to a small open volume and should thus give rise to only a slight increase in the lifetime compared with the bulk (Sect. 3.1.5.3). The dislocation-related lifetime has been theoretically calculated up to now only for metals and was found to be only a few picoseconds higher than the bulk value (e.g. Häkkinen et al. 1990; Kamimura et al. 1995). Thus, open-volume defects on the dislocation line must be responsible for the observed trapping effects. Such vacancy-type defects on the dislocation, called K1, K2, and Y centers, were identified by electron paramagnetic resonance (Kisielowski-Kemmerich 1990). It is likely that one or several of these centers are the observed positron annihilation sites d1.

This interpretation is in agreement with the treatment of positron trapping in dislocations as a cascade process, i.e. the positron is first trapped in the defect-free dislocation line before being trapped in a deep potential of a vacancy-type defect bound to the dislocation (Hashimoto and Kino 1991; Smedskjaer et al. 1980; Xiong and Lung 1989). As is shown in Fig. 3.26, the undisturbed dislocation line acts in this model as a shallow positron trap.

Temperature-dependent measurements of the trapping rates κ_{d1} and κ_{d2} in the

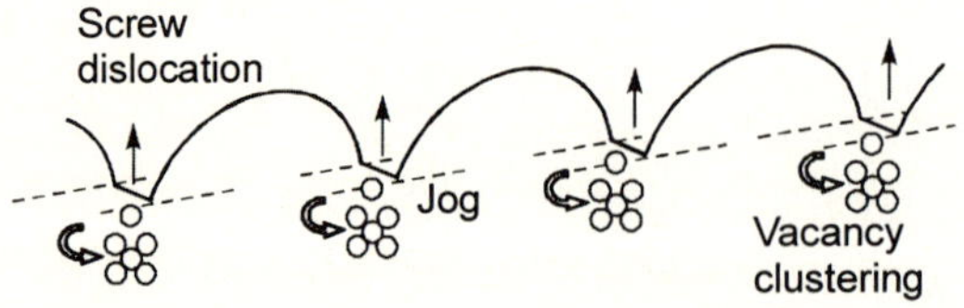

Fig. 4.34. Mechanism of the formation of vacancies by the dragging of non-glissile jogs along screw dislocations. Rows of vacancies are supposed to be unstable, and larger vacancy clusters are formed close to the edge-type jogs. The arrows indicate the direction of the non-conservative motion of the jogs.

as-deformed state yielded a temperature dependence of $\propto T^{-1}$ (Krause-Rehberg et al. 1993a). The authors concluded diffusion-limited positron trapping for both dislocation-related defects (d1) and voids (d2). The deviation from the expected $T^{-1/2}$ dependence of the trapping rate was explained by the charge of the extended defects.

The trapping rate κ_{d2} was calculated as a function of the dislocation density according to (3.13). However, the description of the trapping in dislocations as a transition-limited process with (3.48) was also found to give a satisfactory description of the experimental data. The calculated trapping coefficients are given in Sect. 3.2.5.

Krause-Rehberg et al. (1993a) investigated also p-type silicon with a carrier density of 5×10^{16} cm^{-3} deformed under comparable conditions to sample Si-B of Fig. 4.32. Almost no positron trapping was observed, although the densities of dislocation-related defects and voids are expected to be similar to the deformed n-type Si. It was concluded that the sensitivity of positron lifetime spectroscopy was lost, which was related to band-bending effects at voids or the positive charge of dislocation-related defects.

The temperature dependence of the positron trapping in dislocations was studied in more detail by Kawasuso et al. (1995d). A heat treatment up to 1100 °C was performed after the deformation. A second defect-related lifetime, which decreases around 900 °C from 285 to 240 ps, was separated in addition to a void component of 544 ps. The lifetime of 285 ps was attributed to a mixture of a 300 and a 240 ps lifetime attributed to positron trapping in vacancy-like parts on dislocations and in the dislocation line itself. The 900 °C annealing led to the vanishing of the defect responsible for the 300 ps component. This treatment enabled the separate investigation of the temperature behavior of the trapping to the dislocation line. In Fig. 4.35, the trapping rate corresponding to the 240 ps lifetime is shown versus the temperature for this annealing state.

The obtained trapping rate was fitted as a function of temperature according to the model of a dislocation as a combined trap of a shallow and a deep state. The temperature dependence is in a very good agreement with the model of Manninen and Nieminen (1981) describing the undisturbed dislocation line as a shallow precursor trapping state (Sect. 3.2.1.3). A T^{-1} dependence was found in the temperature range above 100 K, similar to the results of Krause-Rehberg et al. (1993a) discussed above. The fit resulted in a positron binding energy to the shallow state at the dislocation line of $E_b \approx 11$ meV. Such shallow levels may be induced from the deformation potential around dislocations (Kawasuso et al. 1995d). The deformation potential is responsible for shallow electronic levels in the bandgap, which were theoretically predicted (Teichler 1975) and used for the explanation of various photoluminescence lines (Lelikov et al. 1992).

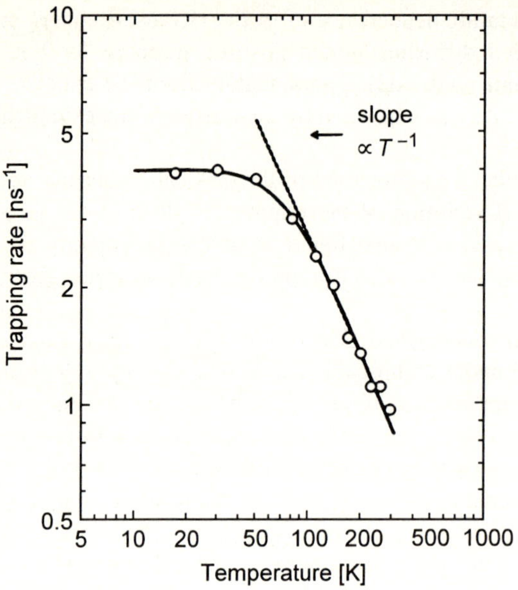

Fig. 4.35. Trapping rate corresponding to the 240 ps lifetime component versus the sample temperature after 900 °C annealing of plastically deformed Si (Kawasuso et al. 1995d). The solid line is a fit according to the model of a dislocation as a combined positron trap of a shallow precursor and a deep state (Sect. 3.2.1.3). The trapping rate follows the curve $\propto T^{-1}$ for temperatures higher than 100 K.

4.1.10 Porous, Amorphous, and Polycrystalline Silicon

Recently, non-standard materials, such as porous and amorphous silicon, have become more and more the subject of experimental investigations because of their unique physical properties. Positron annihilation techniques have been applied as well due to their sensitivity to atomic-scale imperfections. Remarkable changes were found in the annihilation parameters. However, the theory of positron capture in such materials has not been established sufficiently, making the interpretation of the experimental results difficult.

Porous silicon layers with a typical thickness of a few μm can be obtained by anodization of monocrystalline silicon in hydrofluoric acid. The material is sponge-like with a porosity up to about 70 %. The positron annihilation behavior was studied by several groups (e.g. Dannefaer 1995; Itoh et al. 1994; Knights et al. 1995).

Strong positron trapping was observed. It was found by means of slow-positron-beam measurements (Knights et al. 1995) that the positron diffusion length is rather short, indicating saturated positron trapping in the porous layer.

The positron lifetime spectra exhibited two defect-related long-lived components, $\tau_2 = 600$ to 1000 ps and $\tau_3 = 20$ to 90 ns. The latter lifetime arose from the annihilation of free positronium in the open volume of the pores. Such an extremely high lifetime, which comes close to the pick-off-free ortho-positronium lifetime of 142 ns (Sect. 3.1.6), has rarely been found in semiconductors. The existence of positronium was confirmed in measurements of the two-dimensional angular correlation of annihilation radiation, where a narrow peak was observed (Haung et al. 1996). The positron lifetime component τ_3 varied depending on the actual sample preparation conditions, viz. the pore size, surface morphology, and possible adsorbates. It was influenced by annealing under different ambients. The lifetime component τ_2 was attributed to surface-bound Ps and other defects created during anodization. The defect-related S parameter exhibited a weak increase in S_d/S_b up to 3 %. This increase was found to be sensitive to the anodic etching conditions. This may be due to the chemical sensitivity of the S parameter and was related to a different content of SiO_2 in the layer.

Amorphous Si is obtained either as hydrogenated a-Si:H by plasma-assisted film deposition, or molecular beam epitaxy, or as non-hydrogenated a-Si by ion implantation. The structure of pure amorphous Si prepared by sputtering, chemical vapor deposition, molecular beam epitaxy, and ion implantation was investigated by slow-positron-beam measurements (Hakvoort et al. 1994; Roorda et al. 1992). Vacancy clusters with the size depending on the growth temperature were found in a-Si:H. MBE-grown a-Si contained large open-volume defects. The structure of amorphous Si produced by self-implantation and hydrogen co-implantation was studied. The amorphized Si contained small vacancies with a concentration of 1.2 atom %, which decreased upon annealing (relaxation) to 0.4 atom %. The results obtained with the co-implanted hydrogen indicated the passivation of dangling bonds by H.

The role of defects in the amorphization of crystalline Si by ion beams or during chemical vapor deposition was studied by Hiroyama et al. (1993, 1995). The rearrangement of divacancies and interstitials to ring-like structures was described by molecular dynamics simulation (Motooka et al. 1995). The further growth of vacancy-type defects to larger clusters was investigated by positron lifetime measurements, Raman spectroscopy, and electron paramagnetic resonance for different preparation techniques of a-Si (Hiroyama et al. 1993, 1995). Amorphous Si formed by ion implantation was found to contain only small vacancies clusters of a divacancy size, which disappeared during annealing. On the other hand, a-Si and a-Si:H prepared by chemical vapor deposition contained larger vacancy complexes. Vacancy clusters of nanometer size were formed during the annealing. It was concluded that vacancy clusters or voids suppress the structural relaxation in a-Si (Hiroyama et al. 1995).

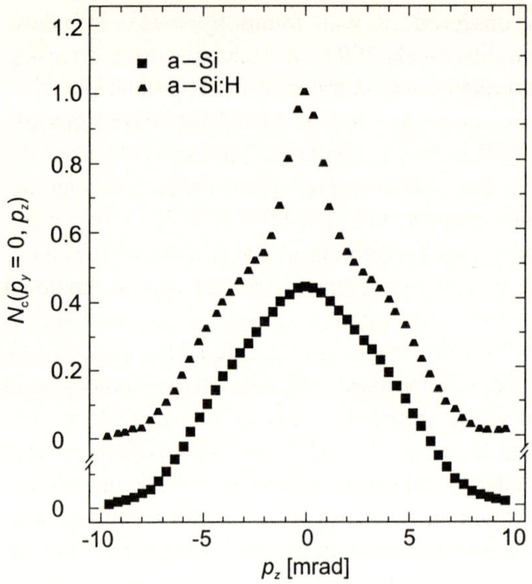

Fig. 4.36. Two-dimensional angular correlation of annihilation radiation (2D-ACAR) spectra of amorphous silicon (squares) and hydrogenated a-Si:H (triangles). The projection $N_c(p_y, p_z)$ of the three-dimensional momentum density $\sigma(\mathbf{p})$ is shown as the cut for $p_y = 0$ (He et al. 1986). The narrow ACAR component of the upper curve from $p_z = -2$ to 2 mrad indicates positronium formation. The spectra were normalized to the maximum peak height of the a-Si:H curve.

The detection of voids in amorphous silicon by positron lifetime spectroscopy is in contrast to the findings in metals (Bhide et al. 1987; Dannefaer et al. 1983b; He et al. 1986; Schaefer et al. 1986). The appearance of voids was concluded from the lifetime component of 400 to 550 ps. The voids are supposed to stabilize the network of the amorphous structure (Motooka et al. 1995). The volume of the voids can be estimated from Fig. 4.4 to be certainly larger than 8 vacancies in accordance with Schaefer et al. (1986), who discussed from their experimental results a size of 10 to 15 vacancies. It should be noted that the defect-related positron lifetime could be saturated, i.e. that this lifetime does not increase further with increasing cluster size. In this case, the void volume could even be much larger.

A pronounced positronium component could be found in a-Si:H (He et al. 1986). This is clearly visible in the spectra of angular correlation of annihilation radiation of Fig. 4.36. The narrow component of the ACAR curve of a-Si:H in connection with the appearance of a lifetime component of 9 ns unambiguously indicates positronium formation. Obviously, hydrogen influenced the structure of the voids and thus promoted the formation of Ps. Nasu et al. (1987) concluded that the size and the concentration of voids increased by hydrogenation. However, other possibilities for the enhanced Ps formation may be the passivation of the surface of the void by hydrogen or the interaction of the positron with hydrogen gas.

Polycrystalline silicon is regarded as a promising material for photovoltaic applications. The grain boundaries are potential positron traps. The grain size must be not much larger than the positron diffusion length in order to obtain a measurable effect (Sect. 3.1.5.8). This condition is not fulfilled in the usual poly-Si grown for photocells. Positron lifetime spectroscopy of such material showed only a one-component lifetime spectrum with $\tau = \tau_b$ (Gebauer et al. 1998a). Thus, only positron trapping effects, e.g. due to impurities such as carbon (Coleman et al. 1995), comparable to monocrystalline bulk Si may be possible.

Nanocrystalline Si is a special case of poly-Si where the crystallite size is of the order of 5 to 10 nm. The material can be prepared by chemical transport and deposition on Si substrates in a hydrogen plasma (Veprek and Marecek 1968). The high density of interfaces gives rise to novel material properties as found for metals (Gleiter and Marquardt 1984). Crystallites of this size are much smaller than the positron diffusion length. Thus, the interface region becomes a highly attractive positron trap. A positron lifetime study of nanocrystalline Si was published by Würschum et al. (1987). Samples with a layer thickness in the order of magnitude of 100 μm exhibited two-component spectra with $\tau_1 = 400$ ps to 450 ps and $\tau_2 = 7$ ns to 12 ns. These results support the hypothesis of the wide interatomic distance distribution in the interfaces. In contrast to metals, in nanocrystalline Si no monovacancy-type defects but voids probably located at the intersections of grains are detected.

4.1.11 Summary

Open-volume defects can hardly be detected in as-grown silicon by means of positron annihilation. Defect investigations are therefore mainly carried out in irradiated material. Systematic studies were presented to characterize the formation and annealing of defects in dependence on the irradiation dose, doping, and oxygen content. The annealing behavior is a rather complex function of these parameters. Monovacancies, divacancies, and vacancy–dopant complexes (*E* centers) were identified by a combination of positron annihilation with other experimental methods such as infrared absorption and electron paramagnetic resonance. The *A* center, as a complex consisting of a vacancy and oxygen, acts as a shallow positron trap with a positron binding energy of 48 meV. A negative charge is not required for the weak binding of the positron, which is related to the smallness of the open volume of the defect. The formation and dissolution of other vacancy–oxygen complexes V_xO_y was observed during thermal treatment of Czochralski-grown Si. The study of ion-implantation-induced defects in silicon has become one of the standard applications of the slow-positron-beam technique. The damage depth and amorphization were investigated after implantation with a broad range of ions. Defect reactions were found to be different in Czochralski-grown and

Table 4.2. Positron annihilation characteristics of defects in silicon.

Structure	τ_d [ps]	τ_d/τ_b	Trap type	Dissolution temperature [K]
Monovacancies	266 to 275[a]	1.22 to 1.26	deep	150[b]
Divacancy	290 to 320[c]	1.33 to 1.47	deep	550[e]
Voids	550 to 600[f]	2.52 to 2.75	deep	1150[f]
VO (*A* centers)	225[g]	1.03	shallow	620 to 670[h]
V_xO_y complexes	240 to 330[i]	1.10 to 1.5[k]	deep	
VOB complexes	270[j]	1.24		1000[j]
Interstitial–oxygen clusters	100 to 113[k]	0.46	deep	920[k]
V–dopant complexes (*E* centers)	270[e]	1.24	deep	200 to 400[e]
Sponge defects	315 to 365[d]	1.44 to 1.67	deep	770[d]
Dislocation-related defects	300[f]	1.38	deep	

τ_d is the defect-related positron lifetime, τ_b the positron lifetime in the defect-free bulk ($\tau_b = 218$ ps).
[a]Fuhs et al. (1978); Mäkinen et al. (1990a); Würschum et al. (1989a), [b]Mäkinen et al. (1989); Watkins (1986), [c]Dannefaer and Kerr (1986a); Fuhs et al. (1978); Hasegawa et al. (1995b); Kawasuso et al. (1994, 1995a); Kwete et al. (1990); Shimotomai et al. (1981); Würschum et al. (1989a); [d]Avalos and Dannefaer (1996), [e]Polity et al. (1998a), [f]Kawasuso et al. (1995d); Krause-Rehberg et al. (1993a), [g]Mascher et al. (1989b), [h]Gregory and Barnes (1968); Mascher et al. (1989b), [i]Kawasuso et al. (1995b), [j]Dannefaer (1990); Dannefaer et al. (1989b), [k]Dannefaer and Kerr (1986a); Puff et al. (1992b).

float-zone Si, indicating the formation of vacancy–oxygen complexes during annealing.

The results of the investigations of plastically deformed silicon gave evidence for the formation of point defects by jog dragging. Defect-related positron lifetimes as high as 600 ps were attributed to the occurrence of vacancy clusters. The results of positron lifetime measurement as a function of temperature were interpreted with the effect of undisturbed dislocation lines as shallow positron traps.

Slow-positron-beam measurements were used to characterize layered structures such as epitaxial silicon, silicon dioxide layers, and metal films. A high density of open-volume defects, giving rise to an extremely high average positron lifetime, was detected by slow-positron lifetime measurements in low-temperature-grown epitaxial Si. Interface states were investigated in the system SiO_2/Si.

Table 4.2 summarizes the experimental data on positron studies of defects in silicon. The defect-related lifetime τ_d at room temperature is given. The character of the positron trap is indicated with respect to the positron binding energy as deep or shallow. The temperature for the dissolution of the respective defect as found from annealing experiments is given.

4.2 Defects in Other Elemental Semiconductors

There exist only a few papers on positron studies of germanium (Sect. 4.2.1), mostly limited to irradiation-induced defects and defects occurring after plastic deformation. The latter ones have been studied in conjunction with deformed silicon and the results are thus presented in Sect. 4.1.9.

Recently, mixed crystals consisting of silicon and germanium (SiGe) as well as diamond crystals and epitaxial layers have found increasing interest. The first positron results on these materials are reviewed in Sects. 4.2.2 and 4.2.3.

4.2.1 Germanium

The first positron studies of Ge were carried out using the angular-correlation technique. Doping with different impurities resulted, in as-grown material, in changes in the angular correlation of annihilation radiation curves, which were interpreted with different chemical species (Arifov et al. 1979).

Irradiation-induced defects were first reported by Bartenev et al. (1979) after irradiation with 1×10^{18} cm^{-2} neutrons. Three annealing stages were reported in correlation with electrical, optical, and electron paramagnetic resonance. Low-temperature electron irradiation is best suited to create Frenkel pairs in a defined manner. Such experiments were carried out by Corbel et al. (1985) on p-type germanium with a resistivity of 2 Ω cm. The annealing curve is shown in Fig. 4.37. The first annealing stage observed at 200 K was interpreted as the transformation of irradiation-induced monovacancies to vacancy–dopant complexes. These complexes disappear at 500 K and the bulk lifetime is recovered.

These results are in agreement with a more recent study of electron-irradiated Ge by Polity et al. (1997b). Undoped Ge samples were irradiated with 2-MeV electrons at 4 K with doses of 10^{15} to 10^{19} cm^{-2}. The isochronal annealing curves are shown in Fig. 4.38. Only one annealing stage in the same range around 200 K as found by Corbel et al. (1985) was observed for irradiation doses in the range from 10^{15} to 10^{18} cm^{-2}. A defect-related lifetime component of 281 ± 5 ps was separated below 200 K. This finding is similar to Corbel et al. (1985), Moser et al. (1985), and Würschum et al. (1989b), who found 290, 292, and 278 ps, respectively. This lifetime can be attributed to positron trapping in monovacancies, taking into consideration the (slightly smaller) value of 263 ps of the theoretical calculations for the monovacancy by Puska et al. (1989b). As the defect-related lifetime around 285 ± 7 ps was experimentally observed for undoped and for different dopants, it is likely that it is thus due to isolated monovacancies.

Additionally, Hall effect measurements were carried out in the annealing experiment of Polity et al. (1997b) for $T_a \geq 300$ K in order to determine the Fermi-level position. It was located at 300 meV above the valence band for all irradiated samples. Puska (1989b) calculated ionization levels of Ge monovacancies in different charge states (Fig. 4.39). It can be concluded from the calculated ionization levels that the monovacancies are in the negative charge state, at least after the annealing stage at 200 K.

The average lifetime measured at 90 K and the corresponding trapping rate are plotted as a function of the irradiation dose in Fig. 4.38b. The average positron lifetime virtually represents the sensitivity curve of the positron lifetime spectroscopy method (Fig. 3.36). The trapping rate (right axis) increases with the irradiation dose. It was obtained, however, that the defect introduction rate slightly decreases with the dose. It could be calculated with an estimate for the trapping coefficient of 3×10^{16} s^{-1} at 90 K to be 0.4 to 0.1 cm^{-1}.

A peculiar behavior was observed for the highest dose of 10^{19} cm^{-2} (Fig. 4.38a). A second annealing stage appeared at 400 K. A defect-related lifetime of 281 ps was obtained in the temperature range between 200 and 350 K. Polity et al. (1997b) interpreted this component as a monovacancy–oxygen complex. The detectability of this defect only at a higher dose was explained by a very low introduction rate of such defects. A similarly low defect introduction rate of about 10^{-4} cm^{-1} was observed for a VO complex by means of deep level transient spectroscopy (Ito et al. 1986). This means that an irradiation dose of at least 10^{19} cm^{-2} is necessary to create a sufficiently high defect concentration for detection by positron lifetime spectroscopy.

No annealing stage at 500 K was found in undoped Ge by Polity et al. (1997b). This confirms the conclusion of Corbel et al. (1985) given above that this stage is related to vacancy–dopant complexes in p-type Ge.

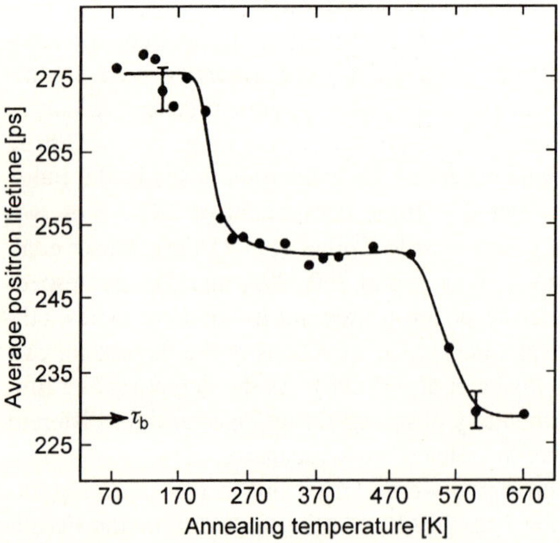

Fig. 4.37. Isochronal annealing of p-type germanium after electron irradiation at 20 K. The bulk lifetime τ_b of 228 ps is indicated (Corbel et al. 1985).

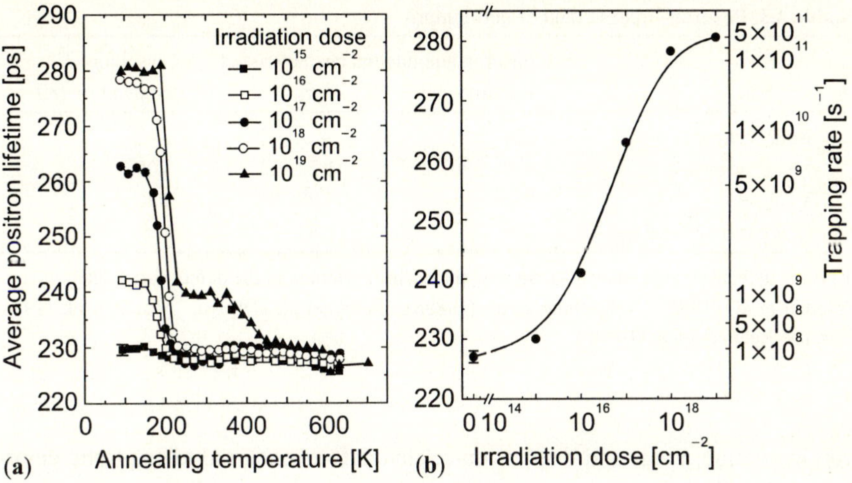

Fig. 4.38. Annealing experiment of electron-irradiated undoped germanium (Polity et al. 1997b). The average positron lifetime is shown as a function of the annealing temperature in (**a**). The lifetime was measured at 90 K. The dose dependence of the average positron lifetime and the trapping rate is shown in (**b**). The lines are drawn to guide the eye.

The annealing of an intrinsic defect was observed by perturbed-angular correlation investigations of electron-irradiated doped germanium (Hässlein and Sielemann 1995; Hässlein et al. 1998). It was interpreted as the vacancy as part of a Frenkel pair. The annealing stage at 200 K is identical to the findings of Corbel et al. (1985) and Polity et al. (1997b) and was attributed to the recombination of the Frenkel defect. In reasonable agreement with the calculations of Puska (1989b), a deep donor (0/−) level for the monovacancy was found at E_{val} + 0.2 eV.

Table 4.3 summarizes positron lifetime parameters of Ge. Only deep positron traps were reported. Both the bulk lifetime τ_b and the defect-related lifetimes τ_d measured at room temperature are given. Monovacancies were found after elec-

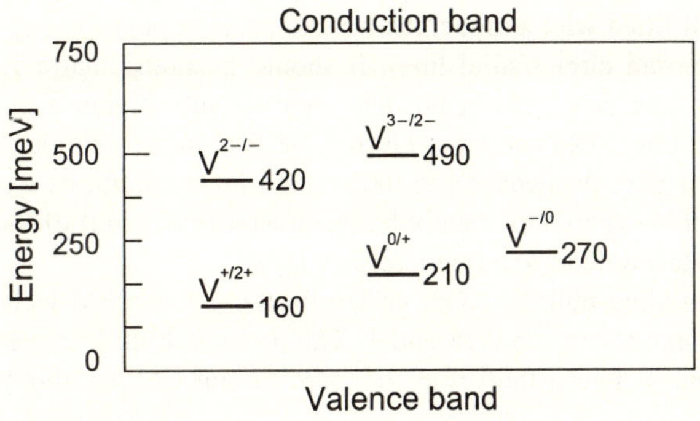

Fig. 4.39. Ionization levels of monovacancies in germanium as calculated by Puska (1989b).

Table 4.3. Positron-lifetime data of germanium.

	Positron annihilation characteristics		Dissolution
	Lifetime [ps]	τ_d/τ_b	temperature [K]
Bulk lifetime	228[a,b]	—	—
Monovacancies	278[c], 290[a], 292[d]	1.22, 1.27, 1.28	200[a,c]
Voids	520[e]	2.28	870[e]
Dislocation-related defects	325[e]	1.43	

τ_d is the defect-related positron lifetime, τ_b the positron lifetime in the defect-free bulk.
[a]Corbel et al. (1985), [b]Würschum et al. (1989b), [c]Polity et al. (1997b), [d]Moser et al. (1985), [e]Krause-Rehberg et al. (1993a).

tron irradiation, voids and dislocation-related defects occurred after plastic deformation (Sect. 4.1.9).

4.2.2 SiGe Mixed Crystals

$Si_{1-x}Ge_x$-based transistors are expected to be well suited for high-frequency applications. Extremely high mobilities for electrons and holes were found in epitaxial SiGe layers (Fitzgerald et al. 1994). Misfit defects between the $Si_{1-x}Ge_x$ layer and the Si substrate are an important technological problem. The lattice strain increases with increasing Ge content in the layer.

Although positrons are not sensitive to the strain field itself, defects in the layer as well as interface defects (Sect. 3.1.5.7) may be detected. Baker and Coleman (1989) and Baker et al. (1989) studied strained and relaxed layers of $Si_{1-x}Ge_x$ with $x = 0.125$ using a slow-positron-beam system. There were practically no dislocations in the strained layer. Misfit dislocations of a density of $10^5\,cm^{-1}$ were present in the interface after relaxation of the layer by thermal treatment. A remarkable difference was found in the measured S parameter for strained and relaxed samples (Fig. 4.40).

A higher S parameter was measured in the relaxed layer containing misfit defects. The data could be well fitted with a model taking into account positron trapping in interface defects (closed circles/solid line). It should be noted that a fit could also be made without interface defects, but with significantly higher S parameter in the layer (dashed line). The authors preferred the first model, although the presence of defects other than dislocations in the relaxed layer cannot be excluded. As a consequence, this experiment cannot be taken as a final proof of the sensitivity of positrons to detect defects in a thin interface layer.

Kawasuso et al. (1997) studied bulk $Si_{1-x}Ge_x$ grown by the Czochralski technique with the alloy composition between 0.18 and 1. The positron bulk lifetimes varied with x and were found between the values for pure Si and Ge. An abrupt

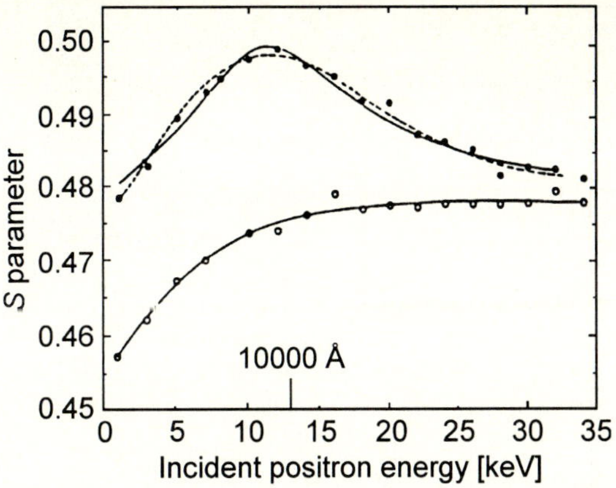

Fig. 4.40. *S* parameter versus incident positron energy for $Si_{0.875}Ge_{0.125}$ overlayers on silicon. The lower curve was measured on a strained epitaxial layer, and the upper curves on a relaxed layer containing misfit defects (Baker and Coleman 1989). The full and broken lines are fits to the data, supposing interface defects and defects in the whole layer, respectively. The thickness of the layer is indicated.

change in the slope of the positron lifetime versus the alloy composition was observed at $x = 0.8$ to 0.83, which seems to correlate with the change in the bandgap energy.

4.2.3 Diamond

Diamond exhibits unique properties, such as extreme hardness and a pronounced high-temperature stability. It represents a promising material for new electronic devices. The vacancy formation energy is about 6 to 7 eV (Bernholc et al. 1981). A prominent feature is the appearance of the so-called GR absorption system that was related to neutral vacancies and their exited states (Clark et al. 1992).

The investigation of the diamond defect structure by means of positron annihilation is just at the beginning. Slow-positron-beam measurements of diamond films grown by chemical vapor deposition (CVD) were reported by Saleh and Rice-Evans (1994). The positron diffusion length was determined from *S* versus *E* plots. Assuming a trapping coefficient of $5\times10^{14}\,s^{-1}$, a defect concentration of 5×10^{-6} was estimated.

The first extended paper on in-grown vacancy-type defects in natural diamond and CVD diamond layers was published by Dannefaer et al. (1996c). The investigated single crystals of IaA, Ib, and IIa types, as well as a boron-doped CVD film, exhibited one dominant lifetime component of about 105 ps with an intensity of 88 % to 98 %. This lifetime was interpreted as positron annihilation from the bulk of the material. Thus, theses samples were essentially free of positron traps. A

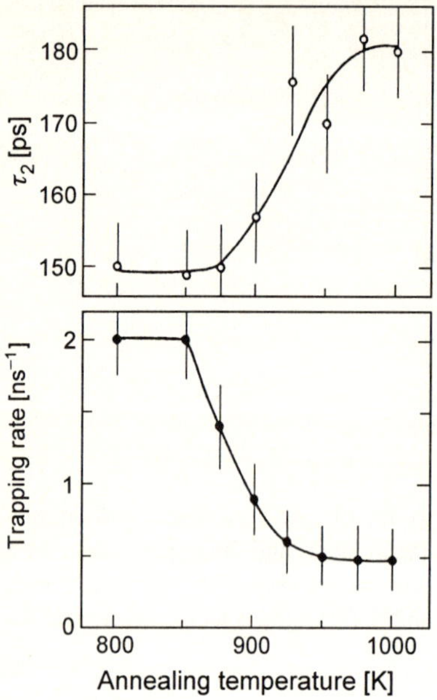

Fig. 4.41. Isochronal annealing experiment of electron-irradiated natural diamond of IaA type. The dose of the 3.5-MeV electron beam was 2.5×10^{18} cm^{-2} (Dannefaer et al. 1996c). The defect-related lifetime component τ_2 and the corresponding trapping rate, resulting from a two-component decomposition of the measured lifetime spectra, are shown. The lines are drawn to guide the eye.

similar bulk lifetime was reported by Uedono et al. (1995a) and Brandes et al. (1992). Material where no positron trapping occurred was used for an electron-irradiation experiment (Fig. 4.41). Monovacancies having a positron lifetime of about 150 ps were found to be stable up to 600 °C. The defect-related lifetime component increased upon annealing above 600 °C. This effect is possibly due to vacancy clustering.

The investigations on layers grown by chemical vapor deposition were performed on thick self-supporting films that were arranged in a stack to form a sandwich for positron lifetime measurements (Dannefaer et al. 1996c). 33 % to 82 % of the positrons annihilated within this stack. The remaining part of the positrons was stopped in the substrate behind the diamond stack. This annihilation fraction was taken into account as a part of the source correction.

The undoped CVD films exhibited very complex lifetime spectra with up to five components. The average lifetime reached a value as high as 630 ps. It was found that various types of open-volume defects, such as monovacancies and di-vacancies, and larger vacancy clusters occurred. Furthermore, positronium formation was observed within the material. Obviously, further experimental effort is needed to clarify the defect structure in diamond layers.

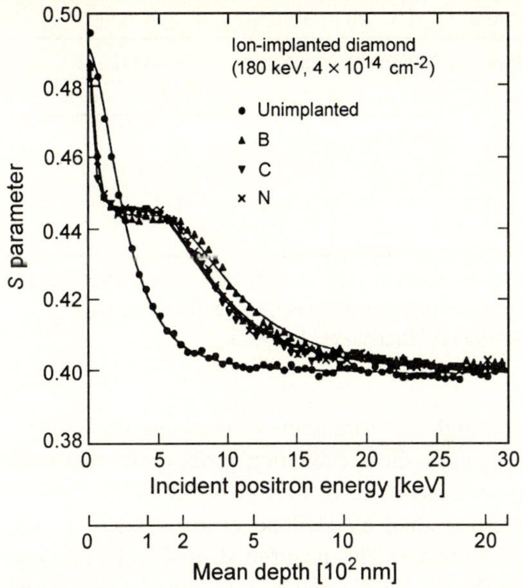

Fig. 4.42. Line shape parameter S as a function of incident positron energy measured in Ib diamond implanted with boron, carbon, and nitrogen ions (Uedono et al. 1995a). The solid curves are fits to the experimental data using the VEPFIT (variable energy positron fit) routine.

Natural diamonds of the Ia and IIa types as well as synthetic IIa and Ib crystals were studied by two-dimensional angular correlation of annihilation radiation (Chiba et al. 1997). All the crystals showed similar anisotropies of the angular correlation of annihilation radiation curves reflecting the covalent character of the binding. A synthetic IIa crystal containing no positron traps was irradiated with 28-MeV electrons (dose 5×10^{18} cm^{-2}). Positron trapping in monovacancies was observed. The anisotropies observed were reproduced by first-principles two-component density functional calculations.

Anisotropies of the electron–positron momentum distribution in diamond were also measured using Doppler-broadening spectroscopy (Nilen et al. 1997a). The single crystalline sample was rotated about the [111] axis with respect to the Ge detector and periodic changes of up to 2 % in the S parameter with the angle were observed.

The age–momentum correlation technique was used to study the effect of light on positron annihilation in natural diamond irradiated with γ-rays (Nilen et al. 1997b). The trapping rate of the defect-related lifetime component was found to increase under illumination. This effect was interpreted as the optical ionization of an open-volume defect, which was stable at least up to 1170 K.

A new model for positron trapping in monovacancies in wide-gap semiconductors and insulators was proposed by Shi et al. (1997). The trapping is mediated,

Table 4.4. Positron lifetime data of diamond.

	Positron annihilation characteristics		Dissolution
	Lifetime [ps]	τ_d/τ_b	temperature [K]
Bulk lifetime	105[a,b,c]	—	—
Monovacancies	150[a]	1.43	
V_3X	217 to 222[b]	2.07 to 2.11	873[b]
Voids	630[a]	6	

τ_d is the defect-related positron lifetime, τ_b the positron lifetime in the defect-free bulk. V_3X is a complex of a vacancy with boron, carbon, or nitrogen, respectively, as the implanted ions.
[a]Dannefaer et al. (1996c), [b]Uedono et al. (1995a), [c]Brandes et al. (1992).

in the framework of this model, through the formation of positronium. The released energy is absorbed by the electron–hole excitation process. In this way, their positron lifetime and *S* parameter measurements may be understood.

Implantation-induced defects were studied by Uedono et al. (1995a). Unimplanted diamond was studied for reference. The positron diffusion length was found to be 52.1 and 100.8 nm for Ib and IIa diamonds, respectively, indicating enhanced positron trapping in the former material. However, the *S* parameter of the Ib sample was smaller by 1.5 %. This was attributed to positron trapping or scattering by nitrogen-related defects. The high chemical sensitivity of the *S* parameter was made responsible for the small *S* value of the Ib sample.

Monocrystalline diamond of IIa and Ib types was implanted at room temperature with B^+, C^+, and N^+ ions up to a dose of 4×10^{14} cm^{-2}. The increase in the *S* parameter up to a mean positron depth of 1 μm indicates the presence of open-volume defects (Fig. 4.42). This is supported by a very short positron-diffusion length of about 3 nm, which suggests saturated positron trapping in these defects. Consequently, one-component positron-lifetime spectra were observed. The defects were supposed to be small vacancy clusters such as V_3 or V_4. The defect-related lifetimes of 217, 234, and 222 ps for B^+, C^+, and N^+ implantation differ more than the standard deviation of 5 ps. It was suggested that complexes with the implanted ions were formed. The difference in the positron lifetime was related to different lattice relaxations with respect to the implanted ion.

Table 4.4 summarizes the positron lifetimes measured at room temperatures for bulk diamond crystals and open-volume defects investigated so far.

5 Defect Characterization in III–V Compounds

A stable existence region around the stoichiometric composition exists, in contrast to elemental semiconductors, in the phase diagram of a compound semiconductor (Jordan et al. 1974; van Vechten 1975). The concentration of point defects in thermal equilibrium can be rather high at the growth temperature, because the deviation from the stoichiometric composition in the existence region of the compound is realized by them. Thus, a larger variety of point defects exists in compounds. Beside interstitials and vacancies in both sublattices, a new class of defects, where the atoms are incorporated on the wrong lattice site, exists. These defects are called antisites. The defects introduced at the growth temperature may be quenched-in during the cooling of the crystal to room temperature.

III–V compound semiconductors crystallize in the sphalerite or wurzite structure, having two sublattices. Compound semiconductors such as $Al_xGa_{1-x}As$ are still binary in the number of sublattices, in such a way that the group III atom sublattice is shared by two atomic species. Such compounds are called pseudo-ternary and pseudo-quaternary with respect to the number of atomic components.

5.1 Defect Characterization in Gallium Phosphide

Gallium phosphide is a basic material for optoelectronic applications and can be used for light-emitting diodes in the visible region. Defects exert a dominant influence on performance and stability of the devices. The ionization levels of point defects are necessary for the interpretation of positron results. The calculation of levels for vacancies and antisite defects in gallium phosphide is presented in Sect. 5.1.1. Experimental results of as-grown GaP, as well as studies of GaP after fast-particle irradiation, are reviewed in Sect. 5.1.2.

5.1.1 Vacancy- and Antisite-Related Defect Levels

Point defects such as vacancies and antisites usually have localized ionization levels in the bandgap of the semiconductor. Depending on the position of the Fermi level, they may exist in different charge states. Calculated ionization levels for vacancies and antisites in GaP (Puska 1989a) are shown in Fig. 5.1. The calculations were carried out without taking into account the lattice relaxation. A negative-U

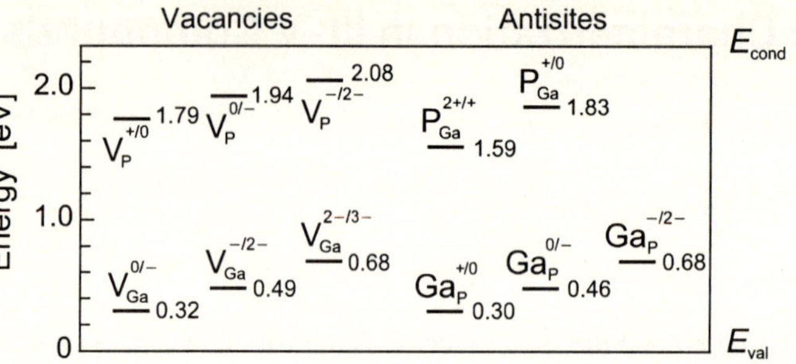

Fig. 5.1. Ionization levels of vacancies and antisites in the bandgap of gallium phosphide calculated without lattice relaxation (Puska 1989a). The positions of the levels are given with respect to the top of the valence band E_{val}. E_{cond} is the minimum of the conduction band.

behavior is thus a priori ruled out. It is expected that the energy levels are shifted in reality towards the corresponding band edges, i.e. the levels of Ga vacancies are closer to the valence band and the levels of phosphorus vacancies closer to the conduction band (Puska 1995). Another theoretical calculation of the vacancy levels was performed by Xu (1990) and provided qualitatively similar results.

Gallium vacancies can be neutral or negatively charged and should thus always be detectable by positrons. In contrast, phosphorus vacancies may appear in negative, neutral, and positive charge states. Depending on the Fermi-level position, phosphorus vacancies are thus positron traps or not.

The negative charge state is essential for the appearance of an antisite as a shallow positron trap (Sect. 3.1.5.2). According to theoretical calculations, P_{Ga} does not occur in a negative charge state and is consequently not a shallow positron trap independent of the Fermi-level position.

5.1.2 Experimental Results of As-Grown and Irradiated Gallium Phosphide

The first positron annihilation study of defects in gallium phosphide was reported by Brudnyi et al. (1983). n-type GaP crystals irradiated at room temperature by 2.3-MeV electrons up to a dose of 1×10^{19} cm^{-2} were electrically characterized and studied by angular correlation of annihilation radiation. It was found that positrons were trapped by irradiation-induced vacancy-type defects of acceptor character. The trapping rate increased with temperature in the range 77 to 300 K. All deep positron traps disappeared after annealing up to 500 °C.

Doppler-broadening experiments were performed by Sen Gupta et al. (1985) in GaP irradiated with 22-MeV α-particles. The increase in the line shape parameter S indicated the clustering of vacancies to larger agglomerates between 200 and

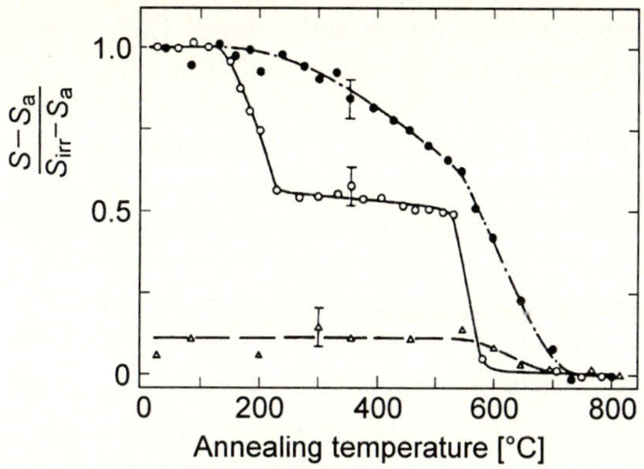

Fig. 5.2. Isochronal annealing curves of irradiated gallium phosphide (Dlubek et al. 1988a). The normalized line shape parameter of angular correlation of annihilation radiation, $(S - S_a)/(S_{irr} - S_a)$, is plotted against the annealing temperature for proton- (○) and neutron-irradiated (●) gallium phosphide samples. S_{irr} and S_a denote the S parameters after irradiation and after annealing at 800 °C, respectively. The irradiation experiments were carried out with doses of 2.3×10^{16} cm^{-2} (protons) and 8×10^{18} cm^{-2} (neutrons). For comparison, the data of an unirradiated reference sample (△) are included.

350 °C. The annealing of these defects was observed in a similar temperature range to that found by Brudnyi et al. (1983).

A distinct difference in the annealing behavior of defects created during room-temperature irradiation by protons or neutrons was found by Dlubek et al. (1986b, 1988a) in GaP (Fig. 5.2). Angular correlation of annihilation radiation and positron lifetime measurements were carried out as a function of the irradiation dose and the annealing temperature. Two annealing stages at about 200 and 600 °C were detected in proton-irradiated samples, whereas neutron-irradiated samples exhibited only one broad annealing range between 300 and 700 °C. The 200 °C stage was attributed to specific defect reactions resulting in more stable complexes rather than defect annealing. This conclusion was drawn in conjunction with internal friction and Rutherford backscattering measurements. The second annealing stage of the proton-irradiated samples is in a reasonable agreement with the stage of the neutron-irradiated sample. Apparently, defect complexes similar to those observed in the proton-irradiated samples at annealing temperatures above 200 °C were formed during the irradiation with neutrons. These complexes were supposed to anneal at the 700 °C stage resulting in the complete disappearance of the defect-specific positron signal.

Low-temperature 3-MeV electron irradiation experiments were performed at 20 K by Sen Gupta et al. (1988b). A first annealing stage in the average positron lifetime was found in the range 200 to 300 K and attributed to the annealing of isolated Ga vacancies. This experimental finding was confirmed by a more recent study of Polity et al. (1995a). Low-temperature electron irradiation was carried out at 4 K in semi-insulating nitrogen-doped and n-type sulfur-doped GaP, respectively. Positron lifetime measurements were performed at 100 K (Fig. 5.3).

The carrier density determined by Hall effect measurements was below 1×10^{10} cm^{-3} after irradiation to a dose of 1×10^{18} cm^{-2}. The sample was hence converted into the semi-insulating state corresponding to a Fermi-level shift close to the midgap position. According to the calculations of Puska (1989a), see Fig. 5.1, only Ga vacancies should be visible for positrons in this conductivity state. It was therefore concluded that annealing of Ga vacancies was observed in this first annealing stage. A defect-related lifetime of $\tau_d = 260$ ps was separated and was thus related to the gallium vacancy. Theoretical calculations provided a positron

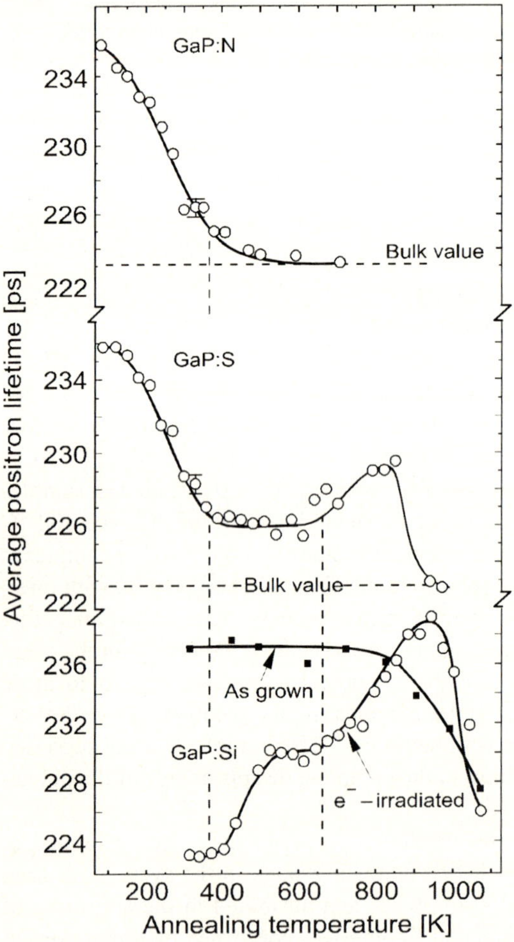

Fig. 5.3. Average positron lifetime as a function of the annealing temperature in nitrogen-, sulfur-, and silicon-doped electron-irradiated gallium phosphide (Polity et al. 1995a).

lifetime of 264 ps for V_{Ga} (Puska et al. 1989b). The observed bulk lifetime of $\tau_b = 223$ ps (Krause-Rehberg et al. 1993c) is in good agreement with the calculations of Puska et al. (1989b).

The first annealing stage was missing in the investigations of Krause-Rehberg et al. (1993c), since the Si-doped GaP specimen (lower panel of Fig. 5.3) was irradiated at room temperature. It was concluded that Ga vacancies had annealed already during the irradiation. Surprisingly, a positron lifetime close to the bulk value was found after irradiation, whereas an increased average lifetime of $\tau = 237$ ps was measured in the non-irradiated sample. This behavior can be understood in conjunction with the results of the correlated Hall effect measurements (Fig. 5.4). The carrier density decreased after irradiation below the detection limit of about $1\times10^{10}\,\text{cm}^{-3}$. Consequently, the Fermi level of the irradiated sample is shifted from a position close to the conduction band prior to irradiation to a mid-gap position. Ga vacancies should be detectable in the irradiated state, and P vacancies should not (Fig. 5.1). Because V_{Ga} annealed during irradiation and V_P were invisible, it was reasonable to observe no vacancy signal and to detect the positron bulk lifetime after irradiation.

A defect-related lifetime of $\tau_d = 295$ ps was observed in the as-grown state. A similar lifetime was found at room temperature by Dannefaer and Kerr (1994). The corresponding vacancy-type defect may be attributed to the phosphorus vacancy, since this positron-lifetime component vanishes with the shift of the Fermi level E_F after irradiation. It should be noted that a defect complex having a level in the upper half of the bandgap and containing a monovacancy may also be the observed trapping center. This possibility will be ignored in the following discussion, since positron lifetime spectroscopy is not able to provide such structural information.

A recovery of the V_P-related lifetime component was observed in two stages in the isochronal annealing experiment (lower panel of Fig. 5.3). This recovery goes along with a continuous increase in the carrier concentration up to a value close to that of the GaP:Si sample prior to irradiation. It was concluded that the Fermi level crosses two ionization levels of the phosphorus vacancy during the annealing experiment. Supposing no negative-U behavior, the first stage must be attributed to the +/0 level, since the average positron lifetime starts to rise above the bulk level. This can be related to the fact that positive vacancies repel positrons. The second stage is then due to the 0/− level. An average lifetime of 239 ps is reached in this second recovery stage. This is only slightly above the value observed in the as-grown state. Obviously, the phosphorus vacancy concentration after 900-K annealing is similar to the concentration prior to irradiation. This finding was explained by Krause-Rehberg et al. (1993c) by the disappearance of irradiation-induced vacancies during annealing below 900 K. The remaining defect signal originates in phosphorus vacancies related to the stoichiometric deviation of the as-grown crystal. The irradiation-induced vacancies are supposed to occur in close Frenkel pairs having a lower annealing temperature than the grown-in defects. The

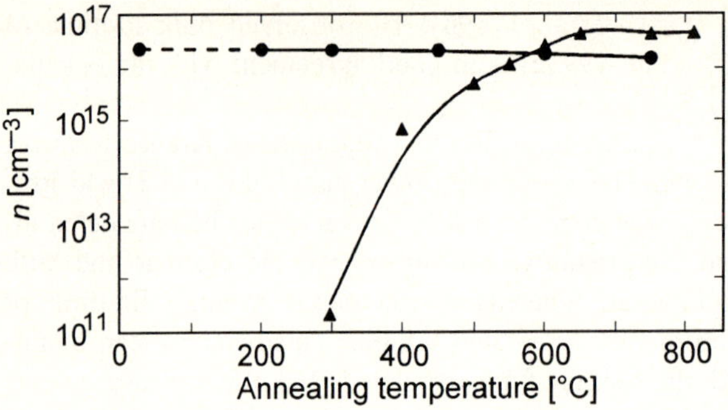

Fig. 5.4. Carrier density n as a function of the annealing temperature in as-grown and electron-irradiated silicon-doped gallium phosphide (Krause-Rehberg et al. 1993c). The circles correspond to the unirradiated reference sample, and the triangles to the sample irradiated with 1-MeV electrons to a dose of 1×10^{18} cm^{-2} at room temperature.

annealing of the grown-in phosphorus vacancies occurs above 900 K in both as-grown and irradiated material.

The position of the Fermi level could be monitored by the Hall effect measurement (Fig. 5.4). Since positron annihilation is sensitive to the charge state, the plot of the average positron lifetime $\overline{\tau}$ versus $E_{\mathrm{cond}} - E_{\mathrm{F}}$ indicates the charge transfer of the phosphorus vacancy, at least for the recovery stage between 700 and 900 K (Fig. 5.5). The position of the Fermi level shifted in the annealing experiment of the electron-irradiated sample due to the vanishing of deep-level defects and the conversion of the sample from semi-insulating to n-type conductivity (lower curve in Fig. 5.5). No Hall effect data were available for the first recovery stage.

A similar charge state transition of the phosphorus vacancy can be achieved by the variation of the temperature for as-grown GaP samples with a density of free carriers of about 10^{14} to 10^{16} cm^{-3}. This is essentially based on the temperature dependence of the density of states. The upper curve in Fig. 5.5 shows the results of such an experiment. The average positron lifetime decreased as a function of the Fermi-level shift indicating the charge state transition of the phosphorus vacancy.

The full lines of Fig. 5.5 correspond to the numerical fit of the data. The position of the level E_{d} can be calculated by a fit of the average positron lifetime taking into account the occupation probability of the level, f, according to the Fermi–Dirac statistics,

$$f = \left[1 + \frac{1}{g}\exp\left(\frac{E_{\mathrm{d}} - E_{\mathrm{F}}}{k_{\mathrm{B}}T}\right)\right]^{-1} . \tag{5.1}$$

g is the degeneration factor. The average positron lifetime is obtained within the framework of the trapping model with one defect type (Sect. 3.2.1.1). The trapping rate has to be modified to

$$\kappa = f\mu^{-}C + (1 - f)\mu^{0}C. \tag{5.2}$$

μ^- and μ^0 are the trapping coefficients of the negative and neutral states of the vacancies with the total concentration C, respectively.

The results of both experiments in Fig. 5.5 are consistent with an ionization level position in the bandgap of $E_{cond} - E_d = (270 \pm 20)$ meV (Krause-Rehberg et al. 1993c). Temperature-dependent positron lifetime measurements by Dannefaer and Kerr (1994) did not show such a dependence in highly S-doped GaP. This is in accordance with the idea of the positron trapping in charged V_P as explained above, since the Fermi level in their samples remained close to the conduction band and did not cross the ionization level of the phosphorus vacancy.

The annealing curves of the GaP:N and GaP:S samples (upper two curves in Fig. 5.3) can easily be understood in the light of the results on GaP:Si. The reason for the increase in the average positron lifetime in the S-doped sample between 700 and 900 K is the same as found for the Si-doped sample, viz. the increase in the positron trapping in native phosphorus vacancies related to the recovery of the n-type conductivity during annealing. This effect could not be observed in the nitrogen-doped sample, since it remained semi-insulating during the annealing.

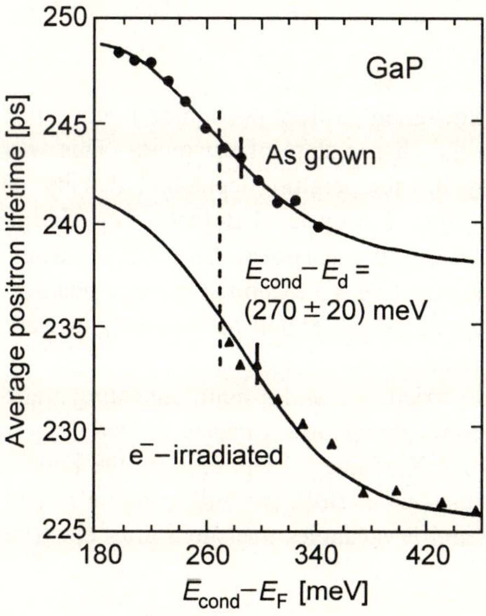

Fig. 5.5. Average positron lifetime in silicon-doped gallium phosphide versus the Fermi-level position, $E_{cond} - E_F$ (Krause-Rehberg et al. 1993c). The Fermi-level shift was obtained for the lower curve by isochronal annealing after electron irradiation. The shift of the Fermi level in a comparable energy range was achieved for a weakly silicon-doped as-grown gallium phosphide sample by the variation of the measurement temperature (*upper curve*). Consistent with both measurements, the 0/− ionization level was determined by a numerical fit (*full lines*) to be at $E_{cond} - E_d = (270 \pm 20)$ meV.

Table 5.1. Positron lifetime data of bulk and vacancy-type positron traps in gallium phosphide.

	Positron annihilation characteristics Lifetime [ps]	τ_d/τ_b	Dissolution temperature [K]
Bulk lifetime	222[a], 223[b,c]	—	—
Ga monovacancy	260[c]	1.17	< 300[b,c]
P monovacancy	295[a,b,c]	1.32	700 to 900[b, c]

τ_d is the defect-related positron lifetime, τ_b the positron lifetime in the defect-free bulk.
[a]Dannefaer and Kerr (1994), [b]Krause-Rehberg et al. (1993c), [c]Polity et al. (1995a).

The concept of the dependence of positron trapping on the charge state of vacancies, i.e. positive vacancies are virtually no positron traps, and neutral and negative vacancies are detected with different sensitivities, could in no other material be demonstrated in such a clear way as in gallium phosphide. In special cases as shown in the example above, it is not only possible to identify defects and to give their concentrations, but also to determine their ionization levels in the bandgap.

5.1.3 Summary

The outstanding result of positron experiments on gallium phosphide is the proof of the sensitivity of positron trapping to the charge state of vacancies. This was achieved by shifting the Fermi level by the temperature-dependent density of states or by introduction of irradiation-induced deep-level defects. The charge state of phosphorus vacancies was changed, and it was shown that positive vacancies are not detected, whereas neutral and negative V_P exhibited distinct positron trapping. The positions of ionization levels in the upper half of the bandgap were estimated from positron lifetime measurements.

Electron irradiation experiments were carried out, and a main annealing stage was found below room temperature. In-grown phosphorus-vacancy-related defects were found to anneal in n-type GaP at 1000 K. Table 5.1 summarizes the known positron-lifetime parameters of gallium phosphide. Both the bulk lifetime τ_b and the defect-related lifetimes τ_d for the Ga and P vacancies measured after electron irradiation are given.

5.2 Gallium Arsenide

Gallium arsenide is the most important compound semiconductor. In contrast to other compounds, the GaAs market has reached a considerable size and is still expanding. The market share of GaAs was 1.6 billion US$ in 1996. An increase to

2.2 billion US$ is expected for the year 2000. Unique physical properties, e.g. the high carrier mobility, make this material interesting for ultra-fast device applications. Field-effect transistors with intrinsic transition frequencies up to 220 GHz have been produced under laboratory conditions with GaAs-based pseudo-ternary III–V compounds (Aniel et al. 1993). In addition, optoelectronic applications, such as light-emitting diodes and lasers based on GaAs pseudo-ternary and quaternary compounds, make the material attractive.

In contrast to elemental semiconductors, a multitude of native defects occur in GaAs, e.g. antisite defects and vacancies in both sublattices. The fact that the presence of these defects dominates the electrical and optical features and, furthermore, the promising properties have stimulated intensive research, making GaAs a model substance in semiconductor physics.

Starting in the 1970s, the first investigations of defects in GaAs by positron annihilation were carried out (Aref'ev et al. 1979; Cheng et al. 1979; Dannefaer 1982; Kerr et al. 1982; Kuramoto et al. 1973; Pogrebnyak et al. 1983). The peculiarities of positron trapping in semiconductors, e.g. the influence of the defect charge, were, however, not sufficiently understood for a reliable interpretation of the experimental data at that time. In the meantime, a large number of results, which have essentially contributed to the understanding of the physics of defects in semiconductors, have become available.

The aim of this section is to give an overview of the most important positron results obtained in as-grown GaAs (Sects. 5.2.1 to 5.2.4), after post-growth annealing (Sect. 5.2.5), fast-particle irradiation, ion implantation (Sect. 5.2.6), plastic deformation (Sect. 5.2.7), and saw-cutting (Sect. 5.2.8).

5.2.1 As-Grown Undoped Gallium Arsenide

There are six elemental point defects in GaAs: Ga and As vacancies, Ga and As interstitials, and, moreover, Ga_{As} and As_{Ga} antisites. A variety of complexes of these defects may be possible. The detectability of these defects by positron annihilation is limited. Only non-positively charged vacancies are detectable and can be unambiguously identified. Additionally, negatively charged non-open volume defects acting as shallow positron traps may be detected at low temperatures.

It is obvious that knowledge of the charge state of the defects under investigation is very important for the interpretation of positron data. The ionization levels of Ga and As vacancies, as well as of antisites, have been calculated in several theoretical studies (Baraff and Schlüter 1985a; Jansen and Sankey 1989; Puska 1989a; Seong and Lewis 1995; Xu and Lindefelt 1990; Zhang and Northrup 1991). It could be shown that lattice relaxation is important, and cannot be ruled out, even for monovacancies (Laasonen et al. 1992). The strongest relaxation effect is found in the first-principles calculations for negatively charged As vacancies.

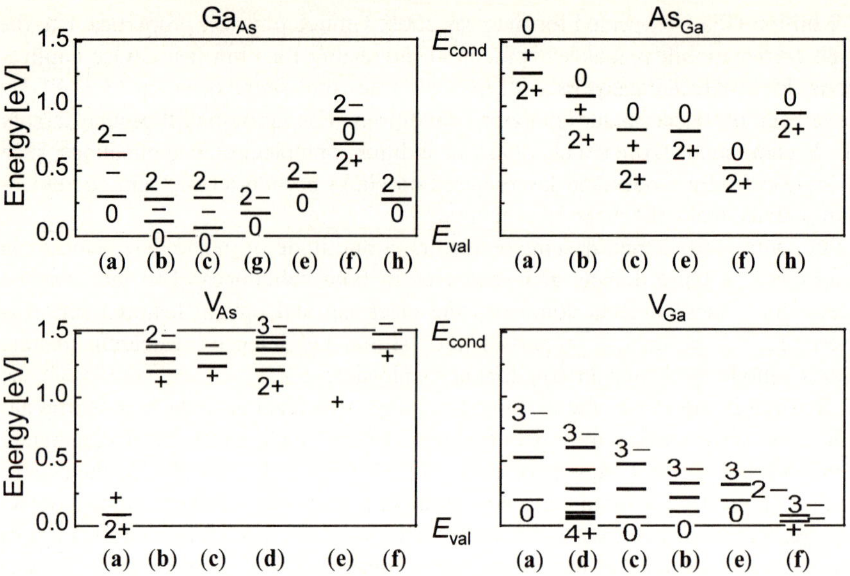

Fig. 5.6. Ionization levels of arsenic vacancies, gallium vacancies, and antisites according to theoretical calculations of (**a**) Baraff and Schlüter (1985a), (**b**) Puska (1989a), (**c**) Jansen and Sankey (1989), (**d**) Xu and Lindefelt (1990), (**e**) Zhang and Northrup (1991), and (**f**) Seong and Lewis (1995), (**g**) Zhang and Chadi (1990), (**h**) Pöykkö et al. (1997). E_{val} and E_{cond} are the edges of the valence and the conduction band, respectively.

The available theoretically calculated ionization levels of As vacancies, Ga vacancies, and antisites are shown in Fig. 5.6. The general trend is that the As vacancy is found in the singly positive charge state and the Ga vacancy is triple negatively charged for the Fermi level in the midgap position.

One has to conclude for positron annihilation measurements that both types of vacancies may be detected in heavily n-doped GaAs. With lowering of the Fermi level to the midgap position, the As vacancy becomes invisible for positrons and the Ga vacancy may still be detected. Most calculations predict for heavily p-doped GaAs that the Ga vacancy should occur in the neutral charge state. Charge state transition to a positive Ga vacancy was, however, found in the calculations of Seong and Lewis (1995). It is therefore not entirely clear from the viewpoint of these calculations whether the Ga vacancy is detectable by positron annihilation in heavily p-doped GaAs or not (see discussion in Sect. 5.2.6.2).

The ionization levels of interstitials in different configurations were also calculated (Baraff and Schlüter 1985a; Zhang and Northrup 1991). It was found that they should always be positively charged or neutral. A more recent calculation of Chadi (1992) revealed, however, a negative charge state for the ⟨110⟩-split As interstitial configuration. Landman et al. (1997) calculated the formation energy of As interstitials for arsenic-rich GaAs, which was found in the same order as for vacancies. Acceptor-like states were predicted for the split As interstitial. As a consequence, it cannot be entirely excluded that As interstitials act as shallow positron traps at low temperatures.

No reliable experimental detection of the ionization levels of the Ga_{As} antisite has been made up to now. The levels at 203 and 78 meV above the valence band edge, previously ascribed to the Ga_{As} antisite, have been attributed recently to boron at an arsenic site (Magerle et al. 1997; Newman et al. 1994). Most of the theoretical calculations provide consistently acceptor levels in the lower half of the bandgap. Two charge state transitions $0/-$ and $-/2-$ were theoretically predicted (Baraff and Schlüter 1985a; Jansen and Sankey 1989; Puska 1989a). The calculations of the fully relaxed antisite configuration by Pöykkö et al. (1997) and Zhang and Northrup (1991) gave a negative-U behavior with a $0/2-$ level (Fig. 5.6). The negative charge state is essential for the Ga_{As} antisite to appear as a shallow positron trap. The transition from the neutral to a negative charge state is expected according to these theoretical calculations between 0.06 and 0.38 eV above the valence band edge. Thus, the Ga_{As} antisite should act as a shallow positron trap, except in heavily p-doped GaAs.

In contrast to earlier theoretical works, ionization levels of Ga_{As} were found at $E_d - E_{val} = 0.706$ eV and 0.896 eV for the $2+/0$ and $0/2-$ transitions, respectively, in the calculations by Seong and Lewis (1995). The $0/2-$ transition, important for the effect of Ga_{As} as a shallow positron trap, is located in the upper half of the bandgap. According to these calculations, Ga_{As} antisites should not act as shallow positron traps in semi-insulating GaAs.

Consistently, the calculations and the experiments provide no negative charge state for the As_{Ga} antisite, which is related to the EL2 defect. Calculations taking into account the lattice relaxation provide a $2+/0$ transition (Pöykkö et al. 1997; Seong and Lewis 1995; Zhang and Northrup 1991). The levels of the EL2 defect identified experimentally by electron paramagnetic resonance (Weber et al. 1982) lie at $E_{cond} - E_d = 770$ meV $(+/0)$ and $E_d - E_{val} = 520$ meV $(++/+)$. As a consequence, this defect is not a positron trap, at least in the stable ground state. The EL2 defect exhibits, however, a metastability, which is characterized by its optical and electrical inactivity. The influence of the transition of EL2 to its metastable state on positron trapping is discussed in Sect. 5.2.1.2.

5.2.1.1 Arsenic and Gallium Vacancies

A vast number of publications dealing with positron annihilation in GaAs have been published since the mid-1980s. Many positron groups previously investigating metals turned to the problems of defects in semiconductors, mainly in Si and GaAs. The experience of the metal investigations was directly transferred to semiconductors, with the consequence that contradictions of results of other methods arose. The densities of vacancy-type defects were found to be too large by at least one order of magnitude. As an example, Dlubek et al. (1986a) studied undoped

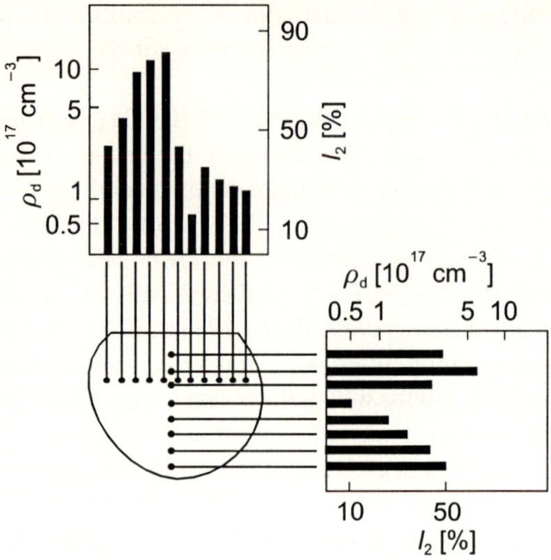

Fig. 5.7. Distribution of the vacancy density ρ_d across an undoped polycrystalline gallium arsenide wafer. The diameter of the wafer is 60 mm. The intensity of the second lifetime component I_2 is given on the second axis. The defect-related lifetime τ_2 was fixed at 295 ps. The vacancy density was calculated using a trapping coefficient of 1×10^{14} s^{-1} (Dlubek et al. 1986a).

polycrystalline GaAs and found a variation of the vacancy distribution between 5×10^{16} and 1×10^{18} cm^{-3} across the wafer (Fig. 5.7). The vacancy densities were calculated with a trapping coefficient of about $\mu = 1\times10^{14}$ s^{-1}, comparable to values obtained in metals. It was realized by the end of the 1980s that the trapping coefficient must be higher (e.g. Dlubek and Krause 1987). One reason is the appearance of a negative charge of the vacancies. Maximum vacancy densities of several 10^{16} cm^{-3}, which are compatible to the results of electrical measurements, are obtained with a more accurate trapping coefficient of about 1×10^{15} s^{-1} in undoped as-grown GaAs.

A few papers dealing with the characterization of vacancies in GaAs by positron annihilation were published at the Seventh International Conference on Positron Annihilation in New Delhi (Dlubek et al. 1985b; Stucky et al. 1985; Uedono et al. 1985). Further positron studies of as-grown GaAs were published in this time (Dannefaer et al. 1984; Dannefaer and Kerr 1986b; Dlubek et al. 1986a; Kerr et al. 1982; Stucky et al. 1986). This first period of the investigation of vacancies in GaAs was summarized in a review by Dlubek and Krause (1987).

The main result of these papers was that a vacancy-related lifetime of $\tau_d = 295$ ps was found in undoped n-type GaAs. The bulk lifetime was determined to be 230 ps. The ratio $\tau_d/\tau_b = 1.28$ indicated the presence of a monovacancy.

Another important result of these papers was that the positron capture in vacancies in semiconductors is ruled by the charge state of the defects. This charge state for defects having states in the bandgap is determined by the position of the Fermi level. The effect of the influence of the vacancy charge on positron trapping was demonstrated by the investigation of undoped GaAs samples of carrier concentrations between 10^9 and 10^{16} cm^{-3} (Dlubek and Krause 1987). The average positron lifetime is plotted in Fig. 5.8 versus the Fermi level position, which was determined by Hall effect measurements. Also, average lifetimes lying below the full line were found. Only the maximum values are shown. These maximum values are controlled by the Fermi-level position and reflect the transition of a vacancy to a positive charge state. It is concluded, according to Fig. 5.6, that the As vacancy is observed. The theoretical calculations have not predicted a negative-U behavior of this vacancy and thus the +/0 level is probably detected. It cannot be excluded from the positron measurements that a defect complex containing a vacancy may be responsible for positron trapping. However, no such defect levels are known in the upper half of the bandgap. It is therefore assumed that indeed the isolated As vacancy is observed, although other transitions were also discussed by Dlubek and Krause (1987). The position of this level E_d can be calculated with a fit of the average positron lifetime (full line in Fig. 5.8) using (5.1). Equation (5.2) has to be modified for the fit to

$$\kappa = f\mu^0 C + (1 - f)\mu^+ C. \tag{5.3}$$

μ^+ is the trapping coefficient to positively charged vacancies. The bulk lifetime is reached for a Fermi level deep in the bandgap (Fig. 5.8), i.e. no positron trapping occurs. The second term in (5.3) can be neglected, since $\mu^0 \gg \mu^+$, and the equation simplifies to $\kappa = f\mu^0 C$. The result of the numerical fit is the ionization level of the

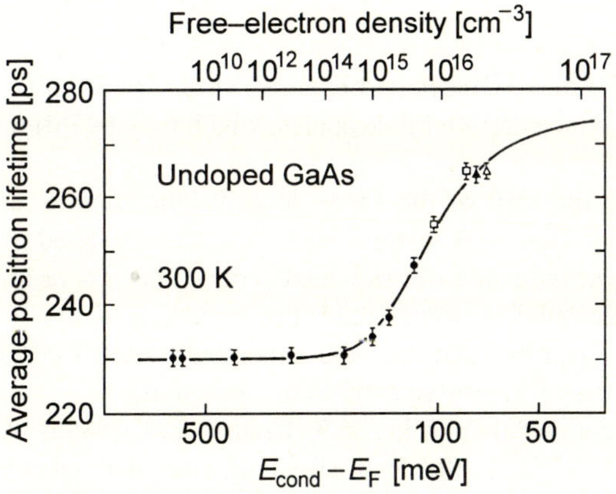

Fig. 5.8. Average positron lifetime in as-grown gallium arsenide samples with free-electron densities versus the Fermi-level position in the bandgap, $E_{cond} - E_F$. In addition to the results of Dlubek and Krause (1987), the values of Stucky et al. (1985, 1986) are included (□). The full line is the fit according to the one-defect trapping model (Sect. 3.2.1.1), using additionally (5.1) and (5.3).

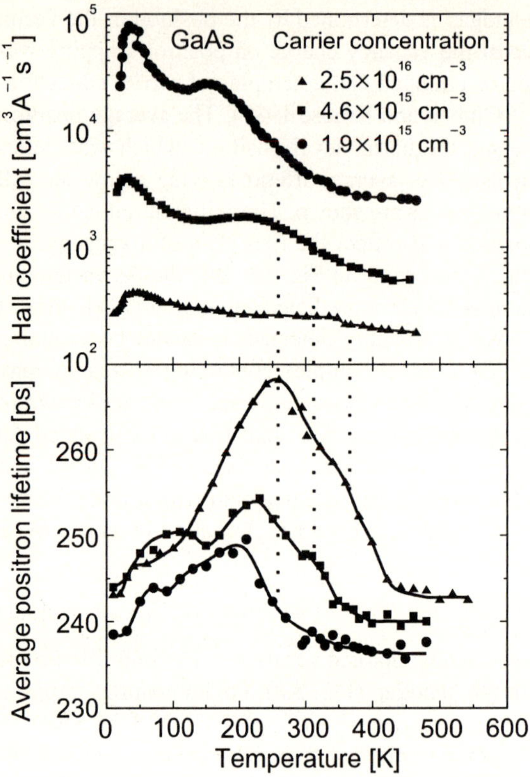

Fig. 5.9. Average positron lifetime and Hall coefficient versus temperature for undoped gallium arsenide with three different carrier densities (Krause et al. 1989a).

As vacancy, $E_{cond} - E_d = (60 \pm 20)$ meV (Dlubek and Krause 1987). The accuracy of this result is influenced by the concentration of vacancies, which may be different in the samples of Fig. 5.8.

A better experiment would be the shift of the Fermi level in one sample, as demonstrated for gallium phosphide (Sect. 5.1.2). Such a shift could be realized by carrier compensation, e.g. via the introduction of irradiation-induced defects or in-diffusion of acceptors. The Fermi-level position could also be changed by the temperature dependence of the density of states in the conduction band by the variation of the temperature in GaAs of a medium carrier concentration.

The result of such an experiment is shown in Fig. 5.9 (Krause et al. 1989a). A charge state transition becomes visible in the Hall effect measurement above 200 K. The three Hall coefficient curves consistently provide a single defect level at $E_{cond} - E_d = (120 \pm 20)$ meV. The decrease in the average positron lifetime in this region is Fermi-level controlled. This is demonstrated in Fig. 5.10, where the lifetime data of Fig. 5.9 are plotted against the Fermi-level position. The values of all three samples lie on a common curve, indicating the same physical mechanism

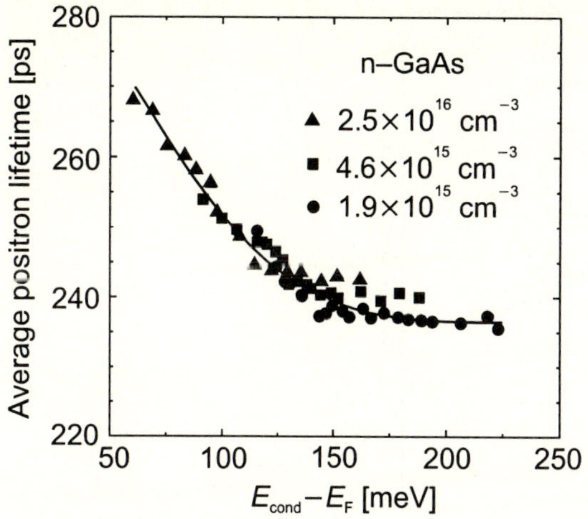

Fig. 5.10. Average positron lifetime as a function of the Fermi-level position $E_{\mathrm{cond}} - E_{\mathrm{F}}$. The Fermi-level variation was achieved by changing the temperature. The lifetime values are taken from Fig. 5.9.

of the temperature dependence of positron trapping. The solid line shown in Fig. 5.10 is the result of a fit similar to Fig. 5.5 providing the defect level in good agreement with the experimental results of the Hall effect measurements as $E_{\mathrm{cond}} - E_{\mathrm{d}} = (100 \pm 30)$ meV. It should be noted that positron trapping in charged defects is a rather complex function of the temperature (Sect. 3.1.5.1) and the multi-component fit is not straightforward. The decrease in the average lifetime in the low-temperature range is attributed to the presence of shallow positron traps (see Sect. 5.2.2).

A similar experiment of a Fermi-level shift by variation of the sample temperature was carried out by Corbel et al. (1988) and Saarinen et al. (1991). Comparable results to those shown in Fig. 5.10 were obtained with undoped and weakly n-doped GaAs (Fig. 5.11). A charge state transition from neutral to negative was postulated in the former paper of Corbel et al. (1988) and later assigned to the +/0 transition (Saarinen et al. 1991). In the latter paper, the position of the ionization level was obtained to be 140 meV below the conduction band edge with the same experimental procedure using a more extended series of GaAs samples with carrier densities ranging from 2×10^{15} to 6×10^{16} cm^{-3}.

Further evidence for the existence of a +/0 charge state transition of the As vacancy in the upper half of the bandgap was found by an annealing experiment of undoped GaAs (Dlubek and Krause 1987). The sample was stepwise annealed in a quartz tube under He atmosphere. The average positron lifetime decreased with increasing annealing temperature (left panel in Fig. 5.12) due to progressive compensation by in-diffusing acceptor-type impurities. The compensation was proven by correlated optical and electrical measurements (Dlubek 1985). The positron

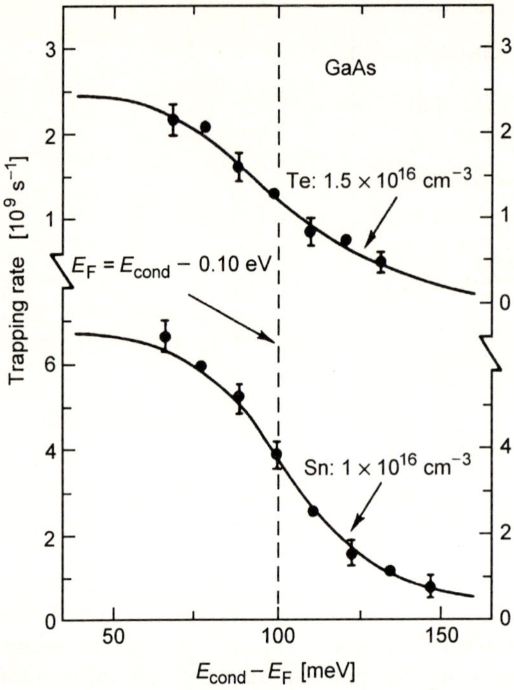

Fig. 5.11. Positron trapping rate of arsenic vacancies in n-type gallium arsenide as a function of the Fermi-level position $E_{cond} - E_F$. The dashed line indicates the ionization level at E_{cond} − 100 meV (Corbel et al. 1988). The solid lines represent the occupation probability of the two ionization states of the arsenic vacancy according to (5.1) and (5.3).

capture was almost eliminated due to the transition to the positive charge state of the As vacancy. The annealing of the defect was excluded, because the positron trapping reappeared after the removal of a near-surface layer by several etching steps (right panel in Fig. 5.12).

A similar compensation to that obtained by the in-diffusion of impurities can be achieved by electron irradiation of undoped or weakly doped GaAs. Such an experiment was carried out by Stucky et al. (1985). They found a pronounced decrease in the average positron lifetime with increasing electron dose. It may be attributed to the 0/+ charge state transition of the As vacancy owing to the Fermi-level shift to the midgap position after irradiation.

The positron lifetime measurements of Dlubek and Krause (1987) and Stucky et al. (1985) were carried out at room temperature. The Fermi level was shifted by the introduction of deep centers compensating the free-carrier concentration, in contrast to the shift of the Fermi level by the variation of temperature (Fig. 5.10 and Fig. 5.11). The measurement of the temperature-dependent positron lifetime alone would allow another interpretation: a similar increase in the average lifetime

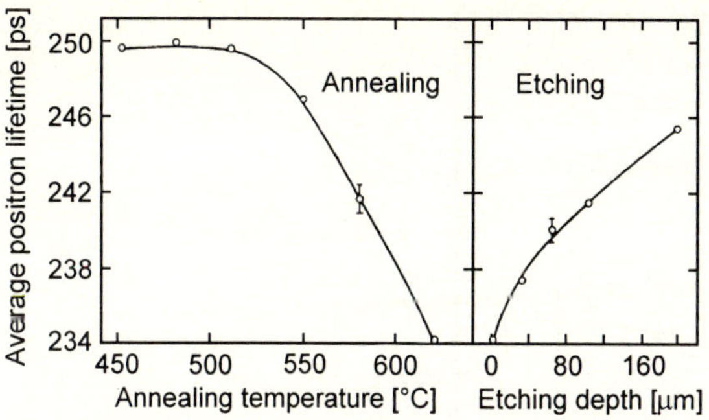

Fig. 5.12. Average positron lifetime as a function of the annealing temperature (*left panel*) measured in undoped gallium arsenide (carrier density $n = 2\times10^{16}$ cm^{-3}). The annealing was carried out in a quartz tube under helium atmosphere (Dlubek and Krause 1987). The right panel shows the recovery of the average positron lifetime after several etching steps.

with falling temperature, as observed in Fig. 5.9 above 200 K, is expected according to (3.17) for a negative vacancy. Such a dependence is observed e.g. for the negative Ga vacancy, without any charge state transitions (compare Fig. 5.34). The compensation experiments are, therefore, of special importance for the identification of the vacancy, as they allow one unambiguously to monitor the transition to a positive charge state in the upper half of the bandgap. Assuming an isolated monovacancy, the only candidate for this defect is the As vacancy (Fig. 5.6).

A similar decrease in the average positron lifetime in the low-temperature range as in Fig. 5.9 was found by Corbel et al. (1988) and Saarinen et al. (1991) in weakly n-doped GaAs. The vacancy-related lifetime component τ_2 exhibited a significant decrease from 295 to 258 ps with lowering the temperature. This decrease is not temperature-controlled but seems to be a function of the Fermi-level position (Fig. 5.13). As several ionization levels of the As vacancy were theoretically predicted, it was supposed that the observed effect was due to one of these levels. It was estimated to be located at 35 meV (Corbel et al. 1988) or 30 meV (Saarinen et al. 1991) below the conduction band edge. It was attributed to the $-/2-$ transition of V_{As} in the former paper and reinterpreted as the $0/-$ level in the latter. The observed change in τ_2 corresponds to a strong change in the open volume of the defect. Indeed, a very strong inward relaxation was found in the theoretical calculations of the $0/-$ transition by Laasonen et al. (1992). The most recent calculations of the ionization levels of the As vacancy by Seong and Lewis (1995) predicted two states in the upper half of the bandgap. Whereas the $+/0$ level at $E_{cond} - 130$ meV is in good agreement with experimental results, the $0/-$ level is a very shallow donor level at $E_{cond} - 2$ meV (Fig. 5.6). With the accuracy of the calculations given as 20 meV (Seong and Lewis 1995), the experimental and theoretical results for this level may be in a reasonable agreement. However, it cannot be excluded from the calculations that the level lies resonant in the conduction band. In this case, no further ionization level should be detectable.

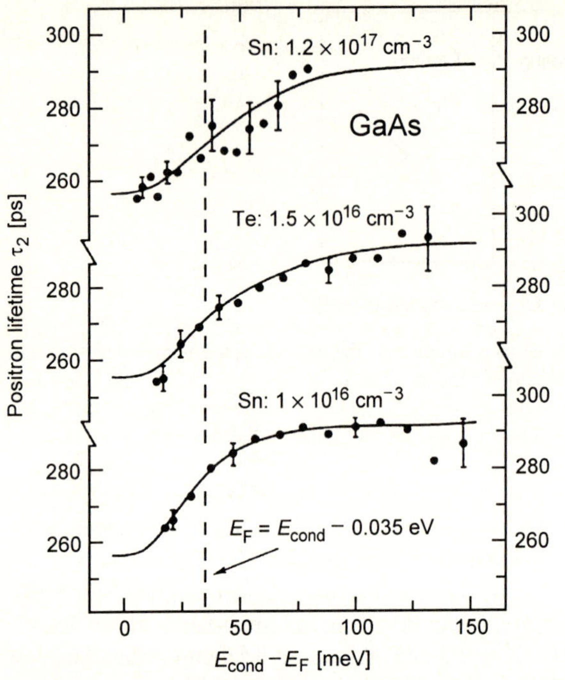

Fig. 5.13. Defect-related lifetime τ_2 separated from positron-lifetime measurements at different temperatures. τ_2 is plotted for differently n-doped gallium arsenide samples versus the position of the Fermi level, $E_{cond} - E_F$, calculated from the carrier concentration and the sample temperature (Corbel et al. 1988). The dashed line indicates the position of the vacancy ionization level at E_{cond} − 35 meV.

Indeed, in contrast to the ionization level of the As vacancy at 120 meV, the 0/− level at about E_{cond} − 30 meV could not be found by Hall effect measurements (Krause et al. 1989a). The identification of this level by positron annihilation was based on the change in the defect-related lifetime component in the two-component decomposition of the spectra (Corbel et al. 1988; Saarinen et al. 1991). However, it has turned out that shallow positron traps are active positron traps in GaAs in the temperature range below 200 K (Saarinen et al. 1989). Trapping in these defects significantly influences the second lifetime component τ_2 if a two-component decomposition of the lifetime spectra is still applied. This has been shown in the analysis of Monte-Carlo simulated lifetime spectra by Somieski et al. (1996), where a decrease in τ_2 was found with the increasing influence of shallow positron traps, e.g. decreasing temperature.

In order to exclude the influence of shallow traps on the decomposition of the positron lifetime spectra, the Fermi-level was shifted not only by temperature variation, but also by the investigation of different carrier concentrations obtained by different doping. The result is shown in Fig. 5.14 (Saarinen et al. 1991). The in-

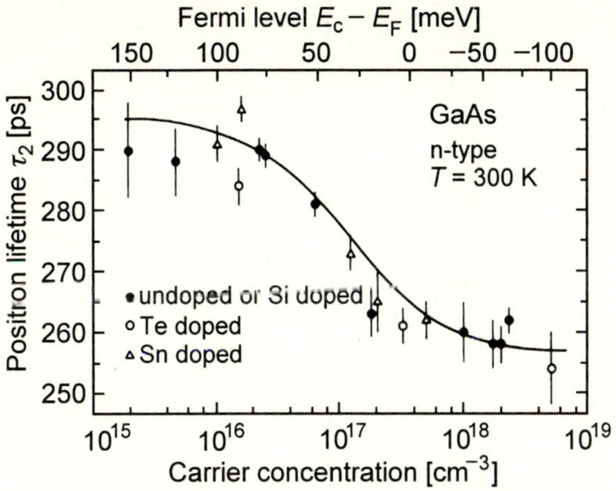

Fig. 5.14. Positron lifetime component τ_2 studied in n-type gallium arsenide as a function of the carrier concentration or the Fermi level $E_{cond} - E_F$, respectively (Saarinen et al. 1991). The solid line corresponds to the fit with the 0/− ionization level of the arsenic vacancy at 30 meV below the conduction band E_{cond}. The measurement temperature was 300 K.

fluence of shallow traps could be excluded at the measurement temperature of 300 K. The data seem to be in agreement with the decrease in the V_{As}-related lifetime due to a pronounced inward relaxation during the 0/− transition (solid line in Fig. 5.14). However, another interpretation for the observed decrease in τ_2 could be the occurrence of another vacancy-type defect. As is explained in Sect. 5.2.3, vacancy–dopant complexes are clearly identified in medium and highly n-doped GaAs (Gebauer et al. 1997c; Krause-Rehberg et al. 1995b; Laine et al. 1996b). These defect complexes exhibit a positron lifetime of 250 to 260 ps, which is reached in the measurements of Fig. 5.14 for high dopant concentrations.

A further indication for the detection of another defect type in highly n-doped GaAs may be the thermal stability of the vacancy-type defect. This is visible in Fig. 5.15 where the annealing behavior of the arsenic vacancy (Corbel et al. 1988) is compared with the stability of the $V_{Ga}Te_{As}$ complex (Krause-Rehberg et al. 1994b). A clear annealing stage at 750 K was found for V_{As}. A decrease in the average lifetime due to a charge-state transition to V_{As}^{+} could be excluded since the samples were still found to be n-type after annealing at 850 and 1000 K (Corbel et al. 1988). The observed defects in the highly doped sample of Fig. 5.15 are stable up to the highest applied temperature of 1050 K. This is a strong indication for the presence of two different vacancy-type defects in the two samples. The defect in the highly doped GaAs was assumed to be the $V_{Ga}Te_{As}$ complex (see discussion in Sect. 5.2.3.2). The small decrease of 2 ps at about 750 K for the sample with $n = 2.5\times10^{18}$ cm^{-3} was attributed to a weak contribution of the annealing of V_{As} at this temperature (Krause-Rehberg et al. 1994b).

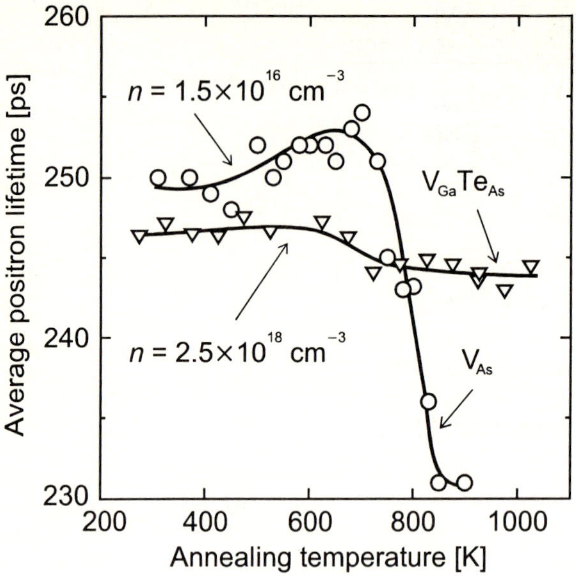

Fig. 5.15. Comparison of the thermal stability of arsenic vacancies (Corbel et al. 1988) in weakly doped n-type gallium arsenide (○) to complexes consisting of gallium vacancies and tellurium (Krause-Rehberg et al. 1994b) in heavily tellurium-doped GaAs (▽).

In order to obtain information on the electronic structure of the As vacancy, angular correlation of annihilation radiation was measured in two dimensions. The perspective and contour plots of the 2D-ACAR distribution measured in weakly n-doped GaAs are shown in Fig. 5.16 (Ambigapathy et al. 1996). It was intended to change the charge state of the As vacancy by changing the sample temperature, similar to the experiments of Corbel et al. (1988) and Saarinen et al. (1991). The ACAR plot of the GaAs sample, exhibiting no positron trapping in defects, was significantly different from the plot of the vacancy (Fig. 5.17). In order to obtain the defect-specific plots, a normalization for saturated positron trapping, i.e. all positrons are captured in defects, is necessary. This can be done by correlated positron lifetime measurements providing the annihilation fraction. The experimental contour plot are compared in Fig. 5.17 to calculations based on the Car–Parinello molecular-dynamics approach (Gilgien et al. 1994). The experimental and theoretical plots of defect-free GaAs are in good agreement. The measured and calculated plots for the negative As vacancy agree satisfactorily.

Similar ACAR plots to those shown in Fig. 5.16 were also taken for the As vacancy in the neutral charge state. The curves of both charge states are rather isotropic. The weak anisotropic parts of the ACAR curves (Ambigapathy et al. 1994, 1996) do not show any particular difference in the structure. This may indicate that the ACAR curves do not differ significantly for both charge states or only the neutral As vacancy is detected in both cases (see discussion above).

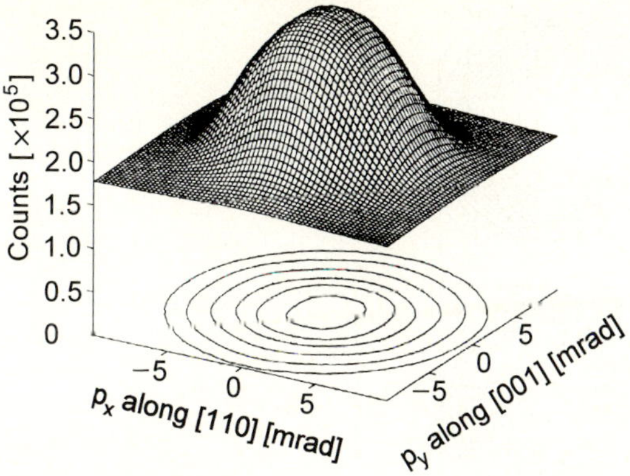

Fig. 5.16. Combined perspective and contour plots of two-dimensional angular correlation of annihilation radiation of negatively charged arsenic vacancies in gallium arsenide (Ambigapathy et al. 1996). The curve was normalized for saturated positron trapping. The perspective plot was vertically shifted by 1.15×10^5. p_x and p_y are the components of the electron momentum.

Positron annihilation experiments were performed with additional monochromatic illumination of the samples (Kuisma et al. 1996b; Saarinen et al. 1993). The GaAs:Zn sample taken as a reference did not show any positron trapping. Semi-insulating GaAs samples already exhibited at low temperatures without illumination an increase in the average positron lifetime, which is attributed to trapping in negatively charged Ga vacancies. This behavior is discussed below in more detail. A new vacancy signal is observed in undoped semi-insulating GaAs illuminated below 150 K with light of 1.42 eV (Fig. 5.18). The additional increase in $\bar{\tau}$ was related to the photoinduced excitation of electrons from the valence band to the As vacancy. It was concluded from the separated vacancy-related lifetime component of 255 ps that the negative charge state of V_{As} is reached (Saarinen et al. 1993). The density of the As vacancies was found in the order of 1×10^{16} cm^{-3}. The difference in the lifetime measured for the samples under illumination and in darkness vanished with increasing temperature at 150 K. This observation was interpreted as thermally stimulated transition of electrons into the conduction band and the reduction of the fraction of negatively charged As vacancies (Saarinen et al. 1993). The ionization level was estimated from this effect to be 50 meV below the conduction band. The level was found at 60 meV in another study (Kuisma et al. 1996b).

The observed near-bandgap absorption in semi-insulating GaAs (Brozel 1995) may be understood as the photoexcitation to this level. The influence of optical excitation on the angular correlation of annihilation radiation was studied by Peng et al. (1994). Significant differences were found in the 2D-ACAR curves between samples measured in the dark and illuminated with 1.4-eV photons.

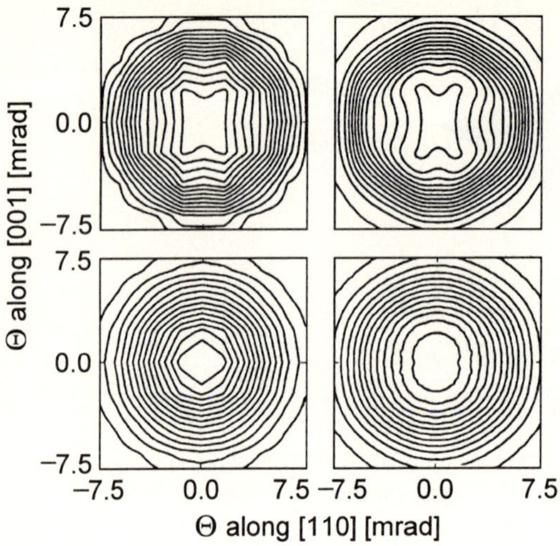

Fig. 5.17. Comparison of the experimental plots (Ambigapathy et al. 1996) of two-dimensional angular correlation of annihilation radiation of the arsenic vacancy (*right panel*) with molecular-dynamics calculations (*left panel*) by Gilgien et al. (1994). The two upper plots represent the measured (*right*) and calculated (*left*) contours of defect-free gallium arsenide. Θ is the angular deviation of the annihilation γ-quanta.

The electrons cannot only be excited from the valence band, but also from occupied states in the bandgap. Such a case was studied for the transition from the EL2 ground-state level to the ionization level of the As vacancy (Kuisma et al. 1996b).

Chen et al. (1996b) found in Doppler-broadening measurements an increase in the S parameter in semi-insulating GaAs under illumination, indicating that the annihilation fraction increased compared with darkness. It was interpreted as the charge state transition of the As vacancy from the positive to the neutral state. However, the 0/− transition could not be ruled out.

To summarize the positron results on the arsenic vacancy, this defect is most probably detectable in undoped n-type or weakly n-doped GaAs. The identification of V_{As} in heavily n-doped GaAs is more difficult, since usually vacancy–dopant complexes occur as competitive trapping centers (see Sect. 5.2.3). The As vacancy exhibits an ionization level between 100 and 140 meV below the conduction band edge that was related to the +/0 transition. Hence, the detectability of V_{As} is lost in semi-insulating GaAs due to its positive charge state. The 0/− level at E_{cond} − 30 meV reported after positron lifetime measurements by Corbel et al. (1988) and Saarinen et al. (1991) cannot be regarded as undoubtedly proven, since the confirmation of such a level by other methods is missing. A defect-related lifetime of 295 ps was found for the neutral As vacancy. Provided that the 0/− transition is observed, it must be accompanied by a pronounced inward lattice relaxation, resulting in a positron lifetime of 265 ps. Some experiments have been

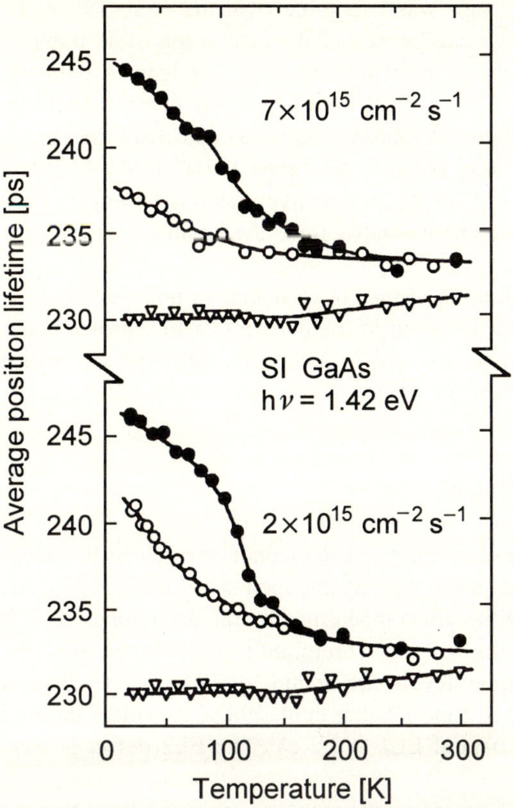

Fig. 5.18. The average positron lifetime $\bar{\tau}$ as a function of the temperature in two undoped semi-insulating gallium arsenide samples. The lifetime was measured in dark (○) and under illumination with 1.42-eV light with the indicated photon flux (•). For reference, the temperature dependence of $\bar{\tau}$ for zinc-doped GaAs (▽) is included (Kuisma et al. 1996b).

reported for the photostimulation of the 0/− transition providing a level of about 60 meV below the conduction band. The annealing stage of the As vacancy centers at 750 K.

No positron trapping is observed in p-type GaAs, in contrast to n-type and semi-insulating material (Dlubek and Krause 1987). A slight increase of about 1 ps is found in the average lifetime in GaAs:Zn with increasing sample temperature (Fig. 5.18). The positrons annihilate from Bloch states in the bulk and the increase is related to the thermal lattice expansion due to the decrease in the electron density (Corbel et al. 1988).

The undoped semi-insulating samples of Fig. 5.18 exhibit the trapping in a monovacancy defect with a defect-related lifetime of 255 ± 5 ps at low temperatures. This behavior could be found only in rare cases, when the influence of positron trapping in shallow positron traps was negligible (Kuisma et al. 1996b; Le Berre et al. 1995). The As vacancy can be excluded as the responsible trapping center, since the Fermi level is located in the middle of the bandgap. A temperature dependence was observed as being typical for a negatively charged vacancy. It was thus proposed that the negative Ga monovacancy was detected (Saarinen et al. 1998). The vacancy densities were found in the range 1×10^{15} to 4×10^{16} cm^{-3} (Le Berre et al. 1995; Saarinen et al. 1998). A negatively charged complex containing a vacancy, however, cannot be excluded from the positron experiments alone.

At first glance, the low concentration of the Ga vacancies in undoped GaAs is surprising, since the stoichiometric deviation of the material usually grown under As-rich conditions may be as high as 10^{18} cm^{-3}. Obviously, other defects realize the stoichiometric deviation (see Sect. 5.2.5).

5.2.1.2 The Metastable EL2 Defect

The EL2 center is certainly the most investigated semiconductor defect. It is technologically very important for the production of undoped semi-insulating GaAs, since residual acceptor-like impurities are compensated by the deep donor EL2. It is a native defect and thus the doping with e.g. chromium is not required to obtain high-resistive GaAs. The ionization levels are found at 770 meV (+/0) and 520 meV (2+/+) above the valence band (Weber et al. 1982). A typical infrared absorption band is found for the neutral EL2 at 1.2 eV. The EL2 defect is thermally stable up to 1000 °C.

It was clearly shown by electron paramagnetic resonance investigations that the As_{Ga} antisite is at least a constituent of the defect (Bäumler et al. 1985; Weber et al. 1982). Remarkably, the EL2 defect has a metastable state (EL2*) with no energy level in the bandgap. The electrical activity, as well as the infrared absorption, vanishes in this state. The transition occurs from the neutral state by illuminating the sample with monochromatic light of 1.1 eV at temperatures below 100 K. The stable ground state reappears by heating the sample above 110 K or by photoinduced recovery with photon energies of 0.8 eV.

A number of different models of the atomic structure were discussed in the past. Reviews of models and properties of EL2 were given e.g. by Manasreh et al. (1989) and Martin and Makram-Ebeid (1986). The model of the center must provide the structure of both the ground and the excited state. Two models were discussed by the end of the 1980s. The model of Baraff and Schlüter (1985b) and von Bardeleben et al. (1986) favored a complex consisting of the antisite and an As interstitial, $As_{Ga}As_i$, whereas Chadi and Chang (1988) and Dabrowski and Scheffler (1989) supposed the isolated As_{Ga} antisite. The metastability is included in the former model by the movement of the interstitial from second to the first nearest-neighbor position or by split interstitial configurations (Manasreh et al. 1989). The latter model describes the metastability by a symmetry-breaking distortion, lead-

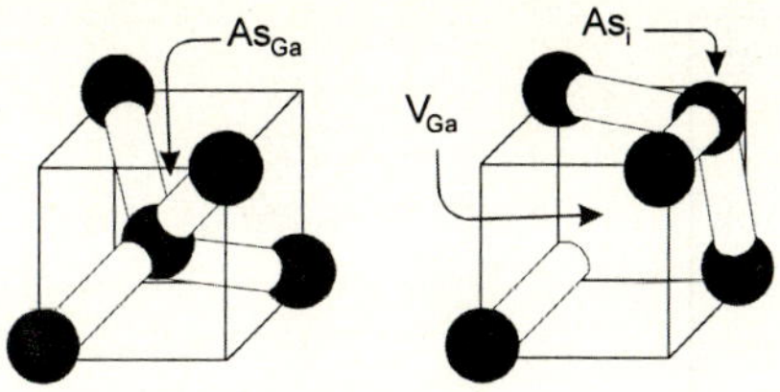

Fig. 5.19. Structure of the EL2 center in the stable (*left*) and metastable (*right*) configuration according to the model of Chadi and Chang (1988) and Dabrowski and Scheffler (1989). The filled circles represent arsenic atoms.

ing to a $V_{Ga}As_i$ configuration (Fig. 5.19). Both models agree in so far that the ground state does not contain vacancies. However, the metastable state contains a vacancy in the model of Chadi and Chang (1988) and Dabrowski and Scheffler (1989) in contrast to the other model.

The possible vacancy-like character of the metastable state can be used for the investigation of EL2 by positron annihilation. The first successful experiment of this kind was reported by Krause et al. (1990b). Two semi-insulating GaAs samples (supplied by VEB Spurenmetalle Freiberg) with different EL2 contents were cooled down to 20 K in darkness. The positron lifetime and the Doppler-broadening *S* parameter were measured in this state (Fig. 5.20). The EL2 was converted thereafter into the metastable state EL2*. An increase in the average lifetime as well as in the *S* parameter occurred. The effect was stronger in the sample with the higher EL2 content, which was determined by infrared absorption according to Martin (1981).

The change in the positron parameters was interpreted as the appearance of a vacancy in the EL2*. The separated defect-related lifetimes of about 255 ps (Krause et al. 1990b) or 245 ps (Le Berre et al. 1994) confirm this assumption, since this value agrees reasonably with that found for Ga monovacancies (Sect. 5.2.6.2).

In order to corroborate this interpretation, the samples were thermally treated in order to study the EL2* → EL2 recovery. The recovery of the ground state was monitored by infrared absorption experiments carried out at the same samples in the same cryostat. The vacancy observed in the metastable state disappears at the transition temperature of 115 K, in good agreement with the recovery of the IR absorption (Fig. 5.20).

In order to exclude any accidental coincidence of the recovery observed in the infrared absorption and positron measurements due to different sensitivities of the methods, the kinetics of the transition was studied (Krause-Rehberg et al. 1994c). Isothermal annealing experiments in the temperature range from 106 to 120 K were performed with both techniques. The results are shown in Fig. 5.21. The recovery of the IR absorption is due to the transition to the ground state of the EL2. It coincides with the disappearance of the metastable vacancy reflected in the decrease in the average positron lifetime. The half-recovery times of both processes are drawn as Arrhenius plots. The slopes agree very well and indicate that the kinetics of the EL2* → EL2 transition and the vanishing of the metastable vacancy

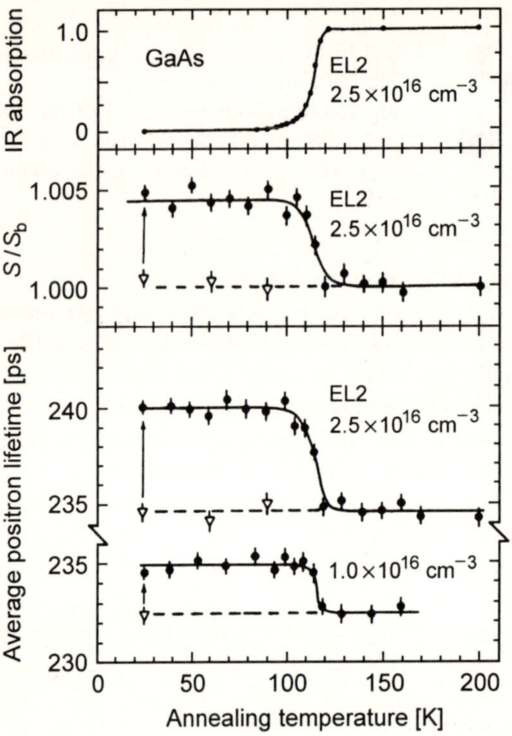

Fig. 5.20. Study of the metastable EL2* state in semi-insulating gallium arsenide by positron annihilation (Krause et al. 1990b). The EL2 density was 1×10^{16} and 2.5×10^{16} cm^{-3}, respectively. The average positron lifetime, the *S* parameter normalized to the bulk value S_b, and the infrared absorption (normalized to the value at 200 K) were measured as a function of the annealing temperature in an EL2* → EL2 recovery experiment. The hollow symbols correspond to the measurement prior to illumination. The arrows indicate the change in the annihilation parameters as a result of the transition to the metastable state by illumination at 25 K. All measurements were carried out in darkness at 25 K.

are identical. The activation energy of the recovery of EL2 was found to be 0.37 ± 0.02 eV, in good agreement with former data (Fischer 1988).

The experiments described above were supplemented by the investigation of three further semi-insulating GaAs samples of different suppliers (Corbel et al. 1992a). The annealing behavior as well as the temperature dependence of the positron trapping in the metastable vacancy was found to be essentially the same. The EL2 concentration was proportional to the trapping rate of the metastable vacancy (Le Berre et al. 1994; Saarinen et al. 1994).

The temperature dependence of positron trapping in the metastable EL2* center was studied by positron lifetime spectroscopy (Krause et al. 1990b). The average positron lifetime was measured in a semi-insulating GaAs sample with EL2 in the

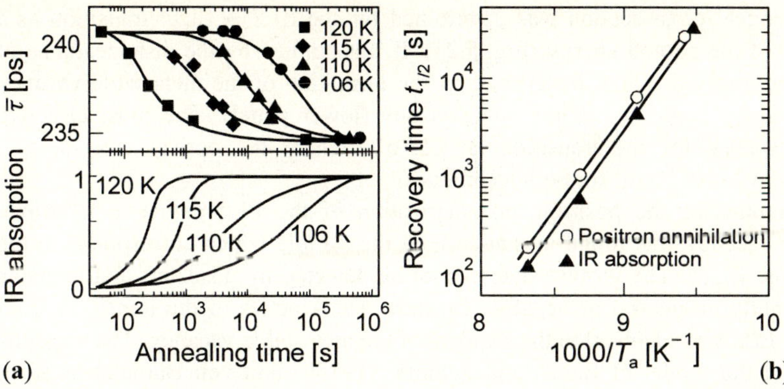

Fig. 5.21. Isothermal annealing experiment to study the transition from the metastable to the stable state of EL2 (Krause-Rehberg et al. 1994c). (**a**) Infrared absorption (IR) recovery in relative units compared with the disappearance of the metastable vacancy monitored by the average positron lifetime. (**b**) Arrhenius plots of the half-recovery time $t_{1/2}$ obtained from (a) for both methods. T_a is the annealing temperature.

ground state (reference) and after photoquenching of EL2 at 25 K (Fig. 5.22). $\bar{\tau}$ reached the reference level of 234 ps at 70 to 100 K, and the positron trapping in the metastable vacancy disappeared. This result proves that the presence of EL2* can be monitored best below 60 K. The change in $\bar{\tau}$ with the temperature is typical for a negatively charged vacancy, as was found for the Ga vacancy (Fig. 5.18).

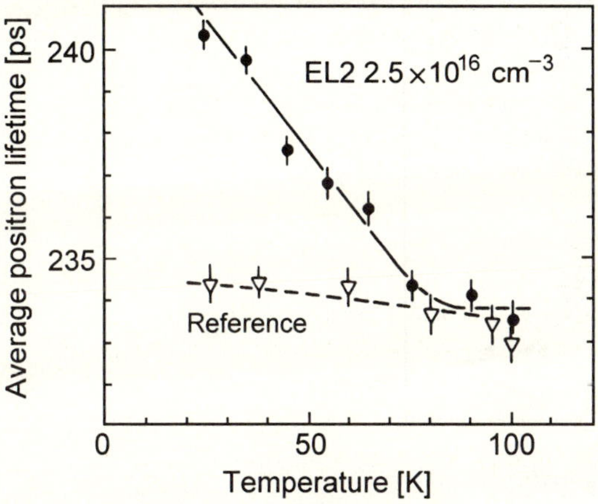

Fig. 5.22. Temperature dependence of positron trapping in the metastable EL2* center (Krause et al. 1990b). The average positron lifetime was measured in semi-insulating gallium arsenide, where the EL2 center was transformed into its metastable state by illumination with light of 1.2 eV before measurement. The reference measurement was carried out with EL2 in the stable state.

The optical cross-section was determined for the EL2 → EL2* transition as a function of the photon energy (Fig. 5.23). It was studied by the disappearance of the infrared signal (upper panel) and by the formation of the metastable vacancy monitored by positron lifetime spectroscopy (lower panel). The most effective photon energy for the transition is 1.15 eV, where the optical cross-section amounts to 2×10^{-18} cm^2 for both techniques.

To summarize the positron characterization of the EL2 defect in as-grown GaAs, a vacancy was found in that state of the sample which corresponds to the metastable EL2*. The ground state cannot be detected by positrons and contains consequently no neutral or negative vacancy. The kinetics of the transition from EL2* to EL2 was identical to the removal of the metastable vacancy. These results contradict the model of Baraff and Schlüter (1985b) and von Bardeleben et al. (1986) but are in a good agreement with the model of Chadi and Chang (1988)

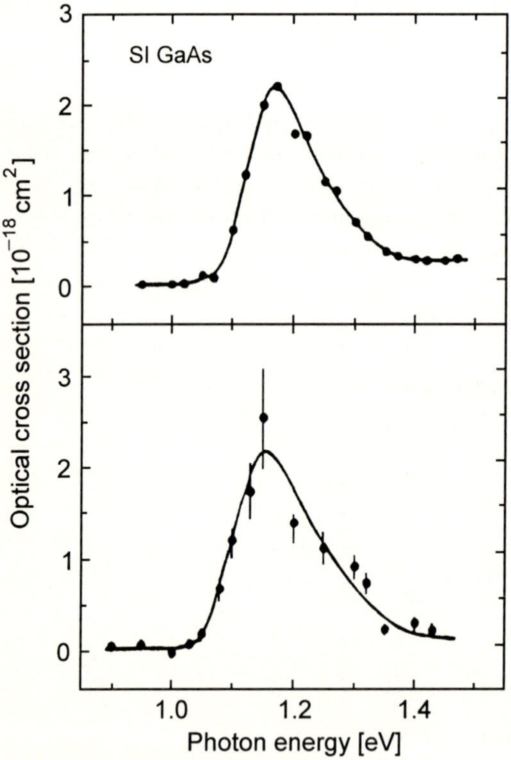

Fig. 5.23. Optical cross-section for the transformation of EL2 into the metastable EL2* in semi-insulating (SI) gallium arsenide as a function of the photon energy (Saarinen et al. 1994). The curve in the upper part was obtained from infrared absorption measurements, and the curve in the lower part from positron lifetime spectroscopy. The measurements were carried out at 25 K.

and Dabrowski and Scheffler (1989). The generally approved structure of the EL2 center is now the isolated As_{Ga} antisite (Wirbeleit and Niklas 1997).

Similarly to the metastable vacancies in as-grown GaAs, such vacancies were also introduced by electron irradiation (Saarinen et al. 1995a). The properties of these defects were essentially the same as in as-grown material. They were thus also attributed to the metastable state of the As_{Ga} antisite. The introduction rate of this antisite defect introduced by 2-MeV electron irradiation at 20 K was estimated to be 0.3 cm^{-1}. This estimate was rather complicated, since three positron traps (Ga vacancies, As and Ga antisites) were already present in the as-grown sample and their concentration increased as a result of the irradiation. The effect of the irradiation on the average positron lifetime was very small and no decomposition of the spectra could be performed. The uncertainty of the introduction rate is thus at least one order of magnitude.

Two properties of EL2* were found to be different to the metastable antisite after electron irradiation (Saarinen et al. 1995a). The conversion to the metastable state required a significantly higher photon flux than for EL2 in as-grown GaAs. The transition from the metastable to the ground state occurred at a temperature 20 K lower than for EL2 in as-grown GaAs. These differences were related to the formation of complexes of the As_{Ga} antisite with other defects. A similar conclusion was drawn from measurements by magnetic circular dichroism of absorption by Hesse et al. (1994). The interpretation was supported by further studies of metastable centers after electron irradiation using combined infrared spectroscopy and positron annihilation (Kuisma et al. 1996a, 1997). The transition temperature to the stable state was found at 70 K. In contrast to EL2 in as-grown GaAs, the disappearance of the irradiation-induced metastable centers occurred after thermal annealing at 520 K. The metastable state can be achieved by 1.075-eV photons and recovered by 1.35-eV illumination. The latter effect and the thermal stability of the metastable state can be seen in Fig. 5.24. The disappearance of the infrared absorption at 0.9 eV and the vanishing of the metastable vacancy indicated by the strong decrease in the positron trapping rate was found consistently at 70 K (full symbols). However, no changes in the lifetime in the range from 30 to 100 K were reported in the previous study by Saarinen et al. (1995a) for a comparable sample irradiated in the same way. Unfortunately, the data of the lifetime experiment were not given in the paper of Kuisma et al. (1997). The strong change in the trapping rate by 2×10^{11} s^{-1} is thus hardly comprehensible.

Fig. 5.24 also shows the results of the measurements after additional illumination with 1.35-eV photons at 25 K (open symbols). Obviously, the stable state of this defect complex can already be reached by this illumination. The infrared absorption increases again at 120 K, indicating the presence of the ordinary EL2, already present in as-grown GaAs. The recovery in the absorption is accompanied by the decrease in the trapping rate, comparable to that in Fig. 5.20.

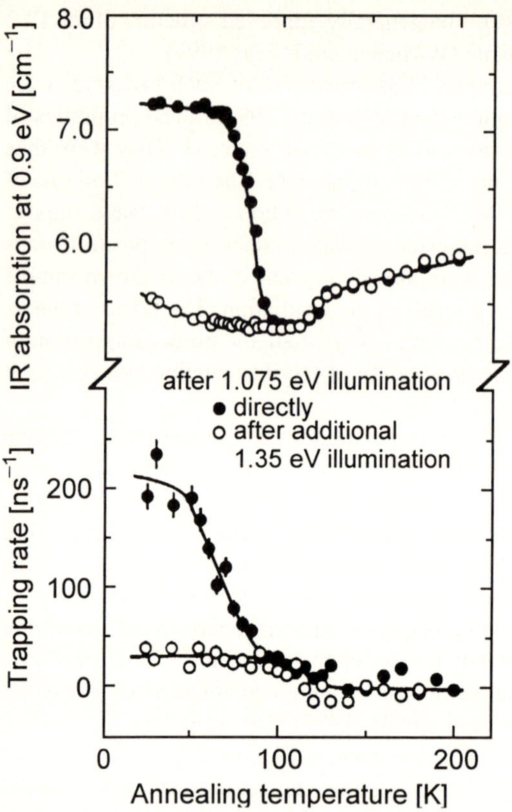

Fig. 5.24. Thermal stability of the metastable state of the electron-irradiation-induced antisite complex in semi-insulating (SI) gallium arsenide (Kuisma et al. 1997). The samples were irradiated with 2-MeV electrons to a dose of 5×10^{17} cm^{-2}. The infrared (IR) absorption at 0.9 eV and the positron trapping rate are shown as functions of the annealing temperature prior to (•) and after (o) 1.35-eV illumination. The measurement temperature was 25 K.

5.2.2 Shallow Positron Traps

The characteristic features of shallow positron traps were described in Sect. 3.1.5.2. The positron binding energy is of the order of some tens of meV, i.e. compared with vacancies rather low. Thermal detrapping of positrons is hence already below room temperatures very likely. Negatively charged non-open volume defects, such as acceptor-type impurities, native defects like antisites, interstitials, or defect complexes may act as shallow positron traps. The positron trapping is

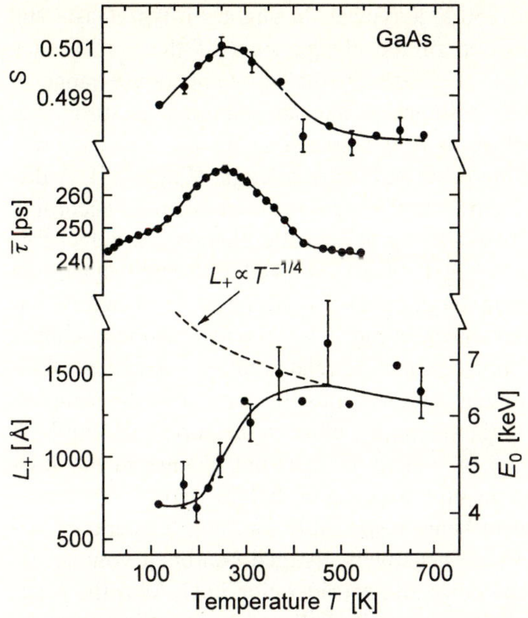

Fig. 5.25. Results of positron lifetime and slow-positron-beam experiments in undoped gallium arsenide (Saarinen et al. 1989). The free-electron density was 2.2×10^{16} cm^{-3}. The positron diffusion length L_+ was determined as a function of the temperature T. The dashed line represents the $T^{-1/4}$ dependence expected for L_+ in case no positron traps are present. The E_0 parameter is shown on the right axis. For reference, the S parameter and the average positron lifetime $\overline{\tau}$ are presented in the upper part as a function of the temperature.

based on the existence of spatially extended Rydberg states owing to the Coulomb potential of the defect.

The occurrence of shallow positron traps in GaAs was first suggested by Dannefaer et al. (1984). Bulk lifetimes ranging from 220 to 232 ps were found in several doped and undoped GaAs samples after positron lifetime measurements with high statistics. The former value was interpreted as due to positron annihilation in delocalized Bloch states. The increase in the bulk value was related to positron trapping in shallow traps. They were attributed to gallium antisite defects or later to substitutional boron or nitrogen impurities (Dannefaer et al. 1989a). The positron binding energy to these defects was determined from the study of the temperature dependence of the lifetime spectra to be about 23 meV.

Bulk lifetimes as low as 220 ps (Dannefaer et al. 1984) have not been reported by other authors. However, the appearance of shallow positron traps was clearly shown by combined positron lifetime and slow-positron-beam experiments (Saarinen et al. 1989). Characteristic courses of the average lifetime in GaAs as a function of the sample temperature are shown in Fig. 5.9. The results of the closer investigation of one of the samples are presented in Fig. 5.25. The increase in the average positron lifetime down to 250 K is typical for slightly n-type GaAs and has been interpreted as the transition of the charge state of the As vacancy (Sect. 5.2.1.1). The average lifetime decreases in the low-temperature range to come close to the bulk value. The assumption that an additional positron trap having a defect-related lifetime close to τ_b is responsible for this decrease was confirmed in the measurement of the positron diffusion length (Fig. 5.25). A distinct decrease in L_+ was found in contrast to the expected increase in L_+ according to the theoretical $T^{-1/4}$ law for the scattering at acoustic phonons in defect-free material (Sect. 3.1.4). Since the investigated GaAs showed a vacancy signal, the same experiment was repeated for material containing no negative or neutral vacancies. The result has already been shown in Fig. 3.14. The identical temperature behavior of L_+ indicates that the course cannot be related to the vacancies. Obviously, a positron trap active at low temperatures must be present in the samples. Its defect-related lifetime and the S parameter are close to the corresponding bulk values. This can be seen in Fig. 3.14, where $\bar{\tau}$ does not change in the low-temperature region with increasing positron trapping in shallow traps.

Possible candidates for the defect being responsible for such temperature dependences of the positron trapping are negatively charged non-open volume defects. A simple native defect which comes into consideration in GaAs is the negatively charged Ga_{As} antisite defect. It is generally believed to act as an acceptor with ionization levels in the lower half of the bandgap. More recent theoretical calculations (Chadi 1992; Landman et al. 1997) predict, in n-type GaAs, the As split interstitials as negatively charged, so that they cannot be excluded as shallow positron traps. Bulk GaAs contains, however, various residual impurities. Among them, acceptors like C_{As}^-, Zn_{Ga}^-, or Mg_{Ga}^- with sufficiently high concentrations may be considered to act as shallow positron traps. The boron content in GaAs grown by the liquid-encapsulated Czochralski method is rather high. Recent studies proved the occurrence of B_{As} acceptors with ionization levels of 203 $(-/2-)$ and 78 meV $(0/-)$ above the valence band edge (Newman et al. 1994). This center is thus a very likely candidate for a shallow positron trap in LEC GaAs. Dislocations have been discussed as shallow positron traps too (Sect. 3.1.5.3). However, their density in as-grown GaAs is below 10^5 cm^{-2} and thus below the detection limit of positron annihilation.

The fit of the temperature-dependent average lifetime can be carried out with the trapping model under consideration of (3.15). The binding energy of the positron is obtained as one of the fitting parameters. It was found to be 43 meV in as-grown undoped GaAs (Saarinen et al. 1989). In order to investigate the nature of the shallow positron traps in GaAs, electron-irradiation experiments were carried out (Corbel et al. 1992b). Two types of irradiation-induced positron traps were found. One of them was identified as the Ga vacancy, either isolated or as part of a complex. The second defect exhibited the properties of a shallow positron trap. A decrease in the average positron lifetime with decreasing temperature similar to the behavior in as-grown GaAs was found. The detrapping rate δ can be calculated from the trapping model taking into account a vacancy and a defect with thermal positron detrapping. The appropriate equations are obtained from the trapping model in Sect. 3.2.1.2 by setting $\kappa_{d3} = 0$.

The detrapping rates calculated in this way are given as Arrhenius plots for three different irradiation doses in Fig. 5.26. The data follow a straight line, indicating that the predicted temperature dependence (Manninen and Nieminen 1981) well describes the thermal detrapping of positrons from the shallow traps. The concentrations obtained from the fit of the average positron lifetime were proportional to the irradiation dose. This clearly indicates that these defects are irradiation-induced and they should be of intrinsic nature. Possible candidates are the

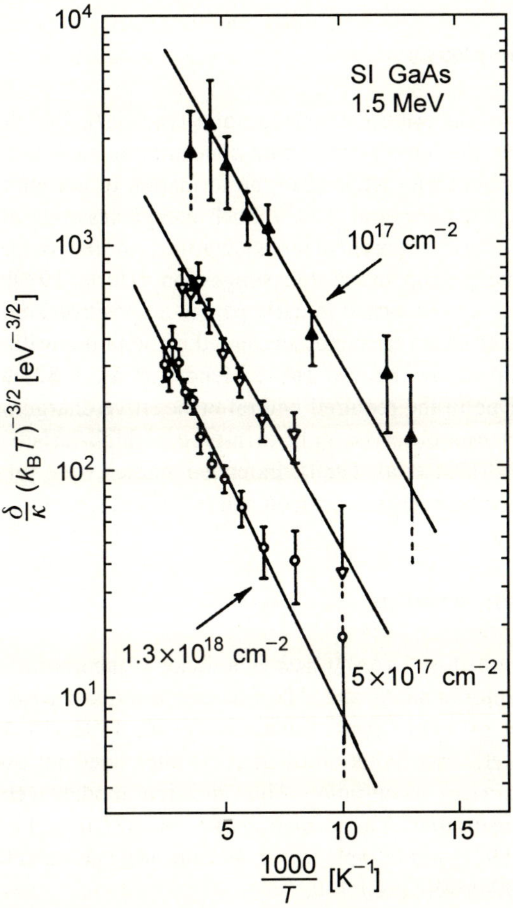

Fig. 5.26. Arrhenius plots of the detrapping rate δ from the Rydberg states of shallow traps in semi-insulating (SI) gallium arsenide (Corbel et al. 1992b). The samples were irradiated with 1.5-MeV electrons at 20 K to doses of 1×10^{17} to 1.3×10^{18} cm^{-2}. The trapping rate κ was determined according to the trapping model and (3.15). T is the sample temperature and k_B the Boltzmann constant.

Ga_{As}^- antisites, negatively charged As interstitials, or negative non-open-volume complexes.

The positron binding energy to the shallow traps may be determined from the plots of Fig. 5.26. It was found to be between 38 and 41 meV, being close to the value obtained in as-grown material (Saarinen et al. 1989). This was taken as an indication that the same defect was detected. However, the properties of a negatively charged defect acting as a shallow positron trap are mainly determined by the extended Rydberg states. The annihilation of the positron occurs far from the central ion due to the Coulomb repulsion. Therefore, there is hardly any information on the chemical nature of the ion.

5.2.3 Vacancy–Dopant Complexes

It was found in the investigations of the electrical behavior of n-type GaAs that the carrier concentration is lower than the introduced dopant concentrations (Whelan et al. 1960). This compensation effect was ascribed to the formation of acceptor-type vacancy–dopant complexes (Maguire et al. 1987). Such complexes were already predicted by Williams (1968). Theoretical considerations of the defect chemistry and the formation energy supported this suggestion (Hurle 1979b; Northrup and Zhang 1993). It was also supposed in early papers on positron annihilation investigations of doped GaAs that vacancies are paired off with impurities (Dannefaer and Kerr 1986b; Dlubek and Krause 1987; Mendez et al. 1988) to form positron trapping centers. Due to the required neutral or negative charge of the defect complexes, this effect was observed only in n-type or high-resistive GaAs (Dlubek and Krause 1987; Takai et al. 1980). Extended studies were presented on the most important donor-type dopants, silicon and tellurium.

5.2.3.1 Silicon-Doped Gallium Arsenide

Silicon is amphoterically incorporated in GaAs. It acts as a donor if incorporated at a Ga lattice site and as an acceptor at an As site. The free-electron density cannot be increased beyond a level of 5×10^{18} cm^{-3} (Schuppler et al. 1993, 1995; Whelan et al. 1960). The amphoteric incorporation of Si at As sites does not explain fully this saturation of the carrier concentration. Thus, different models were developed to explain the autocompensation. The formation of Si pairs (Kolm et al. 1957), Si clusters (Muto et al. 1992), and complexes of Si with intrinsic defects (Maguire et al. 1987; Newman 1994) were proposed.

Highly silicon-doped molecular beam epitaxy GaAs layers were investigated by Laine et al. (1996b) using slow-positron-beam experiments. Measurements of the temperature-dependent S parameter were carried out. Gallium vacancies and shallow positron traps were found. The identification of the Ga vacancies was carried out by the comparison of the high-momentum parts of the Doppler-broadened

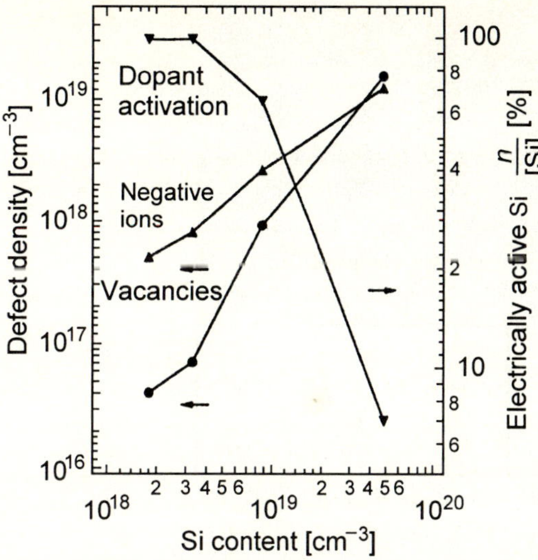

Fig. 5.27. Densities of vacancy-type defects (●) and shallow positron traps (▲) and the degree of silicon donor activation (▼) plotted against the silicon content in highly silicon-doped gallium arsenide layers grown by molecular beam epitaxy. The densities were obtained from the Doppler-broadening *S* parameters of a slow-positron-beam experiment (Laine et al. 1996b). *n* is the free-electron density and [Si] relates to the total silicon content.

annihilation lines of the MBE layers with samples irradiated with fast electrons at low temperatures. Their concentrations could be determined from the fits of the temperature-dependent *S* parameters. Both the concentrations of vacancies and the shallow positron traps are functions of the silicon content (Fig. 5.27). Therefore, it was concluded by Laine et al. (1996b) that the defects are related to silicon and play an important role in the autocompensation. In conjunction with previous studies, it was suggested that the $Si_{Ga}V_{Ga}$ complex and the Si^-_{As} acceptor are observed.

A first microscopic identification of such compensating centers was achieved by atomically resolved scanning tunneling microscopy (Domke et al. 1996). (110) cleavage planes were investigated in ultra-high vacuum. It was found that Si_{Ga} donors were increasingly compensated by Si_{As} acceptors, Si clusters, and $V_{Ga}Si_{Ga}$ complexes. The Si could be imaged and quantitatively determined at any incorporation site. STM images of the $V_{Ga}Si_{Ga}$ complex are shown in Fig. 5.28. The Ga and As sublattice can be imaged separately by reversing the tunneling bias. In this way it could be identified that the complex in Fig. 5.28 consists of a Ga vacancy and a next-nearest neighbor Si_{Ga} acceptor.

However, the situation at the investigated surface may not represent the bulk properties due to the cleavage process. The concentration determined by scanning tunneling microscopy may be wrong. In order to verify the STM results, accom-

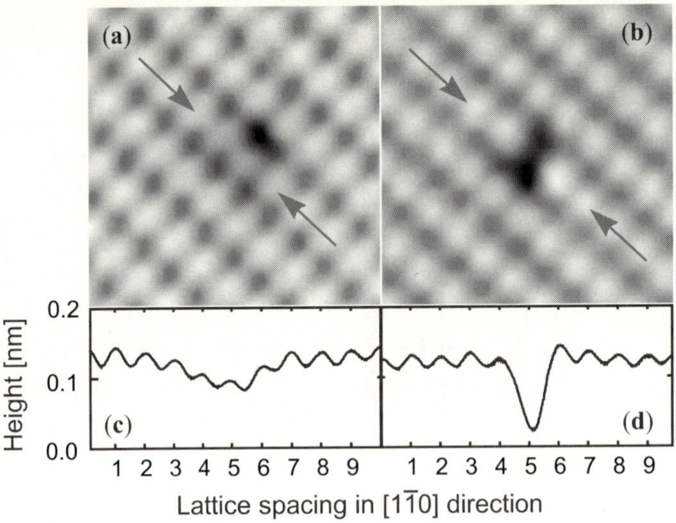

Fig. 5.28. Scanning tunneling microscopy of a $V_{Ga}Si_{Ga}$ complex (Gebauer et al. 1997c). The pictures were taken in constant current mode for occupied (**a**) and empty states (**b**). Line profiles (**c**) and (**d**) were taken along the atomic rows of gallium indicated by arrows in (a) and (b), respectively.

panying positron lifetime experiments were performed. The results are shown in Fig. 5.29. The average positron lifetime increases with increasing Si content. It almost reaches the defect-related lifetime τ_d at the highest Si concentration. The defect-related lifetime of $\tau_d = 262$ ps corresponds to a monovacancy-type defect. No As vacancies were detected by the scanning tunneling microscopy carried out on the same crystals. The concentration of Ga vacancies immediately after cleavage was found by STM to be significantly smaller than that of the $V_{Ga}Si_{Ga}$ complexes. Hence, it was concluded that this complex was observed in the positron measurements.

The density of $V_{Ga}Si_{Ga}$ centers determined by STM is compared with the vacancy density from positron lifetime spectroscopy in Fig. 5.30. A good agreement is obtained. Equation (3.11) was used for the calculation of the monovacancy concentration assuming a trapping coefficient of $1\times10^{15}\,s^{-1}$, which represents the typical value at room temperature for negatively charged monovacancies in semiconductors (Sect. 3.2.4). On the other hand, assuming that identical defects are observed both by scanning tunneling microscopy and POLIS, STM represents a suitable reference method for the determination of the trapping coefficient (Sect. 3.2.4.3).

In addition to the $V_{Ga}Si_{Ga}$ defect, another positron trap appears in measurements of the temperature dependence of the average positron lifetime (Fig. 5.31). The decrease in $\bar{\tau}$ in the low-temperature range is typical for the presence of shallow positron traps (Fig. 3.13). The only kind of defect that comes into consideration

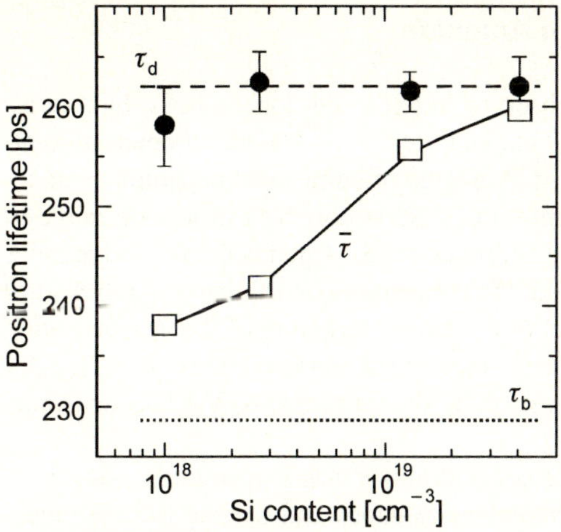

Fig. 5.29. Average positron lifetime $\bar{\tau}$ and defect-related lifetime τ_d measured at room temperature as functions of the silicon content in gallium arsenide (Gebauer et al. 1997c). The silicon content was determined by secondary ion mass spectroscopy. The average value of $\tau_d = 262$ ps (*dashed line*) was used for the determination of the defect concentrations. The bulk lifetime $\tau_b = 228.5$ ps is indicated as a dotted line.

from scanning tunneling microscopy is the Si_{As}^- acceptor, which has a sufficiently high concentration. The density of the shallow positron traps was calculated from the fit of the temperature dependence of the positron lifetime using a three-state trapping model. It is plotted together with the density of the Si_{As}^- acceptor in Fig. 5.31b as obtained by STM. A satisfactory agreement between both types of measurement may indicate that the Si_{As}^- centers are the observed shallow positron traps.

The combination of both methods allows a more reliable identification of the defect type. The drawbacks of STM, i.e. the limitation to the defect detection at or near the surface, and positron annihilation, i.e. the lacking chemical information, can be overcome. Two compensating centers could thus be unambiguously identified as the $Si_{Ga}V_{Ga}$ complex and the Si_{As}^- acceptor. It should be noted that other compensating defects, viz. silicon clusters, have been identified from scanning tunneling microscopy alone (Domke et al. 1996). The role of such clusters on positron trapping is not yet clear.

The use of δ doped layers is a prospective method to achieve high dopant concentrations in semiconductors (Schubert 1993). However, a compensation effect was also observed for such material for high doping concentrations (Newman et al. 1996). Positron Doppler-broadening spectroscopy was carried out at silicon δ-doped GaAs superlattices (Laine et al. 1997a; 1997b). It was found that the δ layers are repulsive for positrons. Vacancies and shallow positrons traps were identified in the intermediate zones between the Si δ layers. They were attributed to Ga vacancies and negatively charged Ga_{As} antisite defects (Laine et al. 1997b).

5.2.3.2 Tellurium-Doped Gallium Arsenide

An autocompensation effect is also observed in tellurium-doped GaAs. In contrast to Si, the Te donor is not incorporated amphoterically. A distinct compensation effect already occurs at a low Te content. A degree of compensation (number of acceptors divided by the number of donors) of 0.25 is observed in a wide range of Te dopant concentrations (Wolfe and Stillman 1975). Chemical thermodynamics was used by Hurle (1979b) to calculate the concentrations of charged and neutral point defects in doped GaAs. A model of the atomic structure of the compensating center was suggested on the basis of these theoretical considerations. The Te_{As} donors are supposed to be partly compensated by the formation of $V_{Ga}Te_{As}$ acceptor-type centers.

A series of GaAs with different Te concentrations was investigated using both Hall effect and positron lifetime measurements (Dlubek and Krause 1987; Krause-Rehberg et al. 1995b). The degree of compensation can be calculated according to Walukiewicz et al. (1979) from the temperature dependence of the Hall coefficient and the mobility. The number of compensating acceptors obtained in this way is plotted versus the number of donors in Fig. 5.32. The degree of compensation is about 0.23 beyond a residual donor density of 2×10^{17} cm^{-3}.

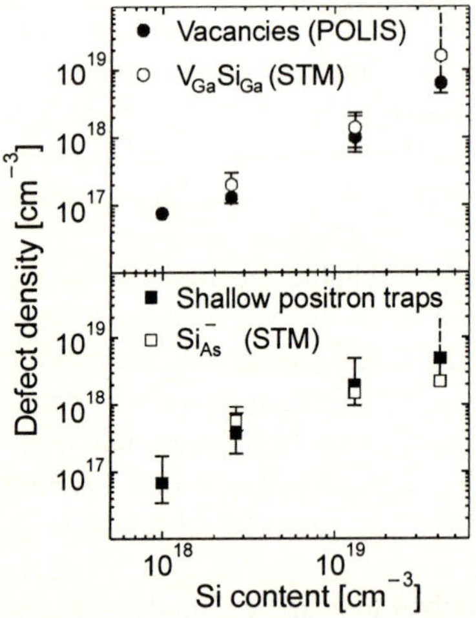

Fig. 5.30. Defect densities in silicon-doped gallium arsenide (Gebauer et al. 1997c). The density of the monovacancy-type defect (*filled circles*) observed by positron lifetime spectroscopy (POLIS) and the density of $V_{Ga}Si_{Ga}$ (*empty circles*) centers determined by scanning tunneling microscopy (STM) in silicon-doped gallium arsenide is plotted in the upper panel versus the silicon content. The density of shallow positron traps (*filled squares*) and the density of Si_{As} acceptors (*empty squares*) determined from STM is shown in the lower panel.

The same set of samples used for the Hall measurements were taken for positron lifetime spectroscopy. The compensating centers are supposed to be vacancy-type defects according to the calculations mentioned above (Hurle 1979b) and should thus be visible in positron lifetime spectroscopy. Indeed, a monovacancy signal with a defect-related lifetime of $\tau_d = 253$ ps was separated. The trapping rate was found to be proportional to the acceptor concentration (Fig. 5.33). Deviations in the range of low concentrations are probably due to residual impurities, acting as additional compensating centers. Using the trapping coefficient for negatively charged vacancies of $1\times10^{15}\,s^{-1}$, the vacancy concentration calculated from (3.9) is close to the concentration of compensating acceptors as found by Hall effect measurements. This may be regarded as a strong indication that the Hurle model of $V_{Ga}Te_{As}$ as the compensating center is correct. In order to exclude the observation of As vacancies instead of $V_{Ga}Te_{As}$ as the positron trapping centers, annealing experiments were carried out. The thermal stability of the observed vacancy-type defect in GaAs:Te was compared with the stability of the As vacancy. The result has already been shown in Fig. 5.15. The defect in Te-doped GaAs is obviously much more stable than V_{As}.

A different series of Te-doped GaAs samples was used to study the temperature dependence of the positron trapping (Polity and Krause-Rehberg 1995). A decrease in the average positron lifetime with increasing temperature above 200 K is shown in Fig. 5.34. This is typical for the presence of negatively charged vacancies. The position of the Fermi level is, in all investigated samples, very close to the conduction band edge and thus a charge state transition of vacancies can be excluded as the cause of the change in $\bar{\tau}$. As a general trend, the average lifetime

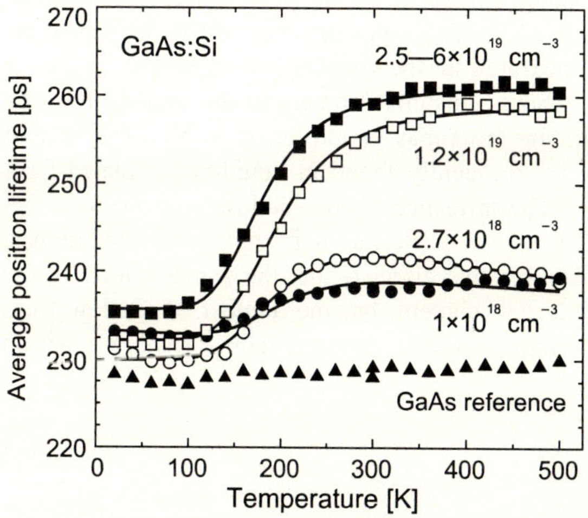

Fig. 5.31. Average positron lifetime as a function of temperature for gallium arsenide of different silicon contents. The solid lines are fits to the experimental data assuming positron trapping in negatively charged vacancies and shallow positron traps (Gebauer et al. 1997c).

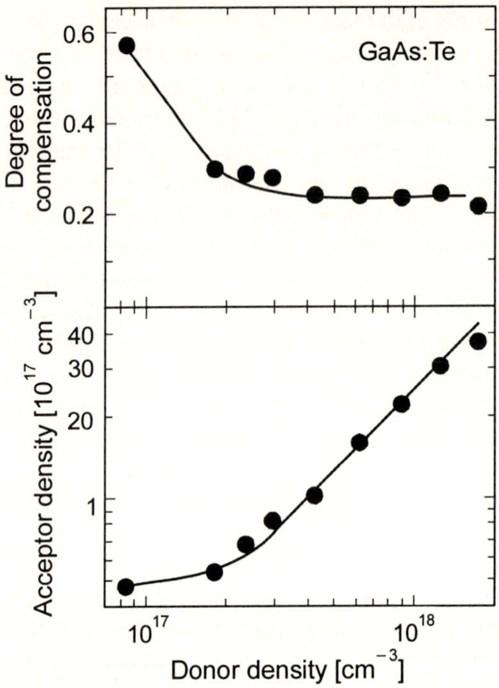

Fig. 5.32. Acceptor density and degree of compensation versus the donor density in gallium arsenide samples with different tellurium contents (Krause-Rehberg et al. 1995b).

increases with increasing Te content, similar to the samples of Fig. 5.33. This result is in accord with the existence of $V_{Ga}Te_{As}$ compensating centers. However, the acceptor concentration was not measured for the samples in Fig. 5.34.

The weakly doped samples exhibit a distinct decrease in the average positron lifetime with decreasing temperature. This may be attributed to the occurrence of shallow positron traps, as they are frequently found in liquid-encapsulated Czochralski-grown GaAs (Fig. 5.9). The influence of the shallow positron traps becomes weaker with increasing Te content. A plateau in $\overline{\tau}$ appears with the highest carrier concentrations. The reason for this effect is that the concentration of the vacancies increases with the tellurium content, but the number of shallow traps does not. In contrast to Si, tellurium obviously does not induce shallow positron traps.

5.2.3.3 Other Dopants

Chromium was used as dopant to produce semi-insulating GaAs. The 3+/2+ acceptor level is situated in the midgap position and is responsible for this behavior.

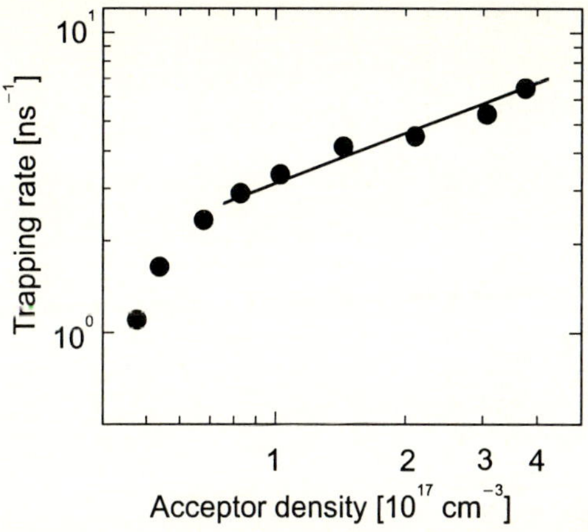

Fig. 5.33. Positron trapping rate in tellurium-doped gallium arsenide as a function of the density of compensating acceptors (Krause-Rehberg et al. 1995b). The acceptor density is obtained from Hall effect measurements (Fig. 5.32). The solid line is the result of the fit according to (3.9), which gives a trapping coefficient of 1.1×10^{15} s^{-1}.

Nowadays, the concentration of residual impurities is low enough that the native EL2 defect can be used for the compensation of the material. The 0.839-eV photoluminescence line of GaAs:Cr was attributed to $V_{As}Cr_{Ga}$ complexes (Fujiwara et al. 1982). Kikoin et al. (1986) also interpreted infrared absorption bands with the occurrence of such centers.

Positron lifetime measurements of chromium-doped GaAs from different suppliers were carried out by Dlubek and Krause (1987). Some of the Cr-doped samples as well as In- and Zn-doped GaAs samples exhibited no positron trapping. However, an increase in the average lifetime between 6 and 12 ps was found in four out of twelve GaAs:Cr specimens. The defect-related lifetime was found to be 250 to 260 ps. Mascher et al. (1987c) separated in Cr-doped GaAs a lifetime component of 250 ps. The trapping rate increased at 260 °C and fell again between 450 and 650 °C. No vacancy clustering was observed during this annealing. The change in the trapping rate was attributed to vacancy migration and further formation and dissolution of $V_{As}Cr_{Ga}$ complexes.

The most important dopant used to achieve p-type bulk GaAs is zinc. Zn_{Ga} represents a shallow acceptor. It was supposed that $V_{As}Zn_{Ga}$ complexes exist in Zn-doped GaAs as compensating centers (Hwang 1968). They could be detected by scanning tunneling microscopy at (110) cleavage planes (Ebert et al. 1996). They are, however, not visible in positron lifetime spectroscopy, as could be seen in Fig. 5.18, where the Zn-doped GaAs sample exhibits only the bulk lifetime. The $V_{As}Zn_{Ga}$ centers in practice do not act as positron traps due to their positive charge. For this reason, Zn-doped GaAs is often used as a reference material exhibiting only positron annihilation from non-localized Bloch states. Such samples

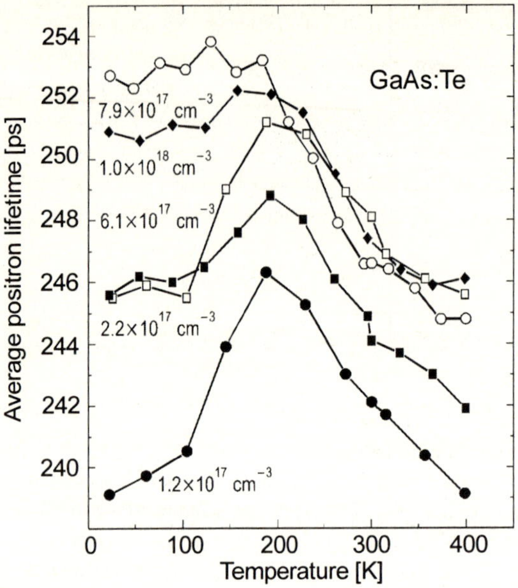

Fig. 5.34. Average positron lifetime as a function of the sample temperature for a series of tellurium-doped gallium arsenide with different carrier densities (Polity and Krause-Rehberg 1995).

are e.g. suitable to carry out the source correction in positron lifetime spectroscopy (Sect. 2.2.2).

5.2.4 Low-Temperature-Grown Gallium Arsenide

GaAs layers can be epitaxially grown by molecular beam epitaxy at temperatures as low as 200 °C. A layer thickness of a few micrometers can be achieved before amorphous growth sets in. The low-temperature-grown (LT) layers exhibit interesting physical properties for device applications. The ultra-short recombination times are interesting for detectors with a sub-picosecond response time. Moreover, the layers can be used after thermal treatment as semi-insulating buffer layers in GaAs devices.

The reasons for the semi-insulating properties of LT GaAs layers are not completely understood. Such layers contain point defects in very high concentrations. They are non-stoichiometric, with an arsenic excess of up to 1 % (Smith et al. 1988). The stoichiometry deviation is essentially controlled by the growth temperature and the beam equivalent pressure (BEP), representing the As/Ga ratio during the MBE growth. The lattice constant is larger than that of bulk GaAs grown from the melt. It is reduced to the value of bulk GaAs after 600 °C annealing by retaining the stoichiometry deviation (Kaminska et al. 1989). The non-stoichiometry can be realized by As_{Ga} antisites, As interstitials, As clusters, or Ga vacancies, or combinations of these.

The neutral arsenic antisite As_{Ga}^0 was detected by infrared absorption measurements. The defect density reduced significantly from 3×10^{19} cm^{-3} in the as-grown state to 3×10^{18} cm^{-3} after annealing at 600 °C (Manasreh et al. 1990). A density of 3×10^{18} cm^{-3} of positively charged As_{Ga}-related point defects in the paramagnetic state was found by electron paramagnetic resonance (Kaminska et al. 1989). The density is reduced to at least 10^{17} cm^{-3} after annealing at 500 °C. The high concentration of these ionized centers must be compensated by a comparable concentration of acceptors, which were proposed to be V_{Ga} (von Bardeleben et al. 1992). Ion channeling experiments indicated the presence of As interstitials, possibly realizing a large fraction of the As excess in the layer. The interstitials vanished after 600 °C annealing (Yu and Liliental-Weber 1992). The observed lattice distortion was related in another interpretation to the presence of a high density of As_{Ga} antisites (Prasad et al. 1995). Transmission electron microscopy showed the presence of As precipitates with diameters of 2 to 10 nm and average distances of 10 to 40 nm after annealing (Liliental-Weber et al. 1991; Melloch et al. 1990).

The high density of native defects made low-temperature-grown GaAs layers interesting for positron annihilation investigations. The first such papers were published at the beginning of the 1990s (Bliss et al. 1992; Hozhabri et al. 1992), applying the slow-positron-beam technique. The results of slow-positron-beam Doppler-broadening measurements indicated an excess concentration of vacancies in the LT layers. They were attributed to the Ga sublattice.

The dependence of the *S* parameter on the growth temperature was studied by Hautojärvi et al. (1993). They found a distinct increase in the *S* parameter when lowering the growth temperature from 225 to 200 °C. The sample grown at 300 °C substrate temperature exhibited no trapping in vacancy-type defects. The defect-related *S* parameter value of 1.037 is higher than that for Ga monovacancies. It was attributed to positron trapping in divacancies or small vacancy clusters. Such high values were found in subsequent investigations only in annealed layers. The concentration of vacancy-type defects was also observed to decrease with increasing growth temperature in a further slow-positron-beam study (Umlor et al. 1993). The same result was obtained in the investigations of Gebauer et al. (1997e), who investigated a series of samples grown at temperatures from 200 to 350 °C with layer thicknesses of 1 to 2 μm and a beam equivalent pressure of 20. The change in the *S* parameter with the growth temperature is plotted in Fig. 5.35.

In order to enable the identification of the observed vacancy-type defect, the *S* parameters found for the GaAs layers grown at different temperatures were plotted against the *W* parameters (Fig. 5.36) in comparison with the results of Si-doped GaAs (Gebauer et al. 1997e). The V_{Ga} paired off with silicon on the next-nearest neighbor site, Si_{Ga}, was unambiguously identified as the dominant positron trap (Gebauer et al. 1997c). The line in the *S* versus *W* plot (Fig. 5.36) runs through the bulk value. This fact may indicate that there is only one dominant defect type. The slope of the line is the same for low-temperature-grown gallium arsenide and for

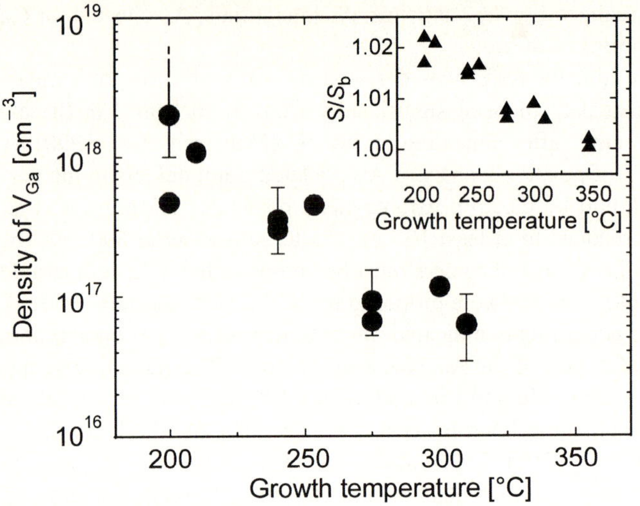

Fig. 5.35. Result of slow-positron-beam Doppler-broadening measurements of low-temperature-grown gallium arsenide (Gebauer et al. 1997e). The change in the line shape parameter S/S_b with the growth temperature is plotted in the inset (S_b bulk S parameter). The density of gallium vacancies was calculated from the average S parameter in the epitaxial layer using (3.11) and (3.42).

GaAs:Si, where the Ga vacancies were identified by scanning tunneling microscopy. Thus, it was concluded that Ga vacancies were detected also in the LT GaAs layers. This is in agreement with the positron lifetime measurements of such layers by Störmer et al. (1996b), where saturated trapping was found with a defect-related lifetime of 275 ps. It cannot be excluded, however, that these vacancies are part of a complex, e.g. with As_{Ga} antisites. This complex could be thermally more stable as the isolated Ga monovacancy.

The S parameter values of the inset in Fig. 5.35 were used to calculate the concentration of Ga vacancies. The trapping rate was obtained via (3.42) with the defect-related S parameter of $S_d = 1.024$ from Fig. 5.36. The concentration was calculated according to (3.11) with a trapping coefficient of $\mu = 1\times10^{15}\ s^{-1}$. The result is given as a function of the growth temperature in Fig. 5.35. The vacancy concentration for the sample with the lowest growth temperature is at the upper limit of the sensitivity range of the S parameter measurement. Ga vacancies are still detected for layers grown at 310 °C. The bulk value is reached, and consequently no vacancies are detected, at growth temperatures of 350 °C.

The S parameter was found to be independent of the measurement temperature in the range of 25 to 300 K (Gebauer et al. 1997e; Hautojärvi et al. 1993). It was concluded that the concentration of shallow positron traps is negligibly small compared with the concentration of Ga vacancies. Ga_{As}^- antisites or residual negatively charged impurities, which act as shallow positron traps (Sect. 5.2.2),

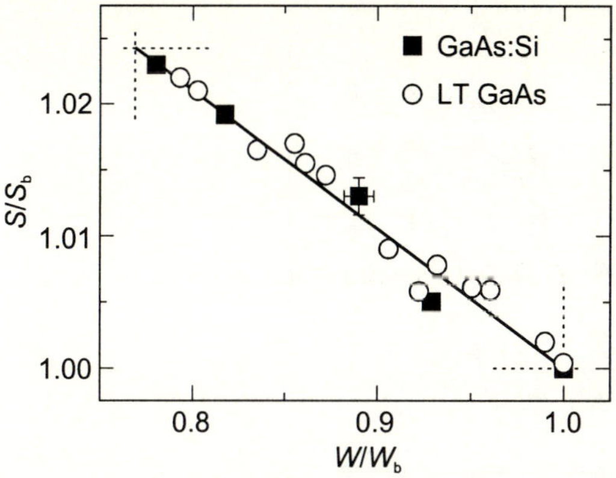

Fig. 5.36. *S* versus *W* plot for low-temperature-grown (LT) gallium arsenide epitaxial layers, grown at substrate temperatures between 200 and 350 °C (Gebauer et al. 1997e). The values are compared with the results obtained in silicon-doped bulk gallium arsenide, where the gallium vacancy paired off with silicon was identified as the positron trap (Sect. 5.2.3.1). The dashed lines indicated the characteristic values for the bulk and for the Ga vacancy, respectively. The solid line is plotted as the linear fit of the values for GaAs:Si. The line shape parameters are normalized to the bulk values S_b and W_b.

are thus not present in significant amounts. Negative As interstitials, which have been discussed as shallow positron traps, are also not detected as traps for positrons.

The influence of doping was the subject of further papers. It was found that beryllium doping decreased the detected concentration of Ga vacancies by at least two orders of magnitude compared with undoped material (Pritchard et al. 1995). This tendency corresponds to the results of study of as-grown LT GaAs layers by Wei et al. (1992). In contrast, Si doping led to an increase in the vacancy concentration. The variation of the vacancy concentration as a function of doping is in agreement with the theoretical considerations of Walukiewicz (1989), relating the formation energies of intrinsic point defects and their concentrations to the position of the Fermi level.

A high resistivity of the low-temperature-grown GaAs layers is obtained only after annealing of the sample at about 600 °C. In order to understand this behavior, experimental investigations were carried out. Positron trapping in LT GaAs layers was studied as a function of annealing temperature (Gebauer et al. 1997d; Hozhabri et al. 1994; Keeble et al. 1993; Störmer et al. 1996b; Umlor et al. 1995). The annihilation parameters, such as the positron lifetime and the *S* parameter, increased significantly after annealing. The change in the *S* parameter with annealing temperature, measured for LT GaAs grown at different temperatures, is shown in Fig. 5.37 (Gebauer et al. 1997d). The measurements were carried out at room temperature after 15 min annealing. The increase in the *S* parameter is less significant for higher growth temperature. The *S* parameter after annealing is, for most

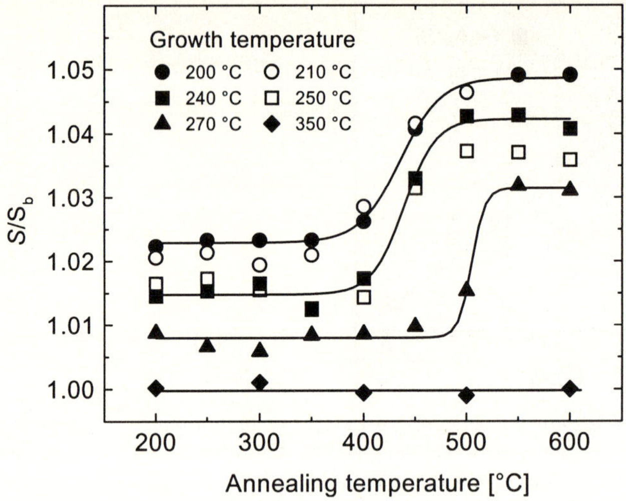

Fig. 5.37. Annealing behavior of the S parameter normalized to the bulk value S_b of low-temperature-grown gallium arsenide, grown at different temperatures (Gebauer et al. 1997d).

samples in Fig. 5.37, much higher than the defect-related value for the as-grown layers. This may prove that another defect is responsible for positron trapping after annealing. The S/S_b and W/W_b values are highest for the lowest growth temperatures and reach 1.05 and 0.67, respectively. Almost saturated trapping was observed and thus these values are close to the defect-related parameters. The open volume of this defect should be larger than that of a monovacancy. This was confirmed by slow-positron-beam lifetime measurements. A defect-related lifetime of 340 to 345 ps was found in the layers (Gebauer et al. 1997d; Störmer et al. 1996b). The theoretical calculations for unrelaxed nearest-neighbor divacancies gave a lifetime of 321 (Puska et al. 1989b) or 317 ps (Pöykkö et al. 1996a). W/W_b was calculated in the latter paper to be 0.59.

While the measured defect-related lifetime indicates that the open volume is slightly larger than a divacancy, the W parameter rather contradicts this interpretation. The atomic structure of the positron trap appearing after annealing is unclear. It was proposed that small vacancy clusters are formed (Keeble et al. 1993; Störmer et al. 1996b). The high thermal stability of such defects up to 700 °C (Umlor et al. 1995) indicates that the observed defects are stabilized by other point defects, e.g. antisites.

Another interpretation is positron trapping related to arsenic precipitates (Gebauer et al. 1997d; Hozhabri et al. 1994). It was discussed in Sect. 3.1.5.5 that precipitates can act as positron traps. The defect-related annihilation parameters can only be understood when vacancies are incorporated either in the interface to the GaAs matrix or in the interior of the precipitates. Indeed, the association of ar-

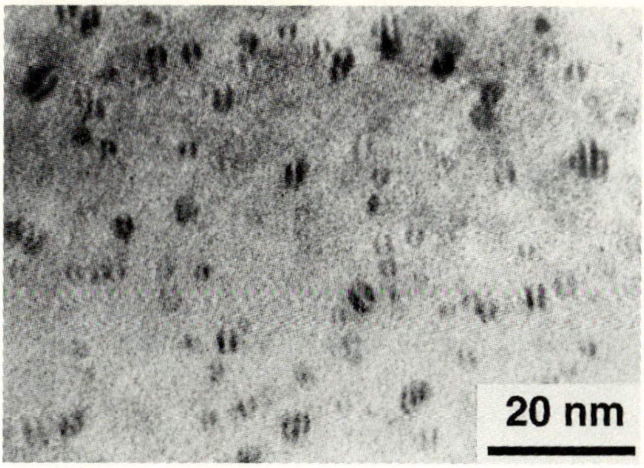

Fig. 5.38. Transmission electron microscopy image of arsenic precipitates in low-temperature-grown gallium arsenide (Luysberg et al. 1998). The layers were grown at 200 °C with a beam equivalent pressure of 30 and annealed at 600 °C.

senic precipitates and vacancy clusters was found by Ruvimov et al. (1998) using transmission electron microscopy. Hozhabri et al. (1994) found a correlation between the depth profile of the As content determined by X-ray photoemission spectroscopy and the depth dependence of the S parameter in the layer. This was taken as a clue for positron trapping in As precipitates. The density of such precipitates was measured together with the positron trapping rate in the same samples by Gebauer et al. (1997d). Densities of 5×10^{17} and 5×10^{16} cm^{-3} and a diameter of about 1.7 nm were found by transmission electron microscopy for samples grown at 210 °C and 250 °C and annealed at 600 °C. A TEM picture of arsenic precipitates in such annealed layers is shown in Fig. 5.38. Assuming diffusion-limited trapping in spherical precipitates, trapping rates of 1.6×10^{12} and 1.7×10^{11} s^{-1} can be estimated according to (3.12) with a diffusion constant of $D_+ = 1.5$ cm^2/s. This should lead to almost saturated positron trapping, as observed in these samples. However, it should be noted that further experiments, especially with non-saturated positron trapping, are required to clarify the nature of the positron trap.

5.2.5 Post-Growth Heat Treatment

Annealing of gallium arsenide at elevated temperatures under variable arsenic pressure is known to change the crystal stoichiometry and hence the electrical and structural properties (Noack et al. 1997; Oda et al. 1992). In order to improve the homogeneity of commercially available GaAs crystals, thermal treatments are nowadays routinely applied after growth (Hurle 1994; Oda et al. 1992). The understanding of this processes requires knowledge of the behavior of native point defects in thermal equilibrium as a function of temperature and stoichiometry.

A compound like GaAs always exhibits a deviation from its ideal stoichiometric composition within the thermodynamically allowed range. This deviation is, by necessity, governed by native defects. Since the crystal is, at a given temperature, in equilibrium with its own vapor phases, the stoichiometry and thus the point defect concentrations depend on the vapor pressure. Only As needs to be considered for GaAs, since the arsenic vapor pressure at high temperatures is usually orders of magnitude higher than the Ga vapor pressure (Arthur 1967). A high As vapor pressure is equivalent to arsenic-rich conditions in the crystal and vice versa. Expressions for the equilibrium concentration of point defects as a function of temperature and As pressure were derived (Tan 1994; Tan et al. 1993) by means of equilibrium thermodynamics. The most striking result was the temperature independence or even negative temperature dependence of the equilibrium concentration of Ga vacancies in heavily n-doped As-rich GaAs. This is essentially based on the so-called Fermi level effect, i.e. the dependence of the concentration of intrinsic point defects on the charge state. The concentration of Ga vacancies should be rather high (10^{17} to 10^{18} cm^{-3} depending on the doping level).

Another way to obtain information on the concentration of intrinsic defects is the calculation of formation energies, carried out in a number of quantum-mechanical first-principles studies (Baraff and Schlüter 1985a; Jansen and Sankey 1989; Zhang and Northrup 1991). Antisites are expected to dominate under most circumstances, i.e. As_{Ga} in As-rich and Ga_{As} in Ga-rich GaAs, respectively. In highly n-doped and As-rich GaAs, however, the Ga vacancy should be the dominant defect due to its low formation energy. Proper values of the point defect equilibrium concentrations at high temperatures are still under debate (Hurle 1995; Wenzl et al. 1990). The GaAs solidus shows a large phase extent corresponding to defect densities of nearly 1×10^{19} cm^{-3} in the model of Hurle (1979a, 1988, 1995). Based on the precision lattice parameter and density measurements of Morozov and Bublik (1986), the defects realizing the non-stoichiometry were supposed to be Frenkel defects in the arsenic sublattice. However, it has not been possible to reproduce these results in the work of Sajovec et al. (1990). Wenzl et al. (1990, 1994) concluded that the phase extent of GaAs should be much smaller, i.e. the defect density should not exceed 10^{18} cm^{-3}.

It is clear from the discussion above that annealing of GaAs under appropriate As vapor pressure should induce changes in the vacancy concentration measurable by positron annihilation. The number of positron studies dealing with vacancies after post-growth annealing is limited. Annealing of as-grown GaAs to study the thermal stability of defects, e.g. As vacancies (Corbel et al. 1988) or vacancy–dopant complexes (Krause-Rehberg et al. 1994b), is clearly also some kind of post-growth annealing. The intention of such treatment is, however, to observe the thermal stability of defects rather than to study the influence of treatment on point defects in equilibrium. Such studies are, therefore, not discussed within this section. One of the first positron annihilation studies regarding post-growth annealing of GaAs was published at the beginning of the 1990s (Clayton et al. 1992). Positron lifetime measurements were performed on semi-insulating GaAs after different heat treatments. A long-lifetime component (≈ 300 ps) was detected after annealing at 950 °C and slow cooling to room temperature. A similar long lifetime (≈ 330 ps) was found in samples quenched from 1000 °C or 1100 °C but with increased intensity. Reannealing of these samples at 950 °C and slow cooling resulted essentially in the same behavior as for the samples without quenching. One may conclude that annealing introduces larger vacancy clusters, which were interpreted to be divacancies (Clayton et al. 1992). Similar studies were reported by Krambrock et al. (1997). They investigated semi-insulating GaAs after inverse thermal conversion (ITC) treatment, i.e. annealing at 1200 °C for 16 h followed by a rapid quench to room temperature. An ITC treatment destroys almost all native EL2 defects present in the as-grown material and leaves the samples slightly p-type (Lagowski et al. 1986).

Positron lifetime measurements as a function of temperature revealed the existence of negatively charged monovacancies and shallow positron traps. The monovacancies had a defect-related lifetime of about 273 ps and were interpreted to be related to Ga vacancies. Their density was estimated to be about 3.5×10^{16} cm^{-3}, one order of magnitude higher than in the as-grown samples. Optically detected magnetic resonance performed on the same samples showed the existence of new acceptor-type defects too, possibly related to the shallow traps observed by positron annihilation. These studies demonstrated the potential of positron annihilation for the study of stoichiometric vacancies in GaAs. Unfortunately, the annealing conditions, most importantly the As vapor pressure, were not specified in more detail in these studies. Thus, it is not possible to relate these results to stoichiometry (i.e. As- or Ga-rich) or to theoretical calculations.

Systematic positron studies of vacancies after post-growth annealing of Te-doped GaAs under controlled vapor pressure were performed by Gebauer et al. (1997f, 1998b). The material under study was device quality GaAs grown by the liquid encapsulated Czochralski method (manufacturer Freiberger Compound Materials GmbH). Annealing was performed in sealed high-purity quartz ampoules in a two-zone furnace. The GaAs samples and a metallic arsenic source were kept at different temperatures. The As vapor pressure in the ampoule was determined from the temperature of the As source according to Gokcen (1989). The ampoules were fast quenched in water after annealing.

The results of such heat treatments on highly Te-doped ($n = 1.5\times10^{18}$ cm^{-3}) GaAs are shown in Fig. 5.39. The average positron lifetime behaves as discussed in Sect. 5.2.3.2 for highly Te-doped GaAs and is well above the reference level obtained in a sample without positron trapping (GaAs:Zn), indicating a relatively high density of vacancies. A defect-related lifetime of about 254 ps was obtained at room temperature from a decomposition of the positron lifetime spectra, compatible with monovacancies and in agreement with the results on as-grown GaAs:Te (Sect. 5.2.3.2). However, a distinct decrease in the average positron lifetime (i.e. of the vacancy density) was observed for the longer annealing time. One may conclude that annealing under the relatively low As pressure of about 0.2 bar reduces the vacancy density, whereas annealing for only 30 min is too short to

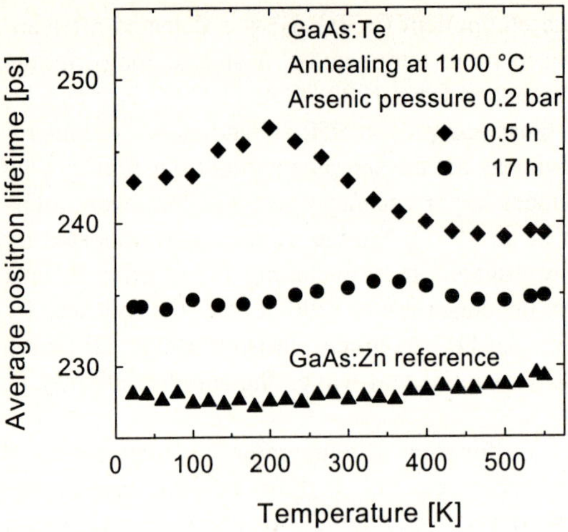

Fig. 5.39. Average positron lifetime measured in tellurium-doped gallium arsenide ($n = 1.5 \times 10^{18}$ cm^{-3}) after thermal treatment as a function of the sample temperature (Gebauer et al. 1998b). The annealing was performed for 0.5 and 17 h at 1100 °C with the arsenic source at 550 °C, corresponding to an arsenic vapor pressure of about 0.2 bar. The measurements are compared with results obtained in a zinc-doped GaAs reference sample.

reach equilibrium between the crystal and gas phase. A variation of both As pressure and annealing time, was performed to confirm these assumptions.

The results are shown in Fig. 5.40. The existence of monovacancies was concluded from the defect-related lifetime, found to be between 251 and 257 ps in all samples. The vacancy density is shown versus the annealing time for GaAs:Te annealed at 1100 °C under different As pressures. A distinct change was not observed for As pressures above 1 bar, i.e. the density of the monovacancies remained constant (about 2 to 3×10^{17} cm^{-3}) regardless of the duration of the annealing. However, a low As pressure reduces the vacancy concentration drastically. A stationary state seemed to be reached for an annealing time of about 1000 min (Fig. 5.40b). The saturation level of the vacancy concentration depends on the annealing temperature. The results show the correlation between As pressure and vacancy concentration, i.e. annealing at a given As pressure and temperature indeed alters the vacancy concentration in a reproducible way. The drop in the vacancy concentration at As pressures below 1 bar is in agreement with the data of Noack et al. (1997) and Oda et al. (1992), who found a strong change in the electrical resistivity in annealed undoped GaAs in the same range of As pressures.

The vacancy concentration increases with the As pressure, i.e. the stoichiometry changes to more As-rich conditions. This proves that Ga vacancies are the dominant stoichiometry-related vacancies in highly n-doped GaAs. This could be independently confirmed by applying the Doppler-broadening coincidence technique to these samples. Weighted difference curves offer advantages over the

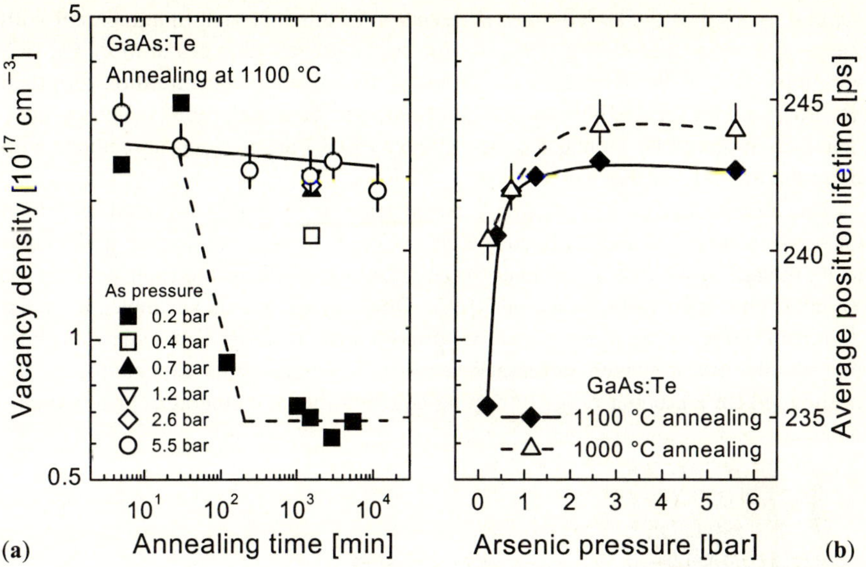

Fig. 5.40. Density of monovacancies in tellurium-doped ($n = 1.5\times10^{18}$ cm^{-3}) gallium arsenide as a function of the annealing time (**a**) and the arsenic vapor pressure (**b**). The samples in (a) were annealed at 1100 °C under arsenic pressures from 0.2 to 5.5 bar, corresponding to a variation of the arsenic source temperature from 550 °C to 700 °C. The lines are drawn to guide the eye. The vacancy densities were calculated from the average positron lifetime at room temperature (*right axes*) using a trapping coefficient of 10^{15} s^{-1} (Gebauer et al. 1998b).

more frequently applied simple ratio (Sect. 3.2.2) in the sense that more reliable information can also be obtained in the low-momentum part of the spectrum. The weighted difference spectra for highly Si-doped GaAs ([Si]=4×10^{19} cm^{-3}) and GaAs:Te shown in Fig. 5.41 were found to be essentially the same. They are clearly different from the spectrum obtained in an undoped n-type GaAs sample, where As vacancies were detected (Sect. 5.2.1.1). V_{As} exhibited a defect-related lifetime of about 290 ps. Complexes of Ga vacancies and Si_{Ga} donors were identified by Gebauer et al. (1997c) in the highly Si-doped sample by the combination of scanning tunneling microscopy and positron annihilation (Sect. 5.2.3.1). The close agreement between the spectra of Te- and Si-doped GaAs suggests that the defects are similar in both samples. The defects in Te-doped GaAs are related to Ga vacancies, probably to $V_{Ga}Te_{As}$ complexes. The minor differences in the spectra around 514 keV were attributed by Gebauer et al. (1998b) to the different chemical surrounding of the Ga vacancies.

The results from positron annihilation may thus be regarded as a proof of the theoretical expectation that Ga vacancies should be the dominant native defects in highly n-doped, As-rich GaAs. Moreover, Ga vacancies can accommodate a substantial fraction of the stoichiometric deviation in this material, in agreement with the results of Sajovec et al. (1990). As the stoichiometry changes to more Ga-rich conditions, i.e. as the As pressure is reduced, the vacancy concentration drops, in accordance with the predictions of Tan (1994) and Tan et al. (1993). The vacancy densities observed by Gebauer et al. (1997c, 1998b) are in good agreement with these theoretical results.

The observation of high vacancy concentrations in highly n-doped as-grown GaAs (Gebauer et al. 1997c; Krause-Rehberg et al. 1994b) can thus be understood to be related to As-rich conditions, since GaAs grown from the melt tends to be arsenic-rich (Hurle 1995; Oda et al. 1992). One may expect Ga vacancies to be the dominant defect types in n-type GaAs, provided the material is arsenic-rich. Further studies are, however, necessary to estimate whether this expectation is entirely valid for a broader range of doping concentrations, temperature, and stoichiometry.

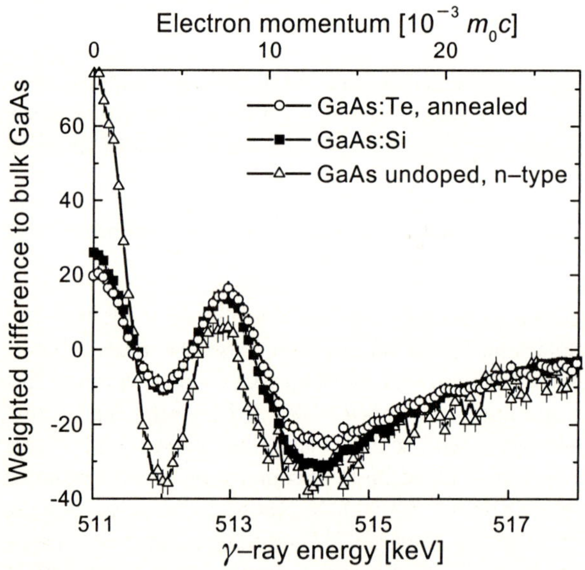

Fig. 5.41. Weighted difference spectra obtained with the Doppler-broadening coincidence technique in highly silicon-doped gallium arsenide ([Si]= 4×10^{19} cm^{-3}, ■) in tellurium-doped gallium arsenide ($n = 5\times10^{18}$ cm^{-3}, o) annealed 1 h at 1000 °C at an arsenic pressure of 5.5 bar, and in undoped n-type GaAs (△). The difference spectra were calculated with respect to zinc-doped GaAs and weighted by the statistical error, which is a function of the γ-ray energy. The spectra were obtained in the doped samples exhibiting saturated positron trapping in vacancy-type defects (Gebauer et al. 1998b). The spectrum of the undoped sample is scaled to saturated positron trapping. The full lines are drawn to guide the eye. The electron momentum (*upper axis*) is given in units of m_0c (m_0 electron rest mass, c speed of light).

5.2.6 Irradiation-Induced Defects

A large number of defect investigations by positron annihilation have been carried out in gallium arsenide irradiated with high-energy particles, such as protons, neutrons, and electrons. The motivation for such studies is not only the technological interest in radiation damage. As the defect identification in as-grown gallium arsenide is not straightforward, particle irradiation is a suitable tool to introduce defects in a defined manner. The early irradiation experiments were carried out with heavy particles (Sect. 5.2.6.1), but of particular importance is irradiation with electrons in the energy range 1 to 2 MeV at low temperatures, where Frenkel pairs are generated, and it is hoped to avoid defect annealing and reactions (Sect. 5.2.6.2).

The defect annealing in ion-implanted layers is of great technological importance. The slow-positron-beam technique is a suitable tool to monitor the depth profile of open-volume defects formed after ion implantation (Sect. 5.2.6.3).

A systematics of the defects found by positron annihilation in GaAs has already been presented above (Sects. 5.2.1 to 5.2.3). The main emphasis is put in the following sections on the annealing of defects and possible defect reactions during heat treatment after irradiation.

5.2.6.1 Heavy-Particle Irradiation

The first heavy-particle irradiation studies on GaAs were performed with protons (Aref'ev et al. 1983, 1985; Pogrebnyak 1984; Pogrebnyak et al. 1983, 1984a), α-particles (Sen Gupta et al. 1984), and neutrons (Dlubek and Krause 1987; Stucky et al. 1986). A special feature is apparently the occurrence of positronium in GaAs irradiated with supercurrent pulsed proton and electron beams (Pogrebnyak 1984). The angular correlation of annihilation radiation showed a narrow peak at low angular deviations, which is typical for positronium. Positron lifetime spectroscopy revealed a lifetime component of $\tau_d > 500$ ps. Quenching experiments with a magnetic field of 0.8 T proved the existence of ortho-positronium in the irradiated GaAs samples. This unusual behavior is similar to the findings in Si and Ge (Sect. 4.1.7.1) after the same irradiation treatment. It was explained by the introduction of large vacancy clusters.

Dlubek et al. (1988b) investigated the formation and annealing of defects in Si- and Te-doped GaAs after neutron irradiation. The irradiation was carried out at 90 °C with fast neutrons of a dose ranging from 5×10^{14} to 4×10^{19} cm^{-2}. The effect of the irradiation on the average positron lifetime, photoluminescence, infrared absorption, and carrier concentration was studied. The behavior of the physical properties shown in Fig. 5.42 is similar for the n-type doping with 4.5×10^{18} cm^{-3} Te or with 1.5×10^{18} cm^{-3} Si. The free carriers are compensated for a neutron dose of 5×10^{17} cm^{-2} and the samples are highly resistive. Simultane-

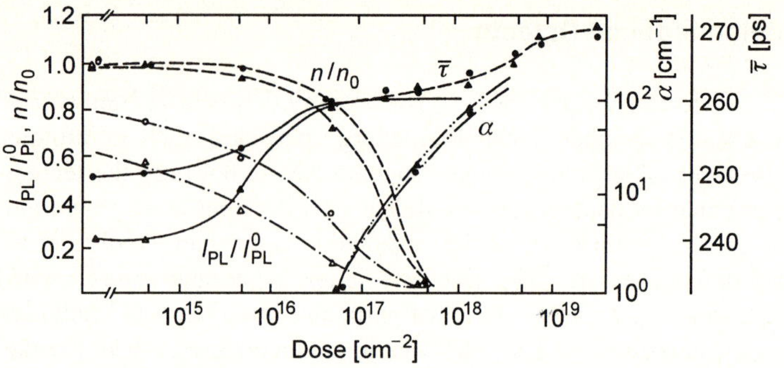

Fig. 5.42. Effect of neutron irradiation on physical properties of gallium arsenide doped with tellurium (*circles*) and silicon (*triangles*). Dose dependence of the average positron lifetime $\bar{\tau}$, the absorption coefficient α measured at the energy of 0.7 eV, the normalized photoluminescence intensity I_{PL} / I^0_{PL} at 300 K, and the normalized carrier concentration n/n_0. I^0_{PL} and n_0 are the photoluminescence intensity and the carrier density before irradiation (Dlubek et al. 1988b).

ously, the photoluminescence intensity vanishes. The absorption coefficient strongly increases for doses higher than 10^{17} cm^{-2}.

Positron lifetime spectroscopy monitored a vacancy signal already in the as-grown samples. This is related to positron trapping in vacancy–dopant complexes (Sect. 5.2.3). The average positron lifetime starts to increase at a dose of 5×10^{15} cm^{-2} and reaches at 5×10^{19} cm^{-2} a level of 270 ps. Although decomposition of the positron lifetime spectra was not carried out, it may be concluded from the $\bar{\tau}$ value and from the expected saturated trapping that monovacancies are the dominant positron traps. The annealing of neutron-irradiation-induced defects is shown in Fig. 5.43. The vanishing of the positron signal was compared with results from other methods, such as conductivity measurements (Lang 1977), deep level transient spectroscopy (Martin and Makran-Ebeid 1983), and electron paramagnetic resonance (Wörner et al. 1982). The annealing stage at 500 °C was found for all methods. The defect in the EPR measurements was identified with the As_{Ga} antisite. The nature of the deep-level defects giving rise to the U and L bands in deep level transient spectroscopy was not specified.

The annealing stage of the average positron lifetime in neutron-irradiated GaAs at 500 °C (Dlubek and Krause 1987; Yu et al. 1995) does not coincide with that found after electron irradiation (see Sect. 5.2.6.2). The higher annealing temperature, which is observed after neutron irradiation, indicates that not isolated vacancies but complexes containing vacancies are detected by positron annihilation. Two annealing stages are usually found in proton-irradiated GaAs (Itoh et al. 1992b, 1995; Pogrebnyak et al. 1983). The main stage at 500 °C is the same as in neutron-irradiated samples. The other annealing stage occurs at 300 °C and is slightly above the stage observed in electron-irradiated GaAs.

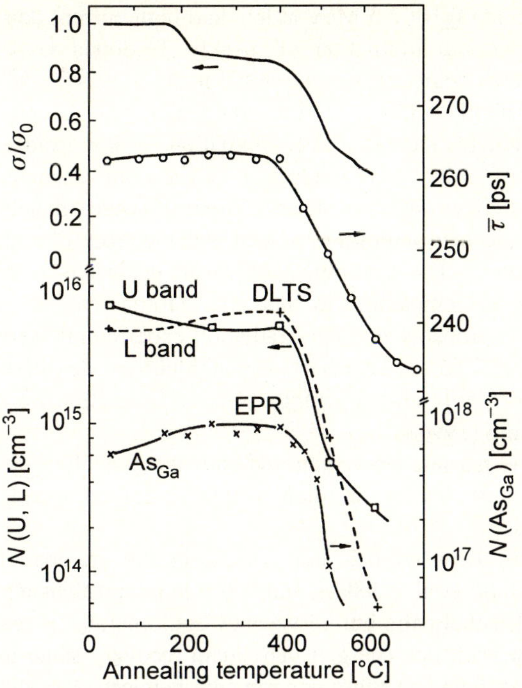

Fig. 5.43. Isochronal annealing of n-type gallium arsenide irradiated with fast neutrons (Dlubek and Krause 1987). The density of As_{Ga} antisites, $N(As_{Ga})$, measured by electron paramagnetic resonance (EPR) is shown together with the density of defects giving rise to the L (E_{val} + 0.7 eV) and U bands (E_{cond} − 0.55 eV) of deep level transient spectroscopy (DLTS), the average positron lifetime $\overline{\tau}$, and the normalized electrical conductivity σ/σ_0 as functions of the annealing temperature. σ_0 is the conductivity after irradiation. E_{val}, and E_{cond} are the edges of the valence band and the conduction band, respectively.

5.2.6.2 Low-Temperature Electron Irradiation

Aref'ev et al. (1979) were the first to report on positron annihilation investigations of electron-irradiated GaAs. The first electron irradiation experiments at low temperatures were carried out by Hautojärvi et al. (1986b) and Mascher et al. (1987c, 1988). A distinct annealing stage in the positron lifetime data was observed around 500 K. The separated defect-related lifetime of 315 ps was attributed by Mascher et al. (1987c, 1988) to di- or trivacancies, possibly in complex with other intrinsic defects. The formation of these defects is probably related to the exceptionally high electron energy of 28 MeV. In contrast, Hautojärvi et al. (1986b) found the formation of Ga monovacancies after irradiation at 20 K with 3-MeV electrons.

A pulsed electron beam of an energy of 2.3 MeV at low and high current densities was used for room-temperature irradiation of weakly Te-doped GaAs (Brudnyi et al. 1989). The dominant component in the initial material was a positron lifetime of 265 ps, but surprisingly a 320 ps lifetime component was also present. The intensity of both components increased after irradiation for low irradiation current densities of 5 mA cm^{-2}. They were interpreted as Ga monovacancies and divacancies. Vacancy agglomerates with a positron lifetime of approximately 590 ps appeared as the primary defects in samples irradiated with electrons of high current densities of 1.5×10^{3} A cm^{-2}. The occurrence of vacancy clusters after high-density electron irradiation is comparable to that after proton irradiation (Pogrebnyak 1984). The reason is probably that the irradiation experiments were carried out at room temperature with pulsed beams of very high current densities. Clustering reactions of vacancies after room-temperature electron irradiation were also reported by Itoh and Murakami (1994).

It is important to carry out the particle irradiation and start the annealing experiment far below room temperature, because irradiation-induced defects already anneal at rather low temperatures.

The irradiation energies were between 0.5 and 2 MeV in the studies of Hautojärvi et al. (1986b), Würschum et al. (1989b), and Würschum and Schaefer (1987). The formation of close Frenkel pairs but no vacancy agglomerates is expected as the primary defects for such low energies. The main recovery stage in the average positron lifetime was found in these experiments between 200 and 350 K. It coincides with the results of electrical measurements (Thommen 1970). It was ruled out that the observed vacancy-type defect was the As vacancy, since the Fermi level was in the midgap position after irradiation and As vacancies should be positively charged. Thus it was assumed that the isolated Ga vacancy was detected. The annealing stage appeared in undoped, p-type (doped with Zn), and n-type (doped with Si or Sn) GaAs. The annealing curves of the irradiated n-type samples show a special feature. The average positron lifetime increases by up to 20 ps at the annealing temperature of 500 K. Würschum et al. (1989b) related this maximum to the shift of the Fermi level, which made residual As vacancies visible (Fig. 5.6).

Defects introduced by electron irradiation of GaAs were investigated in several studies by the groups of Helsinki University of Technology and Laboratoire Positons Saclay (Corbel et al. 1990, 1992b; Saarinen et al. 1995b, 1997a). The appearance of monovacancies and shallow positron traps was proven in n-type and semi-insulating GaAs irradiated with 1.5-MeV to 3-MeV electrons at room temperature and at 20 K (Corbel et al. 1990, 1992b). It was supposed that the irradiation generated negatively charged Ga vacancies and negatively charged Ga_{As} antisites acting as shallow positron traps. The vacancies already present in the non-irradiated material changed their properties under irradiation. The average positron lifetime decreased in weakly Sn-doped GaAs from 265 ps, prior to irradiation, to 240 ps after irradiation to a dose of 1×10^{17} cm^{-2} (Fig. 5.44). This was interpreted as the transition to the positive charge state of the vacancies in as-grown GaAs by the shift of

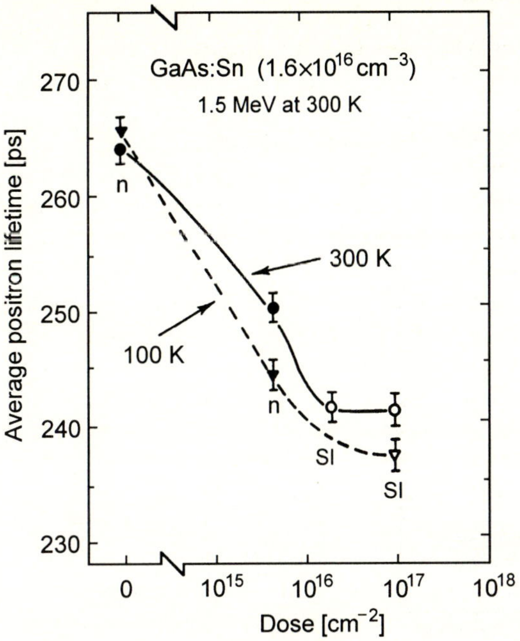

Fig. 5.44. Average positron lifetime as a function of the irradiation dose in tin-doped gallium arsenide (carrier density 1.6×10^{16} cm^{-3}) irradiated with 1.5-MeV electrons at 300 K (Corbel et al. 1990). The samples were still n-type after low-dose irradiation and became semi-insulating (SI) with higher doses. The measurement temperatures were 100 and 300 K.

the Fermi level due to irradiation-induced defects. Obviously, these vacancies have an ionization level in the upper half of the bandgap and were supposed to be As vacancies. Such an interpretation has already been given in Sect. 5.2.1.1.

The annealing of irradiation-induced monovacancies was studied in undoped semi-insulating GaAs. A broad annealing stage was found between 77 and 500 K (Corbel et al. 1992b). The irradiation-induced vacancy must be neutral or negative and cannot be the isolated As vacancy (Fig. 5.6). It was thus identified with the Ga vacancy either isolated or with a vacancy as part of a complex. The Ga_{As} antisites were assumed to be isolated defects and appeared to be stable to at least 440 K. The temperature dependence of the average positron lifetime was extensively studied in order to determine the concentration of shallow positron traps. Under the assumption that the total acceptor concentration is determined by the V_{Ga} and Ga_{As} concentrations, the introduction rate of Ga vacancies for these irradiation conditions was given to be 0.5 cm^{-1} at room temperature.

The same annealing stage of irradiation-induced vacancies was also found in n-type GaAs doped with Si or Te (Saarinen et al. 1995b). In contrast, a high introduction rate for shallow positron traps of 10 cm^{-1} was deduced. This result was

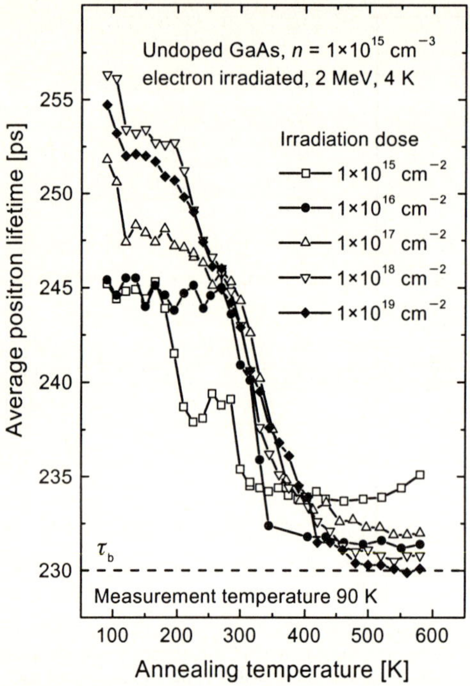

Fig. 5.45. Electron-irradiation experiment in undoped gallium arsenide (Polity 1997; Polity et al. 1997c). The average positron lifetime measured at 90 K is plotted versus the annealing temperature for different irradiation doses. The dashed line indicates the positron bulk lifetime τ_b.

also obtained from the analysis of the temperature-dependent positron lifetime measured after each annealing step. It was concluded that the negative Ga_{As} antisites play a crucial role in the electrical compensation. As has already been stated, it may, however, be difficult to deduce exact quantitative results from multiple-parameter fits to the average positron lifetime measured as a function of the temperature.

Electron-irradiation effects were studied by positron lifetime and Hall effect measurements for GaAs of different carrier concentrations and dopants by Polity et al. (1997c). The irradiation doses were varied between 10^{15} and 10^{19} cm^{-2}, the electron energy was 2 MeV. The average positron lifetime is shown in Fig. 5.45 as a function of the annealing temperature for undoped GaAs. The main annealing stage was found around 300 K, in accordance with the earlier reports. The samples having a carrier density of 1×10^{15} cm^{-3} prior to irradiation were already semi-insulating after irradiation with a dose of 10^{16} cm^{-2} ($E_F = E_{cond} - 570$ meV). An additional weak decrease in $\bar{\tau}$ occurred at 100 K. Although this stage was seen in relation to a possible thermally induced reconversion of the EL2 center, such an effect is, with the results presented in Fig. 5.22, rather unlikely, since the measurement temperature is already too high to detect the metastable vacancy of EL2*. Thus, this stage is rather due to defect annealing.

The sample irradiated with 10^{16} cm^{-2} was still n-type with a carrier density of 7×10^{14} cm^{-3} ($E_F = E_{cond} - 170$ meV measured after the annealing step at 300 K) and showed an additional feature in the annealing curve (Fig. 5.45). This sample makes clearly visible that the main annealing stage around 300 K is composed of two separate stages. Comparing the curves of irradiation with doses of 10^{15} and 10^{16} cm^{-2}, the annealing stage at 200 K is missing for the sample irradiated to a dose of 10^{16} cm^{-2}. As the Fermi level shifted from $E_{cond} - 170$ meV to $E_{cond} - 570$ meV, a charge state transition of a vacancy-type defect was likely to occur. According to Fig. 5.6, the As monovacancy is compatible with this interpretation. The decomposition of the lifetime spectra provided a defect-related lifetime of 260 to 270 ps, indicating the detection of a monovacancy. Presumably, the annealing of the irradiation-induced As vacancy is observed. A defect complex containing a monovacancy with the respective ionization level, however, cannot be excluded as being responsible for the 200-K stage.

The positron lifetime increases significantly with increasing dose from 10^{15} to 10^{16} cm^{-2} in the region of the 300-K stage. Obviously no transition to a positive or neutral charge state can occur by lowering the Fermi level to $E_{cond} - 570$ meV. This is compatible with the Ga vacancy, which has ionization levels rather close to the valence band. The conclusion that the annealing stage at 300 K is related to V_{Ga} is in accordance with former interpretations (Corbel et al. 1990; Hautojärvi et al. 1986b; Saarinen et al. 1995b; Würschum et al. 1989b).

The result of the annealing experiment of electron-irradiated Te-doped GaAs with a carrier density of 6×10^{17} cm^{-3} is shown in Fig. 5.46 (Polity et al. 1997c). An average positron lifetime of 246 ps had already been measured in as-grown samples and was related to vacancy–tellurium complexes (Sect. 5.2.3.2). The annealing stages found at 200 and 300 K are similar to those in undoped GaAs.

A plateau is reached after annealing at 200 K in the low-dose-irradiated samples. This can be understood by positron trapping in the vacancies already present prior to irradiation. The lowering of the plateau with higher doses has to do with the increase in the concentration of shallow positron traps. This is evident, especially in the sample irradiated to 10^{19} cm^{-2}, where the positron lifetime reaches the bulk value. However, complete annealing of the vacancies can be excluded from measurements of the positron lifetime as a function of temperature (Fig. 5.47). The vacancy signal reappears with higher sample temperatures. The positron lifetime of the as-grown sample at 600 K is nearly reached after the annealing step at 580 K, and it can be concluded that the irradiation-induced vacancies have vanished.

A numerical fit was carried out for the temperature-dependent positron lifetime of the irradiated sample after annealing at 580 K (solid line in Fig. 5.47). The fit provided the positron binding energy to the shallow traps of 65 meV and a trapping rate of 8×10^{11} s^{-1} (Polity et al. 1997c). This very high trapping rate is related to a high concentration of irradiation-induced shallow positron traps, which are

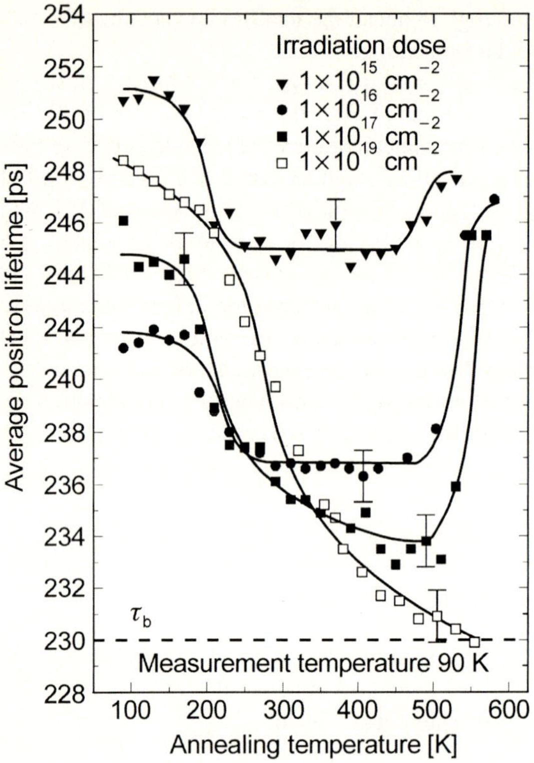

Fig. 5.46. Annealing of tellurium-doped gallium arsenide irradiated with electrons at 4 K with different doses (Polity 1997; Polity et al. 1997c). The average positron lifetime was measured at 90 K. τ_b represents the positron bulk lifetime.

thermally stable at least up to 600 K. This is the reason that the influence of the shallow traps is still visible in the average positron lifetime above room temperature. As a consequence, if the number of shallow positron traps is high enough, it cannot be excluded that these traps influence positron trapping at room temperature and beyond.

The observed shallow positron traps must be intrinsic defects, as these traps occur independently of the dopant. The density increases depending on the irradiation dose far beyond any impurity level. Polity et al. (1997c) estimated approximately 10^{19} cm^{-3} shallow traps after irradiation with 10^{19} cm^{-2} electrons, resulting in an introduction rate of 1 cm^{-1}. Although the positron binding energy was found to be roughly the same as for shallow positron traps in as-grown GaAs (Sect. 5.2.2), it cannot be concluded that the defect types are the same. The simplest intrinsic defects having the necessary negative charge for the shallow trap behavior are the Ga_{As} antisite and possibly the As interstitial. It is generally believed that isolated interstitials already anneal far below room temperature (Hausmann et al. 1996).

A distinct increase in the annealing curve appears above 500 K (Fig. 5.46) for the partially compensated samples, which were irradiated with doses from 10^{15} to 10^{17} cm^{-2}. The partial compensation for low irradiation doses was proven by Hall effect measurement after annealing up to 300 K. The Fermi level was located at 110 meV below the conduction-band edge for the samples irradiated to a dose of 10^{17} cm^{-2}, and the carrier density $n = 6\times10^{15}$ cm^{-3}. The irradiation dose of 10^{19} cm^{-2} leads to complete compensation of the carriers and the Fermi level was found at $E_{\mathrm{cond}} - E_{\mathrm{F}} = 540$ meV ($n = 10^{8}$ cm^{-3}). No change in the carrier concentration was found in a second Hall effect measurement after the annealing at 550 K. This means that the location of the Fermi level did not change, at least during the annealing between these two temperatures. A charge state transition of a defect can thus be excluded as a reason for the observed strong increase in $\overline{\tau}$. A

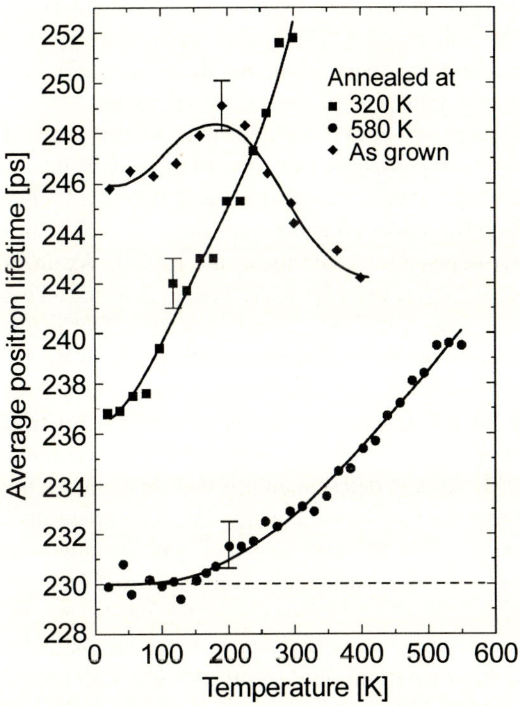

Fig. 5.47. Temperature-dependent average positron lifetime in tellurium-doped gallium arsenide, irradiated with electrons to a dose of 10^{19} cm^{-2} (Polity et al. 1997c). The curves after annealing up to 320 and 580 K are shown and compared with the curve of the as-grown sample. The full line for the curve after annealing at 580 K is a fit to the data according to the trapping model, the other lines are to guide the eye. The dashed line indicates the positron bulk lifetime.

structural change of a defect complex or a defect reaction was concluded. Hausmann and Ehrhart (1995b) found, using magnetic circular dichroism of optical absorption, the annealing of a $As_{Ga}X$ complex at 450 K in electron-irradiated Te-doped GaAs. This defect complex was identified by optical detection of electron nuclear double resonance with the $V_{As}As_{Ga}$ complex having a 2+/+ ionization level at E_{cond} − 220 meV (Wietzke et al. 1995). The positive charge of this defect prevents its detection by positron annihilation before annealing. The dissolution releases a vacancy-type defect, probably the As vacancy either isolated or as part of another complex, which must be neutral or negatively charged and becomes visible for positrons. It should be noted that the isolated As vacancy was probably responsible for the annealing stage at 200 K (Fig. 5.45) in undoped GaAs. In this sense, it may be more likely that a defect complex appears.

A similar increase in the positron lifetime at 500 K has already been observed by Saarinen et al. (1995b, 1997a). They also detected n-type conductivity after annealing at 600 K, but supposed that the samples in the as-irradiated state were compensated. Consequently, they concluded a charge state transition of the isolated As vacancy. The Fermi level did not shift substantially in this annealing range as in the samples measured by Polity et al. (1997c). This was proven by Hall effect measurements after several annealing steps. Thus, the described defect reaction is more likely than a charge state transition of the arsenic vacancy.

The average positron lifetime decreases above the annealing temperature of 600 K (Saarinen et al. 1997a). It should be noted that the annealing stage of the As vacancy was found to be at higher temperatures in as-grown GaAs (Sect. 5.2.1.1).

A similar annealing stage as found in undoped and n-type GaAs at 300 K occurs also in electron-irradiated Zn-doped GaAs (Polity et al. 1997c). Again, a monovacancy with a defect-related lifetime of 260 to 270 ps is detected. This holds only for doses $\geq 5\times10^{18}\ \mathrm{cm}^{-2}$ (Fig. 5.48a). No or only a weak effect is observed for lower doses. A possible explanation is the competing trapping in Zn acceptors acting as shallow positron traps, because the measurement was carried out at 90 K. This possibility can be excluded, since the measurement of the positron lifetime as a function of the temperature showed no vacancies present in these samples. The position of the Fermi level was determined at room temperature for all irradiation doses. It was found that the free hole concentration is compensated. The dose of $1\times10^{18}\ \mathrm{cm}^{-2}$ gave $p = 2.4\times10^{16}\ \mathrm{cm}^{-3}$; $1\times10^{19}\ \mathrm{cm}^{-2}$ resulted in $p = 1.4\times10^{9}\ \mathrm{cm}^{-3}$. The Fermi level shifted consequently from 147 to 578 meV above the valence band edge.

A transition from a positive to neutral or negative charge states seems to be likely to explain the appearance of the irradiation-induced vacancies at high doses. The ionization level of this defect can be obtained from a non-linear fit of the average positron lifetime plotted against the Fermi-level position (Fig. 5.48b) according to (5.1) and (5.2). The fit provides the ionization level of the detected vacancy-type defect to be located at 355 meV above the valence band. The theoretical calculations do not give a uniform picture of the ionization levels of the Ga vacancy. Most calculations provide only neutral or negative charge states, but a re-

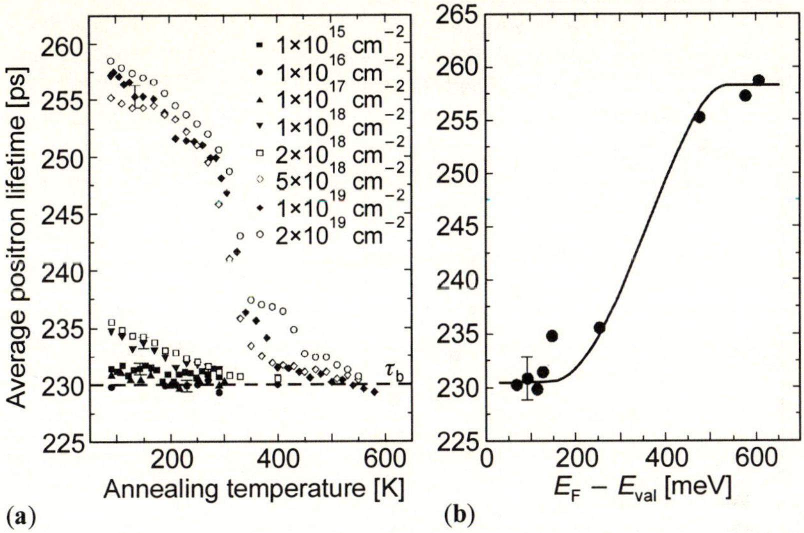

Fig. 5.48. Positron lifetime measurements in zinc-doped gallium arsenide (carrier density $p = 5\times10^{17}$ cm^{-3}) irradiated with 2-MeV electrons (Polity et al. 1997c). Average positron lifetime (measured at 90 K) versus the annealing temperature for different irradiation doses (**a**) and versus the Fermi-level position $E_F - E_{val}$ (**b**). The dashed line in (a) indicates the positron bulk lifetime τ_b. The full line in (b) is the result of a non-linear fit using (5.1) and (5.2).

cent paper does not exclude a positive state (Fig. 5.6). In the latter case, the observed annealing stage at 300 K in electron-irradiated GaAs may be caused by isolated Ga vacancies as suggested by Corbel et al. (1990), Hautojärvi et al. (1986b), Saarinen et al. (1995b), and Würschum et al. (1989b).

Provided that the Ga vacancy cannot be positively charged, the isolated V_{Ga} is not be responsible for the annealing stage at 300 K. Therefore, it must be a defect complex comprising a vacancy, which is positively charged if the Fermi level is close to the valence band edge. Vacancy–dopant complexes, such as V_{Ga}Zn, can be excluded since the same annealing stage also occurs in undoped GaAs. It was also supposed earlier in accordance with this conclusion that the isolated Ga vacancies anneals below 100 K (Pons and Bourgoin 1985). Probably, the complex is composed of a vacancy and an antisite defect.

The result of a low-energy electron irradiation experiment in gallium arsenide is shown in Fig. 5.49 (Polity 1997). The Te-doped and the undoped GaAs samples were found to be compensated after the irradiation and subsequent annealing at 330 K. In contrast, the density of free holes increased in the Zn-doped sample from 5 to 6.9×10^{18} cm^{-3}. The undoped and the n-type samples exhibit the annealing stage at 300 K as found with 2-MeV electron energy (Fig. 5.45 and Fig. 5.46).

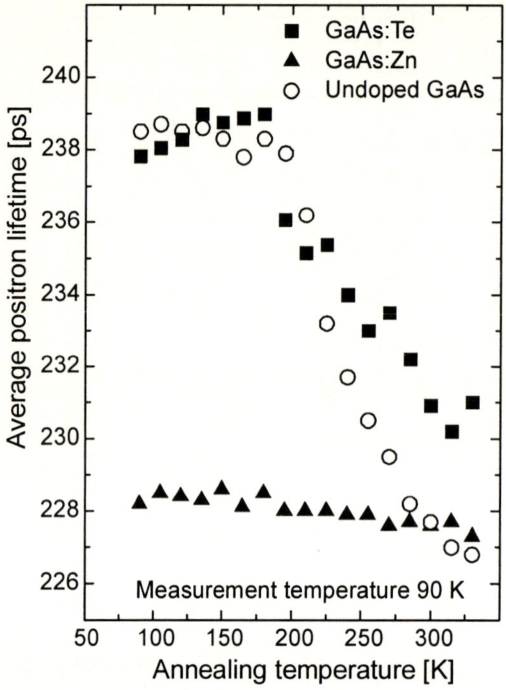

Fig. 5.49. Low-energy electron irradiation experiment in undoped, tellurium-, and zinc-doped gallium arsenide (Polity 1997). The irradiation was carried out at 4 K with 0.55-MeV electrons to a dose of 10^{19} cm^{-2}. The average positron lifetime measured at 90 K is shown as a function of the annealing temperature.

No vacancies are visible in the Zn-doped sample despite the high dose of 10^{19} cm^{-3}. This is in accordance with the interpretation given above that the detected vacancy (or vacancy complex) has a positive/neutral or positive/negative ionization level close to the valence band edge.

To summarize the positron lifetime experiments on differently doped GaAs irradiated at low temperatures, it was found that the annealing of As and Ga vacancies occurs at 300 K. Irradiation-induced vacancies are not responsible for compensation, at least for 2-MeV electron irradiation. This can be concluded, since the electrical compensation is still observed after annealing at 600 K.

5.2.6.3 Implantation-Induced Defects

Ion implantation has found a broad technological application also in GaAs. However, there are special problems with ion implantation in compound semiconductors. A variety of defects with different annealing behaviors is generated in both

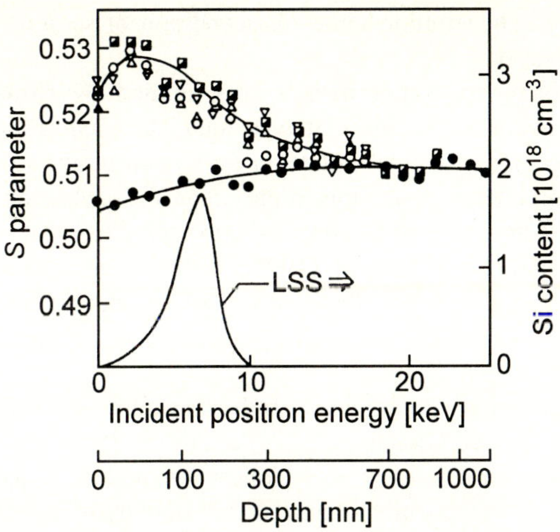

Fig. 5.50. Implantation of silicon in (100) gallium arsenide (Lee et al. 1990). The implantation energy was 120 keV and the dose 3×10^{13} cm^{-2}. The *S* parameter (*left axis*) is plotted as a function of the incident positron energy. A two-step rapid thermal annealing was carried out. The first annealing step was at 900 °C for 5 s. The second step was carried out at 850 °C for 0 s (◪), 10 s (△), 20 s (▽), and 30 s (○), respectively. An unimplanted GaAs sample (●) was included for reference. The profile of implanted silicon (*right axis*) as calculated by the theory according to Linhard, Scharf, and Schiøtt (LSS, Linhard et al. 1963) is shown.

sublattices of GaAs. The heat treatment is difficult, because the arsenic vapor pressure is comparably high, leading to selective evaporation. The recoil of As and Ga by ion implantation creates vacancies near the surface and an excess of interstitials in deeper regions. The knowledge of the defect formation and the annealing in ion-implanted GaAs is rather limited.

The slow-positron-beam technique is a suitable tool for the study of the depth distribution of vacancy-type defects. Starting at the end of the 1980s, it was applied in several studies of implantation with Si, Se, Be, O, and Te ions. Most of the investigations were reported by the positron group of Tsukuba University.

Defect profiles were measured after implantation with Si^+ of 120 keV (Lee et al. 1988a, 1988b, 1989, 1990). They were related to the formation of Ga vacancies either isolated or as part of defect complexes. The observed change in the profile in the first 30 nm between the as-implanted sample and the first annealing step was interpreted as the outdiffusion of Ga vacancies, but can also be simply understood by the change in the surface *S* parameter. The measured profile did not change significantly after several rapid thermal annealing steps (Fig. 5.50). However, the electrical activation of Si changed from below 20 to about 80 %. It was related to the exchange reaction of interstitial Si with Ga_{Ga} rather than the recombination with vacancies. Another possible interpretation is the low sensitivity to moderate changes in the vacancy concentration under the condition of saturated positron trapping in the damaged layer.

The effect of annealing ambients on point defects was studied by slow-positron-beam investigations (Shikata et al. 1992). It was found that annealing with a silicon dioxide cap on top of the sample induced a high concentration of gallium vacancies, which is approximately four times higher than for capless annealing. No effect on the electrical activation of silicon was, however, observed.

Si implantation was also studied by Simpson et al. (1992b) with ion energies of 220 keV. Vacancy-type defects were found in a depth up to 280 nm independent of the dose in the range of 3×10^{13} to 1×10^{15} cm^{-2}. The concentration increased with the dose.

In contrast to Si ions, the implantation with Se^{+} leads to the formation of a high concentration of vacancy clusters after additional annealing at 800 °C (Fujii et al. 1992a, 1992b). The electrical activation efficiency of Se donors was higher, which induced a shift in the Fermi level. It was assumed that a charge state transition of As vacancies from positive to negative occurs, leading to additional positron trapping. The position of the Fermi level also plays an important role for the interpretation of the *S* parameter profiles measured in Be-implanted n-type and p-type GaAs (Uedono et al. 1992b). No change was found in the *S*(*E*) curves after ion implantation in p-type material. A similar effect has already been reported for electron-irradiated p-type GaAs (Fig. 5.48). As long as the sample remains p-type, no positron trapping in open-volume defects can be detected, since they are positively charged. In contrast, open-volume defects were observed after Be implantation in n-type GaAs.

The damage after ion implantation with oxygen was investigated by Eichler (1997). It was found with Rutherford backscattering that the damage was weak in the dose range 1×10^{13} to 1×10^{15} cm^{-2} under investigation. The defect concentration increased with the dose, corresponding to region I of defect generation as a function of the dose in the model of Hecking et al. (1986). With a TRIM simulation, the damage profile for the implantation with 80-keV and 200-keV ions was found to be very broad up to a depth of 500 nm. The Doppler-broadening *S* parameter as a function of the incident positron energy is shown in Fig. 5.52 for several implantation doses. The *S*(*E*) plot could be interpreted with the occurrence of open-volume defects. The *S* versus *W* plots (not shown) indicated that the dominant positron trap is a monovacancy. An additional peak in the *S*(*E*) plot appears in the near-surface region, which becomes more obvious for higher doses. This peak is not related to the damage profile, as no maximum was found by the TRIM simulation in this region. It was concluded, in conjunction with higher *S* and lower *W* parameters compared with monovacancies, that vacancy clusters were formed.

The annealing experiment of the O-implanted samples (Fig. 5.51a) exhibits a constant *S* parameter in the defect-rich layer for a dose of 10^{13} cm^{-2}. The *S* parameter is reduced for annealing temperatures above 500 K. In contrast, a maximum in *S* occurs at 550 K for higher doses. The result cannot be explained by the presence of isolated monovacancies. Arsenic vacancies can be excluded, since the

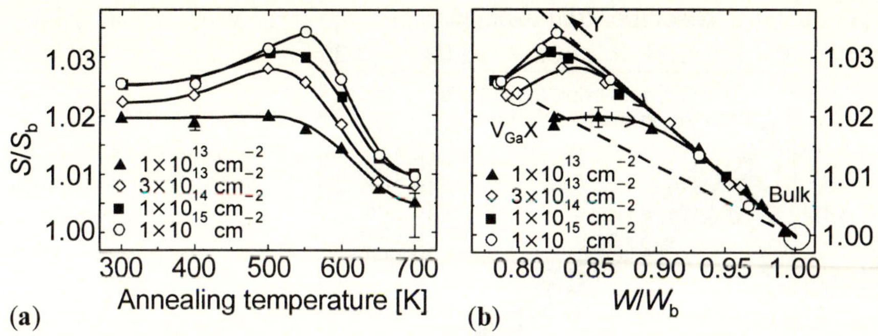

Fig. 5.51. Isochronal annealing experiment for oxygen-implanted gallium arsenide (Eichler 1997). The samples are the same as for Fig. 5.52. (**a**) S parameter in the defect-rich layer versus the annealing temperature for several doses, (**b**) S versus W plot. The arrows on the lines in (b) correspond to increasing annealing temperatures. The annealing time was 10 min. The defect-related value for $V_{Ga}Si_{Ga}$ (Gebauer et al. 1997e) is indicated as $V_{Ga}X$. Two defect-related slopes are drawn as dashed lines ($V_{Ga}X$ and unknown open-volume defect Y). The error of the measurement corresponds to the diameter of the large circles. The line shape parameters S and W are divided by their bulk values, S_b and W_b.

material was semi-insulating, whereas these defects are expected to be positively charged. On the other hand, Ga vacancies are thermally much less stable, as proven by electron irradiation (Sect. 5.2.6.2). It was concluded that the dominant defect is a defect complex containing a vacancy. The typical slope of these complexes in the S versus W plot (Fig. 5.51b) is obtained as the connection of the bulk values (W_b, S_b) to the values in the defect-rich as-implanted layers (dashed line). The defect-related value for $V_{Ga}Si_{Ga}$ (Gebauer et al. 1997e) is indicated as $V_{Ga}X$ in Fig. 5.51b.

The defect clustering is also reflected in the S versus W plot. The (W, S) values leave the dashed line for $V_{Ga}X$, implying a defect reaction to a larger vacancy cluster Y. It is interesting to note that the S parameter remains constant after the annealing at 500 K for the dose of 1×10^{13} cm^{-2}, while the W parameter already shows a clear change. This proves that it is necessary to measure both the S and W parameters to observe defect reactions.

The region of weak damage in the model of Hecking et al. (1986) is followed by the region of constant defect density as a function of implantation dose as the result of the dynamic equilibrium between defect generation and annihilation. The crossing of a threshold dose leads to a sudden inset of amorphization (Eichler 1997). The dose dependence of the implantation damage according to this model is clearly visible in the case of Te implantation in the RBS spectra of Fig. 5.53a. The sample still exhibits a channeling effect up to a dose of 1×10^{13} cm^{-2}. The amorphization at a dose of 3×10^{13} cm^{-2} in the implanted region is obvious. The corresponding $S(E)$ curves are shown in Fig. 5.53b. The S parameter increases significantly for the case of amorphization of the damaged region. The S versus W

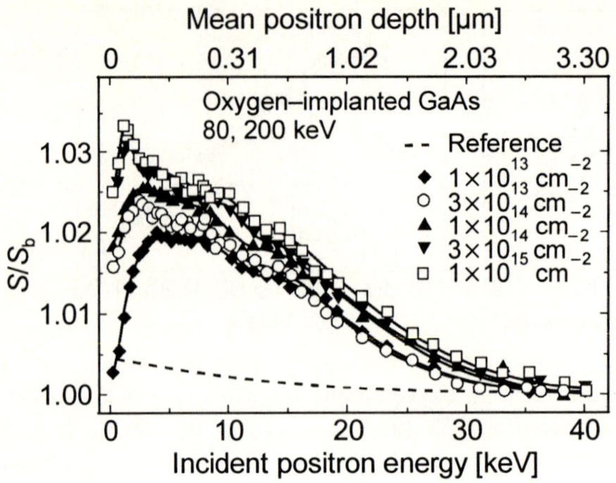

Fig. 5.52. Dose dependence of the defect generation during oxygen implantation in undoped gallium arsenide (Eichler 1997). The *S* parameter, normalized to the bulk value S_b, was measured versus the incident positron energy after oxygen implantation with energies of 80 and 200 keV. The lines are to guide the eye. The result of an unimplanted sample is shown for reference.

plot (Fig. 5.54) gives evidence that this increase is due to the appearance of another defect type with a larger open volume, since two different straight lines are found. In order to exclude the influence of the changing surface *S* parameter, the values for positron energies smaller than 2 keV are not plotted. Similarly to the oxygen-implanted samples, the non-amorphized samples have an *S* versus *W* plot close to a straight line connecting the characteristic pairs of values for the bulk and the $V_{Ga}X$ complex. Interestingly, the vacancy type defects extend over a depth region far below the damaged layer. It may be interpreted as the outdiffusion of vacancy-type defects at room temperature. This seems to be likely with the results from electron irradiation, where the main annealing stage of irradiation-induced defects was found at room temperature (Fig. 5.45). The amorphized samples exhibit in the damaged layer another straight line with a higher slope in the *S* versus *W* plot. These samples also have the monovacancy-type slope for positron penetration depths higher than the thickness of the amorphized layer. This may be a hint that vacancies are emitted in the unimplanted region in the same way as observed for the crystalline samples.

5.2.7 Defects Related to Plastic Deformation

The mechanisms of point defect generation via dislocation motion during plastic deformation of GaAs are basically the same as in elemental semiconductors (Sect. 4.1.9). The first positron annihilation experiments on plastically deformed

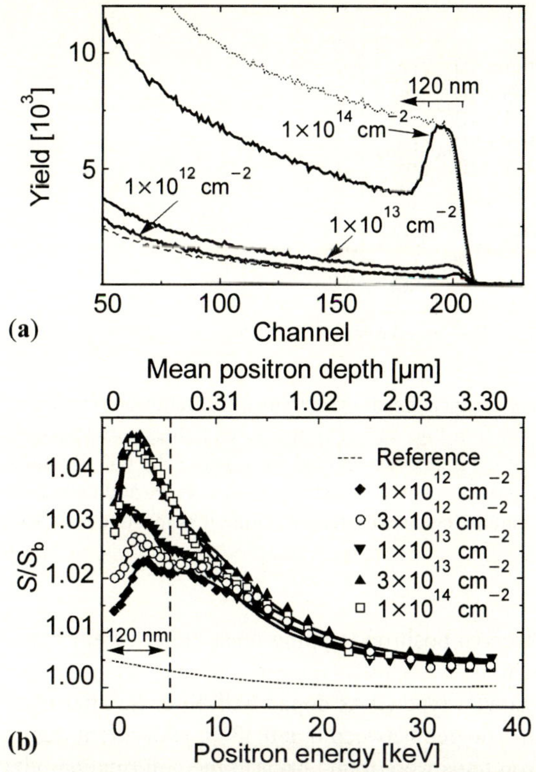

Fig. 5.53. Study of the implantation damage as a function of dose for tellurium-implanted undoped gallium arsenide (Eichler 1997). The implantation energy was 300 keV. (**a**) Rutherford backscattering spectra. The reference spectra for crystalline and amorphous GaAs are given as dashed and dotted lines. (**b**) Doppler-broadening S parameter (normalized to the bulk value S_b) versus the incident positron energy. The curve of the unimplanted GaAs reference sample is shown as a dotted line. The TRIM (transport of ions in matter) simulation gave a damage depth of 120 nm marked in (a) and (b).

GaAs were carried out by Kuramoto et al. (1973). Dislocations were recognized as positron traps, but the presence of point defects after deformation was ignored. This led to the misinterpretation that different types of dislocations, such as α- and β-dislocations, have different positron lifetimes. Generally, edge-type α- and β-dislocations are distinguished by the kind of atom in the dislocation core, either Ga or As (Holt 1962). However, there is no detectable difference in the open volume in the core of both types of dislocations.

Strong positron trapping was also reported by Uedono et al. (1985) in GaAs plastically deformed by bending. They performed Doppler-broadening and positron lifetime measurements. The observed lifetime component of about 400 ps was assigned to dislocations. However, the dislocation density was probably too low to

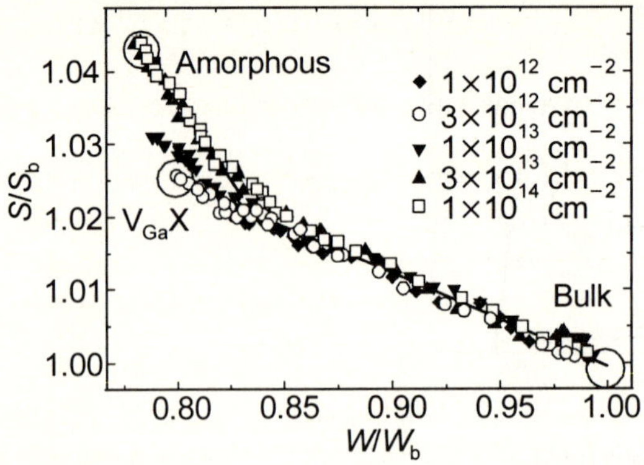

Fig. 5.54. *S* versus *W* plot for the slow-positron-beam measurements of tellurium-implanted undoped gallium arsenide (Eichler 1997) in Fig. 5.53. The measured values corresponding to incident positron energies smaller than 2 keV are omitted for clarity. The (*W*, *S*) values lie close to two straight lines characteristic for two defect types, viz. $V_{Ga}X$ and defects in the amorphous layer. The *S* and *W* parameters are normalized to their respective bulk values, S_b and W_b.

be detected by positrons. The observed positron trapping must rather be attributed to voids as a result of the deformation.

Saarinen et al. (1990) investigated n-type GaAs doped with Si or Se after plastic deformation (compression) to a degree between 6 and 13 % at different deformation temperatures. Two positron traps were found prior to the deformation, viz. a monovacancy-type defect and a shallow positron trap. The vacancy was regarded as the negatively charged isolated As monovacancy. This monovacancy signal in heavily Si-doped GaAs was later on reinterpreted as due to compensating $V_{Ga}Si_{Ga}$ complexes (Laine et al. 1996b). The shallow traps are most probably Si_{As}^{-} acceptors (see Sect. 5.2.3.1), as shown by the correlated study by scanning tunneling microscopy and positron annihilation by Gebauer et al. (1997c).

Both types of defects were also detected after deformation. The concentration of shallow traps increased by a factor of two, and the concentration of monovacancy-type defects by a factor between 1 and 6 depending on the deformation conditions. Another defect with a characteristic positron lifetime of about 490 ps appeared after plastic deformation in addition to the positron traps in as-grown GaAs. This long-lifetime component can only be attributed to rather large open-volume defects, i.e. voids (Saarinen et al. 1990). This long-lived lifetime component does not depend on the deformation conditions, such as the deformation degree or the temperature. Two interpretations are possible: The attained positron lifetime represents a saturation value for large vacancy clusters. A positron lifetime around 500 ps was estimated from empirical considerations of the electron density as the largest possible lifetime in defects in conducting materials such as metals and semiconductors (Seeger et al. 1989). However, positron lifetimes as high as 600 ps were found in deformed Si (Sect. 4.1.9). Another interpretation seems to be possible, viz. voids of a certain size may be preferentially formed during the deformation.

Similar high positron lifetimes also occur if positronium is formed. It is assumed that Ps is normally not formed in metals and semiconductors. The formation of Ps may, however, be possible in voids. It can be investigated by the measurement of the change in annihilation parameters as a function of an external magnetic field. The reduction of a long-lived component should indicate magnetic quenching of ortho-positronium (Sect. 3.1.6). Such an experiment was performed for plastically deformed GaAs containing voids. The presence of positronium could not be observed in these samples (Leipner et al. 1998).

Semi-insulating undoped GaAs was plastically deformed up to 40 % at temperatures between 450 and 600 °C by Dannefaer et al. (1991). They also found a lifetime component of 500 ps, which was attributed to voids. The radius of the voids was determined from the temperature dependence of the corresponding trapping rate to be about 5 nm (Sect. 3.1.5.4). A density of the order of 10^{13} to 10^{14} cm^{-3} was found. This corresponds to a total vacancy generation of 10^{17} to 10^{18} cm^{-3}. A further positron lifetime component between 300 and 320 ps was related to small vacancy clusters comprising two to three vacancies. The density was about 5×10^{16} cm^{-3}. The third dominant positron trap was a shallow one with a positron binding energy of 40 meV. Strains above 20 % resulted in defect concentrations high enough to cause saturated positron trapping.

Highly Te-doped GaAs was investigated for different deformation conditions by Krause-Rehberg et al. (1994b). The samples were compressed in a [213] axis in order to achieve single-slip conditions. The samples were cooled to room temperature after deformation with applied load in order to avoid the relaxation of the defect structure. The average positron lifetime as a function of the sample temperature is shown in Fig. 5.55.

The average positron lifetime at room temperature increases drastically after plastic deformation. A distinct decrease in $\bar{\tau}$ occurs with decreasing temperature. Obviously, the concentrations of both open-volume defects and shallow positron traps become larger as a result of the deformation. The positron trapping in the deformation-induced defects can be well described by the model of Fig. 3.25 including a shallow positron trap d1 and two independent deep traps, d2 and d3. The decomposition was only possible above room temperature and gave two defect-related lifetime components of 255 and 460 ps. These components were attributed, in agreement with the results of Saarinen et al. (1990), to a monovacancy-type defect (d2) and to voids (defect d3). The component of 255 ps was already found in the undeformed state. The annealing behavior of the deformation-induced defects was studied (Fig. 5.56). The average positron lifetime shows one annealing stage beginning at 800 °C. In contrast to elemental semiconductors (Sect. 4.1.9), no maximum occurs during the annealing.

The trapping rates κ_{d2} and κ_{d3} were calculated at room temperature according to a trapping model with two independent open-volume defects (i.e. neglecting the shallow traps in the trapping model of Sect. 3.2.1.2). The trapping rate κ_{d2} in the

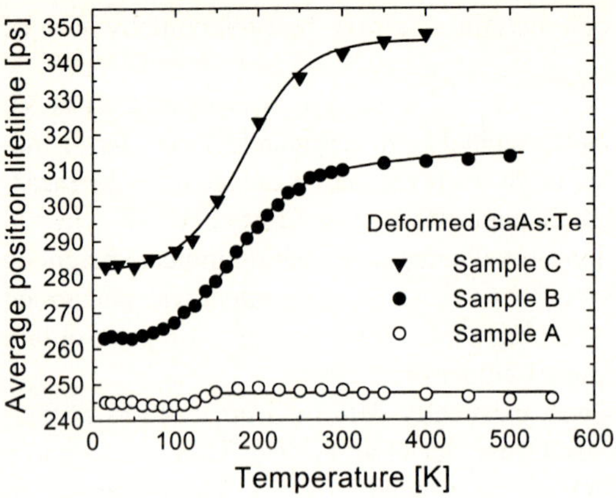

Fig. 5.55. Average positron lifetime as a function of the sample temperature measured in tellurium-doped gallium arsenide after plastic deformation (Krause-Rehberg et al. 1994b). Sample A is the undeformed reference sample. The sample B was deformed up to 0.5 % in compression at 923 K (strain rate $5\times10^{-5}\,s^{-1}$), and sample C was deformed up to 2.9 % at 823 K (strain rate $1\times10^{-4}\,s^{-1}$).

weakly deformed sample B (Fig. 5.56) practically corresponds to that in the undeformed sample. An annealing stage was scarely observed for this defect up to 1075 K. In comparison with the findings in as-grown tellurium-doped GaAs, the defect can be attributed to the $V_{Ga}Te_{As}$ complex (Sect. 5.2.3.2). The void-related trapping rate κ_{d3} exhibits a steep decrease and is thus the reason for the annealing stage in $\overline{\tau}$.

The trapping rate κ_{d2} in the more heavily deformed sample C in Fig. 5.56 is higher by a factor of three in the as-deformed state. The increase cannot be related to dislocations, as the dislocation density determined by transmission electron microscopy is only $5\times10^{8}\,cm^{-2}$ and a trapping coefficient of $1\,cm^{2}\,s^{-1}$ may be assumed (Sect. 3.2.5). An increase in the trapping rate to $1\times10^{10}\,s^{-1}$ cannot be explained in this way. Thus, the formation of additional monovacancies by the plastic deformation must be assumed. They are thermally rather stable and it is possible that they are stabilized by dislocations. The dislocation density does not change significantly upon annealing, but a relaxation of the dislocation network can be observed (Fig. 5.57). Dislocation dipoles are not observed after annealing, but dislocation tangles and small dislocation loops. The rearrangement of the dislocations may lead to the disappearance of dislocation-bound vacancies, which is related to the decrease in κ_{d2}.

The positron lifetime was measured as a function of the sample temperature after certain annealing steps. In this way, the annealing behavior of the shallow positron traps could be studied. A distinct annealing stage in the deformation-induced shallow positron traps was found at 950 K. These shallow traps are likely to be intrinsic defects. A possible candidate is the negatively charged Ga_{As} antisite,

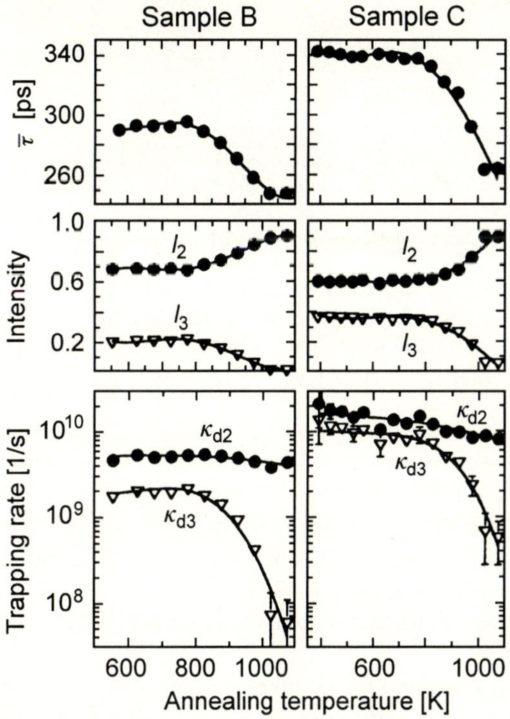

Fig. 5.56. Isochronal annealing experiment of plastically deformed gallium arsenide (Krause-Rehberg et al. 1994b). The trapping rates are obtained at room temperature from the three-component decomposition of the lifetime spectra measured after the respective annealing steps. The samples B and C are the same as in Fig. 5.55.

which was also identified by electron paramagnetic resonance in plastically deformed GaAs (Krüger and Alexander 1992; Leipner et al. 1997). The thermal treatment at 1075 K resulted in positron lifetime spectra comparable to the undeformed reference sample.

In order to understand the influence of the dopant on the generation of the monovacancy-type defects d2 in GaAs:Te, a similar deformation experiment was carried out for undoped GaAs (Hübner et al. 1996). The plastic deformation was performed under identical conditions. Two typical defect-related positron lifetimes of 270 ± 20 and 475 ± 15 ps are found at room temperature in the as-deformed sample (Fig. 5.58). Surprisingly, the average positron lifetime increases after additional annealing of the deformed sample at 725 K. This increase can be understood as the reduction in the concentration of the positron trap with the shorter positron lifetime. A higher intensity is found for the void-related positron lifetime component, and the average positron lifetime increases as a result. As the similar typical lifetimes τ_{d2} are found both in doped and undoped GaAs after deformation,

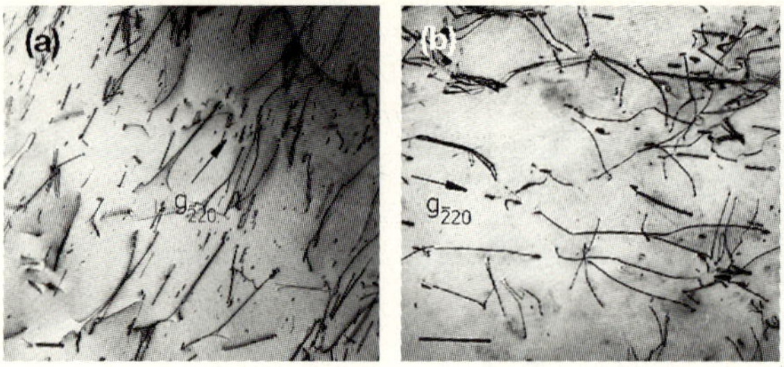

Fig. 5.57. Transmission electron microscopy images of the dislocation structure in plastically deformed tellurium-doped gallium arsenide (sample C in Fig. 5.55 and Fig. 5.56). (**a**) after deformation, (**b**) after annealing up to 650 K, (**c**) after annealing up to 1050 K. The markers correspond to 1 μm. The diffraction vectors **g** of the ⟨220⟩ type are indicated (Leipner et al. 1993).

this monovacancy-like defect is unlikely to be a vacancy–dopant complex. Leipner et al. (1996) argued that this defect may be identical to the deformation-induced EL6 center (Wosinski and Figielski 1989), which exhibited an annealing stage in infrared absorption measurements starting at 673 K.
Another possibility is the detection of dislocation-bound vacancies, which may anneal in this temperature range.

The measurement of the average positron lifetime reveals a typical step at about 100 K. This effect is already visible in the upper curve in Fig. 5.58. It is more pronounced with increasing dislocation density, which is adjusted with a higher strain rate (Fig. 5.59) according to the Orowan equation, $\dot{\varepsilon} = \rho_{\mathrm{dis}} v_{\mathrm{dis}} b$. $\dot{\varepsilon}$ is the strain rate, v_{dis} the average dislocation glide velocity, and b the magnitude of the Burgers vector. The average positron lifetime strongly increases with increasing strain rate as a result of increasing density of open-volume defects. The higher density of generated point defects as a result of the dislocation motion is in accordance with the model of jog dragging (Leipner et al. 1997). Another feature is the appearance of the above-mentioned kink in the curves around 100 K. It is not possible to fit all the curves with the trapping model of Sect. 3.2.1.2 under the assumption of temperature-dependent detrapping to one shallow positron trap.

A possible explanation for the kink is the appearance of another type of shallow positron trap for high strain rates. An attempt was made to fit the curves in the low- and high-temperature parts separately. This fit provided positron binding energies of 12 ± 7 and 38 ± 1 meV, respectively. The former value corresponds to the results in deformed GaAs:Te (Fig. 5.56), where a trapping model with a single shallow trap is sufficient to fit the data. Probably the same point defect is also detected in deformed undoped GaAs. The very small positron binding energy to the second shallow trap may be an indication of the different nature of this trap. It is conceivable that the undisturbed dislocation line represents this second shallow positron trap. This interpretation is supported by the results of Kawasuso et al.

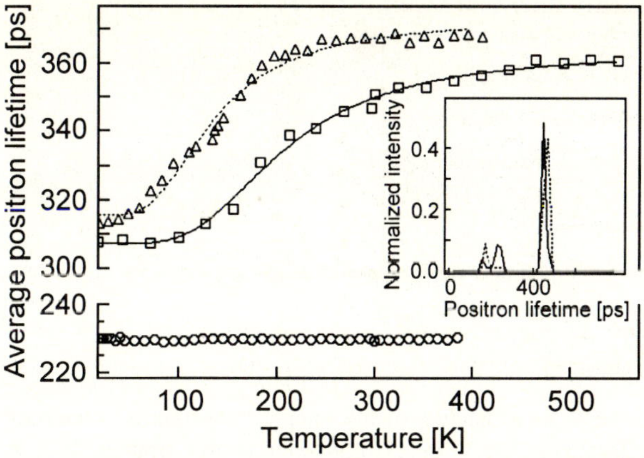

Fig. 5.58. Average positron lifetime measured as a function of the sample temperature after 3 % plastic deformation of undoped gallium arsenide (Hübner et al. 1996). The sample was deformed at 825 K with a strain rate of 1×10^{-4} s^{-1} (□). The lifetime was also measured after additional annealing of the sample at 725 K (△). An undeformed sample was measured for reference (○). The inset shows the result of a maximum entropy of lifetime (MELT) analysis of the positron lifetime spectra measured at room temperature for both sample conditions. The solid and the dotted lines correspond to the as-deformed and the annealed states, respectively.

(1995d), who reported a positron binding energy of 10 meV for dislocations in silicon (Sect. 4.1.9).

The defect densities after plastic deformation were determined for different compression axes. These orientations correspond to a different number of primary slip systems: one dominant slip system for a [213] direction, four systems for [110], and eight systems for [001]. The frequency of intersection processes is higher for a higher number of equivalent slip systems. The intersection of dislocations leads to the formation of jogs, the non-conservative motion of which gives rise to the emission of point defects. The results of the measurements are shown in Fig. 5.60 as the corresponding trapping rates of the defects d1, d2, and d3. The trend of increasing defect concentrations, which are proportional to the respective trapping rates, with a higher number of dominant slip systems is confirmed.

The temperature dependence of the average positron lifetime was studied in deformed GaAs to obtain quantitative results. It should be noted that many factors influence the temperature dependence of positron trapping (Sect. 3.1.5). Beside the fact that the positron diffusion is a function of the temperature, the main factor is the presence of various types of positron traps, which may be charged and may additionally show charge-state transitions. A pronounced temperature dependence exists for the trapping and detrapping in shallow positron traps and for the trapping in charged vacancies. Dislocations may act as shallow traps and exhibit therefore also a strong temperature-dependent trapping. Finally, the trapping in

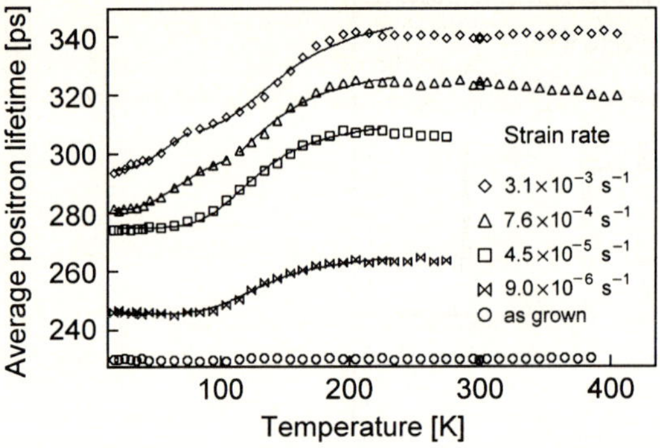

Fig. 5.59. Average positron lifetime as a function of the sample temperature in undoped gallium arsenide plastically deformed at 800 °C in [110] direction up to a strain of 10 % at different strain rates (Leipner et al. 1997). The full lines correspond to fits described in the text.

voids is also temperature dependent. All these defects are found in plastically deformed semiconductors and each require several parameters to describe the positron trapping. Reliable quantitative information can hardly be derived solely from the analysis of the temperature dependence of positron trapping in deformed semiconductors. A more practicable way is the determination of the trapping coefficient with reference methods at a certain temperature, as described in Sects. 3.2.4 and 3.2.5. The combination with other methods providing densities or changes in the charge state of defects is very useful to reduce the number of free parameters. However, despite the uncertainties of the analysis of the temperature dependence of positron trapping in the presence of a multitude of defects, qualitative evidence of the kind of the detected defects may be obtained.

5.2.8 Near-Surface Defects After Saw-Cutting of Gallium Arsenide Wafers

The cutting of wafers from the ingot by diamond saws introduces a damaged layer near the surface with a fissured surface structure. Beneath, various defects, such as extended cracks, dislocations, and point defects, are generated. It is essential for the manufacturer to completely remove this layer by chemo-mechanical polishing in order to guarantee the desired wafer quality. The accurate knowledge of the depth of the damaged layer is required. A microscopic examination of the wafer allows the determination of the etch depth to obtain a sufficiently smooth surface. However, removal of the cracks does not necessarily mean that dislocations and point defects are removed as well. They may extend deeper into the wafer and may have a deteriorating influence on the function of devices.

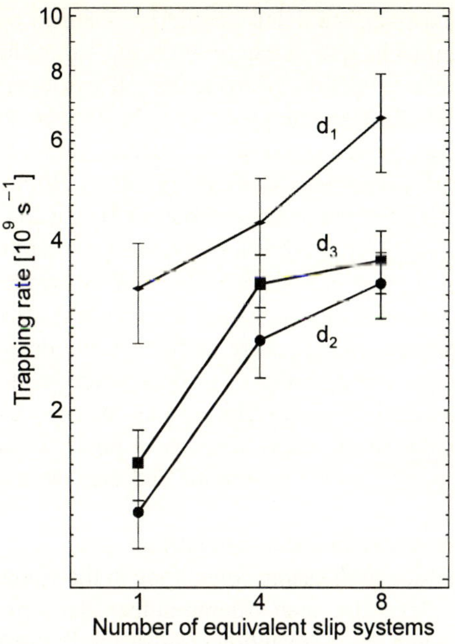

Fig. 5.60. Trapping rates of the shallow traps (d1), vacancies (d2), and voids (d3) determined by temperature-dependent positron lifetime measurements in undoped gallium arsenide plastically deformed up to 3 % in the [213], [110], and [001] directions, respectively. The strain rate was $3\times10^{-4}\,s^{-1}$ and the deformation temperature 500 °C (Hübner et al. 1997b).

The depth profile of defects near as-sawn surfaces of GaAs wafers were investigated by positron annihilation in combination with electron microscopy (Börner et al. 1998; Krause-Rehberg et al. 1996). In the earlier paper, the wafer was stepwise etched, and the defects in the layer were characterized by conventional positron lifetime spectroscopy (Fig. 5.61). The positron lifetime in the as-sawn sample was measurably higher than the lifetime in the bulk, although the effect was very small. The positron lifetime decreased by only 2 ps after the complete removal of the damaged layer. This might indicate that the defect concentration introduced by the saw cutting is rather small. However, one has to take into account that only 18 % of the incident positrons annihilate in the first 10 μm of the GaAs wafer. The measured average positron lifetime is thus the superposition of the lifetime in the defect-rich region and the bulk. The reason is the broad energy distribution of the positrons up to 540 keV from a radioactive ^{22}Na foil source (Fig. 2.17). The average positron lifetime reaches the bulk value at an etch depth of 10 μm, which was taken as the extension of the defect region. This is in fair agreement with cathodoluminescence intensity profiles, indicating a gradual increase in the defect density starting at about 15 μm below the sample surface. Dislocations were observed by transmission electron microscopy up to the same depth.

Depth resolution in the detection of near-surface defects could be obtained by the application of the slow-positron-beam technique. It can only be applied to the study of saw-cutting-induced defects in conjunction with etching, because the mean penetration depth of the positrons having an energy of 40 keV amounts to 3.3 μm, which is significantly smaller than the damaged depth. Such a combination of etching and slow-positron-beam measurements was carried out by Börner et al. (1998) in order to study the dependence of the damage depth on the advance of the saw blade. The *S* parameter is shown in Fig. 5.62 as a function of the depth. The curve comprising the whole damaged layer is assembled by four individual measurements of *S* versus the positron energy *E* in the as-sawn wafer and after three etching steps. The *S*(*E*) values after the etching were shifted on the depth scale by the corresponding layer thickness etched away. The bulk value is approached at a depth of 9.5 μm. The individual curves exhibit a steep increase for small incident positron energies. This is due to the back-diffusion of positrons to the surface. The surface *S* parameter is lower than the bulk value and thus the observed shape of the *S*(*E*) curve is obtained.

All measured (*W*, *S*) pairs were found to lie close to a straight line in an *S* versus *W* plot. This was related to the occurrence of the same defect type in the whole damaged layer. In order to characterize closer the open-volume defect, the *S* parameter measurements were complemented by conventional positron lifetime measurements. In contrast to the first study (Krause-Rehberg et al. 1996), the de-

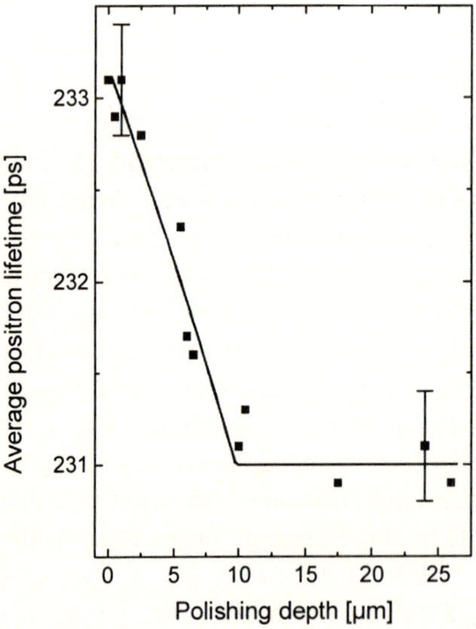

Fig. 5.61. Investigation of near-surface defects in a gallium arsenide wafer cut by a diamond saw. The average positron lifetime was measured after step-by-step polishing. The line is the result of a fit supposing a rectangular defect profile (Krause-Rehberg et al. 1996).

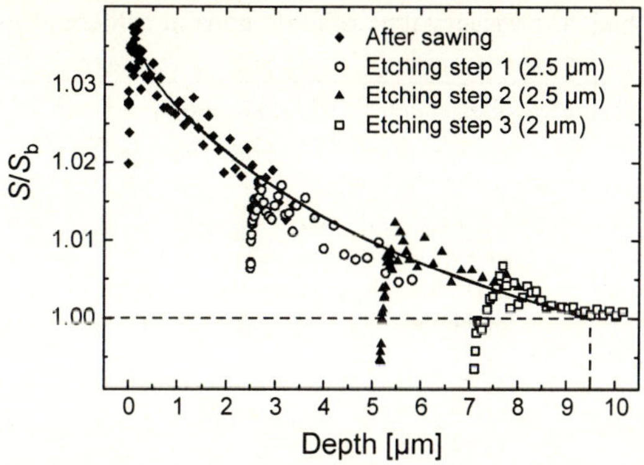

Fig. 5.62. Slow-positron-beam experiment for the determination of the damage depth after saw-cutting of a gallium arsenide wafer (Börner et al. 1998). The *S* parameter normalized to the bulk value S_b is plotted against the depth. The first measurement (◆) was carried out in the as-sawn wafer, the following ones after etching away layers of 2 to 3 µm. The latter values were shifted by the corresponding etch depth. The full line is to guide the eye. The dashed horizontal line represents the bulk value, which is reached at a depth of 9.5 µm.

fect profile was known from the slow-positron-beam experiment and the defect-related lifetime could be determined more accurately to be at least 330 ps. It is higher than the value calculated for divacancies (317 ps, Pöykkö et al. 1996b). It was thus concluded that small vacancy clusters are observed.

The annealing behavior of the sawing-induced defects was studied in an isochronal heating experiment. The average positron lifetime and the *S* parameter increased significantly above 400 K and reached a maximum at 650 K. It was concluded from the annihilation parameters $\tau_d > 400$ ps and $S/S_b = 1.054$ that the vacancy clusters already present after sawing grow further. These clusters did not completely disappear at the highest annealing temperature of 800 K. The defect reaction was confirmed by characteristic changes in the *S* versus *W* plot.

5.2.9 Summary

In contrast to elemental semiconductors, open-volume defects are detected in the as-grown state of gallium arsenide. A vacancy having ionization levels in the upper half of the bandgap was found n-type undoped GaAs. It was related to the arsenic monovacancy. In contrast, a thermally stable complex consisting of a gallium vacancy and a donor atom was unambiguously identified in heavily n-type material using correlated positron lifetime spectroscopy and scanning tunneling microscopy. A high density of Ga vacancies was found in low-temperature-grown arsenic-rich epitaxial layers. Post-growth heat treatments were carried out under defined arsenic vapor pressure in order to investigate stoichiometry-related vacan-

Table 5.2. Positron-lifetime data of bulk and vacancy-type positron traps in gallium arsenide.

	Positron annihilation characteristics		Dissolution
	Lifetime [ps]	τ_d/τ_b	temperature [K]
Bulk lifetime	228 to 232[a,b,c,d]	—	—
As monovacancies	295[e,f]	1.28	750[b]
	255 to 295[b,g]	1.11 to 1.28	
Ga monovacancies	255[h]	1.11	300[s]
	260 to 270[i]	1.13 to 1.17	
$V_{Ga}Te_{As}$	251 to 257[j,k]	1.09 to 1.12	> 1000[k]
$V_{Ga}Si_{Ga}$	262[d]	1.14	
$V_{As}Cr_{Ga}$	250 to 260[l,e]	1.09 to 1.13	
EL2*	255[m]	1.11	
	245[n]		
Dislocation-bound	300 to 320[o]	1.30 to 1.39	
vacancies	270[p]	1.17	
Voids	460 to 500[q]	2 to 2.17	700 to 900[q]
	590[r]	2.56	

τ_d is the defect-related positron lifetime, τ_b the positron lifetime in the defect-free bulk.
[a]Dlubek and Krause (1987), [b]Corbel et al. (1988), [c]Saarinen et al. (1994), [d]Gebauer et al. (1997c), [e]Dlubek and Krause (1987), [f]Krause et al. (1989a), [g]Saarinen et al. (1991, 1993), [h]in as-grown GaAs (Le Berre et al. 1995), [i]after electron irradiation (Polity et al. 1997c), [j]Gebauer et al. (1998b); Krause-Rehberg et al. (1995b), [k]Krause-Rehberg et al. (1994b), [l]Mascher et al. (1987c), [m]Krause et al. (1990b), [n]Le Berre et al. (1994), [o]Dannefaer et al. (1991), [p]Hübner et al. (1996), [q]after plastic deformation (Krause-Rehberg et al. 1994b; Saarinen et al. 1990; Dannefaer et al. 1991; Hübner et al. 1996), [r]after irradiation (Brudnyi et al. 1989). [s]Corbel et al. (1990), Hautojärvi et al. (1986b), Saarinen et al. (1995b), Würschum et al. (1989b).

cies in thermodynamic equilibrium. The results of positron lifetime spectroscopy can be taken as proof that Ga vacancies are the dominant native defects in n-type arsenic-rich GaAs. V_{Ga} was found in this material in a complex with the donor atom.

The structure of the metastable EL2 center was studied in correlated investigations of optical absorption and positron lifetime. The theoretical prediction of a vacancy in the metastable state (EL2*) was verified. The kinetics of the vanishing of the metastable vacancy were in excellent agreement with the transition kinetics of EL2* to EL2 in optical and electrical measurements.

Irradiation-induced defects were extensively studied after irradiation with fast electrons, as well as with heavy particles. The main annealing stage of irradiation-induced vacancies appeared between 200 and 400 K and was attributed to the annealing of vacancies in both sublattices. Defect reactions were observed in Te-doped electron-irradiated GaAs for annealing at temperatures between 500 and 600 K, which lead to open-volume defects giving rise to positron trapping. The generation and annealing of open-volume defects were also studied after ion implantation as a function of the dose and the implanted species. Again, defect reactions starting at 500 K could be manifested with slow-positron-beam measurements.

The point defect formation as a result of dislocation motion was studied in plastically deformed GaAs. In general, independent of the sample conductivity and the deformation conditions, vacancy clusters comprising at least 10 vacancies appeared. Open-volume defects of the size of a monovacancy are probably bound to the dislocation line. The defect profile occurring after saw-cutting of GaAs wafers was determined by slow-positron-beam measurements.

Shallow positron traps appeared in almost all GaAs samples investigated. They were attributed in irradiated or plastically deformed material to the Ga_{As} antisite. The nature of the shallow positron traps in as-grown samples is still under discussion. Acceptor-like impurities are candidates for such defects. In p-type electron-irradiated GaAs, it was proven that the zinc acceptor is the dominant shallow positron trap.

Table 5.2 summarizes the known positron-lifetime parameters of gallium arsenide. Both the bulk positron lifetime τ_b and the defect-related lifetimes τ_d measured at room temperature for the Ga and As monovacancies, vacancy–impurity complexes, dislocation-bound vacancies, and voids are given. Additionally, the positron lifetime of the metastable vacancy of the EL2* center as measured at low temperatures is included in the table.

5.3 The DX Center in III–V Compounds

The DX center is an important defect in n-doped GaAs and related alloys. It exhibits a metastable state which induces persistent photoconductivity. In this way, the defect was discovered in the late 1960s. Cooling the sample below 100 K and illuminating with a photon energy lower than the bandgap causes a strong increase in the conductivity, which persists for hours or days after switching off the light. The metastable state was interpreted by Lang and Logan (1977) as a large atomic rearrangement of the center. The transformation to the stable ground state is thermally activated at about 100 K.

The technological importance of DX centers is related to the metastability, which may cause memory or hysteresis effects in the electrical characteristics of devices. They act as deep centers, trap free electrons and reduce the carrier mobility.

It was supposed that the center consists of a shallow donor D and an unidentified native defect X, resulting in a deep level. Following the theoretical considerations of Chadi and Chang (1989) and Dabrowski and Scheffler (1992), it is now accepted that the DX center is just the distorted state of the substitutional donor, which is displaced in a $\langle 111 \rangle$ direction.

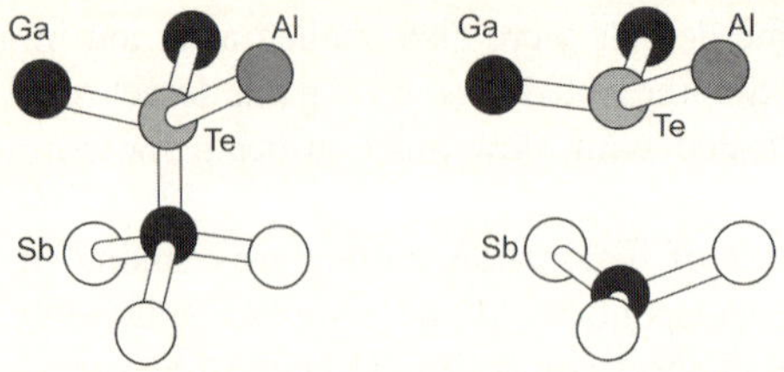

Fig. 5.63. Model of the DX center in tellurium-doped gallium aluminum antimonide in the ground (*right*) and metastable (*left*) state according to Chadi and Chang (1989).

The structure model of the DX center is shown in Fig. 5.63 for tellurium-doped $Al_xGa_{1-x}Sb$ as an example. The bonding between the Te and a Ga atom is broken in the stable state and both atoms are displaced in opposite directions to leave an open volume. The Te donor is incorporated on a group V sublattice site. Donors such as Si and Sn are built in on group III sites.

The metastable state is obtained for tellurium-doped $Al_xGa_{1-x}Sb$ below temperatures of 80 K by illumination with a photon energy larger than 0.6 eV. It is characterized by the restoration of the broken bond and consequently the vanishing of the open volume. The stable state is again established upon heating above 80 K. The metastable state is monitored by the appearance of persistent photoconductivity, as mentioned above.

The DX center is only found under atmospheric pressure in alloys, e.g. in $Al_xGa_{1-x}As$ and $Al_xGa_{1-x}Sb$. Apparently, the formation of the DX centers requires internal strain, which is caused by differently sized atoms in the group III sublattice surrounding the dopant. The DX defect is detected in n-doped binary compounds only under the application of external hydrostatic pressure of the order of 1 GPa (Maude et al. 1989).

It should be possible to verify the model of the DX center as shown in Fig. 5.63 by means of positron annihilation, as the existence of an open volume is predicted. The experiment must be carried out in such a way that at first the open volume in the ground state is detected and than the DX center is transformed into the metastable state by cooling and illumination. In that state the vacancy signal should be diminished. The vacancy must recover on heating.

Such a positron experiment was reported by Krause-Rehberg et al. (1993b). The results are shown in Fig. 5.64. The DX center in the stable and metastable state was investigated in tellurium-doped $Al_{0.5}Ga_{0.5}Sb$ by photoconductivity measurements and positron lifetime spectroscopy. The sample was cooled down to 15 K in darkness. The measured conductivity was in this state below 1×10^{-5} S. The conductivity increased by more than three orders of magnitude after a short illumination with white light, indicating the conversion of the DX into the metastable state. The disappearance of the photoconductivity and thus the transition of metastable DX into the ground state is shown in the upper panel of Fig. 5.64. The low conductivity is reached again upon heating above 100 K.

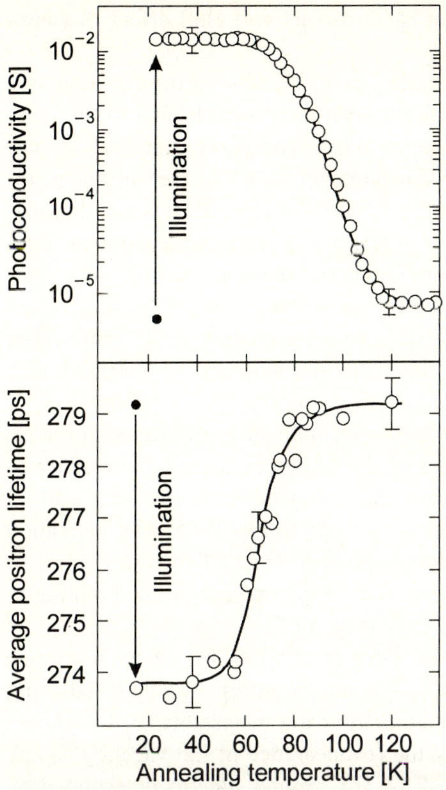

Fig. 5.64. Positron lifetime experiment for the detection of the open volume in the ground state of the DX center in tellurium-doped $Al_{0.5}Ga_{0.5}Sb$ (Krause-Rehberg et al. 1993b). The appearance of the metastable state of this center was monitored by photoconductivity measurements (*upper panel*). Closed circles represent measurements in the stable state of DX, open circles measurements during the annealing experiment up to 120 K performed after illumination with white light.

The result of the corresponding positron lifetime experiment, which was carried out with a conventional positron lifetime spectrometer and the same samples, is shown in the lower panel of Fig. 5.64. An average positron lifetime of 286.5 ps was measured at room temperature. The positron lifetime spectrum consisted of two components with the second lifetime of $\tau_d = 295$ ps. The bulk value was determined to be 270 ps. The ratio $\tau_d/\tau_b = 1.09$, which is smaller than expected, e.g. for monovacancies in GaAs (Table 5.2, p. 270). It is thus reasonable to associate this component with the open volume in the DX center, which is virtually not a vacant lattice site but created by the outward movement of the Te and Ga atoms in [111] directions. The density of the positron trap determined from the corresponding trapping rate with a trapping coefficient of 1×10^{15} s^{-1} amounted to about 3×10^{17} cm^{-3}. This value is in very good agreement with the density of the DX center obtained from deep level transient spectroscopy and Hall effect measurements.

The average positron lifetime after cooling to 15 K prior to illumination was 279.5 ps. This is below the value at room temperature, probably due to the presence of a small number of shallow positron traps. The short illumination performed at that temperature, similarly to the photoconductivity measurement, resulted in a distinct reduction in the average positron lifetime. All positron lifetime measurements in the subsequent annealing experiment were carried out in darkness at 15 K. The average positron lifetime increased upon annealing in the temperature range 20 to 120 K from 274 to 279 ps. The former value is close to τ_b and indicates that the density of vacancy-like defects is lower than $1\times10^{16}\,cm^{-3}$. The recovery of the vacancy component, i.e. the stable ground state, is observed after annealing at 100 K.

The comparison of both curves in Fig. 5.64 may indicate a difference in the recovery process of the positron lifetime and the photoconductivity. In order to study the kinetics more closely, isothermal annealing experiments were carried out. The results of annealing experiments are shown in Fig. 5.65. The activation energies were determined from the Arrhenius plots of the half-recovery times of the photoconductivity and positron lifetime. The activation energies of both methods are within the errors at about 85 ± 10 meV (Fig. 5.66).

To summarize the results of Krause-Rehberg et al. (1993b), an open-volume defect, which vanished after illumination at low temperatures, was detected using positron lifetime spectroscopy. The transition from the metastable to the ground state of the DX center was monitored by the reappearance of the vacancy signal. The activation energies of the recovery of the DX ground state as determined by positron lifetime and photoconductivity were found to be the same within the error limits. All these experimental results are in good agreement with the proposed structure of the DX center (Chadi and Chang 1989; Dabrowski and Scheffler 1992).

The advantage of the choice of $Al_xGa_{1-x}Sb$ in the experiment of Krause-Rehberg et al. (1993b) was the availability of bulk samples, which enabled the use of conventional positron lifetime spectroscopy. However, the main interest in the formation of the DX center lies in the $Al_xGa_{1-x}As$ pseudo-ternary compound, which is only grown as epitaxial layers. It is necessary to apply the slow-positron-beam technique to study open-volume defects in such layers. The DX center was studied extensively in n-type $Al_xGa_{1-x}As$ doped with Si, Sn (Mäkinen et al. 1993, 1995; Umlor et al. 1995), or Te (Laine et al. 1996a) by Doppler-broadening spectroscopy.

The results of such measurements in Te-doped $Al_{0.27}Ga_{0.73}As$ layers (Laine et al. 1996a) are shown in Fig. 5.67. The low *W* parameter in the doped sample is an indication of the presence of an open-volume defect. The *W* parameter increases to the level of the undoped reference sample after illumination at 20 K with 1.32-eV photons. This shows that the concentration of open-volume defects strongly decreases during illumination. The heating of the sample leads to the recovery of the

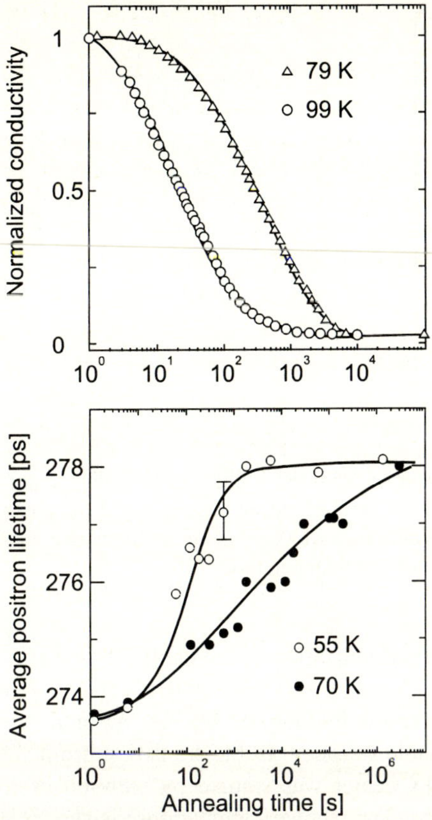

Fig. 5.65. Isothermal annealing experiment for the determination of the activation energy of the recovery of the stable ground state of the DX center (Krause-Rehberg et al. 1993b). The curves for two temperatures are shown for the conductivity and positron lifetime measurements, respectively. The positron measurement was carried out at 15 K after the corresponding annealing step at the indicated temperature.

W parameter to the value measured prior to illumination. This metastable effect, i.e. the appearance of the photoquenchable vacancy, is typical for the existence of the DX center, similar to the positron lifetime measurements in bulk AlGaSb described above. It could thus be proved that the DX centers in epitaxial $Al_{0.27}Ga_{0.73}As$ layers doped with Te contain a vacancy in the stable ground state. Similar results were also obtained in silicon- and tin-doped $Al_xGa_{1-x}As$ layers (Mäkinen et al. 1993, 1995).

The *W* parameter approaches the curve of the undoped reference sample upon increase in the temperature above 300 K (Fig. 5.67). This effect was interpreted as the disappearance of the vacancy. This open volume in the DX ground state van-

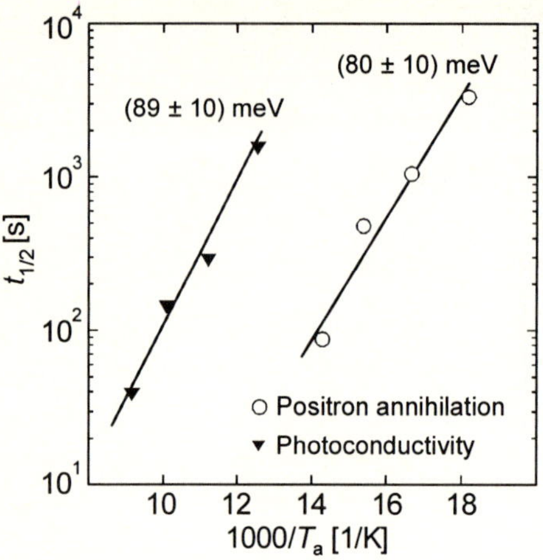

Fig. 5.66. Determination of the activation energies from the isothermal annealing experiment shown in Fig. 5.65. The Arrhenius plots of the half-recovery times $t_{1/2}$ of the photoconductivity and positron lifetime provide, within the error limits, similar activation energies (Krause-Rehberg et al. 1993b). T_a is the annealing temperature.

ished not only by optical but also by thermal ionization of the center (Laine et al. 1996a).

The influence of the alloy composition (mole fraction) of $Al_xGa_{1-x}As$ on the recovery temperature of the photoquenchable vacancy as the transition from the metastable to the ground state of the DX center was studied by slow-positron-beam measurements (Mäkinen et al. 1995). The recovery temperature is shown in Fig. 5.68 as a function of the composition x of $Al_xGa_{1-x}As$. This temperature was determined from plots similar to that of Fig. 5.67. The course of the recovery temperature is compared with the plot of the activation energy for electron capture in the DX center determined by Mooney et al. (1987). Obviously, the ionization of the DX center is in close relation to the recovery temperature, at which the vacancy signal in different $Al_xGa_{1-x}As$ layers appears in positron annihilation.

The available data of Doppler-broadening line shape parameters of the DX center in differently doped $Al_xGa_{1-x}As$ layers are compiled in the W versus S plot of Fig. 5.69. All the (S, W) points in this plot for Te, Si, and Sn doping lie on the same straight line running through the bulk value. This may indicate that the different chemical surrounding of the photoquenchable vacancy, i.e. the Te_{As}, Si_{Ga}, and Sn_{Ga} donors, does not have a large influence on the defect-related annihilation parameters.

The very different core-shell structure of group III and group V atoms in epitaxial $Al_xGa_{0.51-x}In_{0.49}P$ layers may be used for the identification of the sublattice of the photoquenchable vacancy of the DX center (Mäkinen et al. 1996). By comparing the results of Doppler-broadening coincidence measurements in epitaxial

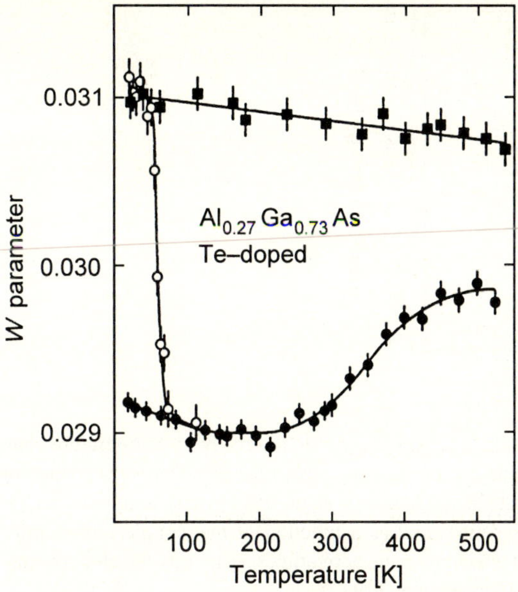

Fig. 5.67. Slow-positron-beam Doppler-broadening measurements of the DX center in tellurium-doped $Al_{0.27}Ga_{0.73}As$ layers (Laine et al. 1996a). The core annihilation *W* parameter was determined as a function of the sample temperature in darkness prior to (●) and after (○) illumination at 20 K with 1.32-eV light. An undoped $Al_{0.25}Ga_{0.75}As$ sample was measured for reference (■).

$Al_xGa_{0.51-x}In_{0.49}P$ layers with those of InP containing vacancies in either sublattice, the vacancy of the DX center was attributed to the group III sublattice.

5.4 Indium Phosphide

Indium phosphide is a III–V compound semiconductor with very promising properties for applications in optoelectronic communication, optoelectronic integrated circuits, and high-speed microwave devices. Optoelectronic components can be designed with optimum lattice and simultaneous bandgap matching by using the InGaAsP pseudo-quaternary alloy. The electron drift velocity at high electric fields is higher in InP than in GaAs. Thus, field-effect transistors with a high cut-off frequency are possible. The thermal conductivity is significantly higher than in GaAs, allowing a better power dissipation in integrated circuits. Other features, such as the large bandgap and higher radiation resistance compared with Si and GaAs, make the technological application of InP-based solar cells conceivable.

On the other hand, there are several disadvantages of InP. The growth of large single crystals is very complicated due to the high partial vapor pressure of phos-

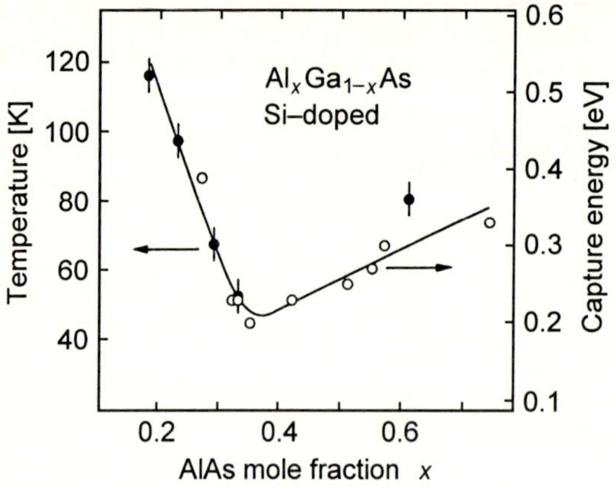

Fig. 5.68. Influence of the mole fraction x of $Al_xGa_{1-x}As$ on the recovery temperature characterizing the appearance of the vacancy signal in slow-positron-beam Doppler-broadening spectroscopy after photoquenching at 20 K (Mäkinen et al. 1995). The temperatures (•) were estimated from the W parameter measured as a function of the sample temperature, similar to that of Fig. 5.67. The right axis shows the activation energy (o) for electron capture at the DX center determined by Mooney et al. (1987).

phorus. Furthermore, the stacking fault energy of InP is the lowest of all III–V compounds, which frequently leads to the occurrence of twins in the crystal.

Similar to other compound semiconductors, intrinsic point defects play an important role in the electrical and optical properties. The ionization levels of vacancies and antisite defects (Fig. 5.70) were calculated by first-principles Car–Parinello methods (Alatalo et al. 1993; Seitsonen et al. 1994). These molecular dynamics calculations were carried out self-consistently and for fully relaxed defects in both sublattices. The appearance of negative-U properties was reported for the phosphorus vacancy, while the lattice coupling of the In vacancy is weaker and the sequence of ionization levels runs from 2+ to 3−. Only a few experimental studies of vacancies in InP exist, which are reviewed in the course of this section.The calculations provide charge states from twofold positive to twofold negative for both the In_P and the P_{In} antisites. They exhibit negative-U properties as well. The P_{In} antisite was successfully identified by electron paramagnetic resonance (Kennedy and Wilsey 1984). However, the energy levels were found by optical detection of magnetic resonance, in contradiction to the theoretical calculations (Fig. 5.70). The +/2+ level of the P_{In} antisite was determined to be at $E_{cond} - 0.23$ eV, while the 0/+ level was suggested to be resonant with the conduction band at $E_{cond} + 0.12$ eV (Chen et al. 1994). No experimental evidence has been found so far for the In_P antisite. Following these experimental results, the P_{In} antisite cannot act as a shallow positron trap.

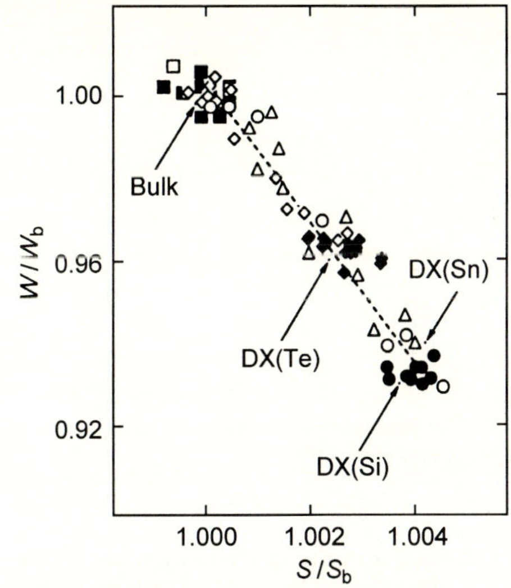

Fig. 5.69. *W* parameter versus the *S* parameter in undoped $Al_xGa_{1-x}As$ layers (*squares*), in tellurium-doped $Al_{0.27}Ga_{0.73}As$ (*diamonds*), and in silicon-doped $Al_{0.29}Ga_{0.71}As$ (*circles*). The line shape parameters are normalized to the bulk values, W_b and S_b. The slow-positron-beam Doppler broadening was measured by Laine et al. (1996a) as a function of the sample temperature in the range 20 to 200 K in darkness (*filled symbols*) and after light illumination (*open symbols*). The data for tin-doped $Al_{0.48}Ga_{0.52}As$ (△) are taken from measurements above room temperature, where the DX center is thermally ionized.

5.4.1 As-Grown Indium Phosphide

The first positron annihilation studies of indium phosphide were reported at the beginning of the 1980s. Brudnyi et al. (1982) studied defects in electron-irradiated InP. Dannefaer (1982) and Iwase et al. (1985a) performed positron lifetime and Doppler-broadening measurements in as-grown material.

Dlubek and Brümmer (1986) and Dlubek et al. (1985a) investigated p-type InP by means of positron lifetime spectroscopy. The main subject was the saturation of the free carrier concentration with a high zinc doping level. It was found that the hole density saturated at a level of 1×10^{18} cm^{-3}, whereas the Zn content can be as high as 10^{21} cm^{-3}. Zn becomes electrically inactive by precipitation (Mahajan et al. 1979). On the other hand, Tuck and Hooper (1975) and Yamada et al. (1983) proposed that the Zn acceptors are bound to complexes with phosphorus vacancies, e.g. $V_PZn_{In}V_P$ or V_PZn_{In}. The latter model could be verified by the measurements of Dlubek et al. (1985a). Samples with dopant contents of 2×10^{18} and 4.5×10^{18} cm^{-3} had the same hole density of 1.7×10^{18} cm^{-3}. The former sample exhibited no positron trapping in defects and the positron bulk lifetime of 242 ps

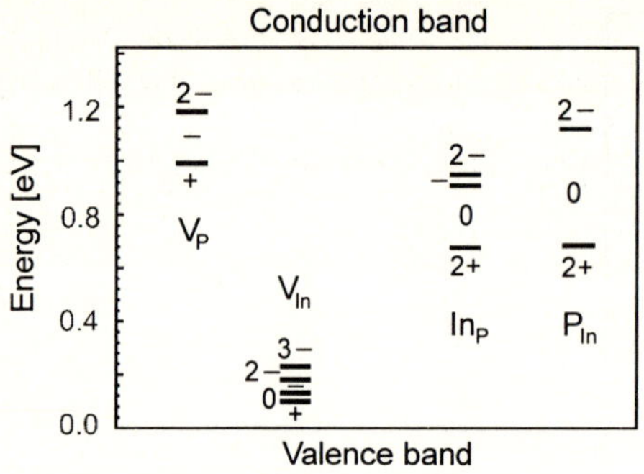

Fig. 5.70. Theoretically calculated ionization levels of vacancies and antisites in indium phosphide. The calculations were performed by first-principles Car–Parinello simulation under consideration of the lattice relaxation (Seitsonen et al. 1994).

was found. The average positron lifetime increased in the latter sample by 35 ps. The decomposition provided a defect-related positron lifetime of 321 ps. This finding is in accordance with the existence of a vacancy–zinc complex. However, it was not possible to distinguish from this lifetime value between different structures proposed for vacancy–zinc complexes. An annealing experiment, which showed a reduction in the vacancy concentration by one order of magnitude, was carried out. The defect-related lifetime component decreased upon annealing at 500 °C, which was interpreted by the structural change of the complexes. The density of the vacancy–zinc complexes was calculated to be 1 to 2×10^{18} cm^{-3} using a trapping coefficient of 2×10^{14} s^{-1}. In this way, the compensation in highly Zn-doped InP was explained. However, the trapping coefficient may be higher at least by a factor of three (Table 3.3, p. 103), provided that the complexes are neutral. If the vacancy–zinc complexes are negatively charged, the trapping rate is even higher and, consequently, the defect concentration lower. It is thus probable that part of the Zn is otherwise inactivated. A similar behavior has already been described in highly Si-doped GaAs (Sect. 5.2.3.1), where different defects contribute to the compensation.

Mahony et al. (1996) also found vacancy–zinc complexes in positron lifetime measurements of highly Zn-doped InP. The defect-related lifetime was 330 ps if the zinc concentration was beyond the saturation level of the carrier concentration. They obtained defect densities below 10^{17} cm^{-3} and came also to the conclusion that these complexes are only partly responsible for compensation. A decrease in the average positron lifetime was found after rapid thermal annealing. $\bar{\tau}$ was close to the positron bulk lifetime after annealing at 700 °C. The decomposition of the spectra was still possible and a defect-related lifetime of 280 ps was separated. This shorter lifetime was related to a proposed −/0 charge state transition of the center, which may be connected to a change in the lattice relaxation.

InP with a Zn content of 6.4×10^{18} cm^{-3} was investigated by Alatalo et al. (1995) and exhibited the same positron lifetime results as reported by Dlubek et al.

Table 5.3. Defect-related positron lifetimes and Doppler-broadening W parameters measured in indium phosphide doped with 6.4×10^{18} cm^{-3} zinc compared with values theoretically calculated for different vacancy–zinc complexes in indium phosphide (Alatalo et al. 1995). The inward relaxation of the corresponding defect taken into account is given.

	Defect	Relaxation [%]	τ_d [ps]	W_d/W_b
Experiment			325	1.05
Theory	V_P	19	323	0.75
	V_PZn_{In}	0	264	1.14
	V_PZn_{In} Zn_IV_P	17	322	0.91
	Zn_{In}	14	319	1.08
	$V_PZn_{In}V_P$	0	327	0.74

τ_d is the defect-related positron lifetime. W_d and W_b are the defect-related and bulk W core annihilation parameters, respectively.

(1985a). The same material was investigated by means of Doppler-broadening coincidence spectroscopy. A value of $W_d/W_b = 1.05$ was experimentally found. The positron lifetime and the W parameter were theoretically calculated for different vacancy–zinc complexes (Table 5.3). The relaxation of the defect structures was chosen for the calculation in such a way as to provide the experimentally obtained defect-related lifetime. The W parameter was calculated for the same relaxation. Comparing these values with the experimental value, it turns out that the best agreement is found for the $Zn_{In}V_PZn_{In}$ structure with a 14 % breathing-mode outward relaxation.

The measurements in the high-momentum parts were compared with theoretical calculations of the phosphorus monovacancy and the $Zn_{In}V_PZn_{In}$ complex (Fig. 5.71). A 3 % lattice outward relaxation was assumed for V_P in order to reproduce the experimental defect-related annihilation parameters of $\tau_d = 267$ ps and $W_d/W_b = 0.94$ (compare Sect. 5.4.2.1). The shape of the Doppler curve is broader for this complex compared with the relaxed isolated phosphorus vacancy. This indicates that the native vacancy observed in heavily Zn-doped InP is different from the isolated V_P. Alatalo et al. (1995) could also exclude the indium vacancy, because the Doppler curve for V_{In} is even narrower than that of V_P. This is indicated by the ratio of $W_d/W_b = 0.7$ for monovacancies.

The calculated shape of the Doppler curve of the $Zn_{In}V_PZn_{In}$ complex is also significantly different from that of V_P (Fig. 5.71b). The broad annihilation line for the vacancy–zinc complex was explained by the high momentum contributions of the 3d electrons of the Zn atom. The features of the calculated curve are in good agreement with the measured ones in Zn-doped InP.

The measurements of Alatalo et al. (1995) clearly show that positron lifetime measurements alone are hardly able to provide information on the chemical surrounding of the detected vacancy. The combination with Doppler-broadening spectroscopy may overcome this disadvantage. In particular, spectra obtained in a

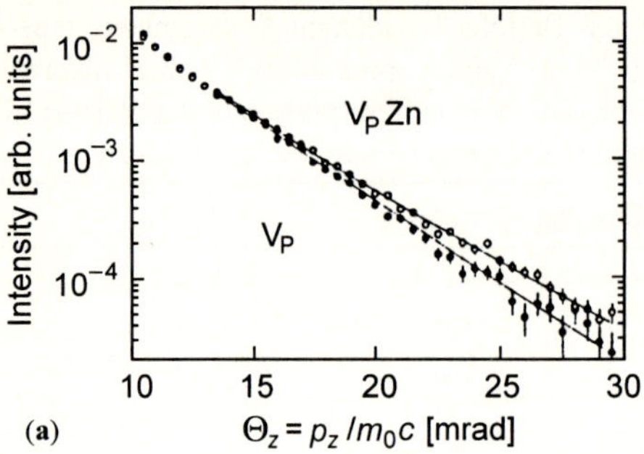

Fig. 5.71. Identification of vacancy–zinc complexes in p-type indium phosphide by means of Doppler-broadening coincidence spectroscopy (Alatalo et al. 1995). (**a**) Doppler-broadening curves measured in zinc-doped (○) and in electron-irradiated (•) indium phosphide. (**b**) Calculated high-momentum parts of the Doppler curves for a phosphorus monovacancy and a $Zn_{In}V_PZn_{In}$ complex. The calculation was carried out for a 3 % outward relaxation for V_P and for a 14 % outward relaxation for the vacancy–zinc complex. Θ_z is the angular deviation related to the component of the electron momentum p_z (m_0 electron rest mass, c speed of light).

coincidence setup can be compared with theoretically simulated ones. In this way, structural information of the vacancy–zinc complex in InP could be obtained.

Further experimental studies were reported (Bretagnon et al. 1993; Chen et al. 1996a; Huang et al. 1995) for differently doped InP (n-type, semi-insulating, and weakly p-type). All samples in the study of Bretagnon et al. (1993) provided the same average positron lifetime of about 242 ps. This value was given as the positron bulk lifetime by other authors (Alatalo et al. 1995; Dlubek et al. 1985a; Leipner et al. 1995; Polity and Engelbrecht 1997a). The spectra measured with high statistics by Bretagnon et al. (1993) were decomposed into three components. The bulk value was than determined to be between 235 and 237 ps. The 242 ps lifetime was interpreted with positron trapping in open-volume defects. The defect-related lifetime τ_2 of 265 to 270 ps was related to trapping in monovacancies. As this signal was also found in p-type and n-type InP, the P vacancy can be ruled out due to its positive charge in p-type samples (Fig. 5.70) and it was thus attributed to the In

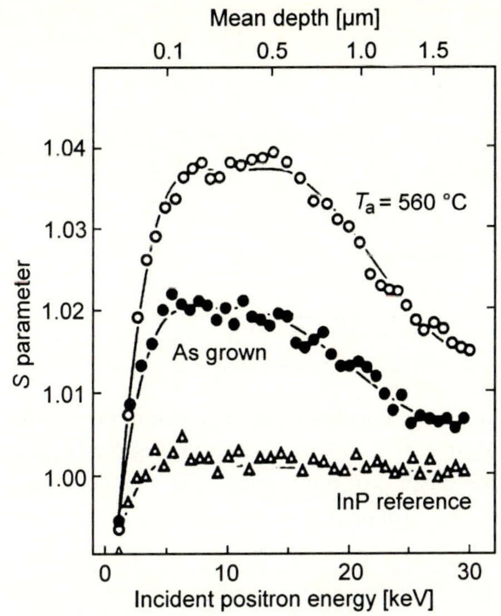

Fig. 5.72. Doppler-broadening line shape parameter S of a slow-positron-beam experiment in low-temperature-grown indium phosphide layers (Hautojärvi et al. 1993). The results of the layer in the as-grown state and after annealing at $T_a = 560$ °C are compared with those of a weakly zinc-doped bulk InP reference sample. The incident positron energy (*lower axis*) is related to the mean stopping depth of the positrons (*upper axis*).

vacancy. No systematic dependence of the positron trapping on the dopant species and doping concentration was found. The separated vacancy-type signal was stable after annealing up to the highest applied temperature of 800 °C. These findings and the fact that the lifetime spectra with $\bar{\tau} = 242$ ps obtained under similar experimental conditions could not be decomposed by other authors may indicate that the obtained vacancy signal is an artificial effect of the decomposition procedure. As a consequence, no vacancies are usually detected in as-grown InP except in highly Zn-doped samples. This is different from the situation in as-grown GaAs, where a vacancy signal is frequently found.

Similar to low-temperature-grown GaAs, it is possible to deposit epitaxial InP layers at fairly low temperatures. These layers are known to be phosphorus-rich and it is assumed that they contain a high number of intrinsic defects. In contrast to LT GaAs, such InP layers were found to be n-conductive, and it was impossible to produce p-type layers. The presence of a high number of P_{In} antisite defects was assumed to be responsible for these properties, as the P_{In} donor level was found to be resonant in the conduction band (Chen et al. 1994; Dreszer et al. 1993). Indium phosphide layers grown at low temperatures by molecular beam epitaxy were investigated by slow-positron-beam Doppler-broadening measurements (Hautojärvi et al. 1993). The growth temperature was 180 °C and the layer thickness 1.5 μm. The S parameter is plotted in Fig. 5.72 as a function of the incident positron en-

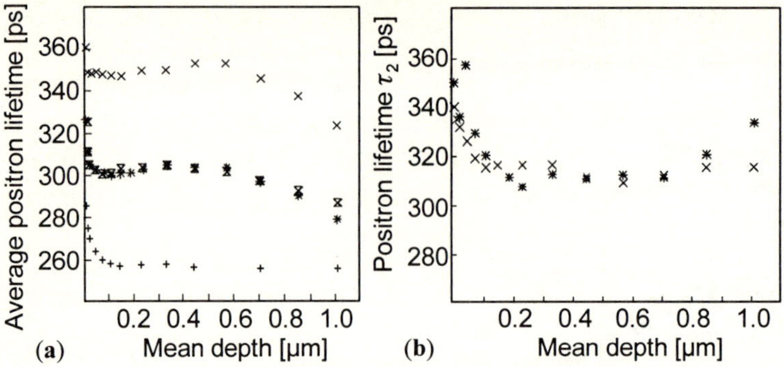

Fig. 5.73. Positron lifetime measurements in low-temperature-grown indium phosphide layers using a slow-positron beam (Willutzki et al. 1995). Average positron lifetime (**a**) and defect-related lifetime τ_2 (**b**) as functions of the mean positron implantation depth for two samples grown at different growth conditions (∗ ⧖), annealed at 560 °C (×), and a Czochralski-grown p-type reference sample (+).

ergy for the LT InP sample in the as-grown state and after 560 °C annealing. A weakly Zn-doped InP sample assumed to be free of positron traps was measured as a reference. The low-temperature-grown sample shows an increase in the *S* parameter by about 2 % in the as-grown layer and a further 2 % increase after the annealing treatment. This result was interpreted by Hautojärvi et al. (1993) by the presence of a high number of In vacancies, which form small vacancy clusters during the annealing.

Furthermore, the appearance of shallow positron traps was proven in as-grown and annealed LT InP samples by the measurement of the *S* parameter as a function of the sample temperature. The shallow positron traps were related to In_P antisites. However, the results of other defect-sensitive methods for the firm identification of these shallow positron traps are required.

The same samples were investigated in positron lifetime measurements using the Munich pulsed positron beam (Willutzki et al. 1995). The results are in agreement with the Doppler-broadening experiments (Hautojärvi et al. 1993). A defect-related lifetime $\tau_d = 312$ ps with a high intensity was detected in the low-temperature-grown InP layers, independent of the PH_3 flow rate during growth (Fig. 5.73). The signal was attributed to the In vacancy, in agreement with the previous Doppler-broadening study. τ_d is slightly above the positron lifetime theoretically calculated by Puska et al. (1989b) for the In monovacancy of 295 ps (Table 3.6, p. 123). The positron bulk lifetime determined in the reference sample was 260 ps and thus higher than usual. However, the ratio τ_d/τ_b equals 1.2 for the measurements and the theoretical predictions. The density of the vacancies was calculated from the decomposition of the spectra. Taking a trapping coefficient for negatively charged vacancies in semiconductors of 1×10^{15} s^{-1} (Sect. 3.2.4), a density of In vacancies of 4×10^{17} cm^{-3} can be calculated.

Only one lifetime component could be resolved in the annealed low-temperature-grown sample. It was interpreted as saturated trapping in divacancies. The density of vacancies was expected to increase upon annealing to at least 1×10^{19} cm^{-3}, which was seen in connection with the precipitation of phosphorus (Claverie et al. 1993). The detailed mechanism of this vacancy formation and clustering was not discussed.

5.4.2 Irradiation-Induced Defects

The investigations on irradiation-induced defects in indium phosphide are less numerous than in GaAs. Systematic studies on low-temperature irradiation with elementary particles (Sect. 5.4.2.1) were carried out. Furthermore, several positron experiments were reported on defects induced by ion-implantation (Sect. 5.4.2.2).

5.4.2.1 Irradiation with Elementary Particles

The first positron annihilation investigations of defects introduced by electron and neutron irradiation of n-type InP were published by Brudnyi et al. (1982, 1986). It was found as a result of Hall effect measurements that the Fermi level shifted only slightly to $E_{\mathrm{cond}} - 0.33$ eV despite very high irradiation doses. The authors reported distinct changes in the positron annihilation parameters measured by angular correlation of annihilation radiation. This was interpreted by the presence of In vacancies. A clustering of vacancies to larger agglomerates appeared during isochronal annealing with a maximum at temperatures between 300 and 500 °C. All irradiation-induced defects, which were similarly observed after electron and neutron irradiation, vanished after 600 °C annealing.

Positron trapping in InP, GaP, and Si irradiated with fast protons was studied as a function of the irradiation dose by Dlubek et al. (1988a). The average positron lifetime and the S parameter exhibited no strong dose dependence in InP and GaP, in contrast to Si. The lifetime $\bar{\tau}$ increased in InP only from 267 to 272 ps by increasing the proton dose from 10^{14} to 10^{16} cm^{-2}. Although the decomposition of the spectra could not be carried out reliably, it was concluded that no larger vacancy clusters were formed.

Bretagnon et al. (1992) studied positron annihilation in differently doped InP irradiated at room temperature with 2.5-MeV electrons to a dose of 8×10^{17} cm^{-2}. They reported the formation of indium vacancies regardless of the doping level and conduction type. Additionally, divacancies were found in semi-insulating and slightly n- or p-type material. The defect-related lifetimes were 265 and 323 ps, yielding $\tau_{\mathrm{d}}/\tau_{\mathrm{b}} = 1.12$ (V_{In}) and 1.37 (divacancies) with the used positron bulk lifetime of 236 ps.

The divacancies did not appear in irradiated heavily p-doped InP, which remained p-type after the irradiation. This was interpreted in conjunction with the charge state of the divacancy, i.e. a charge state transition from positive to neutral or negative should exist. A Fermi-level pinning occurred for low doping levels, resulting in highly resistive n-type samples after irradiation. The divacancies were visible in this state. Surprisingly, the divacancies were not detected in the highly n-doped sample. This was explained by the interaction of vacancies formed by the irradiation with impurities (Bretagnon et al. 1992). The concept of the charge-state transition of the irradiation-induced divacancies was further pursued in a succeeding study (Bretagnon et al. 1994).

Electron-irradiation experiments were also carried out at 100 K by the same authors (Dannefaer et al. 1996a). A vacancy signal detected only in irradiated n-type InP was related to the formation of close Frenkel pairs in the indium sublattice. As this signal was not present in semi-insulating and p-type samples, a +/0 level of $V_{In}In_I$ was assumed to exist in the upper half of the bandgap. It should be noted that the structure of these complexes was not clearly explained. An interstitial in a nearest-neighbor position of a vacancy is hardly conceivable for a stable complex. In contrast to their earlier result (Bretagnon et al. 1994), the divacancy signal was now also reported to exist in semi-insulating and heavily n-doped InP. Differences in the annealing behavior of the divacancy-type defects in both types of InP indicate that the nature of these defects may be different. It was speculated that the divacancies in n-type InP are associated with interstitials but not in semi-insulating InP. Defect-related lifetimes attributed earlier to mono- and divacancies were also reported by Bretagnon et al. (1997) with positron lifetimes of 265 and 338 ps in electron-irradiated n-type InP. Thermal annealing of both defects took place up to 200 °C. These vacancy-type defects in electron-irradiated InP are regarded to be in complex with interstitials. This idea followed from the earlier decomposition of positron lifetime spectra having an average positron lifetime of 242 ps, where a monovacancy-type lifetime was separated (Bretagnon et al. 1993). This defect-related lifetime was attributed to the indium vacancy. This vacancy was found to be stable up to 800 °C. The V_{In}-related defect occurring after irradiation should thus be part of a thermally more unstable complex. However, the low positron bulk lifetime of 236 ps could not be reproduced by other authors (Table 5.4, p. 294), and hence the stable vacancy signal in as-grown InP was not found. Under the assumption of the bulk value of 240 ps, the association of vacancies with interstitials is not necessary to explain the results. Only monovacancy-type positron lifetimes were found by Törnqvist et al. (1994a, 1994b) in Fe- and S-doped InP irradiated with 2-MeV electrons at 20 K. The decomposition of the spectra gave rise to two defect-related lifetimes. The first one was 263 ps, and the second one 297 ps (Törnqvist et al. 1994b) or 283 ps (Törnqvist et al. 1994a). According to theoretical calculations, the positron lifetimes are 275 and 295 ps for the phosphorus and the indium vacancy, respectively (Puska et al. 1989b). Thus, the experimental results are in good agreement with these calculations. No defect-related lifetimes could be identified in Zn-doped InP. This is accordance with the calculation of defect-related ionization levels (Fig. 5.70), from which a positive

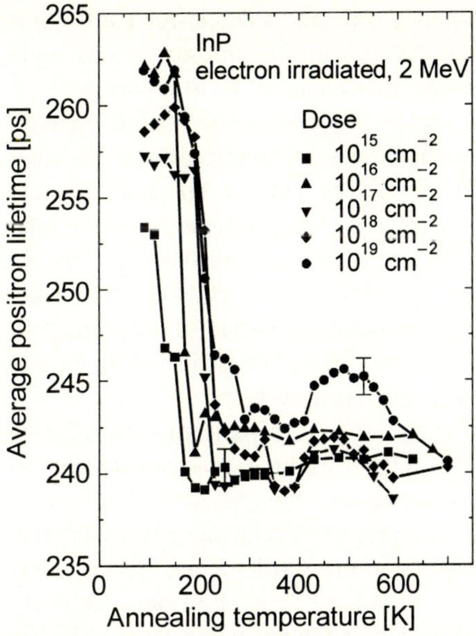

Fig. 5.74. Average positron lifetime as a function of the annealing temperature in electron-irradiated undoped indium phosphide (Polity and Engelbrecht 1997a). The irradiation with 2-MeV electrons was carried out for different doses at 4 K. The positron measurements were performed at 90 K.

charge state is expected for both vacancies in heavily p-doped InP. The monovacancies are expected to be multiply negative in n-type InP.

The average positron lifetime was measured at 80 K as a function of the annealing temperature after the irradiation (Törnqvist et al. 1994a, 1994b). An annealing step in $\bar{\tau}$ was observed between 150 and 200 K. For comparison, Karsten and Ehrhart (1995) observed, by diffuse X-ray scattering experiments, in the same temperature range the annealing of close Frenkel pairs.

The dependence of defect generation on the electron-irradiation dose was studied by Polity and Engelbrecht (1997a) in undoped InP. The irradiation was carried out at 4 K with 2-MeV electrons with doses between 1×10^{15} and 1×10^{19} cm^{-2}. The samples were n-type in the as-grown state. The free carriers were increasingly compensated by irradiation-induced defects and the Fermi level shifted away from the conduction band. The Fermi level was determined at room temperature to be at E_{cond} − 0.08 eV for the as-grown state, and 0.13 eV and 0.33 eV below the conduction band edge for doses of 10^{15} cm^{-2} and 10^{19} cm^{-2}. This means that, according to their charge states (Fig. 5.70), both types of monovacancies should act as positron traps for the whole investigated dose range.

Annealing experiments were carried out after irradiation to different doses (Fig. 5.74). The highest average positron lifetime is reached at 90 K for a dose of 10^{16} cm^{-2}. The lifetime does not increase further, or is even lower, for higher doses. The decomposition of the lifetime spectra provided a defect-related lifetime of $\tau_d = 278$ ps as an average value for the different irradiation doses. The ratio $\tau_d/\tau_b = 1.16$ indicates positron trapping in a monovacancy. Positron trapping in these vacancies is not saturated, since $\bar{\tau}$ is lower than τ_d. The overall positron trapping in irradiation-induced defects may, however, be saturated, as a high number of shallow positron traps is formed (see discussion below). The average positron lifetime is in this case the weighted sum over all contributions from annihilations in vacancies and shallow positron traps and is thus smaller than τ_d. $\bar{\tau}$ in the as-irradiated state varies in the dose range from 10^{16} to 10^{19} cm^{-2} by only 5 ps, although the defect concentration certainly increases by orders of magnitude. It may be concluded that the ratio of the defect introduction rates for vacancies and shallow positron traps does not change much with the dose.

A sharp annealing stage is visible in $\bar{\tau}$ (Fig. 5.74). It shifts from 150 to 250 K with increasing irradiation dose. The defect-related lifetime does not depend on the dose and the annealing temperature, indicating that monovacancies are always detected. It is assumed that these vacancies are present in both sublattices and that the annealing of close Frenkel pairs is observed in the 200-K range. This is in accordance with investigations of the magnetic circular dichroism of the optical absorption (Hausmann and Ehrhart 1995a). The shift in the annealing stage to higher annealing temperatures with increasing dose was supposed to be related to the superposition of the annealing of the Frenkel pairs in both sublattices (Polity and Engelbrecht 1997a).

The average positron lifetime is not higher than 243 ps after annealing at 300 K for all doses. This seems to prove the almost complete annealing of irradiation-induced vacancies. However, there is a slight increase in $\bar{\tau}$ after annealing around 450 K for higher doses. Furthermore, the average positron lifetime measured as a function of the sample temperature shows that vacancies are still present (Fig. 5.75). The slope of the curve above 400 K may indicate that these vacancies are negatively charged.

The abrupt fall of the positron lifetime towards lower temperatures is a sign for the presence of a high number of shallow positron traps. A fit was carried out to the data using a trapping model with a vacancy and a shallow positron trap. This fit provided a positron binding energy of 81 meV in the range of annealing temperatures from 340 to 380 K. A high binding energy of 205 meV was obtained from the fit for annealing temperatures above 420 K. The defect-related lifetime determined in this annealing range was higher than for lower annealing temperatures and was determined to be 310 ps. This finding and the facts of another positron binding energy to shallow positron traps and of a maximum in $\bar{\tau}$ around an annealing temperature of 450 K, give strong evidence of a defect reaction in this temperature range. Annealing at 625 K was sufficient to remove the irradiation-induced defects detected by positron lifetime spectroscopy.

Undoped indium phosphide was investigated by positron lifetime spectroscopy after neutron-transmutation doping (Navarro et al. 1996). The samples were irradiated with thermal neutrons to form Sn donors in the density range between

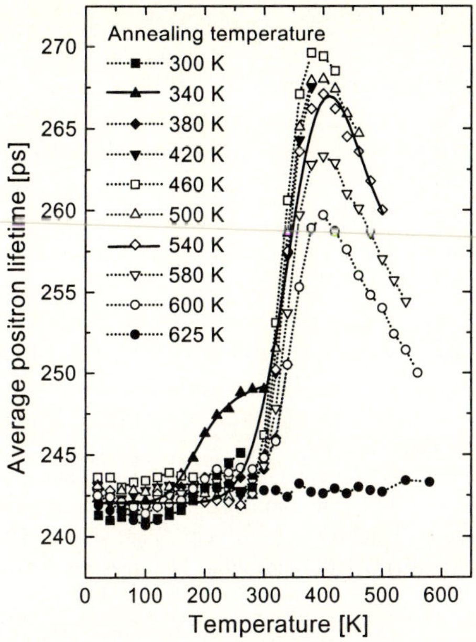

Fig. 5.75. Average positron lifetime as a function of the sample temperature in electron-irradiated undoped indium phosphide measured after different annealing steps (Polity and Engelbrecht 1997a, 1997b). The samples were irradiated at 4 K with 2-MeV electrons to a dose of 1×10^{18} cm^{-2}. The full lines are fits according to the trapping model with a vacancy-type defect and a shallow positron trap.

2×10^{15} and 2×10^{18} cm^{-3} according to the reaction ^{115}In(n, γ)^{116}In → ^{116}Sn + e^{-}. The average positron lifetime measured at room temperature shows a monotonic increase between 246 and 282 ps with the impurity level (Fig. 5.76). The higher lifetime compared with the reference material was attributed to V_{In}-related complexes. This defect annealed almost completely after thermal treatment at 550 °C for 10 min and was thus thermally more stable than defects introduced by electron irradiation.

5.4.2.2 Ion-Implantation-Induced Defects

Defects introduced by ion implantation of indium phosphide have been studied by positron annihilation in only a few papers. Damage induced by ion implantation with Si^{+} in Zn- and S-doped InP was investigated by slow-positron beam and Rutherford backscattering experiments by Schultz et al. (1992) and Simpson et al. (1992a). The implantation dose was varied in a range between 10^{11} and 10^{14} cm^{-2}, the ion energy was 0.6 and 3.0 MeV. A distinct dose dependence was found in the defect depth profiles measured by Doppler-broadening spectroscopy. Partial an-

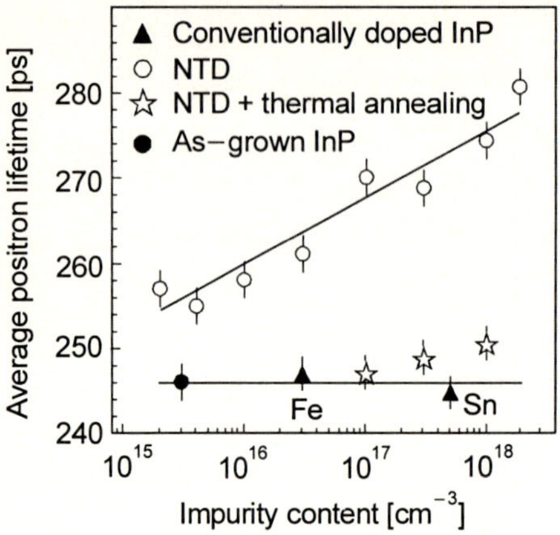

Fig. 5.76. Average positron lifetime versus the impurity content in neutron-transmutation-doped (NTD) indium phosphide (Navarro et al. 1996). For comparison, the results of samples conventionally doped with iron (3×10^{16} cm^{-3}) and tin (4.8×10^{17} cm^{-3}) are shown. The stars represent measurements of NTD samples after 550 °C annealing. The lines are drawn to guide the eye.

nealing of the introduced vacancy-type defects was observed with both techniques after storage for 100 days at room temperature. A full restoration of the lattice was found by RBS after annealing at 375 K. Positron annihilation, however, showed only a moderate reduction in the concentration of vacancy-type defects for isochronal annealing up to 720 K. Moreover, the increase in the *S* parameter monitored the formation of vacancy clusters.

Implantation with beryllium is interesting from a technological point of view for producing p-type layers. The potential of probing vacancy-type defects introduced by low-dose Be^+ implantation in S-doped InP was demonstrated by Uedono et al. (1992b) and Wei et al. (1994). The positron trapping rate was integrated over the depth and thus the total number of introduced vacancy-type defects was found to vary linearly with the implantation dose. The real almost Gaussian-like defect profiles were simulated by step functions. The depth profiles of vacancy-type defects were found to be in agreement with those of the implanted ions for doses higher than 1×10^{11} cm^{-2}.

5.4.3 Defects After Plastic Deformation

The results of positron lifetime investigations of plastically deformed indium phosphide are very similar to those obtained in deformed GaAs. Horii et al. (1992) investigated p-type and n-type InP after compression with up to 8 % strain at 500 °C. A single positron lifetime component of 260 ps was found in the p-type sample. It did not change after deformation. The same value was found in the n-

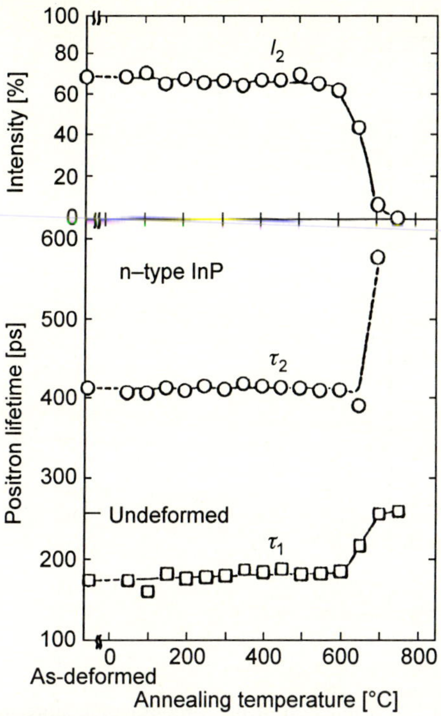

Fig. 5.77. Isochronal annealing experiment of n-type indium phosphide plastically deformed at 500 °C up to 7.6 % (Horii et al. 1992). The positron lifetime spectra were decomposed into two components. The positron lifetime in the undeformed sample is indicated for reference.

type sample prior to deformation and was thus taken as the positron bulk lifetime. The two-component decomposition provided after plastic deformation a defect-related lifetime of about 420 ps, which did not change with the strain. This component was related to vacancy clusters. The trapping rate and hence the defect concentration increased with the strain. The positron lifetime results were confirmed by one-dimensional angular correlation of annihilation radiation measurements. The ACAR curves of deformed and undeformed p-type InP were practically the same. A distinct narrowing of the curve occurred after deformation for the n-type sample, which is typical for the occurrence of open-volume defects.

The thermal stability of deformation-induced vacancy clusters was studied during isochronal annealing (Fig. 5.77). The cluster-related lifetime τ_2 indicated that these vacancies are thermally stable up to about 600 °C.

The generation of point defects during plastic deformation of undoped n-type InP was studied by photoluminescence and positron lifetime spectroscopy by Leipner et al. (1995). In contrast to Horii et al. (1992), the lifetime spectra could be decomposed into three components. Defect-related lifetimes of τ_2 between 329

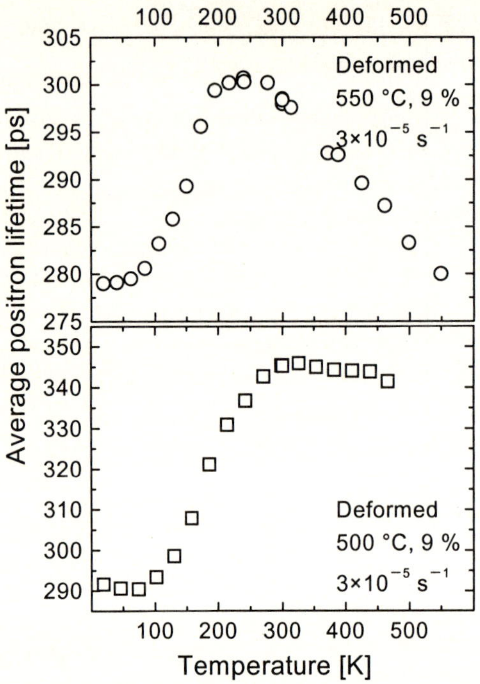

Fig. 5.78. Average positron lifetime measured as a function of the sample temperature in undoped n-type indium phosphide, which was plastically deformed up to 9 % compression at a strain rate of $3\times10^5\ s^{-1}$ (Hübner et al. 1997b; Leipner et al. 1995). The deformation temperatures were 500 °C (□) and 550 °C (○), respectively.

and 340 ps and τ_3 between 486 and 527 ps were obtained for different deformation degrees. Similar to the result in deformed GaAs, the latter component was attributed to positron trapping in voids containing at least 10 vacancies. τ_2 corresponds the theoretically calculated value for nearest-neigbor divacancies of 340 ps (Puska et al. 1989b). It was, however, not excluded that the experimentally found lifetime component is influenced by positron trapping in dislocations and monovacancies.

The temperature-dependent positron lifetime was measured for deformation temperatures in the range 500 to 600 °C. Two curves are shown in Fig. 5.78. The decrease in $\bar{\tau}$ in the low-temperature range was taken as clear evidence for the presence of shallow positron traps. This trap was supposed to be an intrinsic defect introduced by the plastic deformation. Possible candidates could be the negatively charged In_P antisites or interstitials. The only difference in the deformation conditions for the curves of Fig. 5.78 is the deformation temperature. The lower temperature of 500 °C gave a higher average positron lifetime. The trapping rates of the two defects mentioned above, κ_2 and κ_3, were determined at room temperature. It was found that the void-related trapping rate κ_3 decreased significantly for increasing deformation temperatures from 500 to 600 °C. κ_2 decreased only slightly in the same range. A distinct increase in the density of the voids with decreasing deformation temperature can be concluded. This increase in the vacancy concentration with decreasing deformation temperature is in accordance with the proposed model of vacancy generation by dislocation motion and jog dragging (Barrett and Nix 1965; Leipner et al. 1994) and the subsequent vacancy clustering during the deformation.

A peculiar feature of the curves in Fig. 5.78 is the change in the temperature dependence at higher sample temperature. The higher concentration of voids may be the reason that $\bar{\tau}$ stays more or less constant for $T > 300$ K for the sample deformed at 500 °C. The influence of the void component is much weaker for the samples deformed at higher temperatures. Thus, the trapping in divacancies is more dominant in the samples deformed at higher temperatures. Under the assumption that these defects are negatively charged, such a strong decrease in $\bar{\tau}$ may be expected. However, no electrical characterization of the deformed InP samples was carried out and other effects, such as a shift of the Fermi level for different deformation conditions and a resulting charge state transition of the divacancies, cannot be excluded.

5.4.4 Summary

The vacancy–zinc complex was extensively studied in as-grown indium phosphide. For the first time, the comparison of calculated Doppler-broadening spectra with background-reduced experimental curves in the high-momentum region was used to identify the annihilation site. It was found that the phosphorus vacancy is paired off with zinc. The best agreement with the data was found for a $Zn_{In}V_PZn_{In}$ complex with a 14 % breathing-mode outward relaxation (Alatalo et al. 1995). Systematic studies were carried out for the defect generation after low-temperature electron irradiation to different doses. Close Frenkel pairs probably of both sublattices anneal in a sharp stage around 200 K. However, no complete annealing of irradiation-induced defects occurs, and open-volume defects, as well as shallow positron traps, can be found up to the annealing temperature of 600 K. Further positron studies were performed after ion-implantation and plastic deformation, as well as in low-temperature-grown InP.

Table 5.4 summarizes the most important positron lifetime results for indium phosphide. The positron bulk lifetimes of Horii et al. (1992) and Willutzki et al. (1995) are not given, as they are obviously too high. The decomposition of Bretagnon et al. (1992) provided lifetimes too low. Nevertheless, these values were taken for the calculation of the τ_d/τ_b ratio.

Table 5.4. Positron lifetime data of indium phosphide bulk and vacancy-type positron traps in InP.

	Positron annihilation characteristics		Dissolution temperature [K]
	Positron lifetime [ps]	τ_d/τ_b	
Bulk lifetime	242[a, b]	—	—
	241[c]		
	243[d, e]		
	244[f]		
In monovacancies	312[g]	1.2[h]	< 250[c, d, e]
	265[i]	1.12[j]	
	297[d]	1.22	
	283[e, f]	1.16	
P monovacancies	263[d,e,f]	1.08	< 250[c]
V_P-zinc complexes	325[a, f]	1.34	680 to 750[a]
Divacancies	323[i]	1.37[j]	473[k]
Voids	420[l]	1.61[m]	873[l]
	486 to 527[b]	2.00 to 2.18	

τ_d is the defect-related positron lifetime, τ_b the positron lifetime in the defect-free bulk.
[a]Dlubek et al. (1985a), [b]Leipner et al. (1995), [c]Polity and Engelbrecht (1997a), [d]Törnqvist et al. (1994b), [e]Törnqvist et al. (1994a), [f]Alatalo et al. (1995), [g]Willutzki et al. (1995), [h]calculated with the higher positron bulk lifetime of 260 ps given by Willutzki et al. (1995), [i]Bretagnon et al. (1992), [j]with the positron bulk lifetime of 236 ps given by Bretagnon et al. (1992), [k]Bretagnon et al. (1997), [l]Horii et al. (1992), [m]with a positron bulk lifetime of 260 ps (Horii et al. 1992).

5.5 Other III–V Compounds

Compared with the III–V semiconductors treated in Sects. 5.1 to 5.4, most of the other compounds are less important. The only exception is gallium nitride, which has recently gained much attraction due to its promising applications in blue-laser devices. Positron annihilation studies on GaN are reviewed in Sect. 5.5.1. The small-bandgap materials InAs, InSb, and GaSb are treated in Sect. 5.5.2.

5.5.1 Gallium Nitride

Knowledge about point defects is limited in the nitrides. For instance, the reason for the n-type conductivity of undoped GaN was attributed either to the presence of nitrogen vacancies (Tansley and Egan 1992) or to impurities such as silicon (Götz et al. 1996). An important feature of bulk or epitaxially grown GaN is the occurrence of the parasitic yellow luminescence, which has been supposed to be related to Ga vacancies (Neugebauer and Van de Walle 1996) or dislocation-bound defects (Shreter and Rebane 1996). Positron experiments were performed (Saarinen et al. 1997b) to characterize vacancy-type defects in bulk GaN grown under a nitrogen pressure of 1.5 GPa and in 2 μm thick layers grown by metalorganic chemical vapor deposition (MOCVD) at 1000 to 1050 °C on sapphire substrates.

The result of the positron lifetime measurements as a function of the sample temperature is shown in Fig. 5.79. The lifetime spectra could be decomposed above 200 K and a defect-related lifetime of 235 ± 5 ps was found. A positron bulk lifetime of 166 ps was deduced in fair agreement with the theoretical result of $\tau_b = 156$ ps (Puska and Hakala 1997). The density of vacancy-type defects was estimated to be roughly 10^{18} cm^{-3}.

The temperature dependence of positron trapping may be related to the presence of shallow positron traps. These defects have not been closer identified. However, it is known that epitaxial GaN layers contain dislocations with a density of up to 10^9 cm^{-2}. Positron capture in dislocations and dislocation-bound vacancies was neglected in the paper of Saarinen et al. (1997b).

Doppler-broadening annihilation parameters were measured as a function of temperature in four MOCVD layers by slow-positron-beam experiments. The defect-related S parameter $S_d/S_b = 1.038$ and the defect-related lifetime of 235 ps was taken as an indication that a monovacancy was detected both in bulk and in epi-

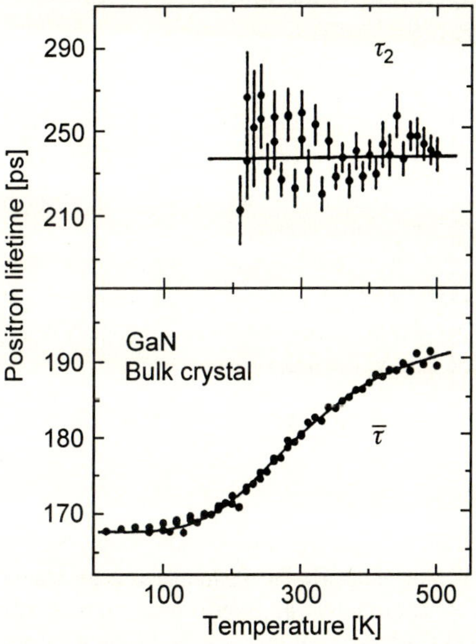

Fig. 5.79. Average positron lifetime $\bar{\tau}$ and defect-related lifetime τ_2 measured as a function of the sample temperature in a bulk gallium nitride crystal (Saarinen et al. 1997b).

taxial GaN. It was concluded from the comparison with the theoretical results of Puska and Hakala (1997) that Ga vacancies are present. The positron lifetime for V_N was calculated to be 169 ps, significantly lower than the experimental value. 213 ps was theoretically obtained for V_{Ga}.

The straight line obtained in the *W* over *S* plot (Fig. 5.80) of the data of the Doppler-broadening measurements at different temperatures is a sign of the predominance of one vacancy-type defect. The uppermost values correspond to the bulk values (S_b, W_b). It is assumed that these values are also valid for annihilation in shallow positron traps. The position on the straight line is determined by the temperature-dependent fraction of positrons annihilating in vacancies.

The vacancy density determined from the *S* parameter measurements was as high as 1.7×10^{18} cm^{-3}. The sample with the highest carrier density of 2×10^{18} cm^{-3} did not show any positron trapping in defects. The vacancy concentration did not vary systematically with the carrier concentration in the other samples. This may be in contradiction to the idea that nitrogen vacancies are responsible for n-type conductivity (Tansley and Egan 1992). Otherwise, these vacancies may also be invisible for positrons, so that no final conclusions can be drawn for the validity of this model.

An attempt was made to correlate the density of Ga vacancies with the intensity of the yellow luminescence as proposed by Neugebauer and Van de Walle (1996). A linear dependence on the vacancy density was given for the photoluminescence intensity of the 2.2-eV band, but no information was provided on the normalization of the PL spectra. Nevertheless, Saarinen et al. (1997b) concluded that the Ga vacancy forms the deep acceptor level responsible for the yellow luminescence.

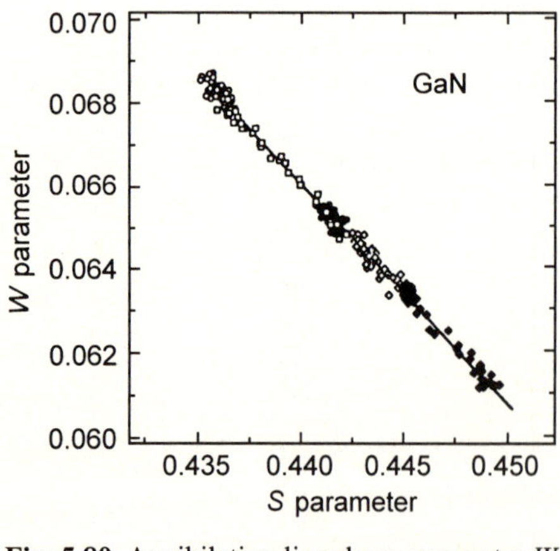

Fig. 5.80. Annihilation-line shape parameter *W* plotted against the *S* parameter measured by Doppler-broadening spectroscopy at different temperatures in as-grown gallium nitride samples. The data from a bulk crystal (□) and from epitaxial layers with different carrier densities (*other symbols*) in the range between 4×10^{17} and 2×10^{18} cm^{-3} are included in the *W* versus *S* plot (Saarinen et al. 1997b).

5.5.2 Small-Bandgap Materials

Only a few positron studies are known on InAs, InSb, and GaSb. Positron annihilation experiments were carried out on electron-irradiated InAs by (Brudnyi (1982). Hall effect measurements proved the increase in the concentration of free electrons by electron irradiation. The dependence of the angular correlation of annihilation radiation on the irradiation dose was studied in the range from 1×10^{17} to 1×10^{19} cm^{-2}. A saturation in the S parameter of the ACAR curve occurred above a dose of 10^{18} cm^{-2}.

A main annealing stage at 100 to 150 °C found in isochronal annealing experiments (Fig. 5.81) was supposed to be related to V_{In}. The maximum around 300 °C was assumed to be caused by vacancy clustering, possibly to divacancies. The S parameter as well as the Hall coefficient R_H recovered completely in the final annealing stage above 300 °C.

Dannefaer (1982) found an average positron lifetime of 247 ps in as-grown InAs. Dlubek and Brümmer (1986) studied the positron lifetime in an InAs polycrystalline sample and obtained an average positron lifetime of 260 and 255 ps before and after heat treatment. A more extended POLIS study of differently doped InAs crystals was presented by Mahony and Mascher (1997). Independent of the carrier density from $n = 5\times10^{18}$ cm^{-3} to $p = 2.2\times10^{18}$ cm^{-3}, the same average positron lifetime of 251 ± 1 ps was measured. However, $\bar{\tau}$ was slightly higher in heavily n-doped material, viz. 257 to 260 ps. The positron bulk lifetime was determined to be 246 ps. The spectra were decomposed into two components with a defect-related component τ_2 of 340 ps for highly n-type InAs and 280 to 295 ps for other samples. The trapping rate was for all samples about 0.8×10^{-9} to 1×10^{-9} s^{-1}, corresponding to $(7 \pm 2)\times10^{16}$ cm^{-3} monovacancies, and did not exhibit any temperature dependence. Rapid thermal annealing up to 850 °C did not show any change in the trapping rate.

Theoretical calculations of the positron lifetimes in vacancies provided 299 and 285 ps for the As and In monovacancies, respectively (Puska et al. 1989b). The ratio of τ_d/τ_b was calculated to be 1.29 and 1.23 for these defects. The experimental value for the highly n-type samples was higher and thus attributed to divacancies. The experimentally detected vacancy-type defects are likely to be monovacancies in the other samples.

In summary, mono- and divacancies were detected, which were exceptionally stable. The temperature dependence of the trapping rates indicated that these defects were neutral. As this was valid for all doping levels in the case of the monovacancy, no ionization level exists in the bandgap. The theoretical calculations of Puska (1989a) predicted that As vacancies are always positive and In vacancies have four charge states between 0 and 3−. Mahony and Mascher (1997) concluded that their detected defects were impurity-bound vacancies. These impurities may be residual impurities, since the defect properties were also found in undoped InAs. It should be noted that the reported properties are rather unusual for

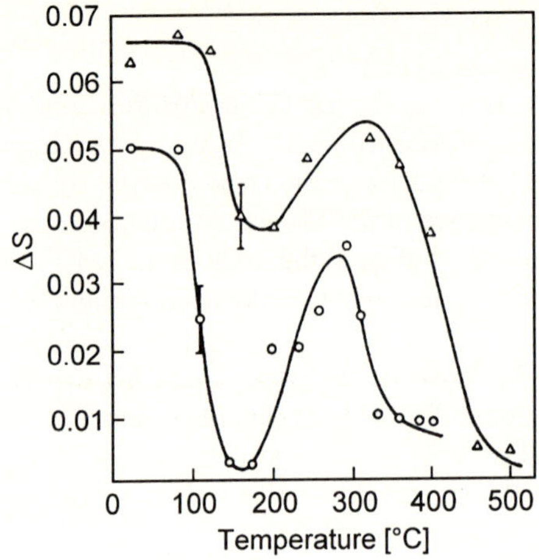

Fig. 5.81. Change in the *S* parameter, ΔS, measured during isochronal annealing by one-dimensional angular correlation of annihilation radiation of indium arsenide irradiated with 2-MeV electrons at room temperature (Brudnyi 1982). The doses are 1.3×10^{18} (○) and 5×10^{18} cm^{-2} (△).

vacancy-like defects in III–V semiconductors. Further data are desirable to exclude artificial effects of the decomposition of the spectra.

The positron lifetime and the annihilation line shape parameter *S* were used to determine the vacancy formation energy in InSb (Sen Gupta et al. 1986). Although the equilibrium vacancy concentration in Si turned out to be too small to be detected by positron annihilation (Sect. 4.1.2), temperature-dependent changes in the annihilation parameters, typical of the occurrence of vacancies in thermal equilibrium, were found in InSb. The formation energy was determined to be 0.24 ± 0.02 eV.

The introduction of defects by low-temperature electron irradiation (20 K, 4×10^{18} cm^{-2}) in InSb was investigated by Sen Gupta et al. (1988a). The average positron lifetime of 282 ps increased by 8 ps after irradiation. This increase recovered on thermal annealing at about 300 K.

Low-temperature electron irradiation experiments were also carried out for a dose of 5×10^{18} cm^{-2} in Te-doped and undoped p-type GaSb. Two annealing stages were identified at about 200 and 400 K. The average positron lifetime of 253 ps of the unirradiated sample was reached again after annealing at 500 K. In addition to this irradiation study, Mahony et al. (1995) investigated deformation-induced defects in GaSb. A positron bulk lifetime of 253 ps and a defect-related lifetime of 300 ps were found in as-grown and in annealed undoped p-type GaSb. The defects were supposed to be related to neutral complexes with a monovacancy-type character of a density of 0.5×10^{17} to 1×10^{17} cm^{-3}. Plastic deformation to the lower yield point was carried out at 430 °C. Small vacancy clusters were only found in

Table 5.5. Positron-lifetime data of GaN, InAs, and GaSb.

Material		Positron lifetime [ps]	τ_d/τ_b
GaN	Bulk	166[a]	—
	Ga vacancies	235[a]	1.4
InAs	Bulk	246[b]	—
	Monovacancies	280 to 295[b]	1.14 to 1.20
GaSb	Bulk	253[c]	—
	Monovacancy-type complexes	300[c]	1.19
	Vacancy clusters	374[c]	1.48

τ_d is the defect-related positron lifetime, τ_b the positron lifetime in the defect-free bulk.
[a]Saarinen et al. (1997b), [b]Mahony and Mascher (1997), [c]Mahony et al. (1995).

the sample, which was pre-deformed at 550 °C up to 5.5 % and then deformed to the lower yield point in the main deformation stage at 430 °C. The defect-related lifetime of the vacancy clusters was 374 ps, while 311 ps was found in the other deformed sample without pre-deformation. The clusters anneal out in a broad stage between 400 and 650 °C. The vacancy clustering was related to differences in the strain rate and temperature dependence of the lower yield stress observed in the main deformation stage with and without pre-deformation.

5.5.3 Summary

Only a few positron studies have been carried out up to now for gallium nitride, indium arsenide, and gallium antimonide. Positron lifetime data are summarized in Table 5.5. The monovacancy-type defects in InAs and GaSb have not been identified.

6 Defect Characterization in II–VI Compounds

II–VI compound semiconductors are considered for applications in fast-particle detectors and can cover the whole wavelength range from the far infrared to the near ultraviolet in optoelectronic devices. The width of the bandgap can be adjusted in pseudo-ternary compounds such as $Hg_{1-x}Cd_xTe$ by varying the composition x. II–VI semiconductors appeared promising for emitter or detector devices due to their excellent optical features together with predicted favorable transport properties. However, no technological breakthrough has been achieved. This is mainly due to the fact that all II–VI compounds except CdTe cannot be amphoterically doped. Bulk crystals of ZnSe, CdSe, ZnS, and CdS are always n-type, independent of impurities (Devlin 1967). ZnTe appears only p-type (Pautrat et al. 1985). A number of theoretical approaches for this behavior exist, but no final experimental proofs of these theories have been given. The reason for the compensation could be the existence of extrinsic defects introduced during crystal growth. However, intrinsic defects or complexes of intrinsic defects with dopants have also been discussed.

Research on II–VI compounds was greatly stimulated by the growth of nitrogen-doped p-type ZnSe layers by molecular beam epitaxy (Park et al. 1990; Taike et al. 1990). p–n junctions were made and the first ZnSe-based lasers could be fabricated (Haase et al. 1991; Ohkawa et al. 1991). Nevertheless, the doping behavior of nitrogen-doped ZnSe layers has also not been fully understood. It is not clear why only nitrogen doping during MBE works for p-type conductivity, and why other epitaxial growth techniques provide hole densities one order of magnitude lower (Ohki et al. 1991).

Obviously, the understanding of point defects in II–VI semiconductors is far from being complete. Vacancy-type defects, e.g. monovacancies and complexes containing a vacancy, play an important role. A prominent defect in doped CdTe is the *A* center, identified by electron paramagnetic resonance measurements as a cadmium vacancy paired off with a dopant atom at the nearest-neighbor site (Hofmann et al. 1992). The dominant defect in $Hg_{1-x}Cd_xTe$ is the mercury monovacancy, acting as an acceptor. A high concentration of V_{Hg} may be the reason for p-conductivity.

Positron annihilation studies of native defects in II–VI semiconductors are reviewed in Sect. 6.1. Irradiation-induced defects are treated in Sect. 6.2.

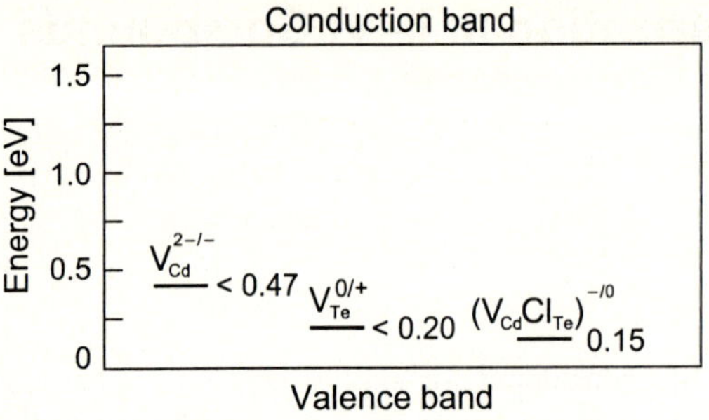

Fig. 6.1. Ionization levels experimentally found for the cadmium vacancy, the tellurium vacancy, and the chlorine-related *A* center ($V_{Cd}Cl_{Te}$) in cadmium telluride (Emanuelsson and Omling 1993; Meyer et al. 1992; Zanio 1972).

6.1 Native Defects in II–VI Semiconductors

Experimental results on defects in as-grown cadmium telluride (Sect. 6.1.1), cadmium mercury telluride (Sect. 6.1.2), and zinc-related II–VI compounds (Sect. 6.1.3) are given in this section. Moreover, the change in the vacancy concentration during the diffusion of silver in CdTe and during post-growth heat treatments of $Hg_{1-x}Cd_xTe$ are discussed.

6.1.1 Cadmium Telluride

Cadmium telluride can be amphoterically doped. However, the doping and compensation behavior are still under discussion. Important defects for the understanding of the compensation are the vacancy–impurity complexes called *A* centers (Kasai and Otomo 1962). These centers consist of a group II vacancy paired off with either a group VII donor (F, Cl, Br, I) on the Te site or with a group III donor (e.g. Ga, In) on the Cd site (Allen 1995). The ionization level of the Cl-related *A* center, $(V_{Cd}Cl_{Te})^{-/0}$, was determined by photoluminescence measurements to be located at 150 meV above the valence band (Zanio 1972). The levels for isolated monovacancies were also investigated experimentally. The 2−/− level of the Cd vacancy was found by electron paramagnetic resonance at $E_d - E_{val} < 470$ meV (Emanuelsson and Omling 1993) and the 0/+ level of the Te vacancy (F center) at $E_d - E_{val} < 200$ meV (Meyer et al. 1992). The experimentally determined level positions of vacancy-related defects are summarized in Fig. 6.1.

6.1.1.1 The A Center

Weakly In-doped cadmium telluride was studied by positron lifetime spectroscopy as a function of temperature (Corbel et al. 1993; Gély et al. 1989; Gély-Sykes et al. 1991). Distinct positron trapping in a monovacancy-type defect was found. The defect-related lifetime was given first as 330 ± 5 ps (Gély et al. 1989), but was corrected later to 320 ± 5 ps (Corbel et al. 1993; Gély-Sykes et al. 1991). The positron lifetime was interpreted to be either due to isolated Cd monovacancies in a doubly negative charge state or due to $(V_{Cd}In_{Cd})^-$ complexes. The average positron lifetime exhibited a distinct decrease with decreasing temperature, which was attributed to the presence of shallow positron traps.

The compensation mechanism in iodine-doped CdTe layers grown by molecular beam epitaxy was investigated by photoluminescence, conductivity measurements, and Doppler-broadening spectroscopy (Fischer et al. 1996). The Doppler-broadening *S* parameter increased significantly with increasing iodine concentration, i.e. with the free-electron concentration. Iodine doping induces obviously vacancy-type defects. This result is in agreement with the proposed microscopic structure of the iodine-related *A* center (Allen 1995).

Kauppinen and Baroux (1997) investigated CdTe crystals doped with In or Cl with positron lifetime and Doppler-broadening spectroscopy. The Doppler measurements were carried out in a background-reducing coincidence setup (Sect. 2.3.2.1). Vacancy-type defects were found in all samples. Defect-related lifetimes of 323 and 370 ps were separated in CdTe:In and in CdTe:Cl, respectively. Indium or chlorine-related *A* centers were assumed to be the defects responsible for positron trapping. This interpretation was supported by Doppler-broadening measurements in the high-momentum range of the spectrum, where the annihilation with core electrons dominates. It was concluded that the annihilation takes place in the cadmium vacancy as part of the *A* center. The distinct difference in the positron lifetime for In- and Cl-related *A* centers was ascribed to different open volumes. A stronger outward lattice relaxation was assumed for $V_{Cd}Cl_{Te}$. However, the observed longer positron lifetime may also be interpreted as the occurrence of an additional defect with larger open volume (see discussion of Fig. 6.3).

In contrast to In doping, chlorine impurities lead to high-resistance CdTe material. A series of CdTe samples containing chlorine in a concentration range from 100 to 3000 ppm was studied by positron lifetime measurements (Krause-Rehberg et al. 1998; Polity et al. 1995b). The average positron lifetime measured as a function of the sample temperature is shown in Fig. 6.2. The reference sample exhibited a single-component spectrum with a temperature-independent positron lifetime of 280 ± 1 ps that was attributed to the positron bulk lifetime. The average positron lifetime increased strongly with increasing Cl content, showing that open-volume defects, probably in a complex with chlorine, were present. It should be noted that the observed change of 100 ps in $\bar{\tau}$ at $T \geq 250$ K is exceptionally large, indicating that the defect-related positron lifetime must be high. The open volume of the defects should thus be significantly larger than that of a monovacancy.

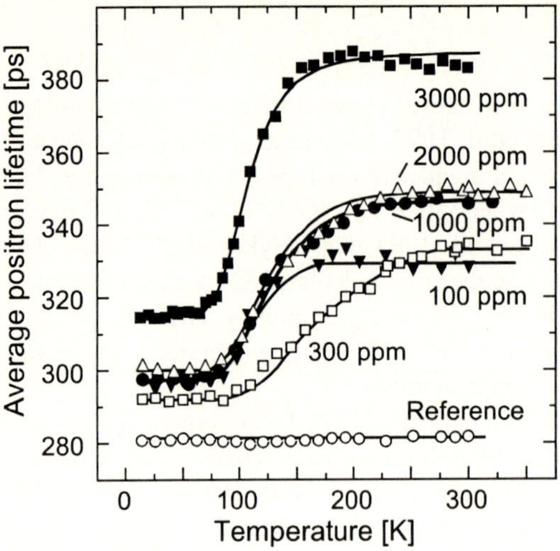

Fig. 6.2. Average positron lifetime as a function of the sample temperature measured in cadmium telluride with a chlorine content in a range from 100 to 3000 ppm (Krause-Rehberg et al. 1998). A nominally undoped sample is included for reference. The full lines are fits according to the trapping model of Sect. 3.2.1.2.

The lifetime spectra were decomposed at first into two components yielding a defect-related positron lifetime of 350 to 395 ps, which increased with increasing Cl content (Polity et al. 1995b). This result corresponds well to the characteristic positron lifetime of 370 ps found in CdTe:Cl by Kauppinen and Baroux (1997). However, the variance of the fit in the experiments by Polity et al. (1995b) was rather poor, indicating the presence of another unresolved lifetime component. Indeed, repeated measurements with a higher figure of 2×10^7 annihilation events allowed the decomposition of three components at temperatures above 250 K for the same samples (Krause-Rehberg et al. 1998). Two positron lifetimes with $\tau_{d2} = (330 \pm 10)$ ps and $\tau_{d3} = (450 \pm 15)$ ps were separated (Fig. 6.3a). Hence, the previously obtained defect-related lifetime of 370 ps must be regarded as an unresolved mixture of τ_{d2} and τ_{d3}. The defect d2 represents a monovacancy-related defect and is attributed to the chlorine *A* center, $V_{Cd}Cl_{Te}$. Defect d3 obviously exhibits an open volume significantly larger than that of a monovacancy. The ratio $\tau_{d3}/\tau_b = 1.6$ indicates that d3 comprises at least the open volume of a divacancy. For comparison, this ratio equals 1.34 for the nearest-neighbor divacancy according to the calculations of Puska et al. (1989b). In contrast to the earlier results of Polity et al. (1995b), the reduced positron bulk lifetimes τ_1 calculated according to a trapping model with two open-volume defects (solid line in Fig. 6.3a) agreed reasonably well with the measured values. This trapping model is obtained from that in Sect. 3.2.1.2 by setting $\kappa_{d1} = 0$, i.e. neglecting the shallow traps in this temperature range.

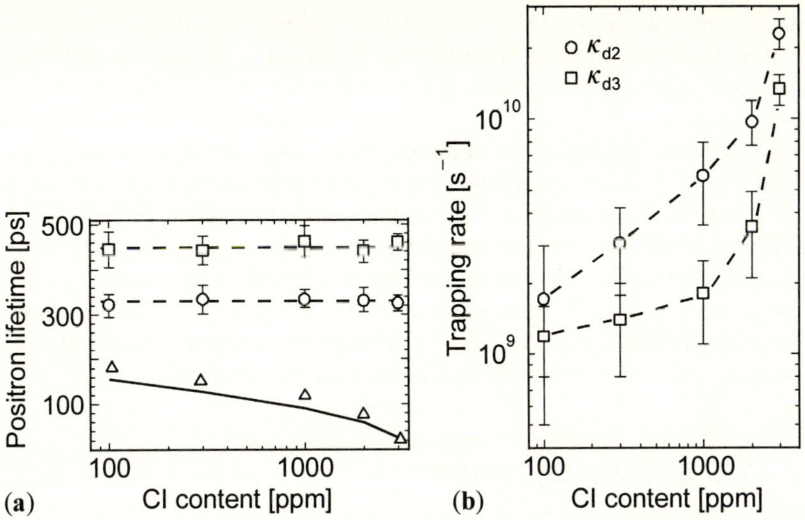

Fig. 6.3. Decomposition of positron lifetime spectra measured in chlorine-doped cadmium telluride at room temperature as a function of the chlorine content (Krause-Rehberg et al. 1998). (**a**) Positron lifetime components. The two long-lived lifetimes (□ and ○) represent the positron lifetimes τ_{d2} and τ_{d3} related to two defects with different open volumes. The shortest lifetime (△) is the reduced positron bulk lifetime τ_1, which corresponds reasonably to the lifetime (full line) calculated from a trapping model with two open-volume defects. (**b**) Trapping rates of the defects d2 (*A* center) and d3 calculated from the decomposition of the spectra. The dashed lines in (a) and (b) are drawn to guide the eye.

The trapping rates κ_{d2} and κ_{d3} calculated from the decomposition of the lifetime spectra are shown in Fig. 6.3b. The trapping rates of both open-volume defects increase with the Cl content, leading to the conclusion that not only d2, but also d3, represents a complex containing Cl. The concentrations C_{d2} and C_{d3} can be calculated from the corresponding trapping rates. When the Cl content is increased from 100 to 3000 ppm, the d2 (*A* center) density increases from 3×10^{16} to 4×10^{17} cm^{-3} and the d3 density from 1×10^{16} to 1×10^{17} cm^{-3}. Hence, the total chlorine content in the defects d2 and d3 amounts to less than 2 % of the Cl added during crystal growth. Trapping coefficients of $\mu = 9\times10^{14}$ s^{-1} and 1.8×10^{15} s^{-1} were taken for these estimates (Krause-Rehberg et al. 1998). Samples from the same set were used for correlated photoluminescence measurements. The concentration of the *A* centers was determined from the shift of the zero-phonon line of the 1.4-eV band, which is characteristic for the *A* center. The concentrations obtained in this way were within the error limits of the positron measurements.

The temperature dependence of the average positron lifetime shown in Fig. 6.2 exhibits a decrease towards lower *T*, indicating the presence of shallow positron traps. The trapping model analysis (solid line in Fig. 6.2), which included the temperature dependence of the detrapping rate δ, revealed that the concentration of the shallow traps did not depend on the chlorine content. The shallow traps were attributed to negatively charged acceptor-type impurities in agreement with photoluminescence results (Krause-Rehberg et al. 1998).

Chlorine-related DX centers were observed in pseudo-ternary compounds $Cd_{0.8}Zn_{0.2}Te$:Cl. The structure was described by a lattice relaxation model (Park and Chadi 1995). Positron annihilation experiments were carried out in order to show the presence of a DX-related open-volume defect (Shan et al. 1997c, 1997d). The transition of the metastable state of the DX center to the ground state was monitored by photoconductivity measurements at a transition temperature of 130 to 150 K. A decrease in the S parameter by 0.3 % was observed by slow-positron-beam Doppler-broadening spectroscopy. Although an increase is expected, the decrease was explained by competitive positron trapping in A centers.

6.1.1.2 Silver Diffusion Experiments

The diffusion of silver in p-type cadmium telluride results in an increase in the degree of compensation as detected by photoluminescence and Hall effect measurements (Zimmermann et al. 1993). This is illustrated in the upper part of Fig. 6.4, where the hole density is plotted versus the time after silver was injected by dipping the crystal into an $AgNO_3$ solution.

When silver diffusion was carried out in p-type CdTe crystals, the concentration of Ag_{Cd} impurities increased. This was indicated by the enhancement of the corresponding (A^0, X) bound-exciton line in the PL spectra. It was supposed that the interstitially diffusing silver interacts with vacancies according to the defect reaction

$$V_{Cd} + Ag_i \rightarrow Ag_{Cd}. \tag{6.1}$$

In order to confirm this assumption, positron lifetime measurements were carried out. As the native concentration of vacancies was too low to be detected by positron annihilation, post-growth annealing at 820 °C under equilibrium conditions in a Te atmosphere was performed in a two-zone furnace over a period of 6 weeks. The annealing conditions were chosen in such a way as to increase the density of Cd vacancies to a level of several 10^{16} cm^{-3}. An average positron lifetime of 294.5 ps was found after this procedure (Krause-Rehberg et al. 1992). The increase in the positron lifetime by about 10 ps compared to the bulk value was attributed to these cadmium vacancies. A silver diffusion experiment was carried out thereafter under conditions comparable to those used by Zimmermann et al. (1993). The result is shown in the lower part of Fig. 6.4. The average positron lifetime decreased markedly during the diffusion experiment carried out at room temperature. This decrease was taken as proof of the dominance of the defect reaction (6.1), resulting in a decrease in the V_{Cd} concentration. A similar experiment was performed by Grillot et al. (1976) in CdS, where cadmium vacancies were also filled by diffusing silver.

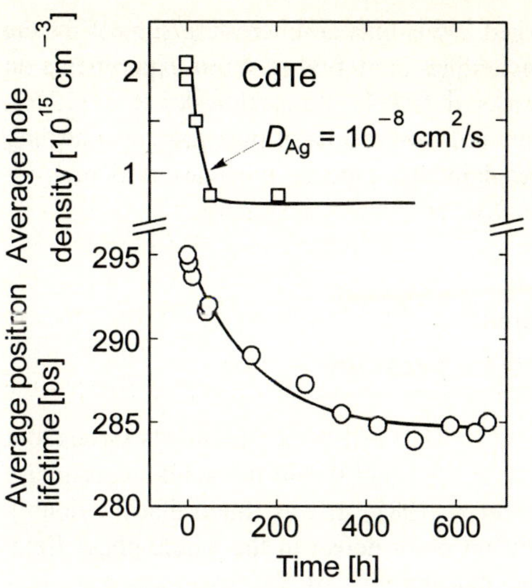

Fig. 6.4. Hole density determined by Hall effect measurements and average positron lifetime as a function of the time after silver injection into a p-type cadmium telluride sample. The upper part of the plot was taken from Zimmermann et al. (1993), the lower part from Krause-Rehberg et al. (1992). The decrease in the hole density corresponds to a diffusion constant D_{Ag} of interstitial silver of 1×10^{-8} cm^2/s (Zimmermann et al. 1993).

However, the time constants of the diffusion process monitored by the change in the hole density and by the change in $\bar{\tau}$ are different (Fig. 6.4). Although the electrical measurement shows the activation of silver interstitials acting as donors in the bulk CdTe, the decrease in the average positron lifetime reflects reaction (6.1). Since the silver diffusion is much faster than the kinetics of (6.1), it was concluded that an additional barrier has to be overcome for the Ag_i in order for cadmium vacancy sites to become occupied (Krause-Rehberg et al. 1992).

6.1.2 Mercury Cadmium Telluride

The intermixing of CdTe with the semimetal HgTe allows the adjustment of the width of the bandgap by variation of the composition x in $Hg_{1-x}Cd_xTe$. The material with a composition of about $x = 0.2$ becomes a narrow-gap semiconductor and is of interest for infrared detector applications in the atmospheric transmission window around 10 µm. The Hg vacancy is the most important point defect because of its electrical activity as an acceptor and the high diffusivity of mercury (Jones 1985; Sher et al. 1991). Furthermore, the Hg partial pressure at low temperatures is already rather high. The stoichiometry, i.e. the content of mercury vacancies, can be influenced by post-growth annealing under defined vapor pressure conditions (Vydyanath 1981).

The Hg vacancy is negatively charged and is thus an interesting subject for the application of positron annihilation techniques. The first positron experiments on $Hg_{1-x}Cd_xTe$ were reported by Dekhtyar et al. (1984), Voitsekhovskii et al. (1986), and Andersen et al. (1987). Positron annihilation results of post-growth annealing and diffusion experiments are reviewed in this section, whereas irradiation-induced defects are the subject of Sect. 6.2.1.

6.1.2.1 Post-Growth Heat Treatment Under Defined Mercury Vapor Pressure

Intrinsic point defects were investigated by Hall effect measurements throughout the existence region of $Hg_{1-x}Cd_xTe$ for $x = 0.2$ and 0.4 in the solid–vapor phase diagram (Vydyanath 1981; Vydyanath et al. 1981). It was found that the doubly negative mercury vacancy is the dominant point defect in the whole phase field. Vydyanath et al. (1981) calculated the concentration of V_{Hg}, assuming a generation of free carriers by band–band transitions and by ionization of the V_{Hg} acceptors,

$$0 \leftrightarrows e^- + h^+, \quad Hg_{Hg} \leftrightarrows V_{Hg}^{2-} + 2h^+ + Hg_{gas} . \tag{6.2}$$

The corresponding equilibrium constants of these reactions are

$$K_i = np, \quad K_{Hg} = [V_{Hg}^{2-}]p^2 P_{Hg} , \tag{6.3}$$

where n and p are the densities of free electrons and holes, respectively. P_{Hg} is the Hg vapor pressure. The neutrality condition $n + 2[V_{Hg}^{2-}] = p$ can be separated into two regions (low and high temperatures corresponding to intrinsic and extrinsic conductivity)

$$n = p, \quad 2[V_{Hg}^{2-}] = p . \tag{6.4}$$

The combination of (6.3) and (6.4) results in a third-order polynomial for the hole density, measured at 77 K after annealing,

$$p_{77}^3 - \frac{K_i^2 P_{Hg}}{2K_{Hg}} p_{77}^2 + 2K_i p_{77} - \frac{2K_{Hg}}{P_{Hg}} = 0 . \tag{6.5}$$

The equilibrium constants in (6.5) were determined as functions of the temperature by Hall effect measurements (Vydyanath et al. 1981). The iso-concentration lines of Hg vacancies in the solid–vapor phase diagram calculated according to (6.5) are shown in Fig. 6.5 within the existence region of $Hg_{1-x}Cd_xTe$.

The concentration of V_{Hg} can be adjusted by the variation of the temperature (Fig. 6.5) during post-growth annealing under a given mercury vapor pressure P_{Hg}. Such an annealing experiment was carried out in a two-zone furnace with $P_{Hg} = 10^5$ Pa (corresponding to a temperature of the Hg reservoir of 650 K) for a

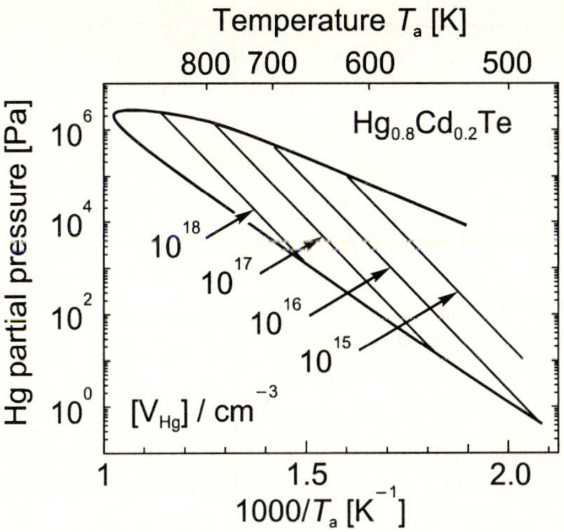

Fig. 6.5. Existence region of $Hg_{0.8}Cd_{0.2}Te$ in a plot of the mercury vapor pressure over the annealing temperature T_a (*thick line*). The thin straight lines indicate the iso-concentration lines of mercury vacancies calculated according to Vydyanath (1981).

series of $Hg_{0.8}Cd_{0.2}Te$ samples. The generated Hg vacancies were studied in correlated positron lifetime and Hall effect measurements (Krause et al. 1989b, 1990a). The average positron lifetime increased significantly with increasing density of Hg vacancies, independently determined as the carrier density p_{77} (Fig. 6.6). The Hg-vacancy density is related to p_{77} via (6.4). The Hall effect measurements showed, however, a compensation of the acceptors by a factor of 0.5 (Neubert et al. 1996), i.e. p_{77} equals the density of V_{Hg}. An important result of the combined positron and Hall measurement is the determination of the trapping coefficient of Hg vacancies, $\mu = 7\times10^{-8}\ cm^3\,s^{-1}$, corresponding to $2\times10^{15}\ s^{-1}$. This has already been described in Sect. 3.2.4.5. The straight line in the inset of Fig. 6.6 is the result of the fit according to (3.11).

The two-component decomposition of the lifetime spectra at first provided a V_{Hg}-related lifetime of $\tau_d = 298$ ps and a positron bulk lifetime of 264 ps (Krause et al. 1990a). These values were later corrected (Krause-Rehberg et al. 1992, 1994a, 1995a) to $\tau_d = 305$ ps and $\tau_b = 275$ ps. The discrepancy was related to problems with the source correction (see Sect. 2.1), which is more difficult for materials with high atomic numbers.

Positron experiments on post-growth-annealed $Hg_{1-x}Cd_xTe$ samples were also carried out in correlation of Hall effect measurements in the Laboratoire Positons CE-Saclay (Baroux et al. 1995; Geffroy et al. 1986; Gély et al. 1990). Positron lifetimes similar to those given above for the bulk and the Hg vacancy were reported. Baroux et al. (1995) studied the temperature dependence of the hole concentration determined by Hall effect measurements, and the positron lifetime as a function of the temperature. The electrical measurements showed evidence of two

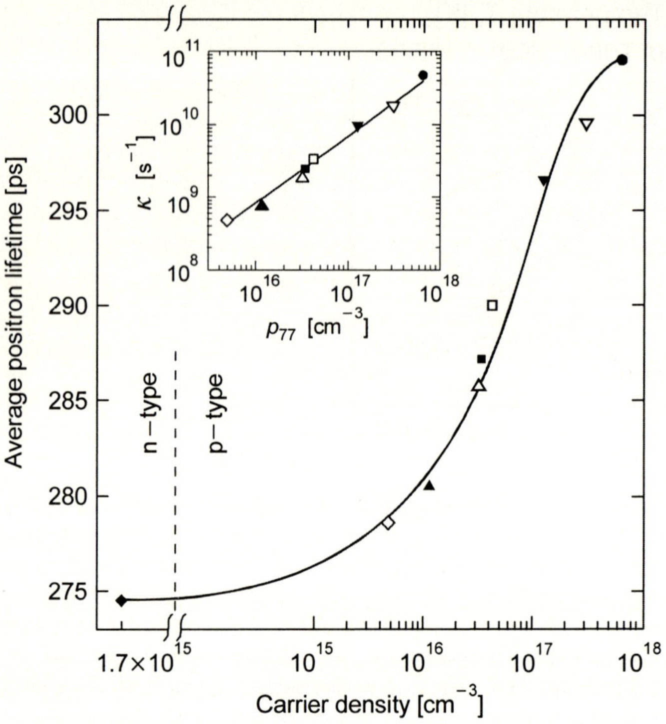

Fig. 6.6. Average positron lifetime in $Hg_{0.8}Cd_{0.2}Te$ versus the carrier density determined by Hall effect measurements at 77 K (Krause et al. 1989b, 1990a). The hole density corresponds to the density of mercury vacancies. The positron trapping rate κ was calculated from the components of the lifetime spectra. It is plotted in the inset against the free-hole density p_{77}.

ionization levels of the mercury vacancy, viz. the 2−/− level located at 41 ± 9 meV above the valence band edge and the −/0 level at 10 ± 1 meV above E_{val}. These results were used for the interpretation of temperature-dependent positron lifetime results. A maximum in the average positron lifetime at about 120 K was found. The decrease in the positron lifetime above this maximum was interpreted with increasing detrapping from Rydberg states, which are connected to the negative charge of the Hg vacancy. The decrease in $\bar{\tau}$ below 100 K was related to two effects: to positron trapping in shallow positron traps and to the two charge-state transitions of the Hg vacancy.

A multiple-parameter fit of the $\bar{\tau}$ curve was carried out using the temperature dependence of the detrapping rate (3.21) and of the trapping rate of the negatively charged mercury vacancies (3.18). The positron trapping in neutral vacancies as well as in shallow positron traps was supposed to be temperature independent. The concentrations of the shallow traps and the Hg vacancies were determined from the fit. Additionally, the donor concentration could be calculated in conjunction with the Hall effect measurements.

The temperature dependence of positron trapping in charged Hg vacancies had already been studied prior to Baroux et al. (1995) by the Doppler-broadening measurements of Smith et al. (1994a). The temperature dependence of the S parameter was similar to the course of the average positron lifetime reported by Baroux et al. (1995). The decline in the S parameter at high temperatures was related to the peculiarities of positron trapping in charged vacancies explained above. The decrease in the S parameter in the low-temperature range was exclusively related to the change in the charge state of the Hg vacancies, and a possible occurrence of shallow positron traps was ignored. Ionization energies of the Hg vacancy higher than those of Baroux et al. (1995) were found. Furthermore, the trapping coefficients of the vacancies were unusually high and had a peculiar temperature dependence, i.e. the trapping coefficient was higher for neutral vacancies than for negative vacancies at room temperature. The contradictions can be resolved by taking into account the occurrence of shallow positron traps (Baroux et al. 1995).

However, it should be noted that the quantitative analysis of the temperature dependence of the positron trapping with the presence of differently charged vacancies and shallow positron traps is not straightforward. The results of the fit are closely related to the temperature dependences of the trapping rates. It was assumed (Baroux et al. 1995) that the positron trapping rates of the singly and doubly charged Hg vacancies were similar to those of charged vacancies in Si (Mäkinen et al. 1992a) or GaAs (Le Berre et al. 1995).

Moreover, the mutual dependence of the fitting parameters gives correlation coefficients very close to one. The real errors of calculated concentrations are much higher than the statistical errors and only rough estimates of the order of magnitude of the concentrations may be possible. The number of fit parameters was eight in the work of Baroux et al. (1995) and must be drastically reduced by data to be provided from other techniques in order to get reliable defect concentrations and ionization levels.

A systematic study of the appearance of mercury vacancies within the existence region of the solid–vapor phase diagram of $Hg_{1-x}Cd_xTe$ was carried out for $x = 0.2$ and 0.3 (Krause-Rehberg et al. 1995a; Neubert et al. 1996). The V_{Hg} concentration was adjusted during annealing in a two-zone furnace under defined Hg vapor pressure. The results of the positron lifetime measurement are shown in Fig. 6.7 for $x = 0.2$. The obtained vacancy concentrations are compared with the results of Vydyanath (1981), calculated as isotherms in the plot of the vacancy concentration versus the vapor pressure according to the defect model given above. The positron lifetime spectroscopy was only sensitive to Hg vacancies, whereas the Hall effect detected all electrically active centers. Since there is a good agreement between both sets of data in Fig. 6.7, the assumptions of the model (6.2) may be fulfilled and it was concluded that the negatively charged Hg vacancies are indeed the dominant point defects.

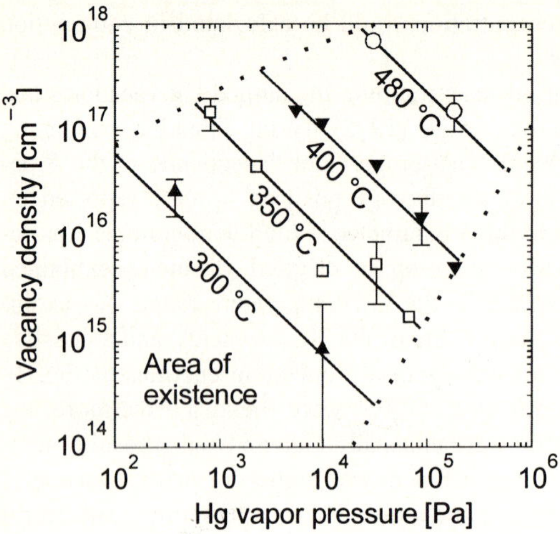

Fig. 6.7. Density of mercury vacancies in $Hg_{0.8}Cd_{0.2}Te$ (Krause-Rehberg et al. 1995a) as determined by positron lifetime spectroscopy (*symbols*). The vacancy densities were adjusted within the region of existence (dotted line) of the compound during post-growth annealing in a two-zone furnace at the given Hg vapor pressure and the indicated sample temperature. The full lines are the isotherms calculated according to (6.5) from the Hall effect measurements of Vydyanath (1981).

Accurate knowledge of the positron trapping coefficient for Hg vacancies in $Hg_{1-x}Cd_xTe$ (Sect. 3.2.4.5) allowed the determination of vacancy profiles in the whole crystal in the as-grown state (Krause et al. 1990a). A traveling-heater grown $Hg_{0.78}Cd_{0.22}Te$ ingot was cut into slices. The positron lifetime was measured after standard polishing procedures at different positions by shifting the positron source across the wafers. The axial position of the wafers is indicated as the third axis in Fig. 6.8. This axis represents the growth direction. The Hg-vacancy concentration was calculated according to (3.11). A low V_{Hg} density of $< 5\times10^{15}$ cm^{-3} was detected near the seed at the start of the growth. The increase in the average positron lifetime and the vacancy density for wafers cut from the end part of the ingot is accompanied by an increase in the hole density p_{77}. Both densities are of the order of 1×10^{17} cm^{-3}. The distribution of Hg vacancies in the ingot was related to the radial and longitudinal temperature variation during the growth. The strong increase at the end of the ingot was attributed to the fast cooling of this part of the crystal after completion of the growth (Krause et al. 1990a). The vacancy profiles observed after growth across the wafer can be homogenized by post-growth annealing in a Hg atmosphere. It could be shown for samples with a pronounced U profile (e.g. the wafer at 51 mm in Fig. 6.8) that the vacancy density could be adjusted according to (6.5) to any desired value between 10^{15} and 10^{18} cm^{-3} (Krause et al. 1992).

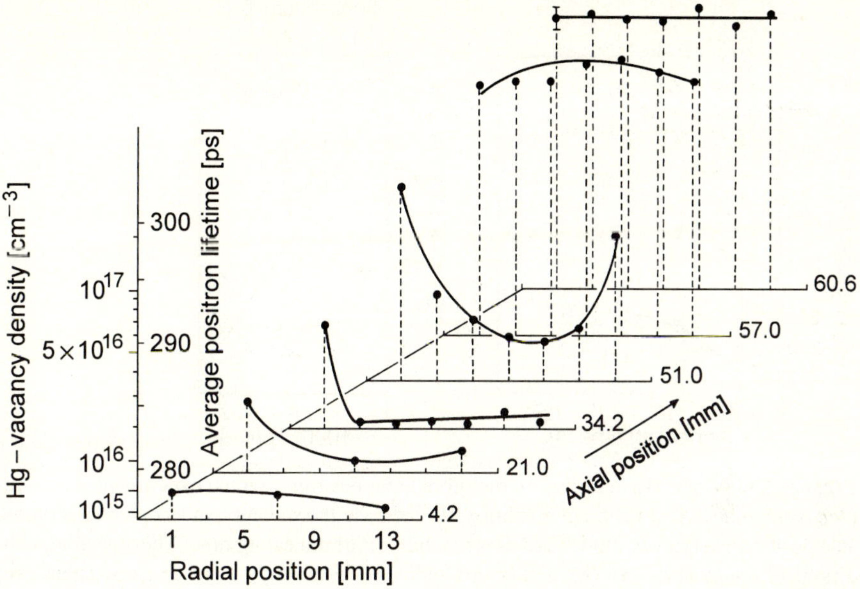

Fig. 6.8. Average positron lifetime $\bar{\tau}$ measured as a function of the position in an ingot of $Hg_{0.78}Cd_{0.22}Te$ (Krause et al. 1990a). In addition, the mercury-vacancy density calculated according to (3.11) is given on the vertical axis. The ingot grown by the traveling-heater method was sliced into wafers, which were measured by placing the positron source at various radial positions. The wafer location in the ingot is indicated as the axial position.

6.1.2.2 Kinetics of Mercury Diffusion

The mercury diffusion in $Hg_{1-x}Cd_xTe$ may be studied by positron annihilation spectroscopy via the detection of the change in the vacancy concentration. The result of such an experiment in samples with $x = 0.2$ is shown in Fig. 6.9 (Krause-Rehberg et al. 1995a). A certain Hg-vacancy concentration was adjusted prior to the diffusion experiment by post-growth annealing according to (6.5). The diffusion conditions are indicated by the arrows a, b, and c in the right-hand panel of Fig. 6.9. The starting conditions for the pre-annealing are the starting points of these arrows. The sample–source sandwich was sealed in a closed ampoule, the other end of which contained the mercury reservoir to control the vapor pressure.

The positron lifetime was measured as a function of the diffusion time. Each positron measurement after a certain diffusion time was carried out after quenching the ampoule in water to room temperature. The measurements could not be carried out at elevated temperatures as originally intended because of the strong dependence of the trapping coefficient on the temperature (Krause et al. 1992).

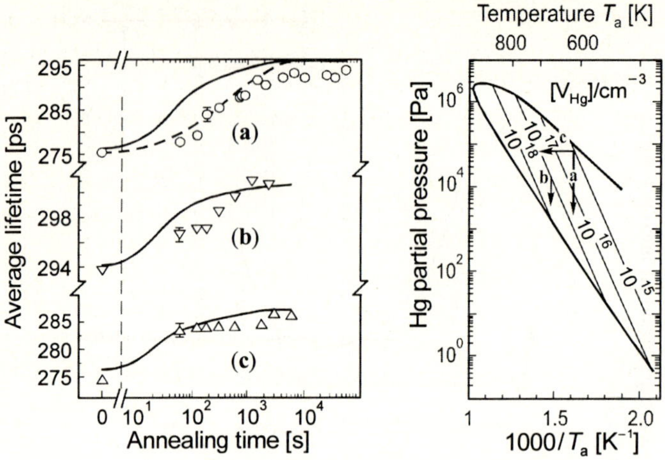

Fig. 6.9. Kinetics of Hg outdiffusion monitored in a positron lifetime experiment. The samples were annealed at different conditions to increase the mercury-vacancy concentration. The positron lifetime was measured as a function of the annealing time. The annealing conditions of the samples (**a**), (**b**), and (**c**) are indicated in the right panel in the existence region of $Hg_{1-x}Cd_xTe$ as the plot of the Hg vapor pressure over the annealing temperature T_a (compare Fig. 6.5). The lines in the left-hand panel are calculated according to a Hg-vacancy diffusion model. The full lines in the left-hand panel are calculated according to a model of mercury diffusion via vacancies. The dashed line corresponds to an extended model additionally taking into account Hg interstitials (Krause-Rehberg et al. 1995a).

This dependence was related to the screening of the negative charge of the Hg vacancy by free carriers in the region of intrinsic conductivity at elevated temperatures (see Sect. 3.1.5.1).

The conditions were chosen in such a way as to realize an outdiffusion of Hg. Correspondingly, the positron lifetime increased during the isothermal (Fig. 6.9a, b) and isobaric (Fig. 6.9c) experiments. The average positron lifetime was calculated using a simple diffusion model supposing mercury diffusion via vacancies and a non-exhausting Hg source outside the sample (full lines in the left panel of Fig. 6.9). The calculation took into account the positron implantation profile as well as the Hg diffusion profile. Obviously, the observed Hg diffusion cannot be described properly by the simple vacancy model used. The kinetics could be better understood by also taking into account Hg interstitials (Krause-Rehberg et al. 1995a).

6.1.3 Zinc-Related II–VI Compounds

ZnSe is one of the promising materials for blue laser diodes. A saturation of the conductivity is not only observed for p-type crystals, but also for n-type material at approximately 1×10^{18} cm^{-3} for Ga doping. MBE-grown ZnSe layers doped with 6×10^{15} to 2×10^{19} cm^{-3} Ga were investigated by slow-positron-beam experiments (Miyajima et al. 1992). It was found that the Zn-vacancy concentration increased with increasing Ga content. The relative vacancy concentration is shown in Fig. 6.10 as a function of the Ga content determined by *S* parameter measurements. The vacancy concentration varies over more than three orders of magnitude. It has been shown for several semiconductors (Sect. 3.2.4.7) that the sensitivity of positron annihilation for the detection of negative vacancies ranges from about 1×10^{-7} to 3×10^{-4}, corresponding to a defect density of 5×10^{15} to 1.5×10^{19} cm^{-3}. One may thus conclude that the relative concentrations given in Fig. 6.10 correspond to that range. Miyajima et al. (1992) suggested that the observed Zn vacancies are bound to Ga atoms. These complexes act as compensating centers and may limit the carrier concentration.

Defects in ZnSe layers grown by metalorganic chemical vapor deposition on GaAs substrates were investigated by Wei et al. (1991). An indication of a high number of open-volume defects near the surface and at the interface with the substrate was found in the depth-dependent slow-positron-beam experiment. Clear differences were detected between the as-grown layer and layers annealed under different atmospheres. It was concluded that heat treatment under zinc vapor is essential to reduce the concentration of native defects in these ZnSe films.

The doping of ZnSe with oxygen resulting in semi-insulating material was also investigated. No effect on positron trapping was observed for oxygen densities lower than 1×10^{18} cm^{-3}. The measured *S* parameter, however, decreased below

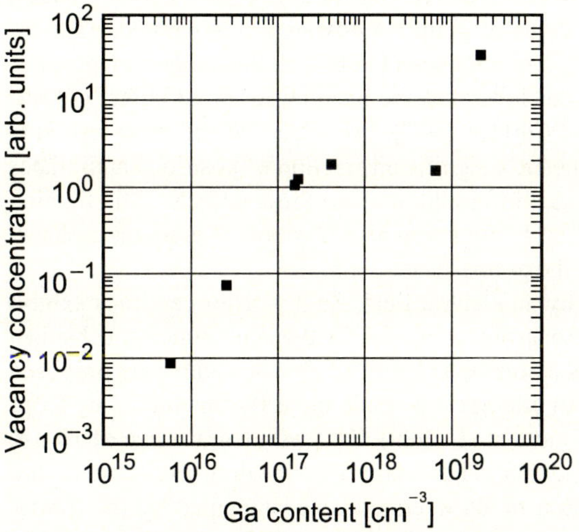

Fig. 6.10. Relative vacancy concentration measured by slow-positron-beam Doppler-broadening measurements as a function of the content of gallium dopants in zinc selenide layers grown by molecular-beam epitaxy (Miyajima et al. 1991).

the bulk value to $S_d/S_b = 0.987$ in ZnSe with an oxygen content of 1×10^{19} cm^{-3} (Miyajima et al. 1992). Such a behavior was observed only rarely, e.g. for oxygen clusters in silicon (Dannefaer and Kerr 1986a). Analogously, the decrease in the S parameter in ZnSe:O was related to positron trapping in oxygen clusters.

Bulk ZnSe crystals were annealed in vacuum and under selenium atmosphere in the experimental investigations of Terashima et al. (1993). The authors found, in positron lifetime measurements, deviations from a three-state trapping model and supposed that a defect having a typical positron lifetime below the bulk value exists. This interpretation was supported by S parameter values being slightly below S_b. It was speculated that the effects were due to positron trapping in Se interstitial clusters, but further evidence with other techniques is required.

There exist only a few positron studies on ZnS crystals. One-component lifetime spectra were obtained in as-grown single crystals (Adams et al. 1995; Pareja and de la Cruz 1993; Pareja et al. 1992). The positron lifetime of 230 ps was interpreted as the positron bulk lifetime in accordance with theoretical calculations (Plazaola et al. 1994, 1995). In contrast, vacancy signals were detected by Doppler-broadening spectroscopy in epitaxial layers of ZnS_xSe_{1-x} layers (Saarinen et al. 1996). Two distinct slopes were found in the (S, W) plots for two types of samples, indicating two different vacancy-type defects. The core annihilation part of the annihilation line could thus be used for the identification of the sublattice of the vacancy. The annihilation with core electrons characterized by the W parameter carries information on the chemical environment of the annihilation site (Sect. 2.3.1). In this way, Zn vacancies were identified in chlorine-doped and Se vacancies in nitrogen-doped material (Saarinen et al. 1996).

The sintering of ZnS polycrystals was investigated by Adams et al. (1995). In addition to a monovacancy-related lifetime component of 290 ps, a longer positron lifetime of 430 ps attributed to voids or grain boundaries was separated. The sintering behavior of ZnO ceramic semiconductors was also the subject of positron studies. Grain boundaries were considered as the location for point defect generation and annihilation (Gupta and Carlson 1987). Provided that the grain size is at most a few micrometers in diameter, a significant fraction of positrons annihilates near grain boundaries and provides information about these defects. This fraction could be calculated by Monte-Carlo simulation as a function of grain size, shape, and defect concentration within the grains (Sect. 3.1.5.8).

When stressed continuously by an electric field, ZnO varistor ceramics exhibit a degradation in the electrical properties. A model for the degradation mechanism involving the grain boundary as a source and sink of Zn vacancies was proposed by Gupta and Carlson (1987). An attempt was made to verify this model by Doppler-broadening spectroscopy (Gupta et al. 1989). ZnO with a grain size of 10 μm sintered at 1200 °C was investigated. The results of annealing experiments are shown in Fig. 6.11. The formation of Zn interstitials accompanied by the formation of negatively charged Zn vacancies near grain boundaries was anticipated in region I. The increase in the S parameter as well as the increase in the photocurrent were taken as proof of the occurrence of both defects. The decrease in region

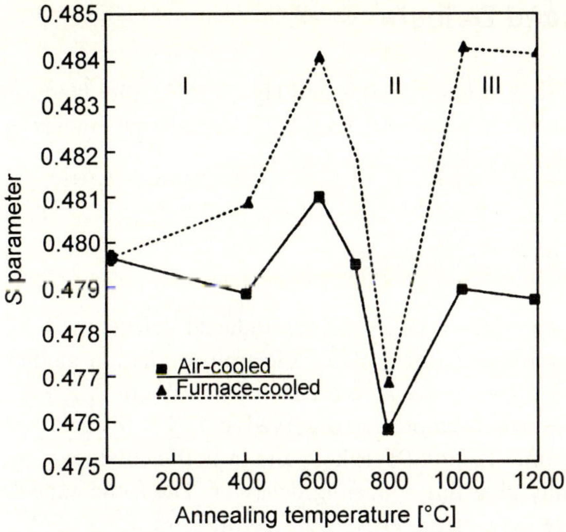

Fig. 6.11. Doppler-broadening S parameter versus the annealing temperature measured in zinc oxide varistor ceramics (Gupta et al. 1989). Different cooling regimes were applied (air- and furnace-cooled, respectively). The different regions I, II, and III are explained in the text.

II was related to the enhanced diffusivity of oxygen and to the reaction of oxygen with the point defects formed in region I. The further increase in the S parameter beyond 800 °C (region III) was interpreted as either the thermal decomposition of ZnO or the generation of Frenkel pairs. The S parameter was a function of the cooling rate after annealing. The quenched air-cooled samples exhibited a lower S parameter than the furnace-cooled samples (Fig. 6.11). This fact was taken as proof that indeed point defects near grain boundaries were investigated.

The sintering of ZnO ceramics has also been studied in more recent investigations by positron annihilation and cathodoluminescence (Mascher et al. 1992; Zhong et al. 1993, 1994). No correlation could be established between the defects acting as positron traps and luminescence centers. Doping with Mn, changing the conductivity from n-type to semi-insulating, did not alter positron trapping significantly (Mascher et al. 1992). The change in positron parameters was studied as a function of the sintering time and temperature. A defect-related lifetime of 255 ps was found after the ZnO was sintered at 1200 °C. It was ascribed to defect complexes involving Zn vacancies. The average positron lifetime attained the positron bulk lifetime of 183 ps after the same sintering conditions. Similarly, an average positron lifetime of 180 ps was given by de la Cruz et al. (1992a) for untreated, high-resistivity as-grown ZnO, in contrast to a one-component spectrum with 169 ps found after thermochemical reduction.

6.2 Irradiation-Induced Defects

Experimental results obtained in cadmium telluride and $Hg_{1-x}Cd_xTe$ after electron irradiation and ion implantation are presented in Sect. 6.2.1. Irradiation studies of zinc-related II–VI semiconductors are dealt with in Sect. 6.2.2.

6.2.1 Cadmium-Related II–VI Compounds

An early positron annihilation experiment on irradiation-induced defects in II–VI compounds was carried out by Moser et al. (1985). Electron-irradiation studies were performed at 20 K with 3-MeV electrons to a dose of 1×10^{18} cm^{-2}. An annealing stage in the average positron lifetime was observed at 130 K for undoped CdTe. It was attributed to the annihilation of Frenkel pairs in both sublattices. Indium doping did not significantly alter this annealing behavior. The same annealing stage was ascribed in a paper of Geffroy et al. (1986) to the annealing of V_{Cd}. The defect-related positron lifetime was found to be 330 ps, which is close to the value for the cadmium vacancy as part of the *A* center (Krause-Rehberg et al. 1998).

Sen-Gupta and Moser (1989) performed correlated Doppler-broadening and positron lifetime measurements of undoped and In-doped CdTe in a temperature range from 20 to 70 K after electron irradiation (2.5 MeV, 4×10^{18} cm^{-2}, 20 K). A weak annealing stage was found at 50 K in correlation with photoluminescence results obtained on the same specimens (Molva and Pautrat 1981).

Extended positron studies of electron-irradiation-induced defects in II–VI semiconductors were carried out by Polity et al. (1997b). The annealing curve of a p-type CdTe sample irradiated at 4 K with 2-MeV electrons (dose 1×10^{18} cm^{-2}) is shown in Fig. 6.12a. The samples were kept, after irradiation, at 77 K and mounted into the cryostat without intermediate warm-up. The positron lifetime was measured at 90 K after every annealing step and at 300 K after the annealing above room temperature. Three distinct ranges can be separated. A pronounced annealing stage is observed in the temperature range below 130 K, similar to the results of Moser et al. (1985). The second range comprises the increase in the average positron lifetime between 130 and 170 K. The positron lifetime reaches the value measured prior to irradiation in the third range up to 600 K.

Correlated Hall effect measurements of the investigated p-type CdTe sample provided a Fermi-level position at $E_F - E_{val} = (215 \pm 25)$ meV, which did not change significantly after irradiation. The electrical measurements could be carried out, however, only after the room-temperature annealing step. Thus, the position of the Fermi level in the intermediate temperature range from 4 to 300 K remains unknown. There are two cases to be discussed: in case (i), the Fermi level stays above the 0/+ ionization level of the Te vacancy (Fig. 6.1), i.e. positron trapping in neutral Te vacancies may be expected during the whole annealing pro-

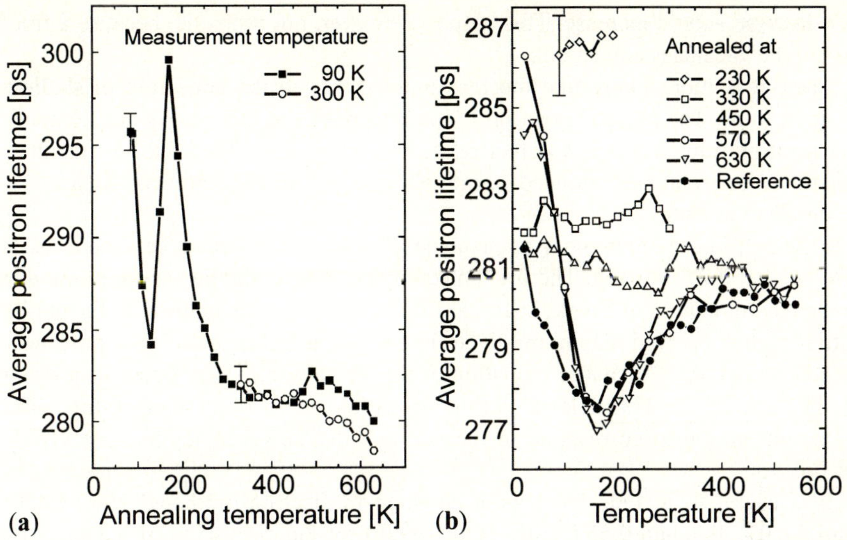

Fig. 6.12. Positron lifetime measurements of p-type cadmium telluride irradiated with 2-MeV electrons at 4 K to a dose of 1×10^{18} cm^{-2} (Polity et al. 1997b). (**a**) Average positron lifetime versus the annealing temperature. The measurement temperatures were 90 and 300 K, respectively. (**b**) Average positron lifetime as a function of the sample temperature after several annealing steps (*open symbols*). The curve of a non-irradiated reference sample annealed at 570 K is added for comparison (*closed circles*).

cedure. In case (ii), the position of the Fermi level is shifted temporarily below the ionization level of the Te vacancy and moves up as a result of defect annealing to the value measured at room temperature, i.e. Te vacancies are not detectable in this state due to their positive charge.

The three annealing ranges are discussed initially under the assumption of the detectability of neutral Te vacancies, case (i). The average positron lifetime measured in the as-irradiated stage indicates the presence of vacancy-type defects. $\bar{\tau}$ reaches after room temperature annealing the value measured before irradiation, indicating the disappearance of both Te and Cd vacancies. This annealing also becomes apparent in Fig. 6.12b, where the average positron lifetime measured after several annealing steps is plotted as a function of the sample temperature. The curves after annealing at 570 and 630 K practically coincide with the curve of the non-irradiated reference sample, also annealed at 570 K.

The recombination of Frenkel pairs in both sublattices is superimposed on the distinct maximum in $\bar{\tau}$ at 170 K. This maximum might be due to clustering of vacancies. It would lead to an increase in the defect-related lifetime. Unfortunately, the positron lifetime spectra could not be reliably resolved to prove this fact. Under the assumption that all vacancies are clustered with no loss due to annealing, the trapping rate should remain constant according to (3.23) and the average positron lifetime should increase. This case is, however, not probable, because a fraction of the vacancies anneal out.

Thus, it is more likely that the maximum is due to the annealing of shallow positron traps. The average positron lifetime increases in this case, since competitive positron traps with a low defect-related lifetime vanish. Such shallow positron traps may be irradiation-induced negatively charged intrinsic defects, such as interstitials or antisites.

Case (ii), i.e. the temporary appearance of positively charged Te vacancies, is discussed in the following. The first annealing stage is, under this assumption, due to the recombination of Frenkel pairs in the Cd sublattice. In addition to the explanation of the observed maximum given for case (i), a further possibility must now be considered. A +/0 charge transition of the Te vacancies may occur as a result of the shift of the Fermi level during annealing up to room temperature. The maximum may then correspond to the fact that the Te vacancies become visible for positrons.

It is useful to include the results obtained in electron-irradiated $Hg_{1-x}Cd_xTe$ with $x = 0.2$, 0.3, and 0.55 (Polity et al. 1997b) for the discussion of the cases (i) and (ii). The irradiation conditions were identical to the treatment of the CdTe sample described above. Only the results of the $Hg_{0.45}Cd_{0.55}Te$ sample are exemplarily described here (Fig. 6.13). The course of the average positron lifetime during the annealing experiment is plotted in Fig. 6.13a. The first annealing stage found in CdTe is clearly visible for $x = 0.2$, but barely detectable for $x = 0.3$ and 0.55. However, the pronounced maximum at 250 K in the average positron lifetime occurs similarly to the case of CdTe in all $Hg_{1-x}Cd_xTe$ samples. The most probable interpretation, which is also supported by the results of Kiessling et al. (1995), is the annealing of shallow positron traps in this range. The appearance of the maximum is thus similar to case (i) discussed above for CdTe.

A two-component decomposition of the positron lifetime spectra was possible in electron-irradiated $Hg_{1-x}Cd_xTe$ below the annealing temperature of 350 K (Polity et al. 1997b). The second lifetime τ_2 equals 303 ps and 317 ps below and above the annealing temperature of 200 K, respectively. This result can be understood under the assumption that the annealing of shallow positron traps is the reason for the maximum in $\bar{\tau}$. The defect-related lifetime is usually too short in a two-component decomposition in the presence of shallow positron traps, as shown by Monte-Carlo simulations of positron lifetime spectra (Somieski et al. 1996).

The defect-related lifetime observed above 200 K lies in between those for a vacancy in the Hg/Cd sublattice and for a Te vacancy. The positron lifetime of a Hg vacancy amounts to 305 ps, as described in Sect. 6.1.2. A positron lifetime of 323 ± 5 ps was experimentally found for the Te vacancy (Kiessling et al. 1995). The Fermi level was found to be shifted upwards after low-temperature electron irradiation, and the samples were heavily n-type (Favre et al. 1989). It was concluded that the Te vacancies are responsible for the Fermi-level pinning. In this case, vacancies in both sublattices are detectable. This is the main difference from the vacancy generation in $Hg_{1-x}Cd_xTe$ during post-growth annealing (Sect.

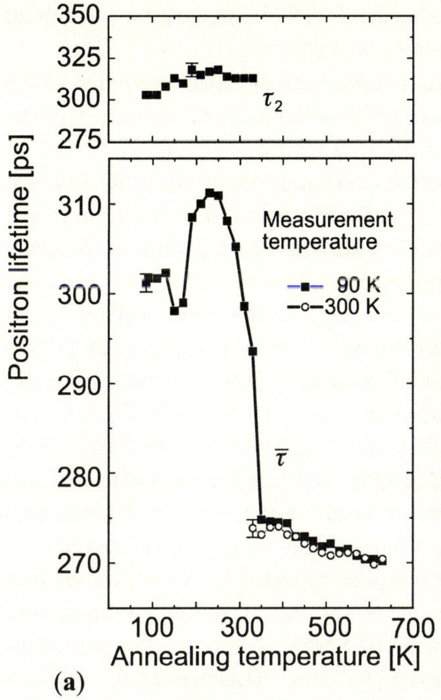

Fig. 6.13. Positron lifetime measurement of $Hg_{0.45}Cd_{0.55}Te$ irradiated at 4 K with 2-MeV electrons to a dose of 1×10^{18} cm^{-2} (Polity et al. 1997b). (**a**) Average positron lifetime $\overline{\tau}$ and defect-related lifetime τ_2 versus the annealing temperature. The measurements were carried out at 90 and 300 K. (**b**) Average positron lifetime as a function of the sample temperature measured after several annealing steps.

6.1.2.1), where the Fermi level is determined by the acceptor-type Hg vacancies. Te vacancies are expected to be positively charged and thus not acting as positron traps.

The results of positron lifetime measurements performed as a function of temperature after certain annealing steps are shown in Fig. 6.13b (Polity et al. 1997b). No influence of shallow positron traps is visible in the low-temperature range after these annealing steps. This is in accordance with the interpretation of the observed maximum given above as the annealing of shallow traps below 200 K. On the other hand, the curves measured after annealing at 230 and 330 K show the typical behavior for negatively charged vacancy-type defects with the steep decrease in $\overline{\tau}$ with increasing temperature. This behavior is most likely caused by the existence of negatively charged Hg vacancies, since Te vacancies should be neutral. However, average positron lifetimes significantly above the defect-related lifetime of the Hg vacancy were reported in another annealing study of electron-irradiated $Hg_{0.78}Cd_{0.22}Te$ by Kiessling et al. (1995). A defect-related lifetime of 323 ps was observed and was attributed to positron trapping in Te vacancies. Combining the results of Polity et al. (1997b) and Kiessling et al. (1995), it may be concluded that, actually, a mixture of Hg and Te vacancies is observed.

To summarize the results obtained in low-temperature electron-irradiated CdTe and $Hg_{1-x}Cd_xTe$, we can say that the annealing behavior is very similar. The annealing of irradiation-induced vacancies in both sublattices occurs in the range from 100 to 400 K leading to a decrease in the average positron lifetime. The observed maximum between 150 and 250 K is likely to be due to the superimposed increase in $\bar{\tau}$ as a result of the annealing of irradiation-induced shallow positron traps.

There have been a few positron studies on ion implantation in Cd-related II–VI compounds (Liszkay et al. 1995; Voitsekhovskii et al. 1995). Liszkay et al. (1995) studied $Hg_{0.78}Cd_{0.22}Te$ implanted with 320-keV Al ions at 100 and 300 K. The defect-related Doppler parameters were found to be $S_d/S_b = 1.05$ and $W_d/W_b = 0.8$. A previous study of Hg monovacancies in this material gave a value of $S_d/S_b = 1.02$ (Smith et al. 1994b). Hence, Liszkay et al. (1995) concluded the existence of implantation-induced small vacancy clusters, at least of the size of divacancies. However, S parameter data of different spectrometers can hardly be compared.

The formation of vacancy agglomerates was also reported by Voitsekhovskii et al. (1995). The divacancy generation reached a saturation at room temperature with a density of 4×10^{16} cm^{-3} at a dose of 3×10^{12} cm^{-2}, whereas after implantation at 100 K the divacancy density exceeded 1×10^{18} cm^{-3} (Liszkay et al. 1995).

6.2.2 Zinc-Related II–VI Compounds

The defect formation after low-temperature electron irradiation was also studied for several Zn-related semiconductors. In practically all studies, the Zn vacancies are supposed to be single or double negatively charged, whereas the anion vacancies are assumed to be positively charged. As a consequence, only the Zn vacancies are detectable by positron annihilation.

One of the first positron studies on Zn-related II–VI compounds was carried out by Moser et al. (1985) for ZnTe irradiated at 20 K with 3-MeV electrons to a dose of 1×10^{18} cm^{-2}. A main annealing stage appeared at 500 K, which was confirmed by the measurements of Brunner et al. (1997).

The average positron lifetime was measured by Abgarjan (1994) in ZnSe electron-irradiated at 4 K (2 MeV, 1×10^{18} cm^{-2}). The results are compared with a sample irradiated with Ne ions of the energy of 200 MeV at room temperature to a dose of 1×10^{14} cm^{-2} (Fig. 6.14). A continuous increase in the average positron lifetime was found in the electron-irradiated sample up to 300 K. The increase was discussed as either resulting from vacancy clustering or annealing of shallow positron traps (Abgarjan 1994). A defect-related component of 300 to 310 ps could be resolved. It can be concluded from the comparison with theoretical calculations (Sect. 3.3.1) that this component may be due to a divacancy or a divacancy-related complex as supposed by Pareja et al. (1992). The case of the an-

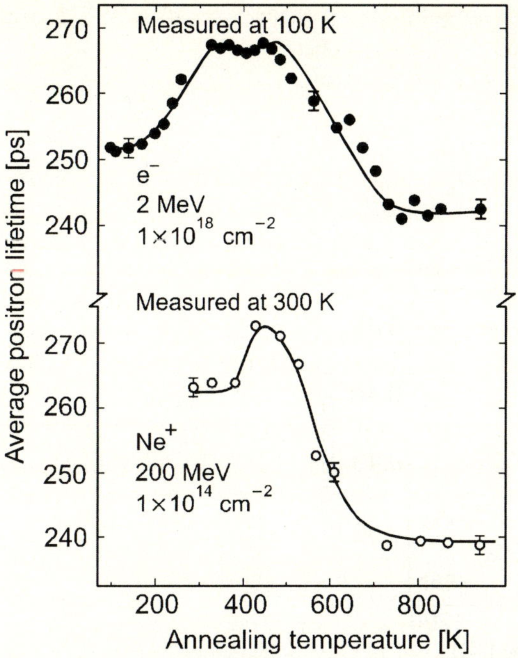

Fig. 6.14. Annealing experiment of electron-irradiated zinc selenide (2 MeV, $1\times10^{18}\,\mathrm{cm}^{-2}$, 4 K) compared with ZnSe crystals irradiated with neon ions (200 MeV, $1\times10^{14}\,\mathrm{cm}^{-2}$, 300 K). The positron lifetime was measured at 100 K in the electron-irradiated sample and at 300 K in the Ne^+-irradiated sample (Abgarjan 1994).

nealing of shallow traps can be excluded with these results, because a comparable increase in $\bar{\tau}$ was found both for low and high sample temperatures.

The ZnSe sample irradiated at room temperature with Ne ions also exhibited a slight increase in the average positron lifetime up to 450 K (Fig. 6.14). Because the same defect-related lifetime of 300 to 310 ps was resolved, and the main annealing stage was also observed between 500 and 700 K, it was concluded that the same defect types dominated the positron trapping in the electron-irradiated and in the Ne-irradiated samples.

In contrast to Abgarjan (1994), Pareja et al. (1992) found an annealing stage in the temperature range between 77 and 125 K that was interpreted as the recombination of close Frenkel pairs in the Zn sublattice. Another difference was the temperature dependence of the positron lifetime. Abgarjan measured an increase by about 5 ps on cooling the sample from room temperature to 100 K. However, a distinct decrease was observed by Pareja et al. (1992) in the same temperature range and was related to the charge-state transition of the divacancy-type defects, which possibly did not occur in the samples of a different supplier used by Abgarjan (1994).

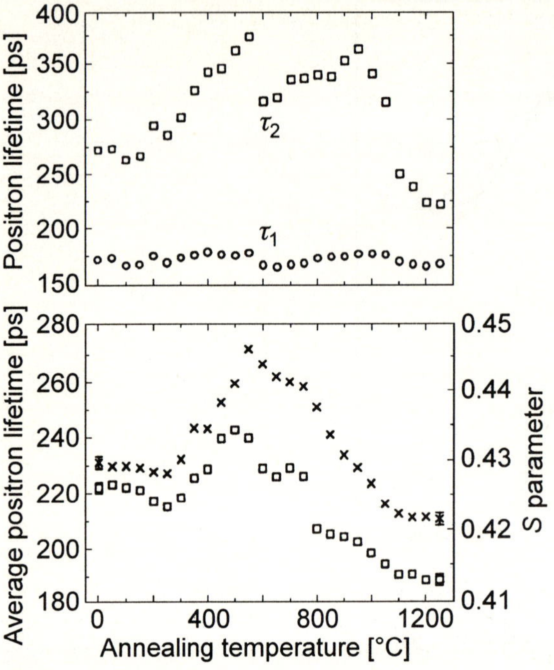

Fig. 6.15. Positron annihilation measurements of an annealing experiment of a zinc oxide single crystal irradiated with 3 MeV protons at 223 K (Puff et al. 1995b). The lifetime components τ_1 (○) and τ_2 (□) of a two-component spectra decomposition are shown in the upper panel. The average positron lifetime (□) and the Doppler-broadening *S* parameter (×) are plotted in the lower panel.

The positron lifetime was measured by Pareja et al. (1992) in ZnS after electron irradiation at 20 and 77 K. The initial annealing stage of close Frenkel pairs observed in ZnSe was not found at the lowest measuring temperature of about 80 K. The average positron lifetime increased from 240 to 280 ps during the annealing procedure up to 700 K. A similar annealing stage to that in ZnSe did not appear in ZnS up to this temperature. The increase in $\bar{\tau}$ was attributed to V_{Zn}-related defects. It was observed that the defect-related lifetime of a sample irradiated at room temperature decreased with decreasing temperature below 300 K (de la Cruz et al. 1992b). This effect was interpreted as the change in the charge state of irradiation-induced vacancy-type defects rather than the appearance of shallow positron traps. However, this interpretation has to be confirmed by electrical measurements.

Electron-irradiated ZnO crystals were investigated by Tomiyama et al. (1992). The irradiation conditions were 28 MeV, 4×10^{18} cm^{-2}, and 77 K. A clear annealing stage was found at 450 K. The interpretation given for this stage as the disappearance of irradiation-induced oxygen monovacancies seems to be unlikely, since

Table 6.1. Positron-lifetime data of the bulk and vacancy-related positron lifetimes of II–VI compound semiconductors.

Material		Characteristic positron lifetime		Defect annealing
		Positron lifetime [ps]	τ_d/τ_b	temperature [K]
CdTe	Bulk	291[a], 285[b,c], 280[d]	—	—
	A center	320[b,c], 330[d,e]	1.10 to 1.18	
HgTe	Bulk	274[e]	—	
$Hg_{0.8}Cd_{0.2}Te$	Bulk	274[e], 275[g], 278[h]		
	V_{Hg}	309[f], 305[g,h]	1.10 to 1.12	< 300[j]
	V_{Te}	325[i,j]	1.17	< 300[j]
ZnO	Bulk	169[k], 183[l], 185[m]	—	—
	V_{Zn}	211[m]	1.15	
	Divacancies	273[m]	1.48	
ZnS	Bulk	230[n]	—	—
	V_{Zn}	290[o]	1.26	
ZnSe	Bulk	240[p]	—	—
ZnTe	Bulk	266[n]	—	—

τ_d is the defect-related positron lifetime, τ_b the positron lifetime in the defect-free bulk.
[a]Dannefaer (1982), [b]Gély-Sykes et al. (1991), [c]Corbel et al. (1993), [d]Krause-Rehberg et al. (1998), [e]Geffroy et al. (1986), [f]Gély et al. (1990), [g]Krause-Rehberg et al. (1992), [h]Baroux et al. (1995), [i]Kiessling et al. (1995), [j]Polity et al. (1997b), [k]de la Cruz et al. (1992a), [l]Zhong et al. (1993, 1994), [m]Puff et al. (1995b), [n]Pareja and de la Cruz (1993), [o]Adams et al. (1995), [p]Pareja et al. (1992).

these defects are expected to be positively charged (Puff et al. 1995b). The annealing behavior was found to be completely different after irradiation with 3 MeV protons at 223 K. A defect-related lifetime of 273 ps was separated after proton irradiation (Fig. 6.15). It was attributed to divacancies, because the τ_d/τ_b ratio reached 1.48. In contrast, this ratio was 1.15 after low-temperature electron irradiation, corresponding to a monovacancy-related lifetime of 211 ps.

The measured average positron lifetime, as well as the Doppler-broadening *S* parameter increased during subsequent annealing in the temperature range of 300 to 550 °C. As the defect-related lifetime τ_2 increased in the same temperature range up to 378 ps, it was concluded that the divacancies grow further to larger agglomerates. These agglomerates annealed during subsequent heat treatment, and the average positron lifetime approached the bulk lifetime at 1100 °C.

6.3 Summary

Native defects in II–VI compound semiconductors, such as the mercury vacancy in $Hg_{1-x}Cd_xTe$, were extensively studied by positron annihilation. This defect is electrically active as acceptor and has a high diffusivity already at low temperatures. The concentration can be adjusted in post-growth annealing experiments under defined vapor pressure. Correlated Hall effect and positron lifetime measurements allowed the accurate determination of the trapping coefficient. In this way, vacancy concentrations within the existence region of the compound were determined in dependence on the Hg partial pressure and sample temperature. Moreover, the distribution of vacancies in a traveling-heater-grown $Hg_{1-x}Cd_xTe$ crystal could be determined.

The *A* center in chlorine-doped cadmium telluride ($V_{Cd}Cl_{Te}$) is an effective positron trap. Positron annihilation shows the presence of two types of open-volume defects in CdTe with varying Cl content. A vacancy–dopant complex was also found in epitaxial zinc selenide layers. Its concentration increased by at least three orders of magnitude with increasing gallium content. A special feature of the annealing curves of electron-irradiated CdTe and $Hg_{1-x}Cd_xTe$ was the appearance of a distinct maximum in the average positron lifetime around 200 K, which was interpreted as the annealing of shallow positron traps.

The published positron lifetime data of the bulk and the vacancy-related lifetimes of II–VI compounds are summarized in Table 6.1. The τ_d/τ_b ratios for monovacancies are found in the range from 1.10 to 1.26, which are similar to the values obtained for other semiconductors. In contrast, the calculated ratios (Sect. 3.3.1) of II–VI compounds are too small compared with the experimental data of Table 6.1.

7 Defect Characterization in Other Compounds

In addition to the semiconductors dealt with in Chaps. 4 to 6, positron investigations of two other less common compounds, viz. SiC and $CuInSe_2$, are treated in this chapter. Silicon carbide has unique chemical and physical properties, which make it interesting for optoelectronics and for high-temperature or high-power applications. The chalcopyrite compound copper indium diselenide is mainly of interest for photovoltaic devices with large conversion efficiency. It holds for both compounds that the characterization of defects is still at the beginning.

7.1 Silicon Carbide

There have been several papers on positron annihilation characterization of native defects in silicon carbide. Girka and Mokhov (1995) investigated the positron lifetimes of various polytypes and growth techniques. They obtained in the 6H polytype a single positron lifetime component of 157 ± 3 ps, which was assumed to be the positron lifetime in the defect-free bulk. Close values of τ were also found in 3C, 4H, and 15R polytypes of SiC, which implied that the density of vacancy-type defects was below the sensitivity limit of about 5×10^{15} cm^{-3} or that these defects were positively charged. An increase in the vacancy density of up to 5×10^{17} cm^{-3} appeared in 6H SiC samples grown with a sublimation sandwich method at a temperature of 1860 to 2000 °C under highly Si-rich conditions. It was concluded that the open-volume defects are carbon vacancies. The excess in V_C was preserved during annealing up to 2400 °C. The comparison with electron irradiation studies led to the conclusion that the high thermal stability cannot be due to isolated carbon vacancies, but rather to vacancy clusters or vacancy–impurity complexes. This conclusion can also be drawn from the results of Puff et al. (1995a). They found in 4H and 6H SiC a second lifetime component of 250 to 300 ps. The theoretical positron bulk lifetime was 150 ps. A defect-related lifetime between 180 and 190 ps was separated by Kawasuso et al. (1996) in n-type 6H SiC. The positron lifetimes in SiC were theoretically calculated by Brauer et al. (1996b). The positron bulk lifetime was 141 ps. The carbon monovacancy had a positron lifetime of 150 ps, the silicon monovacancy about 190 ps, and the $V_{Si}V_C$ divacancy 216 ps. Comparing the experimental results of Kawasuso et al. (1996) with these calculated positron lifetimes, V_C can certainly be ruled out, but not

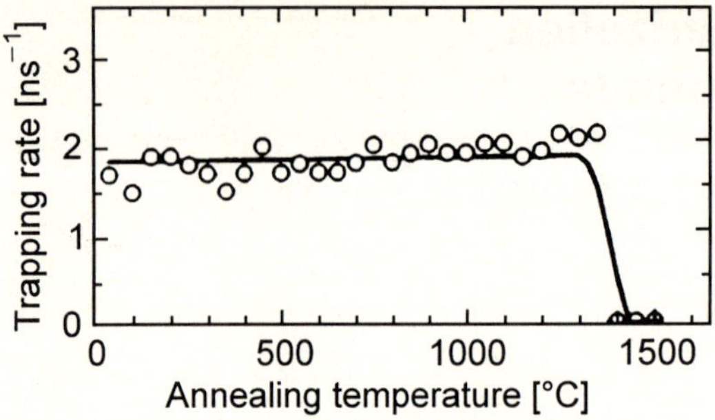

Fig. 7.1. Isochronal annealing of vacancy-type defects in as-grown n-type 6H silicon carbide. The trapping rate was determined from the decomposition of positron lifetime spectra (Kawasuso et al. 1996).

isolated V_{Si}. It was concluded from the thermal stability that the detected defects were vacancy–impurity complexes, probably complexes of silicon vacancies and nitrogen impurities. Their isochronal annealing is shown in Fig. 7.1. The vacancy-type defect disappears at 1400 °C.

The charge states of vacancies were also discussed. A carbon monovacancy is either positively charged in p-type or neutral in n-type SiC (Wang et al. 1988). Si monovacancies should carry a negative charge in highly n-type SiC (Wang et al. 1988), but such defects could not be detected by positron annihilation after growth under excess-carbon conditions (Girka and Mokhov 1995).

In addition to bulk samples, Puff et al. (1995a) investigated nanocrystalline SiC and chemical-vapor-deposited layers. The main fraction of positrons annihilated with a lifetime of about 400 ps and an intensity of 65 % in the nanocrystalline material. The interfaces of the grains having an average diameter of 100 nm were the dominant trapping centers. The CVD SiC layers were characterized by the slow-positron-beam technique. The *S* versus *E* profiles were strongly dependent on the deposition conditions.

Most of the positron studies on SiC were carried out after electron irradiation. Itoh et al. (1992a) combined positron annihilation and electron paramagnetic resonance in order to characterize the defects in n-type 3C SiC after 1-MeV electron irradiation. The *S* parameter in epitaxially grown layers increased linearly by about 4 % with the electron dose varying from 10^{17} to 10^{19} cm^{-2}. The positron diffusion length decreased at the same time from 40 to 5 nm. L_+ was 90 nm in as-grown layers. An EPR spectrum labeled T1, which has an isotropic *g* value of 2.0029, was observed in electron-irradiated 3C SiC. It was interpreted to be related to defects on the Si sublattice. The production rate was in a good agreement with the carrier removal rate, indicating that this center captures electrons from the conduction band. It was concluded that T1 is identical to negatively charged Si vacancies introduced upon irradiation.

The annealing of electron-irradiation-induced defects was studied in 6H SiC (Girka et al. 1989; Kawasuso et al. 1996; Polity 1997; Rempel and Schaefer

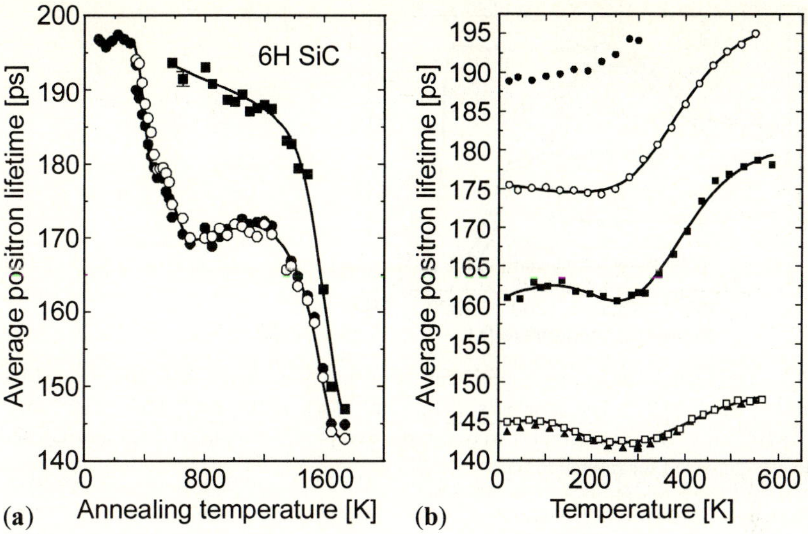

Fig. 7.2. Annealing of electron-irradiation-induced defects in n-type 6H silicon carbide (Polity 1997). The electron irradiation was carried out at 4 K with 2 MeV to a dose of 1×10^{18} cm^{-2}. (**a**) Average positron lifetime measured at several sample temperatures (● 90 K, ○ 300 K, ■ 550 K) as a function of the annealing temperature, (**b**) average positron lifetime versus the sample temperature in the as-grown sample (▲) and after different annealing steps (● 340 K, ○ 580 K, ■ 1490 K, □ 1740 K) of the irradiated sample.

1995). Different annealing stages were reported in these studies. Girka et al. (1989) and Kawasuso et al. (1996) found a first annealing temperature at 420 K, probably related to the annealing of Frenkel pairs. In contrast, the first annealing stage centered at about 650 K in the investigations of Rempel and Schaefer (1995), where p-type samples were used. In this case, the annealing of the Frenkel pairs was possibly hidden for positron annihilation, since the vacancies were positively charged. Furthermore, an increase in the average positron lifetime was found around 1500 K and interpreted as the clustering of vacancies. The defect-related positron lifetime of 210 ps indicated the formation of divacancies. This effect was not observed in n-type SiC by Kawasuso et al. (1996). The final annealing stage at 1650 K led to the vanishing of positron trapping. The annealing curves of n-type 6H SiC, as measured by Polity (1997), show special features in dependence on the measurement temperature (Fig. 7.2a). Two annealing stages were found at 450 and 1600 K when the measurement was carried out at sample temperatures of 90 and 300 K. However, the first stage was missing for the measurement at 550 K. The bulk value of 142 ps measured in as-grown samples is reached again after the annealing step at 1740 K (Fig. 7.2b).

Positron lifetime measurements carried out as a function of the sample temperature indicated the presence of a high density of shallow positron traps. Posi-

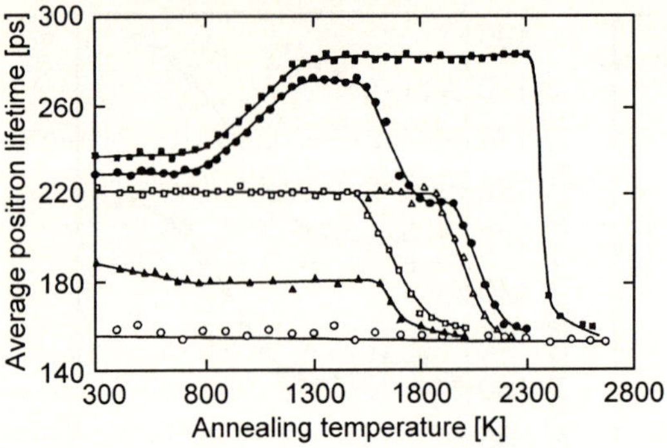

Fig. 7.3. Isochronal annealing experiment of neutron-irradiated 6H silicon carbide (Mokrushin et al. 1991). The dose was 3×10^{16} cm^{-2} (▲), 1×10^{19} cm^{-2} (□), 1×10^{20} cm^{-2} (△●), 1×10^{21} cm^{-2} (■). The samples were n-type, except one sample (△). An unirradiated sample was annealed for reference (○).

tron trapping in these defects is effective up to 600 K. As a consequence, positron lifetime measurements at room temperature may be affected by the presence of shallow positron traps. The determination of vacancy concentrations according to a one-defect trapping model (Sect. 3.2.1.1) should thus be carried out at temperatures above 600 K.

Saturated positron trapping is observed in the irradiated sample after 90 K annealing. Two lifetime components could be separated in the positron lifetime spectrum, τ_{d1} = 178 ps, τ_{d2} = 222 ps. They were attributed to monovacancies and divacancies, in accordance with the theoretical calculations of Brauer et al. (1996b). In accordance with former assignments (Dannefaer et al. 1995; Girka et al. 1989; Kawasuso et al. 1996), the annealing of Frenkel pairs in both sublattices is observed at annealing at 450 K with the disappearance of the positron lifetime component τ_{d1}. Two other lifetimes with τ_{d3} = 145 ps and τ_{d2} = 210 ps appear. The latter lifetime may be still due to positron trapping in divacancies, while τ_{d3} is related to the shallow positron traps. The difference in the curve measured at 550 K from the other two curves in Fig. 7.2a can be well understood with this interpretation. The difference is due to the effect of shallow positron traps leading to the reduction in the average positron lifetime.

Radiation-induced defects in SiC were also studied after neutron or proton irradiation (Hayashi et al. 1995; Mokrushin et al. 1991; Puff et al. 1997). Clustering of irradiation-induced vacancies was observed during thermal annealing of samples irradiated with high-dose neutron beams (Fig. 7.3). A similar effect was already reported after electron irradiation (Rempel and Schaefer 1995). In contrast, the clusters are much larger after neutron irradiation, since defect-related positron lifetimes of 360 or 460 ps were separated. Another difference is that the larger clusters are thermally more stable than the divacancies found by Rempel and Schaefer (1995).

The defects after ion implantation in 6H SiC were studies by Brauer et al. (1996a). 200-keV Ge^+ was implanted in the dose range from 10^{12} to 10^{15} cm^{-2}. Amorphization occurred for doses above 3×10^{13} cm^{-2}. The amorphous layer reached a thickness of 133 nm at the highest dose. The depth of the crystalline damaged layer was up to 600 nm. Slow-positron-beam lifetime measurements identified divacancies as the main positron trap, but Si monovacancies were also present. Larger vacancy clusters consisting of about 4 to 6 vacancies were found in the amorphous layer.

The implantation of 3C SiC films with Al^+ and N^+ was investigated by Uedono et al. (1996, 1997a). Defect profiles and the annealing of implantation-induced defects were studied. Vacancy clusters were identified as the dominant defects in specimens implanted at temperatures higher than 800 °C. In contrast, monovacancies were found after implantation at 400 °C. The profile of vacancy-type defects is shifted towards the surface compared with the sample implanted at lower temperatures. Several annealing stages were observed. Three stages up to 1000 °C were attributed to the agglomeration of vacancy-type defects. Two further stages from 1000 to 1400 °C were assigned to the formation of extended defects.

Saturated positron trapping with a lifetime of 218 ps was found with slow-positron-beam lifetime measurements of 4H SiC implanted with Al^+ of five different energies to an overall dose of 2.3×10^{13} cm^{-2} (Bauer-Kugelmann et al. 1997). This positron lifetime was attributed to nearest-neigbor divacancies in accordance with the calculations of Brauer et al. (1996a), providing a value of 216 ps.

7.2 Copper Indium Diselenide

X-ray photoemission spectroscopy was used by Nelson et al. (1995) in combination with positron annihilation to characterize the composition and defect properties of $CuInSe_2$. The occurrence of different phases with a variation in the composition was found and the observed changes in the near-surface S parameter related to open-volume defects. Al-Kotb et al. (1997) studied the positron lifetime and the Doppler-broadening S parameter in as-grown $CuInSe_2$ samples. The average positron lifetime was in the range 250 to 270 ps. A defect-related lifetime of 285 to 298 ps was separated. The S parameter was found to increase with the Cu/In ratio in the compound. Slow-positron-beam lifetime experiments of thin films grown by molecular beam epitaxy were carried out by Suzuki et al. (1997). Distinct changes in the average positron lifetime measured at an incident positron energy of 10 keV were observed with the variation of the composition (Fig. 7.4). A single-component spectrum with $\tau = 244$ ps was measured for a high Cu/In ratio. The positron bulk lifetime was calculated to be 240 ps (Suzuki et al. 1997). Almost saturated trapping with 319 ps occurred at a Cu/In ratio of 0.8. This corresponds to values calculated theoretically for the divacancies $V_{Cu}V_{Se}$ and $V_{In}V_{Se}$ to be 316.2

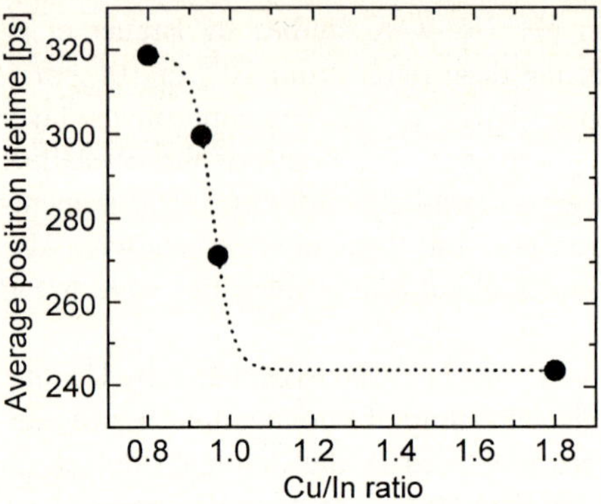

Fig. 7.4. Slow-positron-beam lifetime measurement of copper indium diselenide grown by molecular beam epitaxy (Suzuki et al. 1997). The average positron lifetime measured at an incident positron energy of 10 keV is plotted versus the copper to indium ratio in the compound.

and 328.1 ps, respectively. It was concluded that divacancies are the dominant positron traps in In-rich films.

Low-temperature irradiation with 2-MeV electrons of a dose of 10^{18} cm^{-2} was

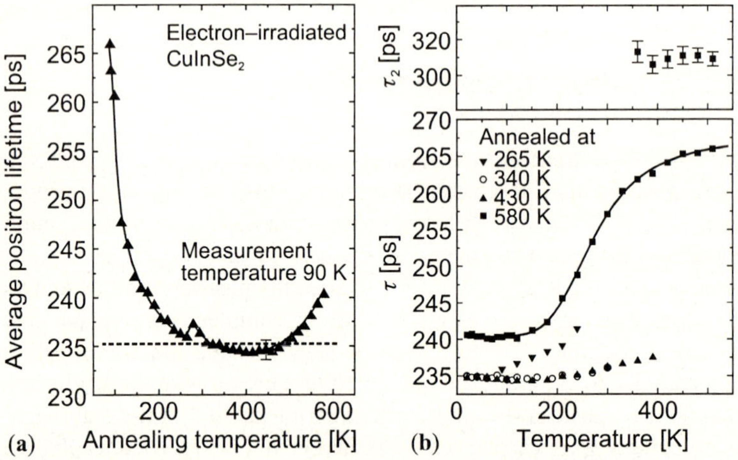

Fig. 7.5. Electron irradiation of copper indium diselenide (Polity et al. 1998b). The irradiation was carried out with 2-MeV electrons at 4 K to a dose of 10^{18} cm^{-2}. (**a**) Average positron lifetime measured at a sample temperature of 90 K as a function of annealing temperature. The dashed line is the positron lifetime measured prior to irradiation. (**b**) Average positron lifetime as a function of the sample temperature after several annealing steps. The solid line corresponds to the fit according to a trapping model with one deep and one shallow positron trap. The upper panel shows the defect-related positron lifetime τ_2.

Table 7.1. Positron-lifetime data for silicon carbide and copper indium diselenide.

Material		Characteristic positron lifetime		Defect annealing
		Positron lifetime [ps]	τ_d/τ_b	temperature [K]
SiC	Bulk	136[a], 142[b], 150[c], 157[d,e], 145[h]	—	—
	Vacancy–impurity complexes	180 to 190[a], 250 to 300[c]	1.32 to 1.39[a], 1.67 to 2.00[c]	1673[a]
	Grain boundaries	400[c]	2.67[c]	
	Divacancies	222[b], 218[h]	1.56[b], 1.50[h]	
	Vacancy clusters	360 to 460[e]	2.29 to 2.93[e]	1650[e]
$CuInSe_2$	Bulk	235[f], 244[g]	—	—
	Divacancies	310[f], 319[g]	1.32[f], 1.31[g]	150[f]

τ_d is the defect-related positron lifetime, τ_b the positron lifetime in the defect-free bulk.
[a]Kawasuso et al. (1996), [b]Polity (1997), [c]Puff et al. (1995a), [d]Girka and Mokhov (1995), [e]Mokrushin et al. (1991), [f]after electron irradiation (Polity et al. 1998b), [g]Suzuki et al. (1997), [h]Bauer-Kugelmann et al. (1997).

carried out by Polity et al. (1998b) in $CuInSe_2$. Prior to irradiation, a single-component spectrum with 235 ps was observed. The composition of the compound was determined by energy-dispersive X-ray analysis. It could be deduced by the consideration of the defect chemistry in the phase diagram of $CuInSe_2$ that Cu_{Se}^{2+} and Cu_{In}^{2-} antisites should be the dominant defects. The concentration of vacancies is lower by orders of magnitude, and it was thus concluded that the measured lifetime corresponds to the positron bulk lifetime, in agreement with theoretical calculations (Polity et al. 1998b).

The results of the electron irradiation and subsequent annealing experiment are shown in Fig. 7.5. The annealing experiment started at a temperature of 90 K, where the average positron lifetime of 267 ps was measured. The steep fall of $\bar{\tau}$ indicated that the onset of the first annealing stage already occurred at lower temperatures. Apparently, all defects disappeared upon annealing at room temperature. An increase in the average positron lifetime, which was measured at 90 K, by 5 ps was observed after annealing at 580 K. The measurement of the positron lifetime as a function of the sample temperature gave evidence of a significantly higher $\bar{\tau}$ at 400 K. The reason for the discrepancy is the activity of shallow positron traps. It could not be decided from the experimental data whether the shallow traps are antisites already present in as-grown material or irradiation-induced defects. The observed vacancy-type defect can be identified with a divacancy, either isolated or as part of a complex. Its density was estimated to be 1×10^{17} cm^{-3}. The defect-related lifetime was separated to be 310 ps, in the temperature range of negligible influence of shallow positron traps (Fig. 7.5). A reference experiment with annealing of an unirradiated sample at rather high temperatures showed that the increase in $\bar{\tau}$ at 580 K of the irradiated sample cannot be related to vacancy generation due to outdiffusion of one component. A Fermi-level shift was excluded as well. Hence, this increase was supposed to be related to a reaction of irradiation-induced defects.

7.3 Summary

Silicon carbide is not only interesting for new device applications, but also from the viewpoint of positron annihilation as a prospective field-assisted positron moderator. Open-volume defects were mainly studied after irradiation. Defect identification in the ternary compound $CuInSe_2$ is just at the beginning, and the first investigations with positron lifetime spectroscopy have been presented. Available positron lifetime data of silicon carbide and copper indium diselenide are summarized in Table 7.1.

8 Applications of Positron Annihilation in Defect Engineering

The specific control of defects is necessary in semiconductor processing to obtain the desired electrical and optical properties of semiconducting materials. The importance of defect characterization increases with the ongoing integration of devices. On the one hand, there are defects which have deteriorating properties and have to be removed from the active region. Ion-implantation-induced defects in planar technology are a typical example. On the other hand, defects are utilized to achieve specific device characteristics. The EL2 defect is used to obtain semi-insulating GaAs. Another example is the use of oxygen-related defects to getter transition metal impurities in silicon. The basic requirement for defect engineering is the application of experimental methods which allow for the identification and determination of defect concentration. Among the different techniques, positron annihilation is especially useful for the identification of open-volume defects, such as vacancies and their agglomerates. There are very few other methods which can specifically identify vacancy-type defects.

It has been demonstrated that positron annihilation can be applied in the field of non-destructive testing of metals (Meyendorf and Krause 1992; Somieski et al. 1994, 1995). The formation of vacancy agglomerates was found during fatigue and high-temperature creep of various steels already in a state where no macroscopic changes could be observed. Positron annihilation may therefore be suited to the prediction of failure of technological components. However, it has not become a routine method in industrial practice. This also holds for semiconductors, although many prospective applications in the field of defect engineering are conceivable. The characterization of defects in semiconductors by means of positron annihilation have already been demonstrated with many examples in Chaps. 4 to 7. In this chapter, possible fields of application of positron annihilation in semiconductor technology are specified without discussing experimental results in detail.

Positron annihilation is able to probe the lateral distribution of vacancy-type defects across a semiconductor wafer as first demonstrated by Dlubek et al. (1986a). The isotope source of conventional positron lifetime spectroscopy must be shifted to different positions. Using a slow-positron beam, the wafer is moved on a x–y stage with the advantage of automation. The lateral resolution corresponds to the size of the source or the beam diameter and amounts to about 1 to 5 mm. Positron microscopes are under development (Sect. 2.5.1) aiming at the improvement of resolution down to the micrometer range.

Technologically important defects in semiconductors, e.g. vacancy–dopant complexes or the metastable DX and EL2 centers, may be profiled in principle in this way. An attempt to correlate the distribution of vacancies with the U- or W-shaped profile of other defects in GaAs was made by Cavallini et al. (1997). If a whole sliced ingot is investigated in this way, the three-dimensional distribution of such defects may be studied. This was demonstrated by Krause et al. (1990a) for the distribution of mercury vacancies in a whole $Hg_{1-x}Cd_xTe$ crystal (Fig. 6.8).

Starting from the ingot, a high number of technological steps are necessary during the fabrication of a device. Additional defects may be introduced during these processes. An obvious case is the saw-cutting of wafers, which generates a near-surface damaged layer. The slow-positron-beam technique may be used to determine the depth profile of point defects (Sect. 2.4.4). Annealing steps during the fabrication process lead to the in-diffusion of impurities. The interaction with vacancies, e.g. the filling of vacancies, may be detected by positron annihilation. The in-diffusion of silver and the vanishing of cadmium vacancies was studied in CdTe by positron lifetime spectroscopy (Fig. 6.4). A vanishing vacancy signal indicated a Fermi level shift in undoped GaAs due to the in-diffusion of acceptor-type impurities (Fig. 5.12).

Post-growth heat treatment is used for whole ingots and wafers to improve the crystal quality (e.g. Mori et al. 1990). It is intended to homogenize the electrical properties of the material, to reduce the concentration of excess point defects, or to dissolve precipitates. It could be shown that the electrical and crystallographic homogeneity of InP wafers improved by annealing in a phosphorus atmosphere (Fornari et al. 1995). Wafer annealing is suited to adjusting the stoichiometry of compound semiconductors after the growth. Stoichiometry-related defects such as vacancies, interstitials, and antisites can be controlled in a defined way using annealing under a fixed vapor pressure of the more volatile component. Post-growth annealing experiments of Te-doped GaAs were carried out by Gebauer et al. (1997f) in a two-zone furnace with defined As vapor pressure (Sect. 5.2.5). Stoichiometry-related Ga vacancies were detected by positron lifetime spectroscopy. Their concentration could be adjusted by the heat treatment.

A very important processing step is ion implantation as the standard method of doping in planar technology. Ions with energies of the order of 100 keV are implanted in a depth of several hundreds of nanometers. They induce a damaged layer, which can be detected using slow positrons (Sects. 4.1.8 and 5.4.2.2). It has become a standard application of the slow-positron-beam technique to non-destructively profile the distribution of vacancy-type defects after ion implantation. This was mainly studied in Si (e.g. Eichler et al. 1997a; Simpson et al. 1991b; Uedono et al. 1994b). The annealing of these defects and occurring defect reactions can be studied in situ.

The growth of thin epitaxial layers also induces defects, especially when the growth temperature is far below the melting temperature of the semiconductor. A prominent example is low-temperature-grown GaAs (Sect. 5.2.4), which is, e.g., used as high-resistive buffer layer. Vacancy densities as high as 2×10^{18} cm^{-3} were detected with the slow-positron-beam technique. The thermal annealing of these defects was studied (e.g. Gebauer et al. 1997d). The combination of molecular-beam epitaxy with a slow-positron-beam system should allow the in-situ observation of the defect generation during growth and the subsequent defect annealing. The configuration of such a combination of MBE and a slow-positron beam was described by He and Hu (1995).

Open-volume defects, which are related to different crystallographic structures at phase boundaries and/or a lattice misfit, may be formed at the interface between the substrate and a cover layer, and they can be detected by positrons. Interface defects were, for example, found in the systems SiO_2/Si and Au/Si (Sects. 4.1.5 and 4.1.6). When the annihilation parameters are significantly different for the substrate and the layer, the thickness can be determined non-destructively using a slow-positron beam. An example is SiO_2 on Si. The thickness of SIMOX or MOS layers could be measured.

The degradation of semiconductor devices is an essential problem of the technology. The lifetime of laser diodes is reduced with the development of defect structures, which increasingly prevent the emission of light. One reason for degradation was found in the development of an extended dislocation network by climb (Ueda et al. 1982). In principle, it should be possible to detect vacancy-type defects connected to this process by positron annihilation. Investigations of point defects generated by dislocation motion have been carried out in plastically deformed semiconductors (Sects. 4.1.9, 5.2.7, and 5.4.3). The degradation of ZnO varistor ceramics was studied by positron lifetime spectroscopy (Fig. 6.11). A degradation model of such devices involving grain boundaries as a source and sink of Zn vacancies was proposed (Gupta and Carlson 1987).

Radiation-induced defects limit the durability of solar panels, especially during applications in space. A higher degree of integration may limit the lifetime of semiconductor devices due to radiation damage. Radiation resistance is an important issue for fast-particle detectors. It was shown that positron annihilation is a suitable tool for investigating defects introduced by irradiation (Sect. 4.1.7 and 5.2.6). The generation of Frenkel pairs after irradiation with different particles and subsequent defect reactions during thermal treatments have been shown in elemental and compound semiconductors. The annealing of irradiation-induced defects can already be monitored in the simplest experimental arrangement of positron annihilation, i.e. Doppler-broadening spectroscopy (Fig. 2.8). The positron techniques may support the development of radiation-resistant semiconductor material.

9 Comparison of Positron Annihilation with Other Defect-Sensitive Techniques

Positron annihilation is limited to the detection of special defect types. On the one hand, the unambiguous identification of open-volume defects may be possible. On the other hand, it is not possible to investigate the whole defect population of a crystal, since several defect types cannot be observed. An information gain is obtained by combination with other defect-sensitive methods. The results of positron annihilation may be complemented in this way, and a more complete picture of the defect chemistry of the crystal may be drawn.

The combination of positron annihilation with electrical or optical techniques is necessary for the interpretation of the results obtained. This becomes obvious from the fact that positrons are sensitive to the charge state of the defects under investigation. The charge state of defects having localized states in the bandgap is determined by the position of the Fermi level (see e.g. Fig. 4.2). Although vacancies are present, they may not be detected by positrons due to their positive charge. So, at least, information on the conductivity type of the semiconductors is required. The position of the Fermi level may change in the course of sample treatment, e.g. by irradiation, deformation, or annealing. The measurement of the carrier concentration is thus recommended in different states of the sample. A suitable method is the Hall effect measurement.

The trapping coefficient μ has been introduced as the sensitivity factor of positron annihilation (Sect. 3.2). A method independent of positron annihilation is necessary for the determination of μ. The trapping coefficient was determined for various defects in the most important semiconductors (Table 3.3, p. 103). Nevertheless, the trapping coefficient of a given defect should be measured anew in the material under investigation for the temperature range of interest. Reference methods, which allow the identification of the defect and the charge state, have to be applied.

Positron annihilation provides information on the size of the open volume of defects, such as vacancies or vacancy clusters. The inherent problem is, however, that only little information is obtained on the chemical surrounding by positron annihilation. This is the case if, e.g., the vacancy is part of a complex. In order to reveal the structure of such a complex, other techniques are useful.

Suitable methods to supplement positron annihilation are presented in this chapter. They are compared with positron annihilation with respect to the kind of defects detected, their sensitivity range, and application limits. The basic measurement principles are outlined; however, a comprehensive treatment is not intended. More specialized techniques are beyond the scope of the book and the reader is referred to other sources. For the sake of comparability, positron annihilation is also included in the list of methods.

Positron annihilation. The detection of defects by positron annihilation is based on their ability to capture positrons. The positrons are implanted in the sample, thermalize, and diffuse. They may be trapped during the diffusion in neutral or negatively charged open-volume defects. Other point defects may be detected at low temperatures if they carry a negative charge. Positron annihilation measurements can be divided into the lifetime and momentum distribution techniques. The positron lifetime is a function of the open volume of the defect and is measured in positron lifetime spectroscopy as the time difference between a start signal (e.g. the 1.27-MeV γ-quantum of the β^+-decay) and the emission of the 511-keV annihilation γ-quanta. The momentum techniques are based on the conservation of momentum during annihilation. The momentum of the annihilating electron–positron pair results in a Doppler broadening of the annihilation line (Doppler-broadening spectroscopy) and a deviation from collinearity of the two annihilation γ-quanta (angular correlation of annihilation radiation). The trapping of positrons in defects is indicated by characteristic changes in annihilation parameters, e.g. the positron lifetime or annihilation line shape parameters. The basic defect types detected are open-volume defects, such as vacancies, vacancy clusters (also as part of defect complexes), dislocations, and grain boundaries. Further defects that may be detected are negative ions and precipitates. The sensitivity range for negatively charged vacancies can be given from about 5×10^{15} to 5×10^{18} cm^{-3} at room temperature. The detection ranges of the other defects and their temperature dependence are discussed in Sect. 3.2. The temperature dependence of positron trapping is the reason that certain defects can only be detected in a limited temperature range. When using unmoderated positrons with a broad energy distribution, only bulk samples can be studied, because of the large penetration depth. The moderation of positrons allows the formation of a monoenergetic beam, which is applicable to the investigation of thin layers and near-surface defects. Defect depth profiling can be carried out by variation of the incident positron energy. Positron annihilation is in practice a non-destructive method and can be applied to a broad range of materials.

Electron paramagnetic resonance. EPR utilizes the resonant absorption of electromagnetic radiation by a system of unpaired electrons placed in a magnetic field (for a review see e.g. Gehlhoff et al. 1992 and Spaeth et al. 1992). The ground state of the system splits in an external static magnetic field into $2S^* + 1$ Zeeman levels, where S^* is the effective spin obtained from the coupling of the spin of the unpaired electron with the orbital angular momentum in the magnetic field. This splitting is detected by electron paramagnetic resonance spectroscopy. The dependence of EPR signals on the direction of the magnetic field with respect to the crystal orientation allows one to determine the symmetry of a defect. The interaction of the unpaired spin with the magnetic moment of the nucleus of the central atom or the magnetic moments of the nuclei of the surrounding atoms leads to the hyperfine or ligand hyperfine interaction, respectively. The chemical nature and the lattice position of the defect may frequently be identified from this information.

The sample is measured in a microwave cavity, and a varying magnetic field and a constant microwave energy are applied. The magnetic field is modulated with an alternating field for the phase-sensitive detection of the impedance of the cavity. In this way, a spectrum of the resonant absorption of the microwave is obtained. There are other techniques, which are based on the same physical principle, but the kind of signal detected is different. The most important variants are the resonance techniques using optical detection, e.g. luminescence and infrared absorption (Spaeth et al. 1992).

The method of electron paramagnetic resonance is limited to systems of unpaired electrons. Defects with a level in the bandgap may be investigated using photoinduced ionization. This technique is called photo-EPR. Electron paramagnetic resonance thus represents a universal method for the investigation of the structure of point defects in semiconductors. Practically all important point defects in silicon could be characterized. The configuration of vacancy-type defects, such as monovacancies, vacancy clusters, and vacancy–impurity complexes, was established by electron paramagnetic resonance (e.g. Corbett and Bourgoin 1975; Watkins 1991). This method is thus a very important reference method for positron annihilation in Si.

In order to give some typical numbers for the detection limit of EPR, the density of unpaired spins must be at least of the order of 10^{14} to 10^{16} cm^{-3}. This number depends on the details of the spin system under investigation. The conductivity of the sample is a limiting factor, because the microwave does not penetrate the sample due to the skin effect at high carrier concentrations. The density of free carriers should be well below 10^{18} cm^{-3}.

Although electron paramagnetic resonance has been very successful in the characterization of defects in Si, its application to III–V compound semiconductors is more difficult. This is mainly due to the high abundance of lattice nuclei with nuclear spin unequal to zero, which causes, via ligand-hyperfine interactions, broad unresolved lines. This limits the possibilities of determination of defect structures, and more sophisticated methods, e.g. electron nuclear double resonance, have to be applied (Spaeth et al. 1992).

Rutherford backscattering. High-energy ions, such as He^+ or H^+, implanted into a solid are scattered mainly at small angles due to the interaction with the nuclei. A small fraction of the probe ions hits the nucleus almost centrally and is scattered at high angles up to 180°. This is called Rutherford backscattering. The number of backscattered ions is significantly smaller if the ion beam is directed parallel to a low-indexed crystallographic direction. Under this channeling condition, lattice defects, especially interstitials, can be detected by the increase in the backscattering yield. RBS represents a suitable method to study ion-implantation-induced defects up to the amorphization of the material. The determination of defect profiles in a depth range of up to several micrometers is possible by the detection of the energy of the backscattered ions (Götz and Gärtner 1988).

The ions are usually accelerated in a van de Graaff generator or an electrostatic Cockcroft–Walton stage. The energies are in a range from several hundreds of keV to some MeV. The backscattered ions are often detected by solid state detectors. The whole energy range can be monitored at once by multi-channel analysis of the detector signal.

The defects which can be primarily observed by Rutherford backscattering are those which occupy interstitial positions in the channel. Open-volume defects are barely detectable, possibly via lattice distortion and the resulting dechanneling. RBS is suited for the detection of heavy dopant atoms. The sensitivity to interstitials, either native interstitials or implanted atoms, ranges in atomic concentrations from 0.01 to 1. The charge state of the defect does not play any role. A closer identification of the defect or defect complex is hardly possible. Its applicability as a supplemental method to positron annihilation is that complementary defects are detected. The depth profiles of open-volume defects obtained from positron annihilation may be compared with RBS spectra. As only very high defect concentrations can be detected, its application in the defect analysis of semiconductors is mainly restricted, e.g., to the investigation of ion-implantation-induced damage or to defect-rich non-stoichiometric layers.

Precision lattice parameter and density measurements. The combination of the measurement of the lattice parameter by X-ray diffraction with the precise determination of the mass density allows the determination of the point defect concentration in a crystal. The reason is that point defects, such as vacancies or interstitials, influence the density of the crystal but have only a small effect on the lattice parameter. The lattice constant is measured by the method of Bond (1960) or in a double-crystal X-ray diffractometer. The density measurement uses the hydrostatic buoyancy measured by a precision balance. The volume measurement is carried out by immersing the sample in liquids of different densities. The main problem is the complete wetting of the surface. Another procedure is the Simmons–Balluffi method, where instead of the density the relative change in the length of the sample is measured, preferably by laser dilatometry (Kluin and Hehenkamp 1991).

The method is suitable for point defects that have an influence on the mass density of the crystal. In this way, Schottky defects, but no Frenkel pairs, should be measured. Antisite defects in compounds are rather unlikely to be found unless the atomic mass difference of the components is distinct. The combination of both techniques has been applied for the measurement of stoichiometry deviations of compound semiconductors, and thus the determination of the existence region in the phase diagram was attempted.

The defect concentration necessary for the detection by precise lattice constant and density measurements was estimated by Morozov and Bublik (1986) to be in the range 5×10^{17} to 3×10^{18} cm^{-3}, depending on the type of the semiconductor and the kind of the defect. This results from the accuracy of the X-ray determination of the lattice constant of 1×10^{-6} nm and of the error of the density measurement of 2×10^{-5} g cm^{-3}.

No direct defect identification can be carried out. Defects are observed as the sum of their influence on the density and the lattice parameter. The presence of several defect types may influence the measurement, e.g. vacancies and interstitials partly cancel out the effect. It is necessary to assume a certain dominant defect type for the numerical evaluation. The defect concentration obtained is very sensitive to the defect structure supposed.

Diffuse X-ray scattering. Elastic X-ray scattering occurs in perfect crystals only if the Bragg condition is fulfilled, i.e. the scattering vector equals a reciprocal lattice vector. A measurable intensity is also detected outside a Bragg reflection in the presence of lattice imperfections. The elastic reflections are superimposed on an inelastic background, which is the result of the incomplete separation of the interfering partial waves of the scattering atoms. The intensity of the diffuse scattering, also called Huang scattering, is proportional to the number of a certain defect. Two contributions determine the intensity, viz. a structure factor, arising from the chemical nature of the lattice imperfection, and the contribution from the displacement field of the lattice surrounding the defect. The measurement of the angular dependence of the scattered intensity can be used in conjunction with theoretical calculations to obtain details of the defect structure (Dederichs 1973).

Such defects, where the structure factor changes or a lattice distortion occurs, can be detected in principle. Diffuse X-ray scattering is one of the few suitable techniques to characterize interstitials. The lattice structure of such defects was investigated in metals and semiconductors. Split interstitial configurations and the arrangement of close Frenkel pairs was revealed in electron-irradiated materials. Moreover, the method is especially sensitive for point defect clusters. The disadvantage is that the defect density must be at least 10^{18} cm^{-3}, which can be reached in semiconductors mostly only after strong damage, e.g. high-dose electron irradiation. No correlated measurements of positron annihilation and diffuse X-ray scattering have been made up to now for semiconductors. However, such a correlation could be useful for the investigation of annealing stages after low-temperature electron irradiation, because in addition to vacancies further defects and their annealing may become visible in X-ray scattering.

Transmission electron microscopy. Imaging with TEM is based on the diffraction of electrons in crystals. Bright or dark field imaging using the direct or one diffracted beam is based on the local deviation from the Bragg condition in the strain field around defects (Bethge and Heydenreich 1987). The visualization of crystal structures in high-resolution transmission electron microscopy is a true Abbe picture as superposition of the diffracted electron beams. The available resolution depends on the accelerating voltage of the electrons and the spherical aberration of the objective lens. Good microscopes achieve a point-to-point resolution of 0.18 nm.

As a reference method for positron annihilation, TEM is applied to the study of extended defects, e.g. dislocations, dislocation loops, voids, and precipitates. It is not suited to point defects. Line defects far below the detection limit of positron annihilation can be imaged. The problem of comparing dislocation densities obtained from other methods with TEM is the statistical relevance of the latter, as only a sample volume of about 10^{-8} cm^3 can be investigated. If defects are inhomogeneously distributed, the defect densities determined by transmission electron microscopy and positron annihilation may differ.

Extended defects with densities that can be detected by positron annihilation should be visible in the TEM diffraction contrast, provided their size exceeds the extinction length. Precipitates may be even smaller if they exhibit an extended strain field, as frequently observed for coherent aggregates. Objects of atomic dimensions can be visualized in high-resolution images, but their density should usually be as high as 10^{17} cm^{-3} to be recognized. The structure of defects is not directly visible, but can only be identified in comparison with image simulation. Even with image processing it is, however, hardly possible to identify small voids in high-resolution images, although such attempts have been made in the past.

In the case of precipitates, it is rather important for positron annihilation to know whether the particle is coherent or incoherent. TEM is a suitable method to directly observe the interface structure of precipitates.

Scanning tunneling microscopy. This method is ideally suited to investigate the surface structure of solids. Individual defects can be imaged with atomic resolution (Chen 1993; Wiesendanger and Güntherodt 1996). Recently, the first experiments have been carried out to characterize lattice defects in the bulk of semiconductors (Feenstra et al. 1993; Zheng et al. 1994). For this purpose, the sample is cleaved and scanned in ultra-high vacuum. It is useful to carry out the investigation at sufficiently low temperatures in order to avoid defect migration on the surface. It could be shown, at least for the GaAs (110) plane, that the defects were not introduced during the cleavage process (Gebauer et al. 1997c). The reversal of the tunneling bias allows the observation of occupied and empty electronic states and thus the identification of the sublattice. Furthermore, this may help to recognize the chemical nature of impurities.

All defects can be in principle studied on an atomic scale. In contrast to high-resolution electron microscopy it is possible to visualize vacancies and to distinguish them from vacancy–impurity complexes. This is the main advantage in comparison with positron annihilation. STM is not restricted to point defects, as small precipitates and dislocations can be imaged as well.

A defect density of at least 10^{17} cm^{-3} is required for the investigation of point defects. Nevertheless, a high effort is necessary, since many pictures have to be evaluated for a statistically reliable value of the defect concentration. The reason is the limited field of view of scanning tunneling microscopy. Another disadvantage is that the technique is only suited for cleavage planes, which are not reconstructed. Such a reconstruction is, however, frequently observed.

Infrared spectroscopy. The optical absorption in semiconductors is determined by the width of the bandgap. In the presence of point defects, there are, in addition to band–band transitions, further transitions from/to localized energy states in the bandgap. The energy position of absorption bands provides information on the location of defect levels in the bandgap. The spectroscopy of the levels can be carried out in two ways (Pankove 1971). The measured quantity in optical absorption spectroscopy, as the first technique, is the absorption coefficient determined by Lambert's law. In many cases, only the absorbance is measured. The second technique is photoluminescence spectroscopy, detecting the emitted light of optical transitions excited by illumination (Bebb and Williams 1972). If the external excitation is carried out by an electron beam inside a scanning electron microscope (cathodoluminescence), the spectral information can be coupled with a lateral resolution as low as 0.1 μm (Yacobi and Holt 1990).

The form of absorption or luminescence bands is mainly determined by the coupling of the electronic states of the defects to the lattice (electron–phonon coupling). The strength of the detected signal is proportional to the defect concentration. The prefactor is the optical cross-section, which is mainly determined by the oscillator strength. It cannot be calculated in many cases, so that a calibration, e.g. by deep level transient spectroscopy, must be carried out for the quantitative measurement of defect concentrations by infrared spectroscopy.

Absorption bands are not only observed for electronic transitions, but also for local vibrational excitations, when the energy of the incident light equals the energy of local vibrational modes (LVM) of defects. This method provides a direct fingerprint for defect identification (Newman 1993). The methods of optical spectroscopy are valuable tools for the detection of impurities or impurity complexes by their LVM or by the chemical shift of the bound-exciton luminescence. Other point defects with levels in the bandgap are detectable without providing closer information on their structure. The sensitivity is very high and the detection limit can be as low as 10^{10} cm^{-3}. The methods have thus found a broad application to many defect issues in semiconductors and they are the most sensitive techniques for investigation of defects. The disadvantage of optical spectroscopy is the necessity of calibration for the quantitative measurement of defect concentrations. A problem in the direct correlation with results of positron annihilation is that the charge state of the defect cannot be directly determined.

Deep level transient spectroscopy. DLTS is the most versatile method for the characterization of defects having levels deep in the bandgap. Defects in the depletion region of a p–n or Schottky junction under reverse bias influence the capacitance of the junction (Bräunlich 1979). A small voltage pulse injects majority or minority carriers, depending on the polarity. This induces a change in the charge state of defect levels in the depletion region. The discharging of the defects after the decay of the pulse depends on temperature due to thermal excitation. It is measured as the transient of the capacitance as a function of temperature. A maximum in the change of the capacitance ΔC occurs at a certain temperature, which is a measure of the position of the ionization level in the bandgap. The magnitude of ΔC is proportional to the defect concentration, which can be calculated directly.

Deep level transient spectroscopy is able to detect deep traps with levels in the bandgap in the range from about 0.15 to 0.8 eV as the distance from one of the band edges. Such traps can be impurities, intrinsic point defects, or dislocations. The energy level, the capture cross-section, and the trap concentration can be measured. Capacitance changes as low as 10^{-5} pF can be measured, corresponding to a lower sensitivity limit of the defect density of about 10^{10} cm^{-3}. The depth of the space charge region can be shifted with the double-pulse technique. A profiling of the defect distribution is possible in a range comparable to the information range of the slow-positron-beam technique. Lateral resolution can be achieved using excitation by an electron beam (scanning deep level transient spectroscopy, Breitenstein and Heydenreich 1985).

The information obtained by DLTS is restricted to electronic properties. The atomic structure of a defect cannot be identified. Such defects, which have only shallow levels in the bandgap or resonant states in the bands, cannot be detected. The method requires the preparation of a Schottky or p–n junction, so the subsequent use of the samples for other characterization techniques is limited. Problems arise if the defect concentration is very high, e.g. after ion implantation, and individual DLTS peaks can no longer be resolved.

Hall effect measurement. The basic principle is the formation of a Hall voltage perpendicular to the direction of the current through the sample and the direction of an external magnetic field. It is a well-established method in semiconductor physics for the determination of the concentration of free carriers. It is frequently carried out for plane-parallel samples of arbitrary shape with the method of van der Pauw (1958). The preparation of Ohmic contacts is required. The position of the Fermi level, which is an important quantity for the interpretation of positron data, can be calculated from the carrier concentration. If the carrier concentration is determined by the ionization of one single defect level, the defect concentration can be directly obtained.

The measurement of the Hall effect as a function of the temperature may provide information on ionization levels and concentration of defects. It is restricted to samples where the Fermi level can be significantly shifted by the temperature, e.g. in GaAs with carrier concentrations in a range of about 10^{14} to 10^{16} cm^{-3}. The shift of the Fermi level across a defect level in the bandgap leads to a change in the measured Hall coefficient. Ionization levels with a distance to the band edge from several to several hundreds of meV can be observed. The method is thus complementary to DLTS, as mainly shallow levels are the subject of the investigation. The determination of defect levels by temperature-dependent Hall effect measurements is, however, not a standard method, since it can only be applied in special cases.

Appendix

A1 Semiconductor Data

Table A.1. Some important data of elemental and compound semiconductors. The values are taken from Lide (1996) and Madelung (1982a, 1982b, 1985).

Material	Structure	a_0 [nm]	ρ [g cm^{-3}]	N_{at} [10^{22} cm^{-3}]	T_m [K]	E_g [eV]	τ_b [ps]
Si	diamond	0.5430	2.329	4.99	1685	1.107 i	218
Ge	diamond	0.5658	5.323	4.42	1210	0.664 i	228
C	diamond	0.3567	3.515	17.62	4100*	5.48 i	105
α-SiC (6H)	wurzite	*a* 0.3082 *c* 1.5118	3.211	9.44	3103	2.86 i	142
β-SiC (3C)	sphalerite	0.4359	3.210	9.44	3070	2.2 i	—
GaAs	sphalerite	0.5653	5.316	4.43	1513	1.425 d	229
GaP	sphalerite	0.5451	4.130	4.94	1740	2.27 i	223
GaN	wurzite	*a* 0.3190 *c* 0.5190	6.095	8.77	1973	3.44 d	166
GaSb	sphalerite	0.6096	5.614	3.53	985	0.70 d	253
InAs	sphalerite	0.6058	5.667	3.60	1215	0.36 d	246
InP	sphalerite	0.5869	4.810	3.97	1335	1.35 d	241
InSb	sphalerite	0.6479	5.775	2.94	800	0.18 d	—
CdTe	sphalerite	0.6481	5.870	2.95	1092	1.43 d	280
HgTe	sphalerite	0.6461	8.070	2.96	943	-0.141 d	274
ZnS	sphalerite	0.5409	4.087	5.05	2103	3.56 d	230
ZnSe	sphalerite	0.5668	5.264	4.39	1793	2.7 d	240
ZnO	wurzite	*a* 0.3252 *c* 0.5213	5.675	4.40	2250	3.2 d	183
$CuInSe_2$	chalcopyrite	*a* 0.5785 *c* 0.1156	5.77	4.13	1260	1.04** d	235

a_0 is the lattice constant at room temperature, ρ the mass density, N_{at} the total number of atoms per cm^3, T_m the melting temperature, E_g the distance between the valence band maximum and the conduction band minimum at room temperature (d stands for direct, i for indirect semiconductor), and τ_b the most reliable experimental positron bulk lifetime.
*at 125 kbar, **at 77 K.

A2 Trapping Model Equations

Trapping models for various types of positron traps have been given in Sect. 3.2.1. A systematic approach for k non-interacting deep positron traps is presented here. Furthermore, the special cases of $k = 1$ and 2 are given separately. It should be noted that a notation different from that used throughout the book is given in this Appendix. If not stated otherwise, the index i runs from 1 to k and the index j from 0 to k throughout the Appendix.

The set of differential equations describing the annihilation from the defect-free bulk state b and from the defects i is

$$\begin{aligned}
\frac{\mathrm{d}}{\mathrm{d}t}n_\mathrm{b}(t) &= -(\lambda_\mathrm{b} + \kappa_1 + \kappa_2 + \ldots + \kappa_k)n_\mathrm{b}, \\
\frac{\mathrm{d}}{\mathrm{d}t}n_1(t) &= \kappa_1 n_\mathrm{b} - \lambda_1 n_1, \\
\frac{\mathrm{d}}{\mathrm{d}t}n_2(t) &= \kappa_2 n_\mathrm{b} - \lambda_2 n_2, \\
&\vdots \\
\frac{\mathrm{d}}{\mathrm{d}t}n_k(t) &= \kappa_k n_\mathrm{b} - \lambda_k n_k,
\end{aligned} \tag{A.1}$$

which can be written in matrix form as $\dot{\mathbf{Y}} = \mathbf{A}\mathbf{Y}$, where

$$\mathbf{Y} = \begin{pmatrix} n_\mathrm{b} \\ n_1 \\ n_2 \\ \vdots \\ n_k \end{pmatrix}, \quad \mathbf{A} = \begin{pmatrix} -\lambda_0 & 0 & 0 & \cdots & 0 \\ \kappa_1 & -\lambda_1 & 0 & \cdots & 0 \\ \kappa_2 & 0 & -\lambda_2 & \cdots & 0 \\ \vdots & \vdots & \vdots & \ddots & \vdots \\ \kappa_k & 0 & 0 & 0 & -\lambda_k \end{pmatrix}. \tag{A.2}$$

The starting conditions at time $t = 0$ for the number of positrons in the states n_b and n_i after the thermalization of the positron are $n_\mathrm{b}(t = 0) = 1$ and $n_i(t = 0) = 0$. For the decay spectrum

$$D(t) = n_\mathrm{b}(t) + \sum_{i=1}^{k} n_i(t), \tag{A.3}$$

one obtains from the solution of the system of linear differential equations (A.1) by standard methods

$$D(t) = \sum_{j=0}^{k} I_j \exp(-\lambda_j t). \tag{A.4}$$

For k defects, the positron lifetime spectrum consists of $k + 1$ components.

Since the characteristic polynomial of (A.2) factorizes, the determination of the eigenvalues is straightforward. Problems arise if detrapping from a shallow positron trap has to be taken into account, leading to additional off-diagonal elements in **A**. The solution of the differential equations with one shallow positron trap has been given in Sect. 3.2.1.2. To obtain the eigenvalues, one has to solve a third-order polynomial if two shallow positron traps have to be considered.

From the general definition of the average positron lifetime,

$$\bar{\tau} = \frac{\int_0^\infty t \frac{\mathrm{d}D(t)}{\mathrm{d}t} \mathrm{d}t}{\int_0^\infty \frac{\mathrm{d}D(t)}{\mathrm{d}t} \mathrm{d}t}, \tag{A.5}$$

one obtains with (A.4)

$$\bar{\tau} = \sum_{j=0}^{k} \frac{I_j}{\lambda_j} = \sum_{j=0}^{k} I_j \tau_j ,$$

which can also be written with the annihilation fraction η_i

$$\bar{\tau} = \eta_0 \tau_\mathrm{b} + \sum_{i=1}^{k} \eta_i \tau_i . \tag{A.6}$$

By determining the eigenvalues and eigenvectors one obtains from the general solution (A.4) the annihilation rates λ_j, which are the reciprocal of the positron lifetimes τ_j, and the corresponding intensities I_j,

$$\tau_0 = 1/\lambda_0 = \left(\lambda_\mathrm{b} + \sum_{i=1}^{k} \kappa_i \right)^{-1}, \quad \tau_1 = 1/\lambda_1, \quad \tau_2 = 1/\lambda_2, \quad \ldots, \quad \tau_k = 1/\lambda_k, \tag{A.7}$$

$$I_0 = 1 - \sum_{i=1}^{k} I_i, \quad I_i = \frac{\kappa_i}{\lambda_0 - \lambda_i} = \frac{\kappa_i}{1/\tau_0 - 1/\tau_i} = \kappa_i \left(\lambda_\mathrm{b} + \sum_{j=1}^{k} \kappa_j - \lambda_i \right)^{-1} . \tag{A.8}$$

The trapping rates κ_i are obtained from the equations above as

$$\kappa_i = I_i(\lambda_0 - \lambda_i) = I_i(1/\tau_0 - 1/\tau_i), \tag{A.9}$$

while the annihilation fractions are

$$\eta_0 = 1 - \sum_{i=1}^{k} \eta_i \tag{A.10}$$

and

$$\eta_i = \kappa_i \tau_b \left(1 + \tau_b \sum_{j=1}^{k} \kappa_j\right)^{-1} = \kappa_i \left(\lambda_b + \sum_{j=1}^{k} \kappa_j\right)^{-1} . \tag{A.11}$$

An important equation for the positron bulk lifetime to be calculated from experimental data is

$$\tau_b^{\text{calc}} = \left(\sum_{j=0}^{k} I_j \lambda_j\right)^{-1} = \left(\sum_{j=0}^{k} I_j / \tau_j\right)^{-1} , \tag{A.12}$$

which can be used for comparison with the independently determined positron bulk lifetime τ_b. A similar equation holds for the reduced bulk lifetime,

$$\tau_0^{\text{calc}} = I_0 \left(\lambda_b - \sum_{i=1}^{k} I_i \lambda_i\right)^{-1} = I_0 \left(1/\tau_b - \sum_{i=1}^{k} I_i / \tau_i\right)^{-1} . \tag{A.13}$$

Both equations can be used to check the validity of the assumptions of the trapping model and the number of components used.

The standard deviation of the trapping rates κ_i and annihilation fractions η_i are obtained via error propagation from the statistical errors of measured values as follows,

$$\sigma_{\kappa_i}^2 = \frac{\kappa_i^2}{I_i^2} \sigma_{I_i}^2 + I_i^2 \left(\sigma_{\lambda_0}^2 + \sigma_{\lambda_i}^2\right), \tag{A.14}$$

$$\sigma_{\eta_i}^2 = \frac{\eta_i^2}{\kappa_i^2} \left[(1 - 2\eta_i)\sigma_{\kappa_i}^2 + \sum_{j=0}^{k} \eta_i^2 \sigma_{\kappa_j}^2\right]. \tag{A.15}$$

Written with the positron lifetimes instead of the annihilation rate one obtains

$$\sigma_{\kappa_i}^2 = \frac{\kappa_i^2}{I_i^2} \sigma_{I_i}^2 + I_i^2 \left(\frac{1}{\tau_0^4} \sigma_{\tau_0}^2 + \frac{1}{\tau_i^4} \sigma_{\tau_i}^2\right). \tag{A.16}$$

The errors of the calculated reduced bulk lifetime τ_0^{calc} and the calculated positron bulk lifetime τ_b^{calc} are

$$\sigma_{\tau_0^{\text{calc}}}^2 = \left(\frac{\left(\tau_0^{\text{calc}}\right)^2}{I_0}\right)^2 \left(\sigma_{\lambda_b}^2 + \sum_{i=1}^{k} I_i^2 \sigma_{\lambda_i}^2 + \sum_{j=0}^{k} \lambda_j^2 \sigma_{I_j}^2\right), \tag{A.17}$$

with $\lambda_0 = 1/\tau_0^{\text{calc}}$, or

$$\sigma_{\tau_0^{\text{calc}}}^2 = \left(\frac{\left(\tau_0^{\text{calc}}\right)^2}{I_0}\right)^2 \left(\frac{1}{\tau_b^4} \sigma_{\tau_b}^2 + \sum_{i=1}^{k} \frac{I_i^2}{\tau_i^4} \sigma_{\tau_i}^2 + \sum_{j=0}^{k} \frac{1}{\tau_j^2} \sigma_{I_j}^2\right), \tag{A.18}$$

and

$$\sigma^2_{\tau_b^{calc}} = \left(\tau_b^{calc}\right)^4 \sum_{j=0}^{k} (\lambda_j^2 \sigma_{I_j}^2 + I_j^2 \sigma_{\lambda_j}^2), \tag{A.19}$$

or

$$\sigma^2_{\tau_b^{calc}} = \left(\tau_b^{calc}\right)^4 \sum_{j=0}^{k} \left(\frac{1}{\tau_j^2} \sigma_{I_j}^2 + \frac{I_i^2}{\tau_j^4} \sigma_{\tau_j}^2 \right). \tag{A.20}$$

The special case of two non-interacting deep positron traps can be deduced from the general expressions (A.7 to A.8),

$$\tau_0 = 1/\lambda_0 = \frac{1}{\lambda_b + \kappa_1 + \kappa_2}, \quad \tau_1 = 1/\lambda_1, \quad \tau_2 = 1/\lambda_2, \tag{A.21}$$

$$I_0 = 1 - I_1 - I_2, \quad I_1 = \frac{\kappa_1}{\lambda_0 - \lambda_1}, \quad I_2 = \frac{\kappa_2}{\lambda_0 - \lambda_2}. \tag{A.22}$$

It follows for the trapping rates

$$\kappa_1 = I_1(\lambda_0 - \lambda_1), \quad \kappa_2 = I_2(\lambda_0 - \lambda_2), \tag{A.23}$$

which are often given in the form

$$\begin{aligned} \kappa_1 &= \frac{I_1(\lambda_b - \lambda_1) + I_1 I_2 (\lambda_1 - \lambda_2)}{I_0}, \\ \kappa_2 &= \frac{I_2(\lambda_b - \lambda_2) + I_1 I_2 (\lambda_2 - \lambda_1)}{I_0}. \end{aligned} \tag{A.24}$$

It holds for the annihilation fractions

$$\eta_0 = 1 - \eta_1 - \eta_2, \tag{A.25}$$

$$\begin{aligned} \eta_1 &= \frac{\kappa_1 \tau_b}{1 + \tau_b(\kappa_1 + \kappa_2)} = \frac{\kappa_1}{\lambda_b + \kappa_1 + \kappa_2}, \\ \eta_2 &= \frac{\kappa_2 \tau_b}{1 + \tau_b(\kappa_1 + \kappa_2)} = \frac{\kappa_2}{\lambda_b + \kappa_1 + \kappa_2}. \end{aligned} \tag{A.26}$$

The solution for the simple case of one deep positron trap is given in the following. In this case, the equations can be partly expressed by the average positron lifetime, which can be experimentally determined with high accuracy.

$$\tau_0 = 1/\lambda_0 = \frac{1}{\lambda_b + \kappa}, \quad \tau_1 = 1/\lambda_1, \tag{A.27}$$

$$I_0 = 1 - I_1, \quad I_1 = \frac{\kappa}{\lambda_0 - \lambda_1} = \frac{\kappa}{\lambda_b + \kappa - \lambda_1}, \tag{A.28}$$

$$\kappa = \frac{I_1}{I_0}(\lambda_b - \lambda_1) = I_1(1/\tau_0 - 1/\tau_1), \tag{A.29}$$

$$\eta = \frac{\kappa}{\lambda_b + \kappa}. \tag{A.30}$$

Using the average positron lifetime one obtains

$$\kappa = \frac{1}{\tau_b}\frac{\bar{\tau} - \tau_b}{\tau_1 - \bar{\tau}}, \quad \eta = \frac{\bar{\tau} - \tau_b}{\tau_1 - \tau_b}. \tag{A.31}$$

This leads to the relation between the trapping rate and the annihilation fraction

$$\kappa = \frac{1}{\tau_b}\frac{\eta}{1-\eta}. \tag{A.32}$$

The error of the trapping rate is

$$\sigma_\kappa^2 = \left(\frac{\kappa}{\tau_b} + \frac{\kappa}{\bar{\tau} - \tau_b}\right)^2 \sigma_{\tau_b}^2 + \left(\frac{\kappa}{\bar{\tau} - \tau_b} + \frac{\kappa}{\tau_1 - \bar{\tau}}\right)^2 \sigma_{\bar{\tau}}^2 + \left(\frac{\kappa}{\tau_1 - \bar{\tau}}\right)^2 \sigma_{\tau_1}^2, \tag{A.33}$$

and for the annihilation rate

$$\sigma_\kappa^2 = \left(\frac{\eta}{\tau_1 - \tau_b} - \frac{\eta}{\bar{\tau} - \tau_b}\right)^2 \sigma_{\tau_b}^2 + \left(\frac{\eta}{\bar{\tau} - \tau_b}\right)^2 \sigma_{\bar{\tau}}^2 + \left(\frac{\eta}{\tau_1 - \tau_b}\right)^2 \sigma_{\tau_1}^2. \tag{A.34}$$

References

Abgarjan T. (1994): Dissertation, Martin-Luther-Universität Halle Wittenberg
Adams M., Mascher P., Kitai A.H. (1995): Appl. Phys. A **61**, 217
Aers G.C. (1990): AIP Conf. Proc. **218**, 171
Akahane T., Chiba T., Shiotani N., Tanigawa S., Mikado T., Suzuki R., Chiwaki M., Yamazaki T., Tomimasu T. (1990): Appl. Phys. A **51**, 146
Akhiezer A.I., Berestetskii V.B. (1965): *Quantum electrodynamics*. Wiley, New York
Alatalo M., Barbiellini B., Hakala M., Kaupinen H., Korhonen T., Puska M.J., Saarinen K., Hautojärvi P., Nieminen R.M. (1996): Phys. Rev. B **54**, 2397
Alatalo M., Kauppinen H., Saarinen K., Puska M.J., Mäkinen J., Hautojärvi P., Nieminen R.M. (1995): Phys. Rev. B **51**, 4176
Alatalo M., Nieminen R.M., Puska M.J., Seitsonen A.P., Virkkunen R. (1993): Phys. Rev. B **47**, 6381
Alexander H. (1986): In *Dislocations in solids*. Nabarro F.R.N. (ed.). North-Holland, Amsterdam, p. 113
Al-Kotb M.S., Puff W., Mohsen M. (1997): Mater. Sci. Forum **255–257**, 509
Allen J.W. (1995): Semicond. Sci. Technol. **10**, 1049
Ambigapathy R., Corbel C., Hautojärvi P., Manuel A.A., Saarinen K. (1996): Appl. Phys. A **62**, 529
Ambigapathy R., Manuel A.A., Hautojärvi P., Saarinen K., Corbel C. (1994): Phys. Rev. B **50**, 2188
Andersen G.Y., Dyakin V.V., Zaitov F.A., Koval V.V., Lyubchenko A.V., Silant'ev V.I., Fedchenko R.G. (1987): Inorg. Mater. **23**, 1149
Andersen J.U., Augustyniak W.M., Uggerhoj E. (1971): Phys. Rev. B **3**, 705
Anderson C.D. (1932): Science **76**, 238
Aniel F., Jin-Delorme Y., Crozat P., de Lustrac A., Adde R. (1993): Electron. Lett. **29**, 1570
Anwand W., Brauer G., Coleman P.G., Goodyear A., Reuther H., Maser K. (1997): J. Phys. Cond. Mat. **9**, 2947
Aref'ev K.P., Brudnyi V.N., Budnitskii D.L., Vorob'ev S.A., Tsoi A.A. (1979): Sov. Phys. Semicond. **13**, 669
Aref'ev K.P., Lopatin V.S., Pogrebnyak A.D. (1983): Sov. Phys. J. **26**, 555
Aref'ev K.P., Vorob'ev S.A., Lopatin V.S., Pogrebnyak A.D., Razov V.I., Rudnev A.S. (1985): Sov. Phys. Sol. State **27**, 1613
Arefiev K.P., Tsoi A.A., Vorobiev S.A. (1978): phys. stat. sol. (a) **47**, K149
Arefiev K.P., Vorobiev S.A., Prokopiev E.P., Tsoi A.A. (1977): Sov. Phys. Sol. State **19**, 1369
Arifov P.U., Arutyunov N.Y., Emtsev V.V., Il'yasov A.Z. (1979): Sov. Phys. Semicond. **13**, 914
Arthur J.R. (1967): J. Phys. Chem. Solids **28**, 2257
Asoka-Kumar P. (1997): Mater. Sci. Forum **255–257**, 166
Asoka-Kumar P., Alatalo M., Ghosh V.J., Kruseman A.C., Nielsen B., Lynn K.G. (1996): Phys. Rev. Lett. **77**, 2097
Asoka-Kumar P., Gossmann H.J., Unterwald F.C., Feldman L.C., Leung T.C., Au H.L., Talyanski V., Nielsen B., Lynn K.G. (1993): Phys. Rev. B **48**, 5345
Asoka-Kumar P., Lynn K.G., Leung T.C., Nielsen B., Rubloff G.W., Weinberg Z.A. (1991): Phys. Rev. B **44**, 5885
Asoka-Kumar P., Lynn K.G., Welch D.O. (1994): J. Appl. Phys. **76**, 4935

Au H.L., Ling C.C., Panda B.K., Lee T.C., Beling C.D., Fung S. (1994): Phys. Rev. Lett. **73**, 2732

Avalos V., Dannefaer S. (1996): Phys. Rev. B **54**, 1724

Baker J.A., Coleman P.G. (1989): In *Positron annihilation.* Dorikens-Vanpraet L., Dorikens M., Segers D. (eds.). World Scientific, Singapore, p. 339

Baker J.A., Coleman P.G., Wakefield B., Gibbings C.J., Tuppen C.G. (1989): Semicond. Sci. Technol. **4**, 815

Baraff G.A., Kane E.O., Schlüter M. (1979): Phys. Rev. Lett. **43**, 956

Baraff G.A., Schlüter M. (1985a): Phys. Rev. Lett. **55**, 1327

Baraff G.A., Schlüter M. (1985b): Phys. Rev. Lett. **55**, 2340

Barbiellini B., Puska M.J., Korhonen T., Harju A., Torsti T., Nieminen R.M. (1996): Phys. Rev. B **53**, 16201

Barbiellini B., Puska M.J., Torsti T., Nieminen R.M. (1995): Phys. Rev. B **51**, 7341

Bardeen J., Shockley W. (1950): Phys. Rev. **80**, 72

Baroux L., Corbel C., Kiessling F.M., Rolland S., Granger R., Hoerstel W., Triboulet R. (1995): Phys. Rev. Lett. **75**, 557

Barrett C.R., Nix W.D. (1965): Acta Metall. **13**, 1247

Bartenev G.M., Bardyshev I.I., Erchak D.P., Stel'makh V.F., Tsyganov A.D. (1979): Sov. Phys. Sol. State **21**, 687

Bartsch H., Hoehl D., Kästner G. (1984): phys. stat. sol. (a) **83**, 543

Bauer-Kugelmann W., Kögel G., Sperr P., Triftshäuser W. (1997): Mater. Sci. Forum **255–257**, 662

Bäumler N., Kaufmann U., Windscheif J. (1985): Appl. Phys. Lett. **46**, 781

Bean A.R., Newman R.C. (1971): Sol. State Commun. **9**, 271

Bebb H.B., Williams E.W. (1972): Semicond. Semimet. **8**, 181

Behringer R., Montgomery C.G. (1942): Phys. Rev. **61**, 222

Beling C.D., Simpson R.I., Charlton M., Jakobsen F.M., Griffith T.C., Moriarty P., Fung S. (1987): Appl. Phys. A **42**, 111

Bernholc J., Antonelli A., Sole T.M.D., Bar-Yam Y., Pantelides S.T. (1981): Phys. Rev. Lett. **61**, 2689

Bertolaccini M., Dupasquier A. (1970): Phys. Rev. B **1**, 2896

Bethge H., Heydenreich J. (eds.) (1987): Electron microscopy in solid state physics. Elsevier, Amsterdam

Bhide V.G., Dusane R.O., Rajarshi S.V., Shaligram A.D., David S.K. (1987): J. Appl. Phys. **62**, 108

Bisi A., Consolati G., Zappa L. (1987): Hyperfine Interact. **36**, 29

Bisson P.E., Descouts P., Dupanloup A., Manuel A.A., Perreard E., Peter M., Sachot R. (1982): Helv. Phys. Acta **55**, 100

Bliss D.E., Walukiewicz W., Chan K.T., Ager J.W., Tanigawa S. (1992): In *Low Temperature (LT) GaAs and Related Materials Symposium.* Witt G.L., Calawa R., Mishara U., Weber E. (eds.). Mater. Res. Soc., Pittsburgh, p. 93

Bond W.L. (1960): Acta Cryst. **13**, 814

Börner F., Eichler S., Polity A., Krause-Rehberg R., Hammer R., Jurisch M. (1998): J. Appl. Phys. **84**, 2255

Boronski E., Nieminen R.M. (1986): Phys. Rev. B **34**, 3820

Bourgoin J., Lannoo M. (1983): *Point Defects in Semiconductors II.* Springer-Verlag, Berlin

Bourgoin J.C., von Bardeleben H.J., Lim H., Stievenard D. (1988): In *Defects in electronic materials.* Stavola M., Pearton S.J., Davies G. (eds.). Mater. Res. Soc., Pittsburgh, p. 411

Brandes G.R., Mills A.P., Zuckerman D.M. (1992): Mater. Sci. Forum **105**, 363

Brandt W. (1974): Appl. Phys. **5**, 1

Brandt W., Dupasquier A. (eds.) (1981): Positron solid state physics. North-Holland, Amsterdam

Brandt W., Paulin R. (1972): Phys. Rev. B **5**, 2430

Brandt W., Paulin R. (1977): Phys. Rev. B **15**, 2511

Brandt W., Waung H.F., Levy P.W. (1968): *Proc. Intern. Symp. Color Centers in Alkali Halides*, Rome, p. 48

Brauer G., Anwand W., Coleman P.G., Knights A.P., Plazaola F., Pacaud Y., Skorupa W., Störmer J., Willutzki P. (1996a): Phys. Rev. B **54**, 3084

Brauer G., Anwand W., Nicht E.M., Kuriplach J., Šob M., Wagner M., Coleman P.G., Puska M.J., Korhonen T. (1996b): Phys. Rev. B **54**, 2512

Brauer G., Anwand W., Nicht E.-M., Coleman P.G., Wagner N., Wirth H., Skorupa W. (1997): Appl. Surf. Sci. **116**, 19

Brauer G., Kerbe F., Kajcsos Z., Ashry A. (1984): phys. stat. sol. (a) **82**, 143

Bräunlich P. (ed.) (1979): Thermally stimulated relaxation in solids. Springer-Verlag, Berlin

Breitenstein O., Heydenreich J. (1985): Scanning **7**, 273

Bretagnon T., Dannefaer A.S., Kerr D. (1997): J. Appl. Phys. **81**, 3446

Bretagnon T., Dannefaer S., Kerr D. (1992): Mater. Res. Soc. Symp. Proc. **262**, 337

Bretagnon T., Dannefaer S., Kerr D. (1993): J. Appl. Phys. **73**, 4697

Bretagnon T., Dannefaer S., Kerr D. (1994): Mater. Sci. Forum **143–147**, 299

Britton D.T., Huttunen P.A., Mäkinen J., Soininen E., Vehanen A. (1989): Phys. Rev. Lett. **62**, 2413

Britton D.T., Störmer J. (1995): Appl. Surf. Sci. **85**, 1

Britton D.T., Willutzki P., Triftshäuser W., Hammerl E., Hansch W., Eisele I. (1994): Appl. Phys. A **58**, 389

Brozel M.R. (1995): Appl. Phys. A **60**, 537

Brudnyi V.N. (1982): phys. stat. sol. (a) **72**, 529

Brudnyi V.N., Charchenko V.A., Kolin N.G., Novikov V.A., Pogrebnyak A.D., Ruzimov S.M. (1986): phys. stat. sol. (b) **93**, 195

Brudnyi V.N., Pogrebnyak A.D., Surov Y.P., Rudnev A.S. (1989): phys. stat. sol. (a) **114**, 481

Brudnyi V.N., Vorobiev S.A., Tsoi A.A. (1982): Appl. Phys. A **29**, 219

Brudnyi V.N., Vorobiev S.A., Tsoi A.A., Shahovtsov V.I. (1983): Rad. Eff. **79**, 123

Brunner S., Puff W., Mascher P., Balogh A.G., Baumann H. (1997): Mater. Sci. Forum **255–257**, 503

Brusa R.S., Duarte Naia M., Galvanetto E., Scardi P., Zecca A. (1992): Mater. Sci. Forum **105–110**, 1849

Canter K.F. (1995): In *Positron spectroscopy of solids*. Dupasaquier A., Mills A.P. (eds.). IOS, Amsterdam, p. 385

Cavallini A., Dupasquier A., Ferro G., Piqueras J., Valli M. (1997): Mater. Sci. Forum **255–257**, 614

Ceperley D.M., Adler B.J. (1980): Phys. Rev. Lett. **45**, 566

Chadderton L.T. (1971): Rad. Eff. **8**, 77

Chadi D.J. (1992): Phys. Rev. B **46**, 9400

Chadi D.J., Chang K.J. (1988): Phys. Rev. Lett. **60**, 2187

Chadi D.J., Chang K.J. (1989): Phys. Rev. B **39**, 10063

Chantre A., Kechouane M., Bois D. (1983): Physica B+C **116**, 547

Chen C.J. (1993): *Introduction to scanning tunneling microscopy.* Oxford University Press, New York

Chen W.M., Dreszer P., Leon R., Weber E.R., Sörman E., Monemar B., Liang B.W., Tu C.W. (1994): Mater. Sci. Forum **143–147**, 211

Chen Z.Q., Hu X.W., Wang S.J., Li S.Q. (1996a): Sol. State Commun. **99**, 745

Chen Z.Q., Hu X.W., Wang S.J., Li S.Q. (1996b): Sol. State Commun. **97**, 951

Cheng L.J., Karins J.P., Corbett J.W., Kimerling L.C. (1979): J. Appl. Phys. **50**, 2962

Cheng L.J., Yeh C.K. (1973): Sol. State Commun. **12**, 529

Cheng L.J., Yeh C.K., Ma S.I., Su C.S. (1973): Phys. Rev. B **8**, 2880

Chiba T., Hasegawa M., Tang Z., Akahane T., Manuel A.A., Saito M., Sumiya H., Kuramoto E., Takenaka M. (1997): Mater. Sci. Forum **255–257**, 521

Chiba T., Kawasuso A., Hasegawa M., Suezawa M., Akahane T., Sumino K. (1995): Mater. Sci. Forum **175–178**, 327

Cho Y.K., Kondo H., Kubota T., Nakashima H., Kawano T., Uedono A., Kurihara T., Tanigawa S. (1992): Mater. Sci. Forum **105–110**, 925
Chu S., Mills A.P., Murray C.A. (1981): Phys. Rev. B **23**, 2060
Clark C.D., Collins A.T., Woods D.S. (1992): In *The properties of natural and synthetic diamond.* Field J.E. (ed.). Academic Press, London, p. 35
Claverie A., Crestou J., Garcia J.C. (1993): Appl. Phys. Lett. **62**, 1638
Clayton J.M., Usmar S.G., Alam A., Hurle D.T.J., Stirland D.J. (1992): Semicond. Sci. Technol. **7**, 233
Clement M., de Nijs J.M.M., Balk P., Schut H., van Veen A. (1996): J. Appl. Phys. **79**, 9029
Clement M., de Nijs J.M.M., Balk P., Schut H., van Veen A. (1997): J. Appl. Phys. **81**, 1943
Coleman C.F. (1979): Appl. Phys. **19**, 87
Coleman P.G., Bless E., Setzler S., LaShell S., Davies G., Sawyer W.D. (1995): Appl. Surf. Sci. **85**, 276
Connors D.C., Crisp V.H.C., West R.N. (1971): J. Phys. F: Met. Phys. **1**, 355
Corbel C., Baroux L., Kiessling F.M., Gely-Sykes C., Triboulet R. (1993): Mater. Sci. Eng. B **16**, 134
Corbel C., Hautojärvi P., LeBerre C., Liszkay L., Saarinen K. (1992a): Mater. Sci. Forum **105–110**, 933
Corbel C., Hautojärvi P., Mäkinen J., Vehanen A., Mathiot D. (1989): J. Phys. Cond. Mat. **1**, 6315
Corbel C., Moser P., Stucky M. (1985): Ann. Chim. Fr. **10**, 733
Corbel C., Pierre F., Hautojärvi P., Saarinen K., Moser P. (1990): Phys. Rev. B **41**, 10632
Corbel C., Pierre F., Saarinen K., Hautojärvi P., Moser P. (1992b): Phys. Rev. B **45**, 3386
Corbel C., Stucky M., Hautojärvi P., Saarinen K., Moser P. (1988): Phys. Rev. B **38**, 8192
Corbett J.W., Bourgoin J.C. (1975): In *Point defects in solids.* Crawford J.H., Slifkin L.M. (eds.). Plenum, New York, p. 1
Corbett J.W., Watkins G.D., Chrenko R.M., McDonald R.S. (1961): Phys. Rev. **121**, 1015
Dabrowski J., Scheffler M. (1989): Phys. Rev. B **40**, 10391
Dabrowski J., Scheffler M. (1992): Phys. Scr. **T45**, 151
Dannefaer S. (1982): J. Phys. C: Semicond. Phys. **15**, 599
Dannefaer S. (1986): Mater. Sci. Forum **10–12**, 103
Dannefaer S. (1990): In *Defect control in semiconductors.* Sumino K. (ed.). Elsevier, Amsterdam, p. 1561
Dannefaer S. (1995): Appl. Phys. A **D61**, 59
Dannefaer S., Bretagnon T. (1995): J. Appl. Phys. **77**, 5584
Dannefaer S., Bretagnon T., Kerr D. (1993a): J. Appl. Phys. **74**, 884
Dannefaer S., Bretagnon T., Kerr D. (1996a): J. Appl. Phys. **80**, 3750
Dannefaer S., Craigen D., Kerr D. (1995): Phys. Rev. B **51**, 1928
Dannefaer S., Dean G.W., Kerr D.P., Hogg B.G. (1976): Phys. Rev. B **14**, 2709
Dannefaer S., Fruensgaard N., Kupca S., Hogg B., Kerr D. (1983a): Can. J. Phys. **61**, 451
Dannefaer S., Hogg B., Kerr D. (1984): Phys. Rev. B **30**, 3355
Dannefaer S., Kerr D. (1986a): J. Appl. Phys. **60**, 1313
Dannefaer S., Kerr D. (1986b): J. Appl. Phys. **60**, 591
Dannefaer S., Kerr D. (1993): Phys. Rev. B **48**, 9142
Dannefaer S., Kerr D. (1994): Phys. Rev. B **50**, 14096
Dannefaer S., Kerr D., Craigen D. (1996b): J. Appl. Phys. **79**, 9110
Dannefaer S., Kerr D., Hogg B.G. (1983b): J. Appl. Phys. **54**, 155
Dannefaer S., Kupca S., Hogg B.G., Kerr D.P. (1980): Phys. Rev. B **22**, 6135
Dannefaer S., Mascher P., Kerr D. (1986): Phys. Rev. Lett. **56**, 2195
Dannefaer S., Mascher P., Kerr D. (1989a): J. Phys. Cond. Mat. **1**, 3213
Dannefaer S., Mascher P., Kerr D. (1991): J. Appl. Phys. **69**, 4080
Dannefaer S., Mascher P., Kerr D. (1993b): J. Appl. Phys. **73**, 3740
Dannefaer S., Puff W., Kerr D. (1977): Phys. Rev. B **55**, 2182
Dannefaer S., Puff W., Mascher P., Kerr D. (1989b): J. Appl. Phys. **66**, 3526

Dannefaer S., Zhu W., Bretagnon T., Kerr D. (1996c): Phys. Rev. B **53**, 1979
David A., Kögel G., Sperr P., Triftshäuser W. (1997): Mater. Sci. Forum **255–257**, 741
de Benedetti S., Cowan C.E., Konnecker W.R., Primakoff H. (1950): Phys. Rev. **77**, 205
de la Cruz R.M., Pareja R., Gonzalez R., Boatner L.A., Chen Y. (1992a): Phys. Rev. B **45**, 6581
de la Cruz R.M., Pareja R., Moser P. (1992b): Mater. Sci. Forum **105–110**, 957
Dederichs P.H. (1973): J. Phys. F: Met. Phys. **3**, 471
Dekhtyar I.Y., Dekhtyar M.I., Dyakin V.V., Zaitov F.A., Vlasenko A.I., Likhtorovich S.P., Lyubchenko A.A., Sakharova S.G., Silant'ev V.I., Fedchenko R.G. (1984): Sov. Phys. Semicond. **18**, 1230
Dekhtyar I.Y., Mikhalenkov V.S., Sakharova S.G. (1969): Fiz. Tverd. Tela **11**, 3322
Deutsch M. (1951): Phys. Rev. **82**, 455
Devlin S.S. (1967): In *Physics and chemistry of II–VI compounds.* Aven A., Prener J.S. (eds.). North-Holland, Amsterdam, p. 551
Dirac P.A.M. (1928): Proc. Roy. Soc. **117**, 610
Dlubek A. (1985): Dissertation A, Martin-Luther-Universität Halle–Wittenberg
Dlubek G. (1986): Mater. Sci. Forum **13–14**, 11
Dlubek G., Ascheron C., Krause R., Erhard H., Klimm D. (1988a): phys. stat. sol. (a) **106**, 81
Dlubek G., Brümmer O. (1986): Ann. Phys. **43**, 178
Dlubek G., Brümmer O., Krause R., Baranowski A., Rozenfeld B. (1983): phys. stat. sol. (a) **78**, 217
Dlubek G., Brümmer O., Plazaola F., Hautojärvi P. (1986a): J. Phys. C: Semicond. Phys. **19**, 331
Dlubek G., Brummer O., Plazaola F., Hautojärvi P., Naukkarinen K. (1985a): Appl. Phys. Lett. **46**, 1136
Dlubek G., Brümmer O., Polity A. (1986b): Appl. Phys. Lett. **49**, 385
Dlubek G., Dlubek A., Krause R., Brümmer O. (1988b): phys. stat. sol. (a) **107**, 111
Dlubek G., Krause R. (1987): phys. stat. sol. (a) **102**, 443
Dlubek G., Plazaola F., Mäkinen J., Hautojärvi P., Brümmer O. (1985b): In *Positron Annihilation.* Jain P.C., Singru R.M., Gopinathan K.P. (eds.). World Scientific, Singapore, p. 684
Domke C., Ebert P., Heinrich M., Urban K. (1996): Phys. Rev. B **54**, 10288
Dreszer P., Chen W.M., Seendripu K., Wolk J.A., Walukiewicz W., Liang B.W., Tu C.W., Weber E.R. (1993): Phys. Rev. B **47**, 4111
Druzhkov A.P., Prokop'ev E.P., Yu N.K., Sidorov A.I., Fedorov V.A. (1980): Sov. Phys. Semicond. **14**, 119
Duke C.B., Paton A., Lazarides A., Vasumathi D., Canter K.F. (1997): Phys. Rev. B **55**, 7181
Dupasquier A., Mills A.P. (eds.) (1995): Positron spectroscopy of solids. IOS Press, Amsterdam
Dupasquier A., Romero R., Somoza A. (1993): Phys. Rev. B **48**, 9235
Ebert P., Heinrich M., Urban K., Chao K.-J., Smith A.R., Shih C.K. (1996): J. Vac. Sci. Technol. A **14**, 1807
Eichler S. (1997): Dissertation, Martin-Luther-Universität Halle–Wittenberg
Eichler S. (1998): unpublished results
Eichler S., Börner F., Gebauer J., Krause-Rehberg R., Wendler E., Weber B., Wesch W. (1998): Phys. Rev. B submitted
Eichler S., Gebauer J., Börner F., Polity A., Krause-Rehberg R., Wendler E., Weber B., Wesch W., Börner H. (1997a): Phys. Rev. B **56**, 1393
Eichler S., Hübner C., Krause-Rehberg R. (1997b): Appl. Surf. Sci. **116**, 155
Eldrup M., Pedersen N.J., Sherwood J.N. (1979): Phys. Rev. Lett. **43**, 1407
Elkin E.L., Watkins G.D. (1968): Phys. Rev. **174**, 881
Emanuelsson P., Omling P. (1993): Phys. Rev. B **47**, 15578
Evwaraye A.O., Sun E. (1976): J. Appl. Phys. **47**, 3776
Fahey P.M., Griffin P.B., Plummer J.D. (1989): Rev. Mod. Phys. **61**, 289
Favre J., Konczykowski M., Blanchard C. (1989): Ann. Physique **14**, 193
Feenstra R.M., Woodall J.M., Pettit G.D. (1993): Phys. Rev. Lett. **71**, 1176
Ferrell R.A. (1958): Phys. Rev. **110**, 1355

Fischer D.A., Lynn K.G., Gidley D.W. (1986): Phys. Rev. B **33**, 4479
Fischer D.W. (1988): Phys. Rev. B **37**, 2968
Fischer F., Waag A., Worschech L., Ossau W., Scholl S., Landwehr G., Mäkinen J., Hautojärvi P., Corbel C. (1996): J. Cryst. Growth **161**, 214
Fitzgerald E.A., Xie Y.H., Monroe D., Watson G.P., Kuo J.M., Silverman P.J. (1994): Mater. Sci. Forum **143–147**, 471
Fluss M.J., Berko S., Chakraborty B., Lippel P., Siegel R.W. (1984): J. Phys. F: Met. Phys. **14**, 2855
Föll H., Gösele U., Kolbesen B.O. (1977): J. Cryst. Growth **40**, 90
Fornari R., Weyher J.L., Krawczyk S., Nuban F., Corbel C., Törnqvist M. (1995): Mater. Sci. Technol. **11**, 1223
Frank W., Seeger A. (1974): Appl. Phys. **3**, 66
Fuhs W., Holzhauer U., Richter F.W., Sturm R., Mantl S. (1978): phys. stat. sol. (b) **89**, 69
Fujii S., Shikata S., Wei L., Tanigawa S. (1992a): J. Appl. Phys. **72**, 1405
Fujii S., Shikata S., Wei L., Tanigawa S. (1992b): Mater. Res. Soc. Symp. Proc. **262**, 343
Fujinami M. (1996): Phys. Rev. B **53**, 13047
Fujinami M., Chilton N.B. (1993a): Appl. Phys. Lett. **63**, 3458
Fujinami M., Chilton N.B. (1993b): J. Appl. Phys. **73**, 3242
Fujinami M., Chilton N.B. (1995): Mater. Sci. Forum **175–178**, 157
Fujinami M., Chilton N.B., Ishii K., Ohki Y. (1993a): J. Physique IV **3**, 169
Fujinami M., Chilton N.B., Ishii K., Ohki Y. (1993b): J. Appl. Phys. **74**, 5406
Fujinami M., Tsuge A., Tanaka K. (1996): J. Appl. Phys. **79**, 9017
Fujiwara Y., Nishino T., Hamakawa Y. (1982): Jpn. J. Appl. Phys. Pt. 2 **21**, L727
Gebauer J., Eichler S., Krause-Rehberg R. (1995): unpublished results
Gebauer J., Eichler S., Krause-Rehberg R., Polity A. (1997a): Appl. Surf. Sci. **116**, 215
Gebauer J., Eichler S., Krause-Rehberg R., Zeindl H.P. (1997b): Appl. Surf. Sci. **116**, 247
Gebauer J., Krause-Rehberg R., Domke C., Ebert P., Urban K. (1997c): Phys. Rev. Lett. **78**, 3334
Gebauer J., Krause-Rehberg R., Eichler S., Bauer-Kugelmann W., Kögel G., Triftshäuser W., Luysberg M., Sohn H., Weber E.R. (1997d): Mater. Sci. Forum **255–257**, 204
Gebauer J., Krause-Rehberg R., Eichler S., Luysberg M., Sohn H., Weber E.R. (1997e): Appl. Phys. Lett. **71**, 638
Gebauer J., Krause-Rehberg R., Lausmann M., Lippold G. (1997f): Mater. Sci. Forum **258–263**, 905
Gebauer J., Lausmann M., Krause-Rehberg R. (1998a): unpublished results
Gebauer J., Rudolf F., Polity A., Krause-Rehberg R., Martin, J., Becker P. (1998b): Appl. Phys. A submitted
Geffroy B., Corbel C., Stucky M., Triboulet R., Hautojärvi P., Plazaola F., Saarinen K., Rajainmaki H., Aaltonen J., Moser P., Sengupta A., Pautrat J.L. (1986): Mater. Sci. Forum **10–12**, 1241
Gehlhoff W., Höhne M., Schmidt J. (1992): In *Hyperfine interaction of defects in semiconductors*. Langouche G. (ed.). Elsevier, Amsterdam, p. 217
Gély C., Corbel C., Triboulet R. (1989): Comptes Rendus de l'Academie des Sciences Serie II **309**, 179
Gély C., Corbel C., Triboulet R. (1990): J. Phys. Cond. Mat. **2**, 4763
Gély-Sykes C., Corbel C., Triboulet R. (1991): Sol. State Comm. **80**, 79
Ghosh V.J. (1995): Appl. Surf. Sci. **85**, 187
Ghosh V.J., Aers G.C. (1995): Phys. Rev. B **51**, 45
Gidley D.W., Frieze W.E. (1988): Phys. Rev. Lett. **60**, 1193
Gilgien L., Galli G., Gygi F., Car R. (1994): Phys. Rev. Lett. **72**, 3214
Girka A.I., Kuleshin V.A., Mokrushin A.D., Mokhov E.N., Svirida S.V., Shishkin A.V. (1989): Sov. Phys. Semicond. **23**, 1337
Girka A.I., Merenkov S.V., Shemaev B.V., Mokrushin A.D. (1992): Mater. Sci. Forum **105–110**, 1017

Girka A.I., Mokhov E.N. (1995): Phys. Sol. State **37**, 1855
Gleiter H., Marquardt P. (1984): Z. Metallkde. **75**, 263
Gokcen N.A. (1989): Bull. Alloy Phase Diagrams **10**, 11
Goldberg R.D., Schultz P.J., Simpson P.J. (1995): Appl. Surf. Sci. **85**, 287
Götz G., Gärtner K. (eds.) (1988): High energy ion beam analysis of solids. Akademie-Verlag, Berlin
Götz W., Johnson N.M., Chen C., Liu H., Kuo C., Imler W. (1996): Appl. Phys. Lett. **68**, 3144
Gregory B.L., Barnes C.E. (1968): In *Radiation effects in semiconductors*. Vook F.L. (ed.). Plenum, New York, p. 124
Greif H., Haaks M., Holzwarth U., Männig U., Tongbhoyai M., Wider T., Maier K., Bihr J., Huber B. (1997): Appl. Phys. Lett. **71**, 2115
Grillot E., Bancie-Grillot M., Egee M. (1976): J. Lumin. **12**, 461
Günther C., Vrijmoeth J., Hwang R.Q., Behm R.J. (1995): Phys. Rev. Lett. **74**, 754
Gupta T.K., Carlson W.G. (1987): J. Mater. Sci. **20**, 3487
Gupta T.K., Straub W.D., Ramanachalam M.S., Schaffer J.P., Rohatgi A. (1989): J. Appl. Phys. **66**, 6132
Haase M.A., Qiu Y., DePuydt J.M., Cheng H. (1991): Appl. Phys. Lett. **59**, 1272
Hahn C., Krause-Rehberg R. (1997): unpublished results
Hakala M., Puska M.J., Nieminen R.M. (1998): Phys. Rev. B **57**, 7621
Häkkinen H., Mäkinen S., Manninen M. (1990): Phys. Rev. B **41**, 12441
Hakvoort R.A., van Veen A., Schut H., van den Boogaard M.J., Berntsen A.J.M., Roorda S., Stolk P.A., Reader A.H. (1994): AIP Conf. Proc. **303**, 48
Halec A., Schultz P.J., Boudreau M., Boumerzoug M., Mascher P., McCaffrey J.P., Jackman T.E. (1994): Surf. Interf. Anal. **21**, 839
Hasegawa M., Chiba T., Kawasuso A., Akahane T., Suezawa M., Yamaguchi S., Sumino K. (1995a): Mater. Sci. Forum **196–201**, 1481
Hasegawa M., Kawasuso A., Chiba T., Akahane T., Suezawa M. (1995b): Appl. Phys. A **61**, 65
Hasegawa M., Tang Z., Chiba T., Saito M., Kawasuso A., Akahane T., Li Z.-Q., Kawazoe Y., Yamaguchi S. (1997): Mater. Sci. Forum **255–257**, 414
Hashimoto E., Kino T. (1991): J. Phys. Soc. Jap. **60**, 3167
Hashimoto E., Ueda Y., Uematsu N., Iwami M., Kino T. (1992): J. Phys. Soc. Jap. **61**, 3799
Hässlein H., Sielemann R. (1995): In *Hyperfine Interactions*. Rots M., Vantomme A., Dekoster J., Coussement R., Langouche G. (eds.). Baltzer Science Publishers, Basel, p. 203
Hässlein H., Sielemann R., Zistl C. (1998): Phys. Rev. Lett. **80**, 2626
Haung C.C., Chang I.M., Chen Y.F., Tseng P.K. (1996): Physica B **228**, 374
Hausmann H., Ehrhart P. (1995a): Phys. Rev. B **51**, 17542
Hausmann H., Ehrhart P. (1995b): Mater. Sci. Forum **196–201**, 1255
Hausmann H., Pillukat A., Ehrhart P. (1996): Phys. Rev. B **54**, 8527
Hautojärvi P. (ed.) (1979): Positrons in solids. Springer-Verlag, Berlin
Hautojärvi P., Huomo H., Lahtinen J., Mäkinen J., Vehanen A. (1986a): Mater. Sci. Forum **10–12**, 527
Hautojärvi P., Mäkinen J., Palko S., Saarinen K., Corbel C., Liszkay L. (1993): Mater. Sci. Eng. B **22**, 16
Hautojärvi P., Moser P., Stucky M., Corbel C., Plazaola F. (1986b): Appl. Phys. Lett. **48**, 809
Hautojärvi P., Tamminen A., Jauho P. (1970): Phys. Rev. Lett. **24**, 459
Hautojärvi P., Vehanen A. (1979): In *Positrons in solids*. Hautojärvi P. (ed.). Springer-Verlag, Berlin, p. 1
Hayashi N., Suzuki R., Hasegawa M., Kobayashi N., Tanigawa S., Mikado T. (1993a): Phys. Rev. Lett. **70**, 45
Hayashi N., Suzuki R., Watanabe H., Sakamoto I., Kobayashi N., Mikado T., Yamazaki T., Kuriyama K. (1993b): Nucl. Instrum. Meth. Phys. Res. **80**, 1006
Hayashi N., Watanabe H., Sakai K., Kuriyama K., Ikeda Y., Maekawa H., Miura T. (1995): Mater. Sci. Forum **196–201**, 1243
He Y.J., Hasegawa M., Lee R., Berko S., Adler D., Jung A.L. (1986): Phys. Rev. B **33**, 5924

He Y.J., Li X.F., Li L.H., Yu W.Z., Xiao J.R. (1989): J. Phys. Cond. Mat. **1**, SA91
He Y.-J., Hu Y. (1995): Mater. Sci. Forum **175–178**, 165
Hecking N., K. F. Heidemann K.F., Te Kaat E. (1986): Nucl. Instrum. Meth. Phys. Res. B **15**, 760
Hehenkamp T. (1985): In *Positron Annihilation.* Jain P.C., Singru R.M., Gopinathan K.P. (eds.). World Scientific, Singapore, p. 856
Hesse M., Koschnick F.K., Krambrock K., Spaeth J.-M. (1994): Sol. State Commun. **92**, 207
Hidalgo C., Linderoth S., González-Doncel G., San Juan J. (1989): In *Positron annihilation.* Dorikens-Vanpraet L., Dorikens M., Segers D. (eds.). World Scientific, Singapore, p. 371
Hill D.E. (1959): Phys. Rev. **114**, 1414
Hirata M., Hirata M., Saito H. (1966): Jpn. J. Appl. Phys. **5**, 252
Hirata M., Hirata M., Saito H. (1969): J. Phys. Soc. Jap. **27**, 405
Hiroyama Y., Motooka T., Tokuyama T., Wei L., Tanigawa S. (1993): Nucl. Instrum. Meth. Phys. Res. B **80**, 982
Hiroyama Y., Suzuki R., Hirano Y., Sato F., Motooka T. (1995): Jpn. J. Appl. Phys. Pt. 1 **34**, 5515
Hodges C.H. (1970): Phys. Rev. Lett. **25**, 284
Hofmann D.M., Omling P., Grimmeiss H.G., Meyer B.K., Benz K.W., Sinerius D. (1992): Phys. Rev. B **45**, 6247
Holt D.B. (1962): J. Phys. Chem. Sol. **23**, 1353
Horii Y., Kawasuso A., Hasegawa M., Suezawa M., Yamaguchi S., Sumino K. (1992): Mater. Sci. Forum **105–110**, 1061
Horsky T.N., Brandes G.R., Canter K.F., Duke C.B., Horng S.F., Kahn A., Lessor D.L., Mills A.P., Paton A., Stevens K., Stiles K. (1989): Phys. Rev. Lett. **62**, 1876
Howell R.H., Fluss M.J., Rosenberg I.J., Meyer P. (1985): Nucl. Instrum. Meth. Phys. Res. B **10**, 373
Hozhabri N., Bailey J.B., Koymen A.R., Sharma S.C., Alavi K. (1994): J. Phys. Cond. Mat. **6**, L455
Hozhabri N., Hyer R.C., Sharma S.C., Ma J.Y., Pathak R.N., Alavi K. (1992): J. Vac. Sci. Technol. B **10**, 788
Huang M., Wang Y., Yang J., He Y., Guo Y., Liu C. (1992): Mater. Sci. Forum **105–110**, 1071
Huang M.-R., Wang Y.-Y., Yang J.-H., Guo Y.-H., He Y.-S., Sun T.-N. (1995): Mater. Sci. Forum **175–178**, 383
Hübner C., Leipner H.S., Krause-Rehberg R. (1997a): Mater. Sci Forum **255–257**, 497
Hübner C., Staab T., Krause-Rehberg R. (1994): Appl. Phys. A **61**, 203
Hübner C.G., Gebauer J., Leipner H.S., Krause-Rehberg R. (1997b): unpublished results
Hübner C.G., Leipner H.S., Storbeck O., Krause-Rehberg R. (1996): In *Physics of Semiconductors.* Scheffler M., Zimmermann R. (eds.). World Scientific, Berlin, p. 2805
Hulett L.D. (1995): Mater. Sci. Forum **175–178**, 99
Huomo H., Jones R., Hurst J., Vehanen A., Throwe J., Usmar S.G., Lynn K.G. (1989): Nucl. Instr. Meth. A **284**, 359
Hurle D.T.J. (1979a): J. Phys. Chem. Sol. **40**, 613
Hurle D.T.J. (1979b): J. Phys. Chem. Sol. **40**, 627
Hurle D.T.J. (1988): In *Semi-insulating III–V Materials, Malmö 1988.* Grossmann G., Ledebo L. (eds.). Adam Hilger, Bristol, p. 11
Hurle D.T.J. (ed.) (1994): Handbook of crystal growth. North-Holland, Amsterdam
Hurle D.T.J. (1995): Mater. Sci. Forum **196–201**, 179
Hutchings S.M., Coleman P.G., West R.N., Stone R.J. (1986): J. Phys. E: Sci. Instrum. **19**, 282
Huttunen P.A., Mäkinen J., Vehanen A. (1990): Phys. Rev. B **41**, 8062
Hwang C.J. (1968): Appl. Phys. **39**, 4307
Hyodo T. (1995): In *Positron spectroscopy of solids.* Dupasaquier A., Mills A.P. (eds.). Ios, Amsterdam, p. 419
Ikari A. (1995): Appl. Surf. Sci. **85**, 253
Ikari A., Haga H., Uedono A., Ujihira Y., Yoda O. (1994a): Jpn. J. Appl. Phys. Pt. 1 **33**, 1723

Ikari A., Haga H., Yoda O., Uedono A., Ujihira Y. (1992): Mater. Res. Soc. Symp. Proc. **262**, 69

Ikari A., Kawakami K., Haga H., Uedono A., Long W., Kawano T., Tanigawa S. (1994b): Jpn. J. Appl. Phys. Pt. 1 **33**, 5585

Ito K., Ito T., Mizuno K. (1986): Rad. Eff. **97**, 101

Itoh H., Yoshikawa M., Wei L., Tanigawa S., Nashiyama I., Misawa S., Okumura H., Yoshida S. (1992a): Mater. Res. Symp. Proc. **262**, 331

Itoh Y., Lee K.H., Murakami H., Iwata R. (1995): Appl. Phys. A **60**, 57

Itoh Y., Murakami H. (1994): Appl. Phys. A **58**, 59

Itoh Y., Murakami H., Kinoshita A. (1994): Hyperf. Interact. **84**, 121

Itoh Y., Murakami H., Suzuki T. (1992b): Mater. Sci. Forum **105–110**, 1085

Itoh Y., Murakami H., Takeno H., Ushio S., Takenaka T. (1997): Mater. Sci. Forum **255–257**, 545

Iwase Y., Uedono A., Tanigawa S., Araki H. (1985a): In *Positron annihilation.* Jain P.C., Singru R.M., Gopinathan K.P. (eds.). World Scientific, Singapore, p. 753

Iwase Y., Uedono A., Tanigawa S., Suzuki T. (1985b): Radioisotopes **34**, 195

Jackman T.E., Aers G.C., Denhoff M.W., Schultz P.J. (1989): Appl. Phys. A **49**, 335

Jackman T.E., Aers G.C., McCaffrey J.P., Britton D., Willutzki P., Schultz P.J., Simpson P.J., Mascher P. (1992): Mater. Sci. Forum **105–110**, 301

Jansen R.W., Sankey O.F. (1989): Phys. Rev. B **39**, 3192

Jastrzebski L. (1982): IEEE J. Sol. State Circuits **17**, 105

Jean Y.C. (1995): Mater. Sci. Forum **175–178**, 59

Jensen K.O., Walker A.B. (1990): AIP Conf. Proc. **218**, 44

Jensen K.O., Walker A.B., Bouarissa N. (1990): AIP Conf. Proc. **218**, 19

Jia Y., Oshima T., Yamada A., Konagai M., Takahashi K., Tanigawa S., Wei L. (1993): Jpn. J. Appl. Phys. Pt. 1 **32**, 1884

Jones C.E. (1985): J. Vac. Sci. Technol. A **3**, 131

Jordan A.S., von Neidar A.R., Caruso R., Kim C.K. (1974): J. Electrochem. Soc. **121**, 153

Jorke H., Herzog H.-J., Kibbel H. (1989): Phys. Rev. B **40**, 2005

Kametani H., Akiyama H., Yamaguchi Y., Koumaru M., Wei L., Tabuki Y., Tanigawa S., Uedono A., Watauchi S., Ujihira Y., Suzuki R., Ohgaki H., Mikado T. (1992): Mater. Res. Soc. Symp. Proc. **262**, 235

Kamimura Y., Tsutsumi T., Kuramoto E. (1995): Phys. Rev. B **52**, 879

Kaminska M., Liliental-Weber Z., Weber E.R., George T., Kortright J.B., Smith F.B., Tsaur B.-Y., Calawa A.R. (1989): Appl. Phys. Lett. **54**, 1881

Karsten K., Ehrhart P. (1995): Phys. Rev. B **51**, 10508

Kasai P.H., Otomo Y. (1962): J. Chem. Phys. **37**, 1263

Kauppinen H., Baroux L. (1997): J. Phys. Cond. Mat. **9**, 5495

Kawasuso A., Hasegawa M., Suezawa M., Yamaguchi S., Sumino K. (1994): Hyperf. Interact. **84**, 397

Kawasuso A., Hasegawa M., Suezawa M., Yamaguchi S., Sumino K. (1995a): Mater. Sci. Forum **175–178**, 423

Kawasuso A., Hasegawa M., Suezawa M., Yamaguchi S., Sumino K. (1995b): Appl. Surf. Sci. **85**, 280

Kawasuso A., Hasegawa M., Suezawa M., Yamaguchi S., Sumino K. (1995c): Jpn. J. Appl. Phys. Pt. 1 **34**, 2197

Kawasuso A., Itoh H., Okada S., Okumura H. (1996): J. Appl. Phys. **80**, 5639

Kawasuso A., Okada S. (1997): Jpn. J. Appl. Phys. Pt. 1 **36**, 605

Kawasuso A., Okada S., Suezawa M., Honda T., Yonenaga I. (1997): J. Appl. Phys. **81**, 2916

Kawasuso A., Suezawa M., Hasegawa M., Yamaguchi S., Sumino K. (1995d): Jpn. J. Appl. Phys. Pt. 1 **34**, 4579

Keeble D.J., Umlor M.T., Asoka-Kumar P., Lynn K.G., Cooke P.W. (1993): Appl. Phys. Lett. **63**, 87

Kennedy T.A., Wilsey N.D. (1984): Appl. Phys. Lett. **44**, 1089

Kerr D.P., Kupca S., Hogg B.G. (1982): Phys. Lett. A **88**, 429

Khatri R., Charlton M., Sferlazzo P., Lynn K.G., Mills A.P.J., Roelling L.O. (1990): Appl. Phys. Lett. **57**, 2374
Kiessling F.M., Corbel C., Baroux L., Rolland S., Triboulet R. (1995): Phys. Rev. B **52**, 9925
Kikoin K., Lyuk P., Pervova L., Vanem R. (1986): J. Phys. C: Semicond. Phys. **19**, 4561
Kimerling L.C. (1977): Inst. Phys. Conf. Ser. **31**, 221
Kimerling L.C., De Angelis H.H., Diebold J.W. (1975): Sol. State Commun. **16**, 171
Kirkegaard P., Pedersen N.J., Eldrup M. (1989): In *Positron annihilation.* Dorikens-Vanpraet L., Dorikens M., Segers D. (eds.). World Scientific, Singapore, p. 642
Kisielowski-Kemmerich C. (1990): phys. stat. sol. (a) **161**, 11
Kitano T., Saito S., Tanigawa S. (1994): Appl. Phys. Lett. **65**, 2434
Kitano T., Wei L., Tabuki Y., Tanigawa S., Mikoshiba H. (1992): Mater. Res. Soc. Symp. Proc. **262**, 707
Kluin J.E., Hehenkamp T. (1991): Phys. Rev. B **44**, 11597
Knights A.P., Kowalski P., Saleh S., Towner A., Patel M.I., Rice-Evans P.C., Moore M., Gledhill G.A. (1995): J. Appl. Phys. **78**, 441
Kögel G. (1995): Mater. Sci. Forum **175–178**, 107
Kögel G. (1996): Appl. Phys. A **63**, 227
Kögel G., Schödlbauer D., Triftshäuser W., Winter J. (1988): Phys. Rev. Lett. **60**, 1550
Kolm C., Kulin S.A., Averbach B.L. (1957): Phys. Rev. **108**, 965
Krambrock K., Corbel C., Spaeth J.-M. (1997): Mater. Sci. Forum **258–263**, 1033
Krause R., Brümmer O., Plazaola F., Dlubek G. (1986a): J. Mater. Sci. **21**, 853
Krause R., Dlubek G., Brümmer O., Skladnikiewitz S., Daut H. (1985a): Cryst. Res. Technol. **20**, 267
Krause R., Dlubek G., Wendrock G. (1985b): Cryst. Res. Technol. **20**, 1495
Krause R., Dlubek G., Wendrock G., Brümmer O. (1986b): Scr. Metall. **20**, 1761
Krause R., Klimakow A., Kiessling F.M., Polity A., Gille P., Schenk M. (1990a): J. Cryst. Growth **101**, 512
Krause R., Neubert M., Drost T., Horstel W., Polity A., Kiessling F.M., Paitz U., Zlomanov V., Mäkinen S. (1992): Mater. Sci. Forum **105–110**, 333
Krause R., Polity A., Dlubek G., Friedland K., Rentzsch R., Siegel W., Kühnel G. (1989a): In *Positron annihilation.* Dorikens-Vanpraet L., Dorikens M., Segers D. (eds.). World Scientific, Singapore, p. 721
Krause R., Polity A., Klimakow A., Schenk M. (1989b): In *Positron annihilation.* Dorikens-Vanpraet L., Dorikens M., Segers D. (eds.). World Scientific, Singapore, p. 684
Krause R., Saarinen K., Hautojärvi P., Polity A., Gärtner G., Corbel C. (1990b): Phys. Rev. Lett. **65**, 3329
Krause-Rehberg R., Abgarjan T., Polity A., Neubert M. (1995a): Mater. Sci. Forum **196–201**, 1285
Krause-Rehberg R., Brohl M., Leipner H.S., Drost T., Polity A., Beyer U., Alexander H. (1993a): Phys. Rev. B **47**, 13266
Krause-Rehberg R., Dlubek G., Polity A. (1995b): Mater. Sci. Forum **196–201**, 1649
Krause-Rehberg R., Drost T., Polity A. (1994a): Mater. Sci. Forum **143–147**, 429
Krause-Rehberg R., Drost T., Polity A., Roos G., Pensl G., Volm D., Meyer B.K., Bischopink G., Benz K.W. (1993b): Phys. Rev. B **48**, 11723
Krause-Rehberg R., Eichler S. (1997): *Poster at International conference on positron annihilation 11,* Kansas City
Krause-Rehberg R., Leipner H.S. (1997): Appl. Phys. A **64**, 457
Krause-Rehberg R., Leipner H.S., Kupsch A., Polity A., Drost T. (1994b): Phys. Rev. B **49**, 2385
Krause-Rehberg R., Leipner H.S., Polity A., Rudolf F., Hammer R., Jurisch M. (1996): phys. stat. sol. (a) **158**, 377
Krause-Rehberg R., Polity A., Abgarjan T., Stadler W., Hofmann D.M., Meyer B.K., Azoulay M. (1998): Phys. Rev. B submitted
Krause-Rehberg R., Polity A., Drost T., Roos G., Pensl G., Volm D., Meyer B.K. (1994c): Mater. Sci. Forum **143–147**, 1099

Krause-Rehberg R., Polity A., Siegel W., Kühnel G. (1993c): Semicond. Sci. Technol. **8**, 290
Krause-Rehberg R., Zimmermann H., Klimakow A., Drost T. (1992): phys. stat. sol. (a) **134**, K45
Krüger J., Alexander H. (1992): Mater. Sci. Forum **83–87**, 899
Kruseman A.C., Schut H., Fujinami M., van Veen A. (1997a): Mater. Sci. Forum **255–257**, 793
Kruseman A.C., Schut H., van Veen A., Mijnarends P.E., Clement M., de Nijs J.M.M. (1997b): Appl. Surf. Sci. **116**, 192
Kuisma S., Saarinen K., Hautojärvi P., Corbel C. (1996a): Phys. Rev. B **53**, R7588
Kuisma S., Saarinen K., Hautojärvi P., Corbel C. (1997): Phys. Rev. B **55**, 9609
Kuisma S., Saarinen K., Hautojärvi P., Corbel C., LeBerre C. (1996b): Phys. Rev. B **53**, 9814
Kulkarni N.M., Kulkarni R., Shaligram A.D. (1992): phys. stat. sol. (a) **133**, 283
Kuramoto E., Takeuchi S., Noguchi M., Chiba T., Tsuda N. (1973): J. Phys. Soc. Jap. **34**, 103
Kuramoto E., Takeuchi S., Noguchi M., Chiba T., Tsuda N. (1974): Appl. Phys. **4**, 41
Kwete M., Segers D., Dorikens M., Dorikens-Vanpraet L., Clauws P. (1990): phys. stat. sol. (a) **122**, 129
Laasonen K., Alatalo M., Puska M.J., Nieminen R.M. (1991): J. Phys. Cond. Mat. **3**, 7217
Laasonen K., Nieminen R.M., Puska M.J. (1992): Phys. Rev. B **45**, 4122
Lagowski J., Gatos H.C., Kang C.H., Skowronski M., Ko K.Y., Lin D.G. (1986): Appl. Phys. Lett. **62**, 3482
Lagowski J., Gatos H.C., Parsley J.M., Wada K., Kaminska M., Walukiewicz W. (1982): Appl. Phys. Lett. **40**, 342
Lahtinen J., Vehanen A., Huomo H., Mäkinen J., Huttunen P., Rytsölä K., Bentzon M., Hautojärvi P. (1989): Nucl. Instrum. Meth. Phys. Res. B **17**, 72
Laine T., Mäkinen J., Saarinen K., Hautojärvi P., Corbel C., Fille M.L., Gibart P. (1996a): Phys. Rev. B **53**, 11025
Laine T., Saarinen K., Hautojärvi P., Corbel C., Ashwin M.J., Newman R.C. (1997a): Appl. Phys. Lett. **71**, 1843
Laine T., Saarinen K., Hautojärvi P., Corbel C., Ashwin M.J., Newman R.C. (1997b): Mater. Sci. Forum **258–263**, 879
Laine T., Saarinen K., Mäkinen J., Hautojärvi P., Corbel C., Pfeiffer L.N., Citrin P.H. (1996b): Phys. Rev. B **54**, R11050
Landman J.I., Morgan C.G., Schick J.T., Papoulias P., Kumar A. (1997): Phys. Rev. B **55**, 15581
Lang D.V. (1977): Inst. Phys. Conf. Ser. **31**, 70
Lang D.V., Logan R.A. (1977): Phys. Rev. Lett. **39**, 635
Lazar M., Wagner G. (1997): Cryst. Res. Technol. **32**, 111
Le Berre C., Corbel C., Brozel M.R., Kuisma S., Saarinen K., Hautojärvi P. (1994): J. Phys. Cond. Mat. **6**, L759
Le Berre C., Corbel C., Saarinen K., Kuisma S., Hautojärvi P., Fornari R. (1995): Phys. Rev. B **52**, 8112
Lee J.L., Shim K.H., Tanigawa S., Uedono A., Kim J.S., Park H.M., Ma D.S. (1988a): J. Physique **49**, 457
Lee J.-L., Kim J.S., Park H.M., Ma D.S., Tanigawa S., Uedono A. (1988b): Appl. Phys. Lett. **53**, 1302
Lee J.-L., Shim K.-H., Kim J.S., Park H.M., Ma D.S., Tanigawa S., Uedono A. (1989): J. Appl. Phys. **65**, 396
Lee J.-L., Uedono A., Tanigawa S., Lee J.Y. (1990): J. Appl. Phys. **67**, 6153
Lee Y.H., Corbett J.P. (1976): Phys. Rev. B **13**, 2653
Leipner H.S., Hübner C., Storbeck O., Polity A., Krause-Rehberg R. (1996): In *Semiconducting and insulating materials 1996.* IEEE, Piscataway, p. 283
Leipner H.S., Hübner C.G., Krüger J., Krause-Rehberg R. (1997): Mater. Sci. Forum **258–263**, 981
Leipner H.S., Hübner C.G., Krüger J., Krause-Rehberg R., Billard I. (1998): phys. stat. sol. (a) in preparation
Leipner H.S., Krause-Rehberg R., Hübner C. (1995): Mater. Sci. Forum **196–201**, 1267

Leipner H.S., Krause-Rehberg R., Kupsch A., Drost T., Polity A. (1993): phys. stat. sol. (a) **138**, 489

Leipner H.S., Krause-Rehberg R., Polity A., Drost T., Alexander H. (1994): Mater. Sci. Forum **143–147**, 1589

Lelikov Y., Rebane Y., Ruvimov V., Tarhin D., Sitnikova A., Shreter Y. (1992): Mater. Sci. Forum **83–87**, 1321

Lessor D.L., Duke C.B., Chen X.M., Brandes G.R., Canter K.F., Ford W.K. (1992): J. Vac. Sci. Technol. A **10**, 2585

Leung T.C., Simpson P.J., Atkinson A., Mitchell I.V., Schultz P.J. (1995): Appl. Surf. Sci. **85**, 292

Lide D.R. (ed.) (1996): Handbook of chemistry and physics. CRC Press, Boca Raton

Liliental-Weber Z., Cooper G., Mariella R., Kocot C. (1991): J. Vac. Sci. Technol. B **9**, 2323

Linhard J., Scharff M., Schiøtt H.E. (1963): Kgl. Dan. Videnskab. Mat. Phys. Medd. **33**, 14

Liszkay L., Corbel C., Baroux L., Hautojärvi P., Bayhan M., Brinkmann A.W., Tatarenko S. (1994): Appl. Phys. Lett. **64**, 1380

Liszkay L., Corbel C., Baroux L., Hautojärvi P., Declemy A., Renault P.O. (1995): J. Phys. Cond. Mat. **7**, 8529

Liszkay L., Huomo H., Rönkä K., Rytsölä K., Hautojärvi P. (1992): Mater. Sci. Forum **105–110**, 1931

Luysberg M., Sohn H., Prasad A., Specht P., Liliental-Weber Z., Weber E.R., Gebauer J., Krause-Rehberg R. (1998): J. Appl. Phys. **83**, 561

Lynn K.G., Asoka-Kumar P. (1992): Mater. Sci. Forum **105–110**, 359

Lynn K.G., Frieze W.E., Schultz P.J. (1984): Phys. Rev. Lett. **52**, 1137

Lynn K.G., MacDonald J.R., Boie R.A., Feldman L.C., Gabbe J.D., Robbins M.F., Bonderup E., Golovchenko J. (1977): Phys. Rev. Lett. **38**, 241

Lynn K.G., Welch D.O. (1980): Phys. Rev. B **22**, 99

MacDonald J.R., Lynn K.G., Boie R.A., Robbins M.F. (1978): Nucl. Instrum. Meth. **153**, 189

MacKenzie I.K., Khoo T.L., McDonald A.D., McKee B.T.A. (1967): Phys. Rev. Lett. **19**, 946

Madelung O. (ed.) (1982a): Landolt-Börnstein. Physics of group II–VI and I–VII compounds, semimagnetic semiconductors. Springer-Verlag, Berlin

Madelung O. (ed.) (1982b): Landolt-Börnstein. Physics of group IV elements and III–V compounds. Springer-Verlag, Berlin

Madelung O. (ed.) (1985): Landolt-Börnstein. Physics of ternary compounds. Springer-Verlag, Berlin

Magerle R., Burchhard A., Forkel-Wirth T., Deichler M. (1997): Mater. Sci. Forum **258–263**, 945

Maguire J., Murray R., Newman R.C., Beall R.B., Harris J.J. (1987): Appl. Phys. Lett. **50**, 516

Mahajan S., Bonner W.A., Chin A.K., Miller D.C. (1979): Appl. Phys. Lett. **35**, 165

Mahony J., Mascher P. (1997): Phys. Rev. B **55**, 9637

Mahony J., Mascher P., Puff W. (1996): J. Appl. Phys. **80**, 2712

Mahony J., Tessaro G., Mascher P., Siethoff H., Brion H.G. (1995): Mater. Sci. Forum **196–201**, 1449

Maier K. (1997): personal communication

Makhov A.F. (1961): Sov. Phys. Sol. State **2**, 1934

Mäkinen J., Corbel C., Hautojärvi P., Mathiot D. (1991a): Phys. Rev. B **43**, 12114

Mäkinen J., Corbel C., Hautojärvi P., Moser P., Pierre F. (1989): Phys. Rev. B **39**, 10162

Mäkinen J., Hautojärvi P., Corbel C. (1992a): J. Phys. Cond. Mat. **4**, 5137

Mäkinen J., Laine T., Partanen J., Saarinen K., Hautojärvi P., Tappura K., Hakkarainen T., Asonen H., Pessa M., Kauppinen J.P., Vanttinen K., Paalanen M.A., Likonen J. (1996): Phys. Rev. B **53**, 7851

Mäkinen J., Laine T., Saarinen K., Hautojärvi P., Corbel C., Airaksinen V.M., Gibart P. (1993): Phys. Rev. Lett. **71**, 3154

Mäkinen J., Laine T., Saarinen K., Hautojärvi P., Corbel C., Airaksinen V.M., Nagle J. (1995): Phys. Rev. B **52**, 4870

Mäkinen J., Palko S., Martikainen J., Hautojärvi P. (1992b): J. Phys. Cond. Mat. **4**, L503
Mäkinen J., Punkka E., Vehanen A., Hautojärvi P., Keinonen J., Hautala M., Rauhala E. (1990a): J. Appl. Phys. **67**, 990
Mäkinen J., Vehanen A., Hautojärvi P., Huomo H., Lahtinen J., Nieminen R.M., Valkealahti S. (1986): Surf. Sci. **175**, 385
Mäkinen S., Puska M.J. (1989): Phys. Rev. B **40**, 12523
Mäkinen S., Rajainmaki H., Linderoth S. (1990b): Phys. Rev. B **42**, 11166
Mäkinen S., Rajainmaki H., Linderoth S. (1991b): Phys. Rev. B **44**, 5510
Manasreh M.O., Fischer D.W., Mitchel W.C. (1989): phys. stat. sol. (b) **154**, 11
Manasreh M.O., Look D.C., Evans K.R., Stutz C.E. (1990): Phys. Rev. B **41**, 10272
Manninen M., Nieminen R.M. (1981): Appl. Phys. A **26**, 93
Mantl S., Triftshäuser W. (1978): Phys. Rev. B **17**, 1645
Manuel A.A. (1995): In *Positron spectroscopy of solids.* Dupasquier A., Mills A.P. (eds.). IOS Press, Amsterdam, p. 155
Marder S., Hughes V., Wu C.S., Bennet W. (1956): Phys. Rev. **103**, 1258
Martin G.M. (1981): Appl. Phys. Lett. **39**, 747
Martin G.M., Makram-Ebeid S. (1986): In *Deep centers in semiconductors.* Pantelides S.T. (ed.). Gordon & Breach, New York, p. 309
Martin G.M., Makran-Ebeid S. (1983): Physica B+C **116**, 371
Mascher P., Dannefaer S., Kerr D. (1989a): Mater. Sci. Forum **38–41**, 1157
Mascher P., Dannefaer S., Kerr D. (1989b): Phys. Rev. B **40**, 11764
Mascher P., Kerr D., Dannefaer S. (1987a): Phys. Rev. B **35**, 4043
Mascher P., Kerr D., Dannefaer S. (1987b): In *18th International Conference on the Physics of Semiconductors.* Engstrom O. (ed.). World Scientific, Singapore, p. 845
Mascher P., Kerr D., Dannefaer S. (1987c): J. Cryst. Growth **85**, 295
Mascher P., Kerr D., Dannefaer S. (1988): Cryst. Res. Technol. **23**, 247
Mascher P., Liu M., Kitai A.H., Puff W. (1992): Mater. Sci. Forum **105–110**, 1145
Massoumi G.R., Lennard W.N., Schultz P.J., Walker A.B., Jensen K.O. (1994): AIP Conf. Proc. **303**, 564
Maude D.K., Portal J.C., Murray R., Foster T.J., Dmowski L., Eaves L., Newman R., Basmaji P., Gibart P., Harris J.J., Beall R.B. (1989): Sol. State Phen. **10**, 121
McKee B.T.A., Saimoto S., Stewart A.T., Stott M.J. (1974): Can. J. Phys. **52**, 756
Melloch M.R., Otsuka N., Woodall J.M., Warren A.C., Freeouf J.L. (1990): Appl. Phys. Lett. **57**, 1531
Mendez B., Piqueras J., Dominguez-Adame F., de Diego N. (1988): J. Appl. Phys. **64**, 4466
Meng X. (1994a): Phys. Scr. **50**, 419
Meng X. (1994b): Phil. Mag. B **70**, 905
Meng X. (1995): Nucl. Instrum. Meth. Phys. Res. B **95**, 65
Meng X.T., Charalambous S., Chardalas M., Dedoussis S.P., Eleftheriadis C.A., Liolios A.K. (1995a): Rad. Eff. Def. Sol. **133**, 97
Meng X.T., Puff W. (1994): J. Phys. Cond. Mat. **6**, 4971
Meng X.T., Zecca A., Brusa R.S. (1995b): Appl. Phys. A **60**, 81
Meyendorf N., Krause R. (1992): Materialprüfung **34**, 188
Meyer B.K., Omling P., Weigel E., Müller-Vogt G. (1992): Phys. Rev. B **46**, 15135
Michel J., Kimerling L.C. (1994): Semicond. Semimet. **42**, 251
Mijnarends P.E. (1979): In *Positrons in solids.* Hautojärvi P. (ed.). Springer-Verlag, Berlin, p. 25
Mills A.P., Jr. (1978): Phys. Rev. Lett. **41**, 1828
Mills A.P.J., Gullikson E.M. (1986): Appl. Phys. Lett. **49**, 1121
Miyajima T., Okuyama H., Akimoto K., Mori Y., Wei L., Tanigawa S. (1991): Appl. Phys. Lett. **59**, 1482
Miyajima T., Wei L., Tanigawa S., Okuyama H., Akimoto K., Mori Y. (1992): Mater. Sci. Forum **105–110**, 1423
Mogensen O., Kvajic G., Eldrup M., Milosevic-Kvajic M. (1971): Phys. Rev. B **4**, 71

Mogensen O.E. (1974): J. Chem. Phys. **60**, 998

Mogensen O.E. (1995): *Positron annihilation in chemistry.* Springer-Verlag, Berlin

Mohorovicic S. (1934): Astron. Nachr. **253**, 93

Mokrushin A.D., Girka A.I., Shishkin A.V. (1991): phys. stat. sol. (a) **128**, 31

Molva E., Pautrat J.L. (1981): Sol. State Commun. **39**, 1151

Mooney P.M., Caswell N.S., Wright S.L. (1987): J. Appl. Phys. **62**, 4786

Mori M., Kano G., Inoue T., Shimakura H., Yamamoto H., Oda O. (1990): In *Semi-insulating III–V materials.* Milnes A.G., Miner C.J. (eds.). Adam Hilger, Bristol, p. 155

Morozov A.N., Bublik V.T. (1986): J. Cryst. Growth **75**, 491

Moser P., Pautrat J.L., Corbel C., Hautojärvi P. (1985): In *Positron annihilation.* Jain P.C., Singru R.M., Gopinathan K.P. (eds.). World Scientific, Singapore, p. 733

Motoko-Kwete, Segers D., Dorikens M., Dorikens-Vanpraet L., Clauws P., Geshef D. (1990): Phys. Lett. A **150**, 413

Motoko-Kwete, Segers D., Dorikens M., Dorikens-Vanpraet L., Clauws P., Lemahieu I. (1989): In *Positron annihilation.* Dorikens-Vanpraet L., Dorikens M., Segers D. (eds.). World Scientific, Singapore, p. 687

Motooka T., Hiroyama Y., Suzuki R., Ohdaira T., Hirano Y., Sato F. (1995): Nucl. Instrum. Meth. Phys. Res. B **106**, 198

Mott N.F., Massey H.S.N. (eds.) (1965): Theory of atomic collisions. University Press, Oxford

Muto S., Takeda S., Hirata M., Fujii K., Ibe K. (1992): Phil. Mag. A **66**, 257

Myler U., Goldberg R.D., Knights A.P., Lawther D.W., Simpson P.J. (1996): Appl. Phys. Lett. **69**, 3333

Myler U., Simpson P.J. (1997): Phys. Rev. B **56**, 14303

Nasu H., Watanabe M., Shizuma K., Imura T., Osaka Y., Hiromi H. (1987): J. Ceram. Soc. Jap. Inter. Ed. **95**, 1095

Navarro F.J., Damonte L.C., Mari B., Ferrero J.L. (1996): J. Appl. Phys. **79**, 9043

Nelson A.J., Gabor A.M., Contreras M.A., Tuttle J.R., Noufi R., Sobol P.E., Asoka-Kumar P., Lynn K.G. (1995): J. Appl. Phys. **78**, 269

Neubert M., Jacobs K., Krause-Rehberg R., Abgarjan T., Gille P., Hoerstel W. (1996): J. Appl. Phys. **79**, 7563

Neugebauer J., Van de Walle C.G. (1996): Appl. Phys. Lett. **69**, 503

Newman R.C. (1993): Semicond. Semimet. **38**, 117

Newman R.C. (1994): Semicond. Sci. Technol. **9**, 1749

Newman R.C., Ashwin M.J., Fahy M.R., Hart L., Holmes S.N., Roberts C., Zhang X., Wagner J. (1996): Phys. Rev. B **54**, 8769

Newman R.C., Davidson B.R., Addinall R., Murray R., Emmert J.W., Wagner J., Götz W., Roos G., Pensl G. (1994): Mater. Sci. Forum **143–147**, 229

Nico J.S., Gidley D.W., Rich A. (1990): Phys. Rev. Lett. **65**, 1344

Nielsen B., Holland O.W., Leung T.C., Lynn K.G. (1993): J. Appl. Phys. **74**, 1636

Nielsen B., Lynn K.G., Chen Y.-C., Welch D.O. (1987): Appl. Phys. Lett. **51**, 1022

Nielsen B., Lynn K.G., Leung T.C., Cordts B.F., Seraphin S. (1991): Phys. Rev. B **44**, 1812

Nielsen B., Lynn K.G., Leung T.C., Welch D.O., Rubloff G. (1988): In *SiO_2 and its interfaces.* Pantelides S.T., Lucovsky G. (eds.). Mater. Res. Soc., Pittsburgh, p. 241

Nielsen B., Lynn K.G., Vehanen A., Schultz P.J. (1985): Phys. Rev. B **32**, 2296

Nielsen B., Lynn K.G., Welch D.O., Leung T.C., Rubloff G.W. (1989): Phys. Rev. B **40**, 1434

Nieminen R.M., Laakkonen J. (1979): Appl. Phys. **20**, 181

Nieminen R.M., Laakkonen J., Hautojärvi P., Vehanen A. (1979): Phys. Rev. B **19**, 1397

Nieminen R.M., Manninen M.J. (1979): In *Positrons in solids.* Hautojärvi P. (ed.). Springer-Verlag, Berlin, p. 145

Nieminen R.M., Puska M.J. (1983): Phys. Rev. Lett. **50**, 281

Nieminen R.M., Puska M.J., Manninen M. (1984): Phys. Rev. Lett. **53**, 1298

Nilen R.W.N., Connell S.H., Britton D.T., Fischer C.G., Sendezera E.J., Schmidt W.G., Sellschop J.P.F., Verwoerd W.S. (1997a): Mater. Sci. Forum **255–257**, 475

Nilen R.W.N., Lauff U., Connell S.H., Stoll H., Siegle A., Schneider H., Castellaz P., Kraft J., Bharuth-Ram K., Sellschop J.P.F., Seeger A. (1997b): Appl. Surf. Sci. **116**, 198

Noack M., Kehr K.W., Wenzl H. (1997): J. Cryst. Growth **178**, 438

Northrup J.E., Zhang S.B. (1993): Phys. Rev. B **47**, 6791

Oda O., Yamamoto H., Seiwa M., Kano G., Inoue T., Mori M., Shikamura H., Oyake M. (1992): Semicond. Sci. Technol. **7**, A215

Ohji Y., Uedono A., Wei L., Tabuki Y., Tanigawa S. (1992): Mater. Res. Soc. Symp. Proc. **262**, 313

Ohkawa K., Karasawa T., Mitsuyu T. (1991): J. Cryst. Growth **111**, 797

Ohki A., Kawaguchi Y., Ando K., Katsui A. (1991): Appl. Phys. Lett. **59**, 671

Oliva J. (1980): Phys. Rev. B **21**, 4909

Omling P., Weber E.R., Samuelson L. (1986): Phys. Rev. B **33**, 5880

Ore A. (1949): Univ. Bergen Aarb. Naturvit. rekke **9**,

Oshima R., Mori M., Hua G.C., Honda S., Kiritani M., Fujita F.E. (1989): Mater. Sci. Forum **38–41**, 1199

Panda B.K., LiMing W., Fung S., Beling C.D. (1997): Phys. Rev. B **56**, 7356

Pankove J.I. (1971): Optical processes in semiconductors. Dover, New York

Pareja R., de la Cruz R.M. (1993): phys. stat. sol (b) **178**, K23

Pareja R., de la Cruz R.M., Moser P. (1992): J. Phys. Cond. Mat. **4**, 7153

Park C.H., Chadi D.J. (1995): Appl. Phys. Lett. **66**, 3127

Park R.M., Troffer M.B., Rouleau C.M., DePuydt J.M., Haase M.A. (1990): Appl. Phys. Lett. **57**, 2127

Pautrat J.L., Francou J.M., Magnea N., Molva E., Saminadayar K. (1985): J. Cryst. Growth **72**, 194

Pedersen M.J., Andersen J.U., Augustyniak W.M. (1972): Rad. Eff. **12**, 47

Peng J.P., Lynn K.G., Asoka-Kumar P., Becker D.P., Harshman D.R. (1996): Phys. Rev. Lett. **76**, 2157

Peng J.P., Lynn K.G., Umlor M.T., Keeble D.J., Harshman D.R. (1994): Phys. Rev. B **50**, 11247

Perkins A., Carbotte J.P. (1970): Phys. Rev. B **1**, 101

Perovic D.D., Weatherly G.C., Noël J.P., Houghton D.C. (1991a): J. Vac. Sci. Technol. B **9**, 2034

Perovic D.D., Weatherly G.C., Simpson P.J., Schultz P.J., Jackman T.E., Aers G.C., Noël J.P., Houghton D.C. (1991b): Phys. Rev. B **43**, 14257

Plazaola F., Seitsonen A.P., Puska M.J. (1994): J. Phys. Cond. Mat. **6**, 8809

Plazaola F., Seitsonen A.P., Puska M.J. (1995): Mater. Sci. Forum **175–178**, 469

Pogrebnyak A.D. (1984): phys. stat. sol. (a) **86**, 191

Pogrebnyak A.D., Boyarko E.Y., Kryuchov, Yu. Yu., Vorobev S.A. (1984a): phys. stat. sol. (a) **81**, 217

Pogrebnyak A.D., Kuznetsov M.F., Surov Y.P. (1984b): Phys. Lett. A **105**, 69

Pogrebnyak A.D., Kuzminikh V.A., Arefiev K.P. (1982): phys. stat. sol. (b) **112**, K79

Pogrebnyak A.D., Lopatin V.S., Ziyakaev R.G., Vorobiev S.A. (1983): Phys. Lett. A **97**, 362

Polity A. (1997): Habilitation, Martin-Luther-Universität Halle–Wittenberg

Polity A., Abgarjan T., Krause-Rehberg R. (1995a): Appl. Phys. A **60**, 541

Polity A., Abgarjan T., Krause-Rehberg R. (1995b): Mater. Sci. Forum **175–178**, 473

Polity A., Börner F., Huth S., Eichler S., Krause-Rehberg R. (1998a): Phys. Rev. B **58**, to be published

Polity A., Engelbrecht T. (1997a): Phys. Rev. B **55**, 10480

Polity A., Engelbrecht T. (1997b): unpublished results

Polity A., Huth S., Krause-Rehberg R. (1997a): Mater. Sci. Forum **255–257**, 602

Polity A., Krause-Rehberg R. (1995): unpublished results

Polity A., Krause-Rehberg R., Staab T.E.M., Puska M.J., Klais J., Möller H.J., Meyer B.K. (1998b): J. Appl. Phys. **83**, 71

Polity A., Krause-Rehberg R., Zlomanov V., Stanov V., Chatchaturov A., Mäkinen S. (1993): J. Cryst. Growth **131**, 271

Polity A., Rudolf F., Abgarjan T. (1997b): unpublished results
Polity A., Rudolf F., Nagel C., Eichler S., Krause-Rehberg R. (1997c): Phys. Rev. B **55**, 10467
Pons D., Bourgoin J.C. (1985): J. Phys. C: Semicond. Phys. **18**, 3839
Pöykkö S., Puska M.J., Alatalo M., Nieminen R.M. (1996a): Phys. Rev. B **54**, 7909
Pöykkö S., Puska M.J., Nieminen R.M. (1996b): Phys. Rev. B **53**, 3813
Pöykkö S., Puska M.J., Nieminen R.M. (1997): Phys. Rev. B **55**, 6914
Prasad A., Liu X., Fujioka H., Jaeger N.D., Nishio J., Weber E.R. (1995): Mater. Sci. Forum **196–201**, 189
Pritchard R.E., McQuaid S.A., Hart L., Newman R.C., Mäkinen J., von Bardeleben H.J., Missous M. (1995): J. Appl. Phys. **78**, 2411
Provencher S.W. (1982): Comp. Phys. Comm. **27**, 213
Puff W., Balogh A.G., Birkholz M., Fiechter S. (1992a): Mater. Sci. Forum **105–110**, 1177
Puff W., Balogh A.G., Mascher P., Baumann H. (1997): Mater. Sci. Forum **248–249**, 285
Puff W., Boumerzoug M., Brown J., Mascher P., Macdonald D., Simpson P.J., Balogh A.G., Hahn H., Chang W., Rose M. (1995a): Appl. Phys. A **61**, 55
Puff W., Brunner S., Mascher P., Balogh A.G. (1995b): Mater. Sci. Forum **196–201**, 333
Puff W., Mascher P., Hahn S., Cho K.H., Lee B.Y. (1992b): Mater. Sci. Forum **105–110**, 1185
Puff W., Mascher P., Kerr D., Dannefaer S. (1989): Mater. Sci. Forum **38–41**, 225
Puska M. (1978): Elinaikaspektrin sovitusohjelman LIFSPECFIT kehittäminen. Teknillinen Korkeakoulu, Otaniemi
Puska M.J. (1987): phys. stat. sol. (a) **102**, 11
Puska M.J. (1989a): J. Phys. Cond. Mat. **1**, 7347
Puska M.J. (1989b): In *Positron Annihilation.* Dorikens-Vanpraet L., Dorikens M., Segers D. (eds.). World Scientific, Singapore, p. 101
Puska M.J. (1995): private communication
Puska M.J., Corbel C. (1988): Phys. Rev. B **38**, 9874
Puska M.J., Corbel C., Nieminen R.M. (1990): Phys. Rev. B **41**, 9980
Puska M.J., Hakala M. (1997): unpublished results
Puska M.J., Jepsen O., Gunnarsson O., Nieminen R.M. (1986): Phys. Rev. B **34**, 2695
Puska M.J., Lanki P., Nieminen R.M. (1989a): J. Phys. Cond. Mat. **1**, 6081
Puska M.J., Mäkinen S., Manninen M., Nieminen R.M. (1989b): Phys. Rev. B **39**, 7666
Puska M.J., Nieminen R.M. (1983): J. Phys. F: Met. Phys. **13**, 333
Puska M.J., Nieminen R.M. (1994): Rev. Mod. Phys. **66**, 841
Read W.T. (1954): Phil. Mag. **45**, 775
Rempel A.A., Schaefer H.E. (1995): Appl. Phys. A **61**, 51
Rich A. (1981): Rev. Mod. Phys. **53**, 127
Roorda S., Hakvoort R.A., Veen A.v., Stolk P.A., Saris F.W. (1992): J. Appl. Phys. **72**, 5145
Rosenberg I.J., Weiss A.H., Canter K.F. (1980): Phys. Rev. Lett. **44**, 1139
Rubaszek A., Kiejna A., Daniuk S. (1993): J. Phys. Cond. Mat. **5**, 8195
Rubloff G.W., Nielsen B., Lynn K.G., Welch D.O., Leung T.C. (1990): Vacuum **41**, 790
Ruvimov S., Dicker C., Washburn J., Liliental-Weber Z. (1998): Appl. Phys. Lett. **72**, 226
Saarinen K., Corbel C., Hautojärvi P., Lanki P., Pierre F., Vignaud D. (1990): J. Phys. Cond. Mat. **2**, 2453
Saarinen K., Hautojärvi P., Corbel C. (1997a): Nucl. Instrum. Meth. Phys. Res. A **388**, 434
Saarinen K., Hautojärvi P., Corbel C. (1998): In *Identification of defects in semiconductors.* Stavola M. (ed.). Academic Press, New York
Saarinen K., Hautojärvi P., Lanki P., Corbel C. (1991): Phys. Rev. B **44**, 10585
Saarinen K., Hautojärvi P., Vehanen A., Krause R., Dlubek G. (1989): Phys. Rev. B **39**, 5287
Saarinen K., Kuisma S., Hautojärvi P., Corbel C., LeBerre C. (1993): Phys. Rev. Lett. **70**, 2794
Saarinen K., Kuisma S., Hautojärvi P., Corbel C., LeBerre C. (1994): Phys. Rev. B **49**, 8005
Saarinen K., Kuisma S., Mäkinen J., Hautojärvi P., Tornqvist M., Corbel C. (1995a): Phys. Rev. B **51**, 14152

Saarinen K., Laine T., Kuisma S., Nissilä J., Hautojärvi P., Dobrzynski L., Baranowski J.M., Pakula K., Stepniewski R., Wojdak M., Wysmolek A., Suski T., Leszczynski M., Grzegory I., Porowski S. (1997b): Phys. Rev. Lett. **79**, 3030
Saarinen K., Laine T., Skog K., Mäkinen J., Hautojärvi P., Rakennus K., Uusimaa P., Salokatve A., Pessa M. (1996): Phys. Rev. Lett. **77**, 3407
Saarinen K., Seitsonen A.P., Hautojärvi P., Corbel C. (1995b): Phys. Rev. B **52**, 10932
Sadamitsu S., Okui M., Sueoka K., Marsden K., Shigematsu T. (1995): Jpn. J. Appl. Phys. Pt. 2 **34**, L597
Sadamitsu S., Umeno S., Koike Y., Hourai M., Sumita S., Shigematsu T. (1993): Jpn. J. Appl. Phys. **32**, 3675
Saito H., Nagashima Y., Hyodo T., Chang T.B. (1995): Mater. Sci. Forum **175–178**, 769
Saito M., Oshiyama A. (1996): Phys. Rev. B **53**, 7810
Saito M., Oshiyama A., Tanigawa S. (1991): Phys. Rev. B **44**, 10601
Saito M., Tang Z., Chiba T., Hasegawa M. (1997): Mater. Sci. Forum **255–257**, 184
Sajovec F., Wolf R., Fattah A., Bickmann K., Wenzl H., Nagel G., Rufer H., Tomzig E., de Bievre P. (1990): phys. stat. sol. (a) **122**, 139
Saleh A.S., Rice-Evans P.C. (1994): Diam. Rel. Mater. **3**, 1293
Schaefer H.E., Würschum R., Schwarz R., Slobodin D., Wagner S. (1986): Appl. Phys. A **40**, 145
Schröter W., Labusch R. (1969): phys. stat. sol. **36**, 539
Schubert E.F. (1993): In *Doping in III–V semiconductors.* Schubert E.F. (ed.). University Press, Cambridge, p. 433
Schultz P.J., Logan L.R., Jackman T.E., Davies J.A. (1988a): Phys. Rev. B **32**, 6369
Schultz P.J., Lynn K.G. (1988): Rev. Mod. Phys. **60**, 701
Schultz P.J., Simpson P.J., Akano U.G., Mitchell I.V. (1992): In *Materials Modification by Energetic Atoms and Ions Symposium.* Grabowski K.S., Barnett S.A., Rossnagel S.M., Wasa K. (eds.). Mater. Res. Soc., Pittsburgh, p. 319
Schultz P.J., Tandberg E., Lynn K.G., Nielsen B., Jackman T.E., Denhoff M.W., Aers G.C. (1988b): Phys. Rev. Lett. **61**, 187
Schuppler S., Adler D.L., Pfeiffer L.N., West K.W., Chaban E.E., Citrin P.H. (1993): Appl. Phys. Lett. **63**, 2357
Schuppler S., Adler D.L., Pfeiffer L.N., West K.W., Chaban E.E., Citrin P.H. (1995): Phys. Rev. B **51**, 10527
Schut H., van Veen A., van de Walle G.F.A., van Gorkum A.A. (1991): J. Appl. Phys. **70**, 3003
Seeger A. (1992): Mater. Sci. Forum **105–110**, 821
Seeger A., Banhart F., Bauer W. (1989): In *Positron annihilation.* Dorikens-Vanpraet L., Dorikens M., Segers D. (eds.). World Scientific, Singapore, p. 275
Seeger A., Swanson M.L. (1968): In *Lattice defects in semiconductors.* Hasiguti R.R. (ed.). University of Tokyo Press, Tokyo, p. 93
Seitsonen A.P., Virkkunen R., Puska M.J., Nieminen R.M. (1994): Phys. Rev. B **49**, 5253
Sen Gupta A., Moser P., Corbel C., Hautojärvi P. (1988a): Cryst. Res. Techn. **23**, 409
Sen Gupta A., Moser P., Corbel C., Hautojärvi P., Sen P. (1988b): Cryst. Res. Techn. **23**, 243
Sen Gupta A., Naidu S.V., Bhandari R.K., Sen P. (1984): Phys. Lett. A **104**, 117
Sen Gupta A., Naidu S.V., Roy R., Sen P. (1986): Sol. State Commun. **58**, 219
Sen Gupta A., Naidu S.V., Sen P. (1985): Phys. Lett. A **112**, 399
Sen-Gupta A., Moser P. (1989): In *Positron annihilation.* Dorikens-Vanpraet L., Dorikens M., Segers D. (eds.). World Scientific, Singapore, p. 731
Seong H., Lewis L. (1995): Phys. Rev. B **52**, 5675
Shan Y.Y., Asoka-Kumar P., Lynn K.G., Fung S., Beling C.D. (1996): Phys. Rev. B **54**, 1982
Shan Y.Y., Au H.L., Ling C.C., Lee T.C., Panda B.K., Fung S., Beling C.D., Wang Y.Y., Weng H.M. (1994): Appl. Phys. A **59**, 259
Shan Y.Y., Ling C.C., Deng A.H., Panda B.K., Beling C.D., Fung S. (1997a): Phys. Rev. B **55**, 7624
Shan Y.Y., Lynn K.G., Asoka-Kumar P., Fung S., Beling C.D. (1997b): Phys. Rev. B **55**, 9897

Shan Y.Y., Lynn K.G., Szeles C., Asoka-Kumar P., Thio T., Bennett J.W., Beling C.B., Fung S., Becla P. (1997c): Phys. Rev. Lett. **79**, 4473
Shan Y.Y., Lynn K.G., Szeles C., Asoka-Kumar P., Thio T., Bennett J.W., Beling C.D., Fung S., Becla P. (1997d): Mater. Sci. Forum **255–257**, 506
Sharma S.C., Hozhabri N., Hyer R.C., Hossain T., Kim S., F. O. Meyer, Pas M.F., Stephens A.E. (1992): Mater. Res. Soc. Symp. Proc. 45
Sher A., Berding M.A., van Schilfgaarde M. (1991): Semicond. Sci. Technol. **6**, C59
Shi M., Waeber W.B., Triftshauser W. (1997): Appl. Surf. Sci. **116**, 203
Shikata S., Fujii S., Long W., Tanigawa S. (1992): Jpn. J. Appl. Phys. Pt. 1 **31**, 732
Shimotomai M., Ohgino Y., Fukushima H., Nagayazu Y., Mihara T., Inoue K., Doyama M. (1981): Inst. Phys. Conf. Ser. **59**, 241
Shimura F. (ed.) (1994): Oxygen in silicon. Academic Press, Boston
Shreter Y.G., Rebane Y.T. (1996): In *The physics of semiconductors*. Scheffler M., Zimmermann R. (eds.). World Scientific, Singapore, p. 2937
Shukla A., Peter M., Hoffmann L. (1993): Nucl. Instrum. Meth. A **335**, 310
Siegle A. (1997): Dissertation, Universität Stuttgart
Siethoff H. (1992): Semicond. Semimet. **37**, 143
Siethoff H. (1998): phys. stat. sol. (b) **205**, R3
Simpson P.J., Akano U.G., Schultz P.J., Mitchell I.V. (1992a): Mater. Sci. Forum **105–110**, 1435
Simpson P.J., Schultz P.J., Jackman T.E., Aers G.C., Nöel J.-P., Houghton D.C., Perovic D.D., Weatherly G.C. (1991a): AIP Conf. Proc. **218**, 125
Simpson P.J., Schultz P.J., Lee S.T., Chen S., Braunstein G. (1992b): J. Appl. Phys. **72**, 1799
Simpson P.J., Vos M., Mitchell I.V., Wu C., Schultz P.J. (1991b): Phys. Rev. B **44**, 12180
Smedskjaer L.C., Chason M.K., Siegel R.W. (1985): In *Positron Annihilation.* Jain P.C., Singru R.M., Gopinathan K.P. (eds.). World Scientific, Singapore, p. 549
Smedskjaer L.C., Manninen M., Fluss M.J. (1980): J. Phys. F: Met. Phys. **10**, 2237
Smith C., Rice-Evans P.C., Shaw N. (1994a): AIP Conf. Proc. **303**, 78
Smith C.D., Rice-Evans P., Shaw N. (1994b): Phys. Rev. Lett. **72**, 1108
Smith D.L., Smith C., Evans H.E., Rice-Evans P.C., Evans J.H. (1991): AIP Conf. Proc. **218**, 134
Smith F.W., Calawa A.R., Chen C.L., Manfra M.J., Mahoney L.J. (1988): IEEE Electron. Dev. Lett. **9**, 77
Snead C.L., Lynn K.G. (1985): In *Positron annihilation.* Jain P.C., Singru R.M., Gopinathan K.P. (eds.). World Scientific, Singapore, p. 720
Soininen E., Huomo H., Huttunen P.A., Mäkinen J., Hautojärvi P., Vehanen A. (1990): Phys. Rev. B **41**, 6227
Soininen E., Mäkinen J., Beyer D., Hautojärvi P. (1992): Phys. Rev. B **46**, 13104
Soininen E., Schwab A., Lynn K.G. (1991): Phys. Rev. B **43**, 10051
Somieski B., Krause-Rehberg R., Salz H., Meyendorf N. (1995): J. Physique IV **5**, C1/127
Somieski B., Salz H., Krause-Rehberg R. (1994): NDT&E Intern. **27**, 235
Somieski B., Staab T.E.M., Krause-Rehberg R. (1996): Nucl. Instrum. Meth. Phys. Res. A **381**, 128
Sørensen H., Uggerhøj E. (1987): Nature **324**, 311
Spaeth J.-M., Niklas J.R., Bartram R.H. (1992): Structural analysis of point defects in solids. Springer-Verlag, Berlin
Sperr P., Kögel G. (1997): Mater. Sci. Forum **255–257**, 109
Staab T.E.M., Leipner H.S., Hübner C.G., Haugk M., Krause-Rehberg R. (1998): phys. stat. sol. (a) submitted
Staab T.E.M., Somieski B., Krause-Rehberg R. (1996): Nucl. Instrum. Meth. Phys. Res. A **381**, 141
Stadler W., Hofmann D.M., Meyer B.K., Krause-Rehberg R., Polity A., Abgarjan T., Salk M., Benz K.W., Azoulay M. (1995): Acta Phys. Polon. A **88**, 921
Stein T.S., Kauppila W.E., Roelling L.O. (1974): Rev. Sci. Instrum. **45**, 951

Steindl R., Kögel G., Sperr P., Willutzki P., Britton D.T., Triftshäuser W. (1992): Mater. Sci. Forum **105–110**, 1455
Stoll H., Koch M., Lauff U., Maier K., Major J., Seeger A., Wesolowski P., Billard I., Abbe J.C., Duplatre G., Connell S.H., Sellschop J.P.F., Sideras-Haddad E., Bharuth-Ram K., Haricharun H. (1994): AIP Conf. Proc. **303**, 179
Stoll H., Koch M., Maier K., Major J. (1991): Nucl. Instrum. Meth. B **56-57**, 582
Störmer J., Goodyear A., Anwand W., Brauer G., Coleman P.G., Triftshäuser W. (1996a): J. Phys. Cond. Mat. **8**, L89
Störmer J., Triftshäuser W., Hozhabri N., Alavi K. (1996b): Appl. Phys. Lett. **69**, 1867
Störmer J., Willutzki P., Britton D.T., Triftshäuser W., Kiunke W., Hansch W., Eisele I. (1995): Mater. Sci. Forum **175–178**, 213
Stucky M., Corbel C., Geffroy B., Moser P., Hautojärvi P. (1986): Mater. Sci. Forum **10–12**, 265
Stucky M., Paulin R., Geffroy B., Corbel C., Suski J. (1985): In *Positron annihilation.* Jain P.C., Singru R.M., Gopinathan K.P. (eds.). World Scientific, Singapore, p. 714
Sugiura J., Ogasawara M., Uedono A., Wei L., Tanigawa S. (1992): Mater. Res. Soc. Symp. Proc. **262**, 1055
Suzuki R., Kobayashi Y., Mikado T., Ohgaki H., Chiwaki M., Yamazaki T., Tomimasu T. (1991): Jpn. J. Appl. Phys. **30**, L532
Suzuki R., Kobayashi Y., Mikado T., Ohgaki H., Chiwaki M., Yamazaki T., Tomimasu T. (1992): Mater. Sci. Forum **105–110**, 1459
Suzuki R., Ohdaira T., Uedono A., Ishibashi S., Matsuda A., Yoshida S., Ishida Y., Niki S., Fons P.J., Mikado T., Yamazaki T., Tanigawa S., Cho Y.K. (1997): Mater. Sci. Forum **255–257**, 714
Szeles C., Asoka-Kumar A.P., Lynn K.G., Gossmann H.J., Unterwald F.C., Boone T. (1995): Appl. Phys. Lett. **66**, 2855
Szpala S., Asoka-Kumar P., Nielsen B., Peng J.P., Hayakawa S., Lynn K.G., Gossmann H.-J. (1996): Phys. Rev. B **54**, 4722
Taike A., Migita M., Yamamoto H. (1990): Appl. Phys. Lett. **56**, 1989
Takai O., Hisamatsu Y., Owada N., Ishimura H., Hinode K., Tanigawa S., Doyama M. (1980): Phys. Lett. A **76**, 157
Tan T.Y. (1994): J. Phys. Chem. Solids **55**, 917
Tan T.Y., You H.M., Gösele U.M. (1993): Appl. Phys. A **56**, 249
Tandberg E., Schultz P.J., Jackman T.E., Denhoff M.W., Aers G.C., Lynn K.G., Nielsen B. (1989): In *Positron annihilation.* Dorikens-Vanpraet L., Dorikens M., Segers D. (eds.). World Scientific, Singapore, p. 708
Tang Z., Hasegawa M., Chiba T., Saito M., Kawasuso A., Li Z.Q., Fu R.T., Akahane T., Kawazoe Y., Yamaguchi S. (1997): Phys. Rev. Lett. **78**, 2236
Tanigawa S., Hinode K., Nagai R., Doyama M. (1979): Appl. Phys. **18**, 81
Tanigawa S., Tabuki Y., Wei L., Hinode K., Kobayashi N., Onai T., Owada N. (1992a): Mater. Res. Soc. Symp. Proc. **262**, 307
Tanigawa S., Uedono A., Wei L., Suzuki R. (1995): In *Positron spectroscopy of solids.* Dupasquier A., A. P. Mills J. (eds.). IOS Press, Amsterdam, p. 729
Tanigawa S., Watanabe K., Kurihara T., Kubota T. (1990): In *Defect control in semiconductors.* Sumino K. (ed.). Elsevier, Amsterdam, p. 1573
Tanigawa S., Wei L., Tabuki Y. (1992b): Mater. Res. Soc. Symp. Proc. **262**, 319
Tansley T.L., Egan R.J. (1992): Phys. Rev. B **45**, 10942
Teichler H. (1975): Inst. Phys. Conf. Ser. **23**, 374
Terashima K., Tokizaki E., Uedono A., Tanigawa S. (1993): Jpn. J. Appl. Phys. Pt. 1 **32**, 736
Thommen K. (1970): Rad. Eff. **2**, 201
Throwe J., Leung T.C., Nielsen B., Huomo H., Lynn K.G. (1989): Phys. Rev. B **40**, 12037
Tomiyama N., Takenaka M., Kuramoto E. (1992): Mater. Sci. Forum **105–110**, 1281
Törnqvist M., Corbel C., Liszkay L., Saarinen K., Hautojärvi P. (1994a): Mater. Sci. Eng. B **28**, 126

Törnqvist M., Nissilä J., Kiessling F., Corbel C., Saarinen K., Seitsonen A.P., Hautojärvi P. (1994b): Mater. Sci. Forum **143–147**, 347
Triftshaeuser W. (1995): J. Physique IV **5**, 217
Triftshäuser W., Kögel G., Sperr P., Britton D.T., Uhlmann K., Willutzki P. (1997): Nucl. Instrum. Meth. Phys. Res. B **130**, 264
Troev T., Dimova V., Zolov R., Levay B. (1979): Radiochem. Radioanalyt. Lett. **37**, 61
Tuck B., Hooper A. (1975): J. Phys. D **8**, 1806
Ueda O., Yamakoshi S., Sanada T., Umebu I., Kotani T. (1982): J. Appl. Phys. **53**, 9170
Uedono A., Cho Y., Tanigawa S., Ikari A. (1994a): Jpn. J. Appl. Phys. Pt. 1 **33**, 1
Uedono A., Fuji S., Wei L., Tanigawa S. (1994b): AIP Conf. Proc. **303**, 92
Uedono A., Itoh H., Ohshima T., Aoki Y., Yosaikawa M., Nashiyamai I., Okumura H., Yoshida S., Moriya T., Kawano T., Tanigawa S. (1996): Jpn. J. Appl. Phys. Pt. 1 **35**, 5986
Uedono A., Itoh H., Ohshima T., Suzuki R., Ohdaira T., Tanigawa S., Aoki Y., Yoshikawa M., Nashiyama I., Mikado T., Okumura H., Yoshida S. (1997a): Jpn. J. Appl. Phys. Pt. 1 **36**, 6650
Uedono A., Iwase Y., Tanigawa S. (1985): In *Positron annihilation.* Jain P.C., Singru R.M., Gopinathan K.P. (eds.). World Scientific, Singapore, p. 711
Uedono A., Kawano T., Long W., Tanigawa S., Ikari A., Kawakami K., Itoh H. (1994c): Jpn. J. Appl. Phys. Pt. 2 **33**, L1131
Uedono A., Kawano T., Tanigawa S., Suzuki R., Ohdaira T., Mikado T., Fujii S., Shikata S. (1995a): Jpn. J. Appl. Phys. Pt. 1 **34**, 1772
Uedono A., Kitano T., Hamada K., Moriya T., Kawano T., Tanigawa S., Suzuki R., Ohdaira T., Mikado T. (1997b): Jpn. J. Appl. Phys. Pt. 1 **36**, 2571
Uedono A., Kitano T., Watanabe M., Moriya T., Komuro N., Kawano T., Tanigawa S., Suzuki R., Ohdaira T., Mikado T. (1997c): Jpn. J. Appl. Phys. Pt. 1 **36**, 969
Uedono A., Long W., Tanigawa S., Ohji Y. (1994d): Jpn. J. Appl. Phys. Pt. 1 **33**, 3330
Uedono A., Moriya T., Tanigawa S., Kawano T., Ohji V. (1995b): J. Appl. Phys. **78**, 3269
Uedono A., Tanigawa S., Sugiura J., Ogasawara M. (1988): Appl. Phys. Lett. **53**, 25
Uedono A., Tanigawa S., Suzuki R., Ohgaki H., Mikado T. (1994e): Jpn. J. Appl. Phys. Pt. 1 **33**, 6286
Uedono A., Ujihira Y., Wei L., Tabuki Y., Tanigawa S., Sugiura J., Ogasawara M., Tamura M. (1992a): Mater. Res. Soc. Symp. Proc. **262**, 1061
Uedono A., Ujihira Y., Wei L., Tabuki Y., Tanigawa S., Wade K., Nakanishi H. (1992b): Mater. Res. Soc. Symp. Proc. **262**, 325
Uedono A., Watauchi S., Ujihira Y., Wei L., Tanigawa S., Suzuki R., Ohgaki H., Mikado T., Kametani H., Akiyama H., Yamaguchi Y., Koumaru M. (1993a): Hyperf. Interact. **79**, 621
Uedono A., Watauchi S., Ujihira Y., Wei L., Tanigawa S., Suzuki R., Ohgaki H., Mikado T., Kametani H., Akiyama H., Yamaguchi Y., Koumaru M. (1994f): AIP Conf. Proc. **303**, 101
Uedono A., Wei L., Li X., Tanigawa S. (1993b): J. Physique IV **3**, 209
Uedono A., Wei L., Tabuki Y., Kondo H., Tanigawa S., Ohji Y. (1992c): Mater. Sci. Forum **105–110**, 1475
Uedono A., Wei L., Tabuki Y., Kondo H., Tanigawa S., Sugiura J., Ogasawara M. (1992d): Mater. Sci. Forum **105–110**, 1479
Umlor M.T., Asoka-Kumar P., Keeble D.J., Cooke P.W., Lynn K.G. (1995): Appl. Surf. Sci. **85**, 295
Umlor M.T., Keeble D.J., Cooke P.W., Asoka-Kumar P., Lynn K.G. (1993): J. Electron. Mater. **22**, 1405
Valkealahti S., Nieminen R.M. (1983): Appl. Phys. A **32**, 95
Valkealahti S., Nieminen R.M. (1984): Appl. Phys. A **35**, 51
van der Pauw L.J. (1958): Philips Res. Rep. **13**, 1
van House J., Rich A. (1988): Phys. Rev. Lett. **61**, 488
van Vechten J.A. (1975): J. Electrochem. Soc. **122**, 423
van Vechten J.A. (1986): Phys. Rev. B **33**, 2674
van Vechten J.A. (1987): In *March Meeting of the American Physical Society,* New York, p. BI 1

van Veen A., Schut H., Clement M., de Nies J.M.M., Kruseman A., Ijpma M.R. (1995): Appl. Surf. Sci. **85**, 216

van Veen A., Schut H., Haakvoort J., de Vries R.A., Ijpma M.R. (1990): AIP Conf. Proc. **218**, 171

Vehanen A. (1989): In *Positron annihilation.* Dorikens-Vanpraet L., Dorikens M., Segers D. (eds.). World Scientific, Singapore, p. 39

Vehanen A., Lynn K.G., Schutz P.J., Eldrup M. (1983): Appl. Phys. A **32**, 163

Vehanen A., Saarinen K., Hautojärvi P., Huomo H. (1987): Phys. Rev. B **35**, 4606

Veprek S., Marecek V. (1968): Sol. State. Electron. **11**, 683

Voitsekhovskii A.V., Kokhanenko A.P., Lilenko Y.V., Pogrebnyak A.D., Ruzimov S.M., Babadzhanov R.D. (1986): Sov. Phys. Semicond. **20**, 514

Voitsekhovskii A.V., Korotaev A.G., Kokhanenko A.P. (1995): Rus. Phys. J. **38**, 1007

von Bardeleben H.J., Stievenard D., Deresmes D., Huber A., Bourgoin J.C. (1986): Phys. Rev. B **34**, 7192

von Bardeleben J.J., Manasreh M.O., Look D.C., Evans K.R., Stutz C.E. (1992): Phys. Rev. B **45**, 3372

Vydyanath H.R. (1981): J. Electrochem. Soc. **128**, 2609

Vydyanath H.R., Donovan J.C., Nelson D.A. (1981): J. Electrochem. Soc. **128**, 2625

Waeber W.B. (1990): Helv. Phys. Acta **63**, 448

Walukiewicz W. (1989): Appl. Phys. Lett. **54**, 2094

Walukiewicz W., Lagowski L., Jastrzebski L., Lichtensteiger M., Gatos H.C. (1979): J. Appl. Phys. **50**, 899

Wang C., Bernholc J., Davies R.F. (1988): Phys. Rev. B **38**, 12752

Watkins G.D. (1964): In *Radiation damage in semiconductors.* Baruch P. (ed.). Dunod, Paris, p. 97

Watkins G.D. (1986): In *Deep centers in semiconductors.* Pantelides S.T. (ed.). Gordon & Breach, New York, p. 147

Watkins G.D. (1991): In *Materials science and technology.* Cahn R.W., Haasen P., Kramer E.J. (eds.). VCH, Weinheim, p. 105

Watkins G.D., Corbett J.W. (1961): Phys. Rev. **121**, 1001

Watkins G.D., Corbett J.W. (1965): Phys. Rev. **138**, A543

Watkins G.D., Troxell J.R. (1980): Phys. Rev. Lett. **44**, 593

Weber E.R., Ennen H., Kaufmann U., Windscheif J., Schneider J., Wosinski T. (1982): J. Appl. Phys. **53**, 6140

Wei L., Lee J.L., Tanigawa S., Kawabe M. (1992): Mater. Sci. Forum **105–110**, 1487

Wei L., Tanigawa S. (1994a): AIP Conf. Proc. **303**, 107

Wei L., Tanigawa S. (1994b): Jpn. J. Appl. Phys. Pt. 1 **33**, 290

Wei L., Tanigawa S., Uedono A., Wada K., Nakanishi H. (1994): Jpn. J. Appl. Phys. Pt. 1 **33**, 33

Wei L., Yang-Koo C., Dosho C., Tanigawa S., Yodo T., Yamashita K. (1991): Jpn. J. Appl. Phys. Pt. 1 **30**, 2442

Weiss A. (1995): In *Positron spectroscopy of solids.* Dupasquier A., Mills A.P. (eds.). IOS Press, Amsterdam, p. 259

Weiss A., Jibaly M., Lei C., Mehl D., Mayer R., Lynn K.G. (1989): In *Positron annihilation.* Dorikens-Vanpraet L., Dorikens M., Segers D. (eds.). World Scientific, Singapore, p. 357

Weiss A.H., Rosenberg I.J., Canter K.F., Duke C.B., Paton A. (1983): Phys. Rev. B **27**, 867

Wenzl H., Mika K., Henkel D. (1990): Cryst. Res. Technol. **25**, 699

Wenzl H., Oates W.A., Mika K. (1994): In *Handbook of Crystal Growth.* Hurle D.T.J. (ed.). North-Holland, Amsterdam

Wertheim G.K. (1958): Phys. Rev. **111**, 1500

West R.N. (1973): Adv. Phys. **22**, 263

West R.N. (1979): In *Positrons in solids.* Hautojärvi P. (ed.). Springer-Verlag, Berlin, p. 89

Whelan J.M., Struthers J.D., Ditzenberger J.A. (1960): In *Proceedings International Conference on Semiconductor Physics.* Academic Press, New York, p. 943

Wiesendanger R., Güntherodt H.-J. (eds.) (1996): Scanning tunneling microscopy. Springer-Verlag, Berlin

Wietzke K.H., Koschnick F.K., Spaeth J.M. (1995): Mater. Sci. Forum **196–201**, 1061

Williams E.W. (1968): Phys. Rev. **168**, 922

Willutzki P., Stoermer J., Britton D.T., Triftshaeuser W. (1995): Appl. Phys. A **61**, 321

Wirbeleit F., Niklas J.R. (1997): Mater. Sci. Forum **258–263**, 987

Wissman B.D., Gidley D.W., Frieze W.E. (1992): Phys. Rev. B **46**, 16058

Wolfe C.M., Stillman G.E. (1975): Appl. Phys. Lett. **27**, 564

Wörner R., Kaufman U., Schneider J. (1982): Appl. Phys. Lett. **40**, 141

Wosinski T., Figielski T. (1989): Inst. Phys. Conf. Ser. **104**, 151

Würschum R., Bauer W., Maier K., Major J., Seeger A., Stoll H., Carstanjen H.D., Decker W., Diehl J., Schaefer H.E. (1989a): In *Positron annihilation.* Dorikens-Vanpraet L., Dorikens M., Segers D. (eds.). World Scientific, Singapore, p. 671

Würschum R., Bauer W., Maier K., Seeger A., Schaefer H.E. (1989b): J. Phys. Cond. Mat. **1**, SA33

Würschum R., Kümmerle E.A., Badura-Gergen K., Seeger A. (1996): J. Appl. Phys. **80**, 724

Würschum R., Schaefer H.E. (1987): phys. stat. sol. (a) **103**, 101

Würschum R., Scheytt M., Schaefer H.E. (1987): phys. stat. sol. (a) **102**, 119

Xiong L.Y., Lung C.W. (1989): In *Positron Annihilation.* Dorikens-Vanpraet L., Dorikens M., Degers D. (eds.). World Scientific, Singapore, p. 374

Xu H. (1990): J. Appl. Phys. **68**, 4077

Xu H., Lindefelt U. (1990): Phys. Rev. B **41**, 5979

Yacobi B.G., Holt D.B. (1990): *Cathodoluminescence in inorganic solids.* Plenum, New York

Yamada M., Tien P.K., Martin R.J., Nahory R.E., Ballman A.A. (1983): Appl. Phys. Lett. **43**, 594

Yamagishi H., Fusegawa I., Fujimaki N., Katayama M. (1992): Semicond. Sci. Technol. A **7**, 135

Yamauchi C., Tsumori Y., Nakashizu C., Esaka H., Takao S., Shinoyama S. (1992): Jpn. J. Appl. Phys. **31**, L439

Yang G., Kim J.H., Yang S., Weiss A.H. (1996): Surf. Sci. **367**, 45

Yonenaga I., Sumino K. (1985): In *Dislocations in solids.* Suzuki H., Ninomiya T., Takeuchi S. (eds.). University of Tokyo, Tokyo, p. 385

Yu K.M., Liliental-Weber Z. (1992): Mater. Res. Soc. Symp. Proc. **241**, 145

Yu W.-Z., Fu D.-Q., Xu W.-Y., Pang F.-C. (1995): Mater. Sci. Forum **175–178**, 597

Zanio K. (1972): Semicond. Semimet. **13**, 129

Zhang S.B., Chadi D.J. (1990): Pgys. Rev. Lett. **64**, 1789

Zhang S.B., Northrup J.E. (1991): Phys. Rev. Lett. **67**, 2339

Zheng J.F., Liu X., Newman N., Weber E.R., Ogletree D.F., Salmeron M. (1994): Phys. Rev. Lett. **72**, 1490

Zhong J., Kitai A.H., Mascher P., Puff W. (1993): J. Electrochem. Soc. **140**, 3644

Zhong J., Mascher P., Puff W., Kitai A.H. (1994): Mater. Sci. Forum **143–147**, 465

Zhu S., Li A., Li D., Huang H., Zheng S., Du H., Ding H., Gou Z., Iwata T. (1995): Mater. Sci. Forum **175–178**, 609

Ziegler J.F., Biersack J.P., Littmark U. (1985): The stopping and range of ions in solids. Pergamon Press, New York

Zimmermann H., Boyn R., Rudolph P., Bollmann J., Klimakow A., Krause R. (1993): Mater. Sci. Eng. B **16**, 139

Index

Printing: Druckhaus Beltz, Hemsbach
Binding: Buchbinderei Schäffer, Grünstadt

Springer Series in Solid-State Sciences

Editors: M. Cardona P. Fulde K. von Klitzing H.-J. Queisser

47 **Statics and Dynamics of Nonlinear Systems**
Editors: G. Benedek, H. Bilz, and R. Zeyher

48 **Magnetic Phase Transitions**
Editors: M. Ausloos and R. J. Elliott

49 **Organic Molecular Aggregates**
Electronic Excitation and Interaction Processes
Editors: P. Reineker, H. Haken, and H. C. Wolf

50 **Multiple Diffraction of X-Rays in Crystals**
By Shih-Lin Chang

51 **Phonon Scattering in Condensed Matter**
Editors: W. Eisenmenger, K. Laßmann, and S. Döttinger

52 **Superconductivity in Magnetic and Exotic Materials** Editors: T. Matsubara and A. Kotani

53 **Two-Dimensional Systems, Heterostructures, and Superlattices**
Editors: G. Bauer, F. Kuchar, and H. Heinrich

54 **Magnetic Excitations and Fluctuations**
Editors: S. W. Lovesey, U. Balucani, F. Borsa, and V. Tognetti

55 **The Theory of Magnetism II** Thermodynamics and Statistical Mechanics By D. C. Mattis

56 **Spin Fluctuations in Itinerant Electron Magnetism** By T. Moriya

57 **Polycrystalline Semiconductors**
Physical Properties and Applications
Editor: G. Harbeke

58 **The Recursion Method and Its Applications**
Editors: D. G. Pettifor and D. L. Weaire

59 **Dynamical Processes and Ordering on Solid Surfaces** Editors: A. Yoshimori and M. Tsukada

60 **Excitonic Processes in Solids**
By M. Ueta, H. Kanzaki, K. Kobayashi, Y. Toyozawa, and E. Hanamura

61 **Localization, Interaction, and Transport Phenomena** Editors: B. Kramer, G. Bergmann, and Y. Bruynseraede

62 **Theory of Heavy Fermions and Valence Fluctuations** Editors: T. Kasuya and T. Saso

63 **Electronic Properties of Polymers and Related Compounds**
Editors: H. Kuzmany, M. Mehring, and S. Roth

64 **Symmetries in Physics** Group Theory Applied to Physical Problems 2nd Edition
By W. Ludwig and C. Falter

65 **Phonons: Theory and Experiments II**
Experiments and Interpretation of Experimental Results By P. Brüesch

66 **Phonons: Theory and Experiments III**
Phenomena Related to Phonons
By P. Brüesch

67 **Two-Dimensional Systems: Physics and New Devices**
Editors: G. Bauer, F. Kuchar, and H. Heinrich

68 **Phonon Scattering in Condensed Matter V**
Editors: A. C. Anderson and J. P. Wolfe

69 **Nonlinearity in Condensed Matter**
Editors: A. R. Bishop, D. K. Campbell, P. Kumar, and S. E. Trullinger

70 **From Hamiltonians to Phase Diagrams**
The Electronic and Statistical-Mechanical Theory of sp-Bonded Metals and Alloys By J. Hafner

71 **High Magnetic Fields in Semiconductor Physics**
Editor: G. Landwehr

72 **One-Dimensional Conductors**
By S. Kagoshima, H. Nagasawa, and T. Sambongi

73 **Quantum Solid-State Physics**
Editors: S. V. Vonsovsky and M. I. Katsnelson

74 **Quantum Monte Carlo Methods in Equilibrium and Nonequilibrium Systems** Editor: M. Suzuki

75 **Electronic Structure and Optical Properties of Semiconductors** 2nd Edition
By M. L. Cohen and J. R. Chelikowsky

76 **Electronic Properties of Conjugated Polymers**
Editors: H. Kuzmany, M. Mehring, and S. Roth

77 **Fermi Surface Effects**
Editors: J. Kondo and A. Yoshimori

78 **Group Theory and Its Applications in Physics**
2nd Edition
By T. Inui, Y. Tanabe, and Y. Onodera

79 **Elementary Excitations in Quantum Fluids**
Editors: K. Ohbayashi and M. Watabe

80 **Monte Carlo Simulation in Statistical Physics**
An Introduction 3rd Edition
By K. Binder and D. W. Heermann

81 **Core-Level Spectroscopy in Condensed Systems**
Editors: J. Kanamori and A. Kotani

82 **Photoelectron Spectroscopy**
Principle and Applications 2nd Edition
By S. Hüfner

83 **Physics and Technology of Submicron Structures**
Editors: H. Heinrich, G. Bauer, and F. Kuchar

84 **Beyond the Crystalline State** An Emerging Perspective By G. Venkataraman, D. Sahoo, and V. Balakrishnan

85 **The Quantum Hall Effects**
Fractional and Integral 2nd Edition
By T. Chakraborty and P. Pietiläinen

86 **The Quantum Statistics of Dynamic Processes**
By E. Fick and G. Sauermann

87 **High Magnetic Fields in Semiconductor Physics II**
Transport and Optics Editor: G. Landwehr

88 **Organic Superconductors** 2nd Edition
By T. Ishiguro, K. Yamaji, and G. Saito

89 **Strong Correlation and Superconductivity**
Editors: H. Fukuyama, S. Maekawa, and A. P. Malozemoff

Springer Series in Solid-State Sciences

Editors: M. Cardona P. Fulde K. von Klitzing H.-J. Queisser

1 **Principles of Magnetic Resonance** 3rd Edition By C. P. Slichter
2 **Introduction to Solid-State Theory** By O. Madelung
3 **Dynamical Scattering of X-Rays in Crystals** By Z. G. Pinsker
4 **Inelastic Electron Tunneling Spectroscopy** Editor: T. Wolfram
5 **Fundamentals of Crystal Growth I** Macroscopic Equilibrium and Transport Concepts By F. E. Rosenberger
6 **Magnetic Flux Structures in Superconductors** By R. P. Huebener
7 **Green's Functions in Quantum Physics** 2nd Edition By E. N. Economou
8 **Solitons and Condensed Matter Physics** Editors: A. R. Bishop and T. Schneider
9 **Photoferroelectrics** By V. M. Fridkin
10 **Phonon Dispersion Relations in Insulators** By H. Bilz and W. Kress
11 **Electron Transport in Compound Semiconductors** By B. R. Nag
12 **The Physics of Elementary Excitations** By S. Nakajima, Y. Toyozawa, and R. Abe
13 **The Physics of Selenium and Tellurium** Editors: E. Gerlach and P. Grosse
14 **Magnetic Bubble Technology** 2nd Edition By A. H. Eschenfelder
15 **Modern Crystallography I** Fundamentals of Crystals Symmetry, and Methods of Structural Crystallography 2nd Edition By B. K. Vainshtein
16 **Organic Molecular Crystals** Their Electronic States By E. A. Silinsh
17 **The Theory of Magnetism I** Statics and Dynamics By D. C. Mattis
18 **Relaxation of Elementary Excitations** Editors: R. Kubo and E. Hanamura
19 **Solitons** Mathematical Methods for Physicists By. G. Eilenberger
20 **Theory of Nonlinear Lattices** 2nd Edition By M. Toda
21 **Modern Crystallography II** Structure of Crystals 2nd Edition By B. K. Vainshtein, V. L. Indenbom, and V. M. Fridkin
22 **Point Defects in Semiconductors I** Theoretical Aspects By M. Lannoo and J. Bourgoin
23 **Physics in One Dimension** Editors: J. Bernasconi and T. Schneider
24 **Physics in High Magnetics Fields** Editors: S. Chikazumi and N. Miura
25 **Fundamental Physics of Amorphous Semiconductors** Editor: F. Yonezawa
26 **Elastic Media with Microstructure I** One-Dimensional Models By I. A. Kunin
27 **Superconductivity of Transition Metals** Their Alloys and Compounds By S. V. Vonsovsky, Yu. A. Izyumov, and E. Z. Kurmaev
28 **The Structure and Properties of Matter** Editor: T. Matsubara
29 **Electron Correlation and Magnetism in Narrow-Band Systems** Editor: T. Moriya
30 **Statistical Physics I** Equilibrium Statistical Mechanics 2nd Edition By M. Toda, R. Kubo, N. Saito
31 **Statistical Physics II** Nonequilibrium Statistical Mechanics 2nd Edition By R. Kubo, M. Toda, N. Hashitsume
32 **Quantum Theory of Magnetism** 2nd Edition By R. M. White
33 **Mixed Crystals** By A. I. Kitaigorodsky
34 **Phonons: Theory and Experiments I** Lattice Dynamics and Models of Interatomic Forces By P. Brüesch
35 **Point Defects in Semiconductors II** Experimental Aspects By J. Bourgoin and M. Lannoo
36 **Modern Crystallography III** Crystal Growth By A. A. Chernov
37 **Modern Chrystallography IV** Physical Properties of Crystals Editor: L. A. Shuvalov
38 **Physics of Intercalation Compounds** Editors: L. Pietronero and E. Tosatti
39 **Anderson Localization** Editors: Y. Nagaoka and H. Fukuyama
40 **Semiconductor Physics** An Introduction 6th Edition By K. Seeger
41 **The LMTO Method** Muffin-Tin Orbitals and Electronic Structure By H. L. Skriver
42 **Crystal Optics with Spatial Dispersion, and Excitons** 2nd Edition By V. M. Agranovich and V. L. Ginzburg
43 **Structure Analysis of Point Defects in Solids** An Introduction to Multiple Magnetic Resonance Spectroscopy By J.-M. Spaeth, J. R. Niklas, and R. H. Bartram
44 **Elastic Media with Microstructure II** Three-Dimensional Models By I. A. Kunin
45 **Electronic Properties of Doped Semiconductors** By B. I. Shklovskii and A. L. Efros
46 **Topological Disorder in Condensed Matter** Editors: F. Yonezawa and T. Ninomiya

Springer Series in Solid-State Sciences

Editors: M. Cardona P. Fulde K. von Klitzing H.-J. Queisser

Managing Editor: H. K.V. Lotsch

90 **Earlier and Recent Aspects of Superconductivity**
Editors: J. G. Bednorz and K. A. Müller

91 **Electronic Properties of Conjugated Polymers III** Basic Models and Applications
Editors: H. Kuzmany, M. Mehring, and S. Roth

92 **Physics and Engineering Applications of Magnetism** Editors: Y. Ishikawa and N. Miura

93 **Quasicrystals** Editors: T. Fujiwara and T. Ogawa

94 **Electronic Conduction in Oxides**
By N. Tsuda, K. Nasu, A. Yanase, and K. Siratori

95 **Electronic Materials**
A New Era in Materials Science
Editors: J. R. Chelikowsky and A. Franciosi

96 **Electron Liquids** 2nd Edition By A. Isihara

97 **Localization and Confinement of Electrons in Semiconductors**
Editors: F. Kuchar, H. Heinrich, and G. Bauer

98 **Magnetism and the Electronic Structure of Crystals** By V.A. Gubanov, A.I. Liechtenstein, and A.V. Postnikov

99 **Electronic Properties of High-T_c Superconductors and Related Compounds**
Editors: H. Kuzmany, M. Mehring, and J. Fink

100 **Electron Correlations in Molecules and Solids** 3rd Edition By P. Fulde

101 **High Magnetic Fields in Semiconductor Physics III** Quantum Hall Effect, Transport and Optics By G. Landwehr

102 **Conjugated Conducting Polymers**
Editor: H. Kiess

103 **Molecular Dynamics Simulations**
Editor: F. Yonezawa

104 **Products of Random Matrices** in Statistical Physics By A. Crisanti, G. Paladin, and A. Vulpiani

105 **Self-Trapped Excitons**
2nd Edition By K. S. Song and R. T. Williams

106 **Physics of High-Temperature Superconductors**
Editors: S. Maekawa and M. Sato

107 **Electronic Properties of Polymers**
Orientation and Dimensionality of Conjugated Systems Editors: H. Kuzmany, M. Mehring, and S. Roth

108 **Site Symmetry in Crystals**
Theory and Applications 2nd Edition
By R. A. Evarestov and V. P. Smirnov

109 **Transport Phenomena in Mesoscopic Systems** Editors: H. Fukuyama and T. Ando

110 **Superlattices and Other Heterostructures**
Symmetry and Optical Phenomena 2nd Edition
By E. L. Ivchenko and G. E. Pikus

111 **Low-Dimensional Electronic Systems**
New Concepts
Editors: G. Bauer, F. Kuchar, and H. Heinrich

112 **Phonon Scattering in Condensed Matter VII**
Editors: M. Meissner and R. O. Pohl

113 **Electronic Properties of High-T_c Superconductors**
Editors: H. Kuzmany, M. Mehring, and J. Fink

114 **Interatomic Potential and Structural Stability**
Editors: K. Terakura and H. Akai

115 **Ultrafast Spectroscopy of Semiconductors and Semiconductor Nanostructures**
2nd Edition By J. Shah

116 **Electron Spectrum of Gapless Semiconductors**
By J. M. Tsidilkovski

117 **Electronic Properties of Fullerenes**
Editors: H. Kuzmany, J. Fink, M. Mehring, and S. Roth

118 **Correlation Effects in Low-Dimensional Electron Systems**
Editors: A. Okiji and N. Kawakami

119 **Spectroscopy of Mott Insulators and Correlated Metals**
Editors: A. Fujimori and Y. Tokura

120 **Optical Properties of III–V Semiconductors**
The Influence of Multi-Valley Band Structures
By H. Kalt

121 **Elementary Processes in Excitations and Reactions on Solid Surfaces**
Editors: A. Okiji, H. Kasai, and K. Makoshi

122 **Theory of Magnetism**
By K. Yosida

123 **Quantum Kinetics in Transport and Optics of Semiconductors**
By H. Haug and A.-P. Jauho

124 **Relaxations of Excited States and Photo-Induced Structural Phase Transitions**
Editor: K. Nasu

125 **Physics and Chemistry of Transition-Metal Oxides**
Editors: H. Fukuyama and N. Nagaosa

Springer and the environment